T0132573

POLITICAL ESSAY ON THE ISLAND OF CUBA

ALEXANDER
VON HUMBOLDT
IN ENGLISH

A series edited by Vera M. Kutzinski and Ottmar Ette

POLITICAL ESSAY ON THE ISLAND OF CUBA

A Critical Edition

ALEXANDER VON HUMBOLDT

Edited with an Introduction
by Vera M. Kutzinski and Ottmar Ette

Translated by J. Bradford Anderson,
Vera M. Kutzinski, and Anja Becker

With Annotations by Tobias Kraft, Anja Becker,
and Giorleny D. Altamirano Rayo

THE UNIVERSITY OF CHICAGO PRESS CHICAGO AND LONDON

Vera M. Kutzinski is the Martha Rivers Ingram Professor of English and professor of comparative literature at Vanderbilt University. Ottmar Ette is professor of Romance literatures at University of Potsdam, Germany. Giorleny D. Altamirano Rayo is a research coordinator at Vanderbilt University and HiE's assistant editor. Anja Becker is an independent scholar. Tobias Kraft is an instructor in Romance literatures at the University of Potsdam, Germany. J. Bradford Anderson is an independent scholar.

The University of Chicago Press, Chicago 60637
The University of Chicago Press, Ltd., London

Printed in the United States of America

20 19 18 17 16 15 14 13 12 11   1 2 3 4 5

ISBN-13: 978-0-226-46567-8 (cloth)
ISBN-10: 0-226-46567-5 (cloth)

Library of Congress Cataloging-in-Publication Data

Humboldt, Alexander von, 1769–1859.
[Essai politique sur l'ile de Cuba. English]
Political essay on the island of Cuba / Alexander von Humboldt ; edited with an introduction by Vera M. Kutzinski and Ottmar Ette ; translated by J. Bradford Anderson, Vera M. Kutzinski, and Anja Becker ; with annotations by Tobias Kraft, Anja Becker, and Giorleny D. Altamirano Rayo.
p. cm.
The translation in this edition is based upon the full text of the Alexander von Humboldt's two-volume *Essai politique sur l'Île de Cuba* from 1826.
Includes index.
ISBN-13: 978-0-226-46567-8 (cloth : alk. paper)
ISBN-10: 0-226-46567-5 (cloth : alk. paper) 1. Cuba—Description and travel. 2. Slavery—Cuba—History. I. Kutzinski, Vera M., 1956– II. Ette, Ottmar. III. Title.
F1763.H913 2011
972.91′05—dc22

2010015028

♾ The paper used in this publication meets the minimum requirements of the American National Standard for Information Sciences—Permanence of Paper for Printed Library Materials, ANSI Z39.48-1992.

CONTENTS

Inventories and Inventions: Alexander von Humboldt's Cuban Landscapes.

AN INTRODUCTION BY
VERA M. KUTZINSKI AND OTTMAR ETTE

The popular images of Alexander von Humboldt are those of a traveling adventurer and a collector of all sorts of scientific data, ranging from a bewildering array of physical measurements to boxes of plant specimens. Humboldt was both a hardened traveler and an enthusiastic collector, but he was also so much more than that: in addition to being a meticulous empirical scientist who pioneered fieldwork as we know it, he was also an imaginative thinker of the first order, and of global proportions. It is for good reason that the venerated Cuban anthropologist Fernando Ortiz, in his introduction to the 1930 Libros Cubanos edition of the *Political Essay on the Island of Cuba*, emphasized that Humboldt's various honorary epithets—Simón Bolívar called him the "re-discoverer of America," José de la Luz y Caballero the "second discoverer of Cuba"—hardly do him justice. According to Ortiz, Alexander von Humboldt was a "bold inventor" of the island of Cuba and of the Spanish Americas as a whole, someone who recognized both the uniqueness of the western hemisphere and its growing importance to the rest of the world. Scores of natural and social scientists, from ecologists, mineralogists, and geographers to anthropologists, economists, and political scientists, followed in Humboldt's footsteps. Along with not a few politicians, notably Charles IV of Spain, Simón Bolívar, and Thomas Jefferson, they mined Humboldt's writing for his data and built on his scientific insights.

While Humboldt's *opus americanum* has long been an indispensible compendium for historians of science, many aspects of his *Voyage to the Equinoctial Regions of the New Continent*—which included the initial versions of the *Political Essay on the Island of Cuba*—have long remained underappreciated. It is to these features of Humboldtian writing that we would like to draw special attention here. There are of course Humboldt's unshakable democratic convictions, inspired by the French Revolution, and his incisive, impassioned criticisms of slavery and other forms of colonial exploitation. No less important is his comparative global perspective on politics, economics, and science, along with a discourse on the tropics that revolutionized the ways in which

Europeans thought about the New World. This discourse is as characteristic a part of Humboldt's unorthodox, fragmented travelogues as are his metaphoric combinations of natural and cultural imagery and his use of different media and languages. All these come together in a distinctive narrative voice that has been obscured, either unwittingly or deliberately, in earlier English translations of his writing.

Most remarkably, there is an alertness in Humboldt's writing, an intellectual and emotional energy that springs from the desire for "full impressions" [Gesamteindrücke]. Such impressions, for Humboldt, are the result of assembling myriad minute details into larger pictures—he typically calls them *tableaux*—that would show how everything, absolutely everything, is interrelated. Rendering intelligible the forms and shapes of these interrelations was as important to Humboldt as was his beloved work in the field. Showing how differing and seemingly incompatible relations come together in patterns and networks on both a hemispheric and a global scale lends his work the imaginative dimensions that make him truly an inventor. This imaginative dimension is perhaps the most significant and lasting part of his intellectual legacy.

Invention in this sense pervades Humboldt's writing at all levels, from the thematic to the aesthetic. Humboldt's works are not mere scientific reports; they are also works of art that range across many different genres, notably those of the essay and the travelogue. And because they are works of art, they do what academic science writing typically does not: incorporate multiple perspectives and integrate information from diverse fields of knowledge in sometimes surprising ways to keep readers' minds open, receptive, and attentive to new inputs and ideas. To create such an openness of understanding, which some have erroneously taken as evidence that Humboldt did not know how to write books, is what the Prussian cared about above all else in his work. This aspect of his thinking and writing makes Alexander von Humboldt's influence not just the stuff of history but something that extends well into the future.

Traveling to the Americas

Alexander von Humboldt's writings on the Americas form a vast corpus. His *Voyage to the Equinoctial Regions to the New Continent* encompasses no less than thirty volumes. The *Political Essay on the Island of Cuba* is a fairly small but central part of this oeuvre. It occupies a prominent place in Humboldt's writings, because Cuba held a special position in his scientific imagination. Since Cuba was the first major island on which Christopher Columbus had set foot in 1492, Humboldt, the "Columbus of Science," decided to follow the same route to the Indies. (His mother's maiden name being Colomb, Humboldt's name in Spanish would actually have been Alejandro de Humboldt y Colón.) When Humboldt set sail for the Americas from the Spanish port of La Coruña on June 5, 1799, he intended to go to Cuba first. But his travels rarely unfolded according to his plans. Prior to undertaking the voyage to the New World, for instance, Humboldt and the French botanist and physician Aimé Bonpland, who accompanied him, had hoped to join Nicolas Thomas Baudin's expedition to Australia, which ran into financial difficulties and had to be aborted. Being also unable to visit North Africa because of Napoleon's invasion of Egypt, Humboldt decided to go to Spain and organize an expedition to the Americas instead. With the help of Phillip von Forell, Saxony's Ambassador in Madrid, Humboldt achieved the nearly unthinkable for a non-Catholic foreigner at the time. Not only did Charles IV grant the Prussian explorer unrestricted access to Spain's American colonies; he also issued him a Spanish passport, ensuring the full cooperation of the colonial authorities. It probably helped that Humboldt did not need any money from the king. Unlike most other scientific explorers before and even after him, he could finance his own expeditions from a considerable personal inheritance from his mother, freeing him from allegiance to a country's commercial and political interests.

After months of intense preparations and further delays because of the British blockade of La Coruña, Humboldt and Bonpland were finally aboard the frigate *Pizarro*, beginning a journey that would make Humboldt one of the most famous men of his times. Some of his contemporaries regarded him as second in importance only to Napoleon Bonaparte, while others, such as Ralph Waldo Emerson, even compared him to Aristotle. Following Columbus's route, Humboldt's transatlantic voyage stopped first in the Canary Islands (June 19–25, 1799). Tenerife, the largest island of the Canaries, was the first non-European island that Humboldt had ever visited. It was there that he learned to describe a place in all its different dimensions: anthropological, botanical, economical, geographical, geological, historical, political, and sociocultural. The island of Tenerife, where he climbed the first of many volcanoes, became both the theoretical and the practical model that Humboldt would apply elsewhere in his travels. Humboldt had hoped to cross from the Canaries directly to Cuba, the major island of the Caribbean. But an outbreak of "fever" on his ship foiled these plans, and the captain of the *Pizarro* decided instead to divert to Cumaná in today's Venezuela. In Cumaná, Humboldt first became acquainted with the "equinoctial regions," the tropics of which he had dreamed ever since he had developed an interest in plants as a boy. Explorations on the mosquito-infested Orinoco and Casiquiare rivers followed in short order. The Casiquiare is a natural link between the Amazon and the Orinoco, whose disputed existence Humboldt confirmed. These adventurous sojourns would become the best-known part of his American voyage.

After his return to the Venezuelan coast, Humboldt could at last proceed to the island of Cuba. His first visit to the island lasted from December 19, 1800 to March 15, 1801; though brief, it was very productive. In Havana, Humboldt met many eminent politicians, local scientists, and scholars, who became not only friends but crucial parts of his growing network of informants and collaborators. After a nearly three-month stay on the island of Cuba, Humboldt traveled on to the regions of South America that later became Colombia, Ecuador, and Peru. These visits became the basis for his *Tableau physique des Andes*, sketches in which he brought together information about the material and cultural worlds of the Cordilleras. There, Humboldt climbed (or attempted to climb) some of the world's loftiest volcanoes, reaching the highest point anyone had measured up to that time on Chimborazo. From Lima, Peru, Humboldt sailed north to the Ecuadorian port of Guayaquil on the waters that would later be named after him: the Humboldt Current. He continued on to Acapulco in New Spain (now Mexico), where he remained for a year. From Mexico City, he made excursions to many of the colony's different regions and climbed more volcanoes, this time the Jorullo and the Nevado de Toluca. On these excursions, Humboldt also spent much time studying the working conditions of miners, most of whom were descended from the same pre-Columbian peoples whose languages and cultures he researched in the archives of New Spain. This research became the foundation for his *Vues des Cordillères et monumens des peuples indigènes de l'Amérique* [Views of the Cordilleras and of the monuments of the indigenous peoples of America] and his *Political Essay on the Kingdom of New Spain*.

Departing New Spain via the port of Veracruz, Humboldt returned to Cuba to retrieve the portion of the botanical collection that he had stored in Havana. During this second stay in Cuba (March 19 to April 29, 1804), he completed his Cuban data gathering, a task for which the social network he had built up during his initial visit proved indispensable. He would maintain this network for years to come, drawing from it valuable information up to and during the time when he was writing the *Political Essay on the Island of Cuba* in the 1820s.

Thinking Globally: *Weltbewusstsein*

Alexander von Humboldt prepared for and carried out his scientific voyages—both to the Americas and, later, to Central Asia—in times of worldwide military and political conflicts. The Seven Years' War had barely ended by the time he was born in 1769, and the U.S. Revolution was not far off. The revolutions in Europe and in the Americas were not just background noise to Humboldt's scientific pursuits; they became a vital part of his political consciousness. His first visit to Paris in 1790, a year after the French Constituent Assembly had adopted the Declaration of the Rights of Man, left an indelible impression on the twenty-one-year-old student. Humboldt enthusiastically embraced the egalitarian principles of the French Revolution and would remain true to them for the rest of his life. When the March Revolution broke out in Berlin in 1848, the seventy-eight-year-old Humboldt, wearing the garb of the French *Directoire*, joined the workers, who hailed him as one of them. Humboldt had worn these clothes as a sign of his deeply held democratic convictions ever since his somewhat grudging return to the Prussian court, and to the city of birth, in 1827. At the same time that his political ideas were profoundly influenced by the French Revolution, Humboldt abhorred the violence with which revolutions were fraught. One year after his visit to Paris, the first large-scale slave uprisings erupted on the island of Saint-Domingue. The Haitian Revolution, which would produce the world's first independent black republic in 1804, was one of the most important historical events to reach its final stage while the Prussian explorer was visiting the New World. Humboldt's local contacts apprised him of these events, and he mentions what he calls "the troubles in Saint-Domingue" repeatedly in the *Political Essay on the Island of Cuba*. Even in the mid-1820s, he still warned that the colonial practice of slavery might unleash similar outbreaks of violence elsewhere in the Caribbean and in the United States, appealing to the "restorative forces in any intelligently lead social body . . . with [which] one can eradicate even the most ingrained of evils" (69). Some historians credit him with having foreseen the U.S. Civil War.

Independence revolutions throughout the Spanish colonies followed in the wake of the successful Haitian Revolution. By the time that Humboldt sat down to write about his travels in 1807, the Spanish empire was coming apart at the seams. By the mid-1820s, when he published the *Political Essay on the Island of Cuba*, both New Granada and New Spain had gained their sovereignty, with only Cuba, Puerto Rico, and the Philippines remaining under Spanish rule. Given his preference for gradual institutional reform and his dislike for violence, Humboldt would be an unlikely candidate for the position of "father of independence movements" in Spanish America, which some historians have awarded him. He was no more that than he was the figure of the lone heroic scientist. But these political revolutions had a dramatic impact on Humboldt in that they made him rethink political developments on a global scale. Rather than as isolated events, he saw them as part of a larger fabric of international politics and economics.

Another, more practical, effect of the tumultuous times during which Humboldt lived was that not all of the expeditions he had planned came to fruition. For one, Napoleon Bonaparte's Egyptian campaign frustrated his plans to travel to North Africa. For another, once Humboldt became known for his critical stances toward colonialism, the British Empire closed its doors, preventing him from visiting India, Tibet, and other British colonies in Asia. But Humboldt's studies of Egyptian and Near Eastern cultures, and of Arabic, Persian, and other languages, still enabled him to develop comparative concepts and approaches in his work, highlighting an area's distinctive characteristics by embedding it within a worldwide network of information, dependencies, exchanges, and correspondences. Humboldt's transareal ideas come into view with particular clarity in his comparisons of ideas of time in the radically different cultures of the pre-Columbian Americas, China, Japan, Tartary, India, Greek and Roman Antiquity, Judaism, and of the Catholic Middle Ages in *Views of the Cordilleras*. Always at the core of Humboldt's studies was a profound sense of cultural relativity and relatedness. Acutely aware of processes of globalization in his own time, Humboldt understood those processes in relation to what one might call the first period of accelerated globalization, during which Iberian imperial expansion reached its first peak with the expeditions by Columbus, Vasco da Gama, and Ferdinand Magellan. In his *Examen critique de l'histoire de la géographie du Nouveau Continent* [Critical examination of the historical development of the geographical knowledge of the New World], he singled out the period between 1492 and 1498 as decisive for world history, arguing that these six years determined how political and economic power was to be distributed across the surface of the globe.

For Humboldt, slavery and the slave trade were undeniable consequences of the two different stages of globalization. In the process of invigorating commerce between Europe and the American territories, Spain and Portugal's early overseas exploits spawned the transatlantic slave trade. Humboldt saw both the positive and the negative effects of the trade triangle between Europe, Africa, and the Americas, which, before long, came to involve Britain and other European countries. Whereas in New Spain Humboldt had focused on indigenous miners, in Cuba his keen critical eye was trained on the deplorable conditions of rural and urban slaves and of the island's growing free population of color. He criticized the slave trade as "barbaric" and "unreasonable" and slavery as "possibly the greatest evil ever to have afflicted humanity" (118 and 144). That he thought slavery wrong both in moral and in economic terms is a clear sign of what he called *Weltbewusstsein* [world consciousness]—a combination of humanistic principles and scientific insight. In the same way that he tied the "happiness" of New Spain's white population to that of the marginalized and exploited natives, the "indios," Humboldt saw the fate and fortunes of Cuba's Spanish and Creole elites as inextricably bound to that of the enslaved Africans and their progeny. Humboldt recognized that, in the end, all larger geopolitical developments in the Americas would depend crucially on how, and how equitably, these diverse populations would be able to live together in the future.

Revolutionizing Discourses

While Humboldt's political philosophy was no doubt a product of Enlightenment thought, his views of the Americas as a natural and cultural space radically diverged from the dominant discourse about the New World as inferior to Europe. Similarly, Humboldt's view of the Americas reflects an early understanding of science not as an exclusively European project but as a globalized practice. His awareness not just of the work of other scientists but of scientists in the Americas, on whose research he relied heavily, is in evidence throughout his writing, in which the tropics occupy center stage.

The tropics, which Humboldt called his "real element," were at the core of all of his scientific investigations and emotional investments. His person and his image would remain forever connected with what he also calls the "Torrid Zone" or the "equinoctial region," which was, for him, the very heart of America. As a young man, Humboldt was very conscious of the fact that, shortly before his birth, the so-called Berlin Debate on the New World had reached its first culmination point after the publication of the initial volume of Cornelius de Pauw's *Réflections philosophiques sur les Américains* [Philosophical reflections about the Americans] in 1768. Taking the work of the Count of Buffon a step further, De Pauw categorically separated the Old World from the New, characterizing America as weak, immature, and incapable of independent progress. In this schema, the Europeans were humanity's true, and indeed only, representatives. Other thinkers, notably Guillaume-Thomas Raynal and, later, Georg Wilhelm Friedrich Hegel sided with de Pauw, setting off a heated worldwide controversy. Humboldt was vehemently opposed to De Pauw's ideas and launched a bitter polemic against all those "schematic thinkers" who had never even set foot in the Americas.

Around 1800, Humboldt proposed a paradigm shift in the empirical foundations of knowledge: "the happy revolution," as he called it in his introduction to *Views of the Cordilleras*. In keeping with this "happy revolution," Humboldt himself began to articulate a new discourse on the New World.

Only five weeks after his departure from Spain, in a letter dated July 16, 1799, he wrote to his brother Wilhelm von Humboldt about his first impressions of the tropics.

> What trees! Coco palms, 50 to 60 feet high! Poinciana pulcherrima, with foot-tall bushes of the most magnificent bright-red blossoms; Pisange, and a cluster of trees with huge leaves and aromatic blossoms the size of a hand, of which we know nothing. Just think that the land is so little known that a new genus (s. *Cavanilles iconus, tom.* 4) which Mutis publicized only two years ago, is a 60-feet tall tree with a broad canopy. We were so happy that we already found this splendid plant (with its inch-long stamens) yesterday. How numerous may be the smaller plants, which are not visible to the observer's eye? And what colors the birds have, and the fish, even the crabs (sky-blue and yellow)! We run around like fools; in the first three days, we could not categorize anything because we would always toss aside one object in order to pick up another. Bonpland assures me that he will lose his mind if the wonders do not soon cease. But even more beautiful than these individual marvels is the impression of the entirety of the potent, profuse, and yet also so light, uplifting, mild nature of the plants. I feel that I will be very happy here and that the impressions around me will continue to lift my spirits frequently. (Humboldt, *Briefe aus Amerika 1799–1804*, ed. Ulrike Moheit [Berlin: Akademie Verlag, 1993, 42])

No other passage from Humboldt's letters from the Americas renders his discourse on the tropics quite so explicit. From the very start of his visit to the New World, Humboldt combines his representations of directly experienced knowledge with scientific discourses, especially in the areas of botany and zoology. At the same time, his lines express both *motion* and *emotion*, the aesthetic equivalent of his elation at the deeply pleasurable flood of sensual impressions. Even decades later, in the *Political Essay on the Island of Cuba*, Humboldt had lost none of his enthusiasm for the "organic vigor typical of the Torrid Zone" (26).

The discourses of Buffon, de Pauw, and Raynal emphasized America's weakness and decay. Humboldt's own writing revolutionized these images of the New World by offering depictions of the tropics that focused on the size and abundance of the plant life, its striking colors, and the multitude of mysteries that the landscape still held. Everything in the tropics seemed to await discovery and study. In his letter, botanical classifications and detailed measurement alternate with expressions of sheer rapture. Language can barely express the intensity and profusion of sensual impressions. Even the brief references to the scientific publications by José Celestino Mutis and Antonio José Cavanilles are pulled into a veritable maelstrom of colors, forms, and smells, which draws everything into the realm of the marvelous. Here, Humboldt writes himself into the tradition of *lo maravilloso*, in which nature impresses and stirs humans through its marvels. In the Americas, this tradition began in the late fifteenth century with the accounts of the first Spanish conquerors and chroniclers. One can still sense in Humboldt's writing the adventure of discovery, much as one can, a century and a half later, in Claude Lévi-Strauss's *Tristes Tropiques* and, in fictionalized form, in Alejo Carpentier's novel *The Lost Steps*.

Humboldt's letter from the American tropics is like a letter from paradise. Its pervasive language of excitement and "happiness" leaves no doubt that the tropics are as much a mixture of Eden and El Dorado for Alexander von Humboldt as they once were for Sir Walter Raleigh. It is no coincidence that the *Atlas physique et géographique des régions équinoxiales du Nouveau Continent* [Geographical and physical Atlas of the equinoctial regions of the New Continent] from 1814 includes many cartographic representations of El Dorado, the legendary city of gold that sixteenth- and seventeenth-century explorers believed to be situated near the Lago Manoa (or Lago Parime) in parts of the Amazon jungle that today are part of Brazil.

Moving Parts: Humboldt's Travelogue

During the early years of European conquest and colonialism, Cuba, as part of the Spanish Caribbean, quickly emerged as a global space because of its geostrategic value. Already Juan de la Cosa, in his beautiful world map—*Mapamundi*—from 1500, which is the centerpiece of Humboldt *Atlas*, had recognized Cuba's importance on this count. At the beginning of the nineteenth century, Humboldt saw Cuba as even more of a global island. By then, the destruction of most of Saint-Domingue's sugar mills during the Haitian Revolution had propelled Cuba onto the international economic stage as the world's foremost sugar producer and thus as a major trade partner both of Europe and of other countries in the hemisphere, notably the United States. Humboldt recalls these aspects of Cuba in the metaphors that frame his account of sailing into the harbor of Havana.

> Havana's appearance from the entrance of the port is one of the most pleasing and picturesque on the coastline of tropical America north of the equator. Celebrated by travelers of all nations, this site has neither the luxurious vegetation that lines the banks of the Guayaquil, nor the wild majesty of Rio de Janeiro's rocky shoreline, two ports in the southern hemisphere. But it has

> the grace that, in our climates, adorns scenes of cultivated nature, blending the majesty of vegetal forms with an organic vigor typical of the Torrid Zone. In this mixture of gentle impressions, the European forgets the dangers that threaten him at the heart of the Antilles' populous cities. He tries to take in the diverse elements of a vast landscape: the fortified castles that crown the rocks to the left of the port, which is an interior basin surrounded by villages and farms; the palm trees that grow to a prodigious height; and the city half-hidden behind a forest of masts and sails. (26)

In this picture of giant palm trees and a "forest" of ships' masts, Humboldt intertwines nature and culture, weaving both into the meshes of the net of global political and economic relations that he will spread out before the reader in the pages of the *Political Essay on the Island of Cuba*. In the same way that Havana harbor is a microcosm of the whole island, this passage is a microcosm of the entire text.

Where the texture of Humboldt's travel narrative is most condensed, as is the case in this passage, different discourses interrupt and transect each other. Among these discourses are those associated with different academic disciplines, such as nautical astronomy, climatology, cultural history, demography, economics, geology, philology, philosophy, plant geography, statistics, and zoology, to name but a few. In the above lines from the opening of the *Political Essay on the Island of Cuba*, these discourses cross each other's paths much like the ships entering and leaving Havana's harbor do. Whenever the travel narrative pauses, the island of Cuba itself comes into view as a dense and dynamic conceptual space. In this space, as in the image of Havana itself, the "diverse elements of a vast landscape" mix and mingle. As Humboldt moves on to describe both the port and the city in more detail, he combines precise scientific observations about the harbor's dimensions with accounts of the deplorable condition of the city streets. More than simply surprising evidence of earlier, ill-fated paving projects (the "beautiful logs of *Cahoba*" that he finds buried in Havana's knee-deep mud) tie the narrative back to the metaphoric "forest of masts"—until sensual impressions intrude. The rancid smell of poorly cured meat interrupts suddenly to introduce another discourse: that of slavery. This meat, called *tasajo*, was a staple in the diet of the African slaves, who are as much part of the city as are its graceful plants and picturesque promenades. Shortly after, we encounter what the smell has already announced: next to the beautiful botanical garden, there is "something else altogether, whose appearance at once aggrieves and appalls: the barracks [barracones] in front of which the pitiable slaves are exposed for sale" (27). The effects of the transatlantic slave trade, which the "forest of masts" also represents, become unexpectedly visible here. The references to Columbus and Fernando Cortés, which follow in the very next sentence as a seeming non sequitur, add to these impressions and observations the discourse of discovery and conquest. From this far-flung context of global history, the narrative shifts just as abruptly to a detailed botanical *excursus* on palm trees to draw us back to the importance of the tropics.

Unlike traditional travel narratives, then, Humboldt's text does not have a unilinear structure; no simple chronology builds in his long descriptive passages. Instead, Humboldtian writing generates an interplay of scientific details, sensual impressions, and historical accounts. Humboldt's sentences take us back and forth between stasis—the moments when the traveler pauses to reflect—and movement. In this way, spatial history (geography) becomes vectorial history (travel and migration).

Because Humboldt's preferred mode of writing is impersonal, something to which French lends itself much more readily than does English, even the figure of the traveler, which is the centerpiece of typical travelogues, gradually fades away, leaving behind little more than a scanty trail of personal pronouns. This effacement across long stretches of the narrative makes the traveler's reappearance all the more startling. One such reappearance occurs right after Humboldt marks the provisional "end" of his *Political Essay on the Island of Cuba*, when the traveler steps out of the narrative to expound on the explorer's social and political responsibilities. A first-person singular pronoun promptly follows. Humboldt rarely uses the "I," reserving it for utterances that he attributes directly to himself and setting it off from the "we" that also includes Bonpland: "It befits the traveler who witnessed up close the torment and degradation of humanity to bring the laments of the wretched to the ears of those who have the power to allay them. I have observed the conditions and circumstances of blacks in countries where the laws, religion, and national customs tend to soften their lot. Nevertheless, when leaving America, still I harbored the same hatred for slavery with which I had left Europe" (143).

The art of writing, for Humboldt, is always a form of time travel. As this last citation shows, he combines at least three temporal levels in the narrative of the *Political Essay on the Island of Cuba*: first, the times of his two visits to Cuba, in 1801 and 1804; second, the time of the writing itself, the mid-1820s; and third, the level between the two, that is, between 1804 and the 1820s, when he collected additional information through his worldwide network of correspondents and updated earlier versions of his publications. A fourth level would be the time prior to his voyage. "The relationships that I have kept up with America since my return to Europe," Humboldt points out, "have moved me to complete the materials that I had gathered on the region"

(1.8). Writing, for Humboldt, is a process of never-ending adjustments and revisions, in which diverse pasts and presents permeate each other. His writing is restless, projecting events backward into the past and forward into the future; like his mind, his prose is always in motion.

Unlike the classical travelogue, which recounts past events in more or less orderly chronological succession, Humboldt's narrative is both retrospective and prospective at the same time. In this way, the *Political Essay on the Island of Cuba* can point toward the future horizons of Cuba and of the Americas more broadly. The prospective dimension of his writing is already implicit, in his above statement on slavery, in the notion that slavery is a hateful condition that can and must be remedied. It is quite explicit in his pointed remarks about political nomenclature in the "two Americas":

> To avoid fastidious circumlocutions, I continue in this study to designate the countries inhabited by *Spanish-Americans* by the name *Spanish-America*, despite the political changes that the colonies have undergone. I call the *United States*—without adding *north America*—the country of *Anglo-Americans*, although other *United States* have formed in south America. It is awkward to speak of peoples who play such an important role on the world scene, but who lack collective names. The word *American* may no longer be applied exclusively to the citizens of the United States of North America, and it would be desirable if this nomenclature for the independent nations of the New Continent could be fixed in a way that would be at once convenient, consistent, and precise. (209)

Tellingly, Humboldt bases his case for why different names are needed on the fact that, by the 1820s, the United States was no longer the only sovereign nation state in the New World. His own inclusive use of "America" anticipated later disputes, in academia and elsewhere, about the appropriation of the term "America" by the United States of America. Humboldt points to linguistic imprecisions and inconsistencies that have momentous cultural and political consequences for the citizens of the Americas.

That Humboldt calls himself "a historian of America" (142) in no way keeps him from speculating and hypothesizing. His many "reasoned" forecasts throughout the *Political Essay on the Island of Cuba* are as integral a part of his hemispheric historiography as are his export and population statistics. The note on which he ends the essay suggests that one is, in fact, already contained in the other: "The language of numbers, the only hieroglyphics maintained amidst the signs of thought, does not require interpretation. There is something gravely prophetic in these inventories of the human species: the New World's entire future seems inscribed within them" (323).

Humboldt's somewhat overstated emphasis here on the purported objectivity of statistics—"inventories"—does not, however, detract from the highly subjective and political nature of his vision in this long essay. The many echoes of this prophetic note throughout the *Political Essay on the Island of Cuba* serve as constant reminders of the openness of the form of the essay, which Humboldt's title already announces. An essay is, after all, an attempted, or provisional, articulation, not a conclusive or even final inscription.

Fragmentation is another key aspect of Humboldtian writing. The *Political Essay on the Island of Cuba* is a part of Humboldt's actual travel narrative, the so-called *Relation historique,* or *Personal Narrative*, which began to appear in Paris in 1814 and would not be complete until 1831. This three-volume travelogue covers merely one-third of Humboldt's entire voyage to and in the New World. It goes up to April 1801, that is, just after Humboldt had departed Cuba for the Andes region. Although Humboldt had contemplated a final volume for years, he never actually wrote it, and many of the *Voyage*'s missing portions have been reconstructed from his travel diaries. The *Personal Narrative*, then, is effectively a fragment, in which the movement of travel ceases with the *Political Essay on the Island of Cuba* in volume three. The *Personal Narrative*, in its turn, is a fragment of the thirty volumes of the *Voyage to the Equinoctial Regions of the New Continent.* The *Voyage*, which itself remains incomplete, was published over the course of more than thirty years (1805 to 1839) by a number of different publishers. Several of them went bankrupt in the process; Humboldt himself almost did as well. In fact, Humboldt rarely ever completed a book. Like the *Relation historique*, the *Examen critique* (1834–38), *Asie Centrale* [Central Asia] (1843), and *Kosmos* (1845–58 and 1862) all remained incomplete, and quite intentionally so. Their incompleteness reflects Humboldt's aversion to narrative closure and his insistence that knowledge and scientific understanding always remain open. He regarded the individual volumes of his writing as parts of a larger work-in-progress—his own work and that of the many scientists with whom he collaborated during his lifetime and others, who would read his work after his death.

Mapping Cuba

Since Cuba is geographically located at the center of the American hemisphere, it is not surprising that the island was correspondingly central to Alexander von Humboldt's interests and that he would refer back to it throughout his later writings. Cuba occupies a prominent place in the *Examen critique*; it appears in *Kosmos* and even, rather unexpectedly, in *Asie Centrale.* Cuba is no less present in many of the letters Humboldt wrote and received from the Americas, especially in the continued exchanges with his Cuban hosts, notably with the landholder and statesman Francisco Arango y Parreño and the philosopher Luz y Caballero. Their correspondence continued well after Humboldt's departure and is part of an astonishingly vast corpus of almost one hundred thousand letters, which he wrote and received during his lifetime. Only about eight thousand of these letters have been preserved. This continuing flood of information made Humboldt's portrait of Cuba not just a record of the actual experiences he had had during his visits; nor are his observations and projections in the *Political Essay on the Island of Cuba* limited to research he himself had conducted in relatively few of the island's locales. While scientific inventories supplied the empirical underpinnings of Humboldt's multifaceted vision of Cuba, his research went far beyond his actual fieldwork, extending to various archives in the Americas, especially in Mexico, and in Europe. From the foundations of this extensive research, a modern image of Cuba emerges in the *Political Essay on the Island of Cuba*—both from Humboldt's narrative and from his two maps.

Alexander von Humboldt reinvented the island of Cuba through cartography, determining, for the first time, the exact geographical positions of Cuba's towns and cities and charting its coastline with unprecedented precision. By updating and correcting earlier maps, he gave Cuba the shape that it still has on maps today. Although only one map was included with the *Political Essay on the Island of Cuba* in 1826, Humboldt actually drew two maps of Cuba: the first is dated 1820; the second is a corrected version from 1826. That he printed both in his *Atlas*, a version of which would be reissued in connection with the *Examen critique*, emphasizes once again the importance of ongoing research and revisions to Humboldt's writing and thinking. (We include both maps for the same reason.) Humboldt's cartographic images of Cuba constitute a radical break with earlier maps, such as the atlas that was part of Raynal's *Philosophical and Political History of the Settlements and the Trade of the Europeans in the East and West Indies*. Better known as *The History of the Two Indies*, this colonial encyclopedia from 1772 was a bestseller in its time. Notably, both of Humboldt's maps feature the same close-up inset of the Havana harbor, a visual representation of the very scene with which he opens the *Political Essay on the Island of Cuba*, establishing an immediate dialogue between image and text. The entwinement of two different media recalls the juxtapositions of word and image that Humboldt had used to such striking effect in *Vues des Cordillères*. Now, he employs it to reframe the island of Cuba. The scale and the visual border he selected for these two maps show Cuba as separate from the rest of the Antilles, underscoring the island's importance and distinctiveness. The narrative discursively enhances this effect and projects it into the future.

That Humboldt, in the 1820s, chose for his essay on Cuba a title that parallels the one of his earlier writings on Mexico suggests that he already envisioned Cuba as a future independent nation. At the time that Humboldt drew his maps and published his work on Cuba, Mexico had already gained its political independence. In the decades leading up to and including the Mexican War, Humboldt's *Political Essay on the Kingdom of New Spain* came to play a fundamental role in shaping U.S.-Mexican relations. Likewise, for Cuban intellectuals, such as Fernando Ortiz and Vidal Morales y Morales, both Humboldt's maps and the text of his *Political Essay on the Island of Cuba* would be key components in the ongoing nation-building process that led up to the Spanish-Cuban-Philippine-American War in 1898 and the Cuban revolution in 1959.

For Humboldt, Cuba was a true microcosm of the New World, a place in which the most diverse peoples had been thrown together into an uneasy coexistence. Most precarious of all were Cuba's race relations, which slavery and racial mixing continued to exacerbate. Many thought an escalation of racial conflict on this "sugar island" virtually inevitable after the Haitian Revolution, and indeed, there were two major slave insurrections in Cuba by the mid-1840s. Both were violently put down. Humboldt's characteristic dislike for such violence shows in the daring projection of an "African Confederation of the Free States of the Antilles," which he posts as a warning to Cuba's all-too complacent slaveholders:

> If the laws in the Antilles and the legal status of people of color do not change for the better soon and if we continue to talk without acting, political supremacy will pass into the hands of those who have the power of labor, the will to emancipate themselves, and the courage to endure long privations. This bloody catastrophe will occur as a necessary result of circumstances without any involvement on the part of Haiti's free blacks, who will not have to abandon the isolation that they have practiced up to this day. Who would dare predict the impact that an *African Confederation of the Free States of the Antilles*, situated between Colombia, North America, and Guatemala, would have on the politics of the New World? (68)

No such confederation came about either in the nineteenth or in the twentieth century, and the Caribbean has continued to be linguistically and politically balkanized even after European colonialism had ended on most islands, on some as late as the 1960s. But attempts at unifying the region, such as the short-lived West Indies Federation (1958–62) and especially the more broadly based Caribbean Community (CARICOM) that succeeded it in 1973, do lend some credence to Humboldt's vision.

The explicit stand against slavery and the slave trade in the *Political Essay on the Island of Cuba* made it arguably the most controversial of Humboldt's publications—arguably because *The Political Essay on the Kingdom of New Spain* had not been received warmly in Spain due to Humboldt's pointed criticisms of colonial abuses and the exploitation of the native populations. But slavery was a far more volatile political topic in Humboldt's day, for the European colonial powers as much as for America's Creole elites, and his arguments and adamant pleas fell largely on deaf ears, at least in Spanish America. The colonial authorities in Havana promptly banned the Spanish translation of Humboldt's text, José de Bustamante's *Ensayo pólitico sobre la Isla de Cuba* from 1827, because of its "dangerous" sentiments, paying little attention to statistically supported analyses that showed how neither slavery nor the slave trade, which continued illegally after 1820, could be justified even on economic grounds.

Barely three decades later, these same sentiments and arguments propelled Humboldt to instant fame in U.S. abolitionist circles. The catalyst for this recognition was a public letter, in which Humboldt complained that one John Sidney Thrasher, in his English translation of the *Political Essay on the Island of Cuba* from 1856, had suppressed "the very portion of my work to which I attach far greater importance than to the arduous labors of determining locations through astronomical observations, experiments with magnetic intensity, or statistical statements."

Translating Humboldt

Although John Thrasher's *The Island of Cuba* was not the first but, in fact, the third English translation of the *Political Essay on the Island of Cuba*, it has had a disproportionate impact on how Humboldt's writings on Cuba have been received in the English-speaking world since the mid-nineteenth century. The first English version of Humboldt's text had appeared in 1829 as part of Helen Maria Williams's seven-volume *Personal Narrative of Travels to the Equinoctial Regions of the New Continent* (her version of the *Relation historique*). The poet Williams lived in Paris at the time and was in close contact with Humboldt, who spoke quite highly of her translation—to the diplomatic and, at times, ironic Humboldt, it was "très jolie" [very nice]. Thomasina Ross's abridged three-volume version of the *Personal Narrative*, which saw print in London in 1852–53, proved far more marketable than Williams's idiosyncratic rendition, but it included only portions of the work on Cuba. Ironically, what has been deemed the most readable English translation of the *Political Essay on the Island of Cuba* was Thrasher's. It was also by far the most unreliable. For one, Thrasher did not know any French and based his translation on the Spanish edition of Humboldt's text (the one that had been banned in Cuba). For another, he deliberately and systematically distorted the Spanish text to create the impression that Humboldt, like Thrasher himself, supported slavery.

Thrasher, who had spent considerable time in Cuba, favored the island's annexation to the United States as another slave state and would stop at little to promote his cause. Thrasher's attempt at turning public sentiment in the United States in favor of annexationism, a political scheme that would collapse with the outbreak of the U.S. Civil War, backfired when Humboldt himself issued a letter of protest only a few months after Thrasher's book had appeared in New York. In this letter, which was first published in the *Berlinische Nachrichten von Staats-und gelehrten Sachen* in Germany and reprinted widely in abolitionist journals from its English version in the *New York Daily Times* (August 12, 1856), Humboldt called Thrasher on his "politically motivated" cannibalization of the *Political Essay on the Island of Cuba.* What particularly provoked Humboldt's ire was that Thrasher had excised an entire section, which the Spanish version titled "Esclavos" (slaves) (it has no title in Humboldt's French text). "As an unrelenting advocate for the most unfettered expression of opinion in speech and in writing," Humboldt wrote, "I should never have thought of complaining, even if I were to be harshly attacked on account of my statements; I do, however, think I am entitled to demand that in the Free States of the continent of America, people should be allowed to read what has been permitted to circulate in the Spanish translation from the first year of its appearance." With this remonstrance, Humboldt decisively distanced himself from Thrasher's fraudulent fabrication.

No English translation of the *Political Essay on the Island of Cuba* since 1856 has endeavored to restore Humboldt's full text. In this vacuum, Thrasher's version, which was reprinted in 1966 and in 2001, has continued to supply a gravely skewed image of Alexander von Humboldt for English-speaking readers, especially in the United States. It is of little surprise, then, and highly ironic given Humboldt's recorded feelings and writings on the subject, that postcolonial critics from the English-speaking world have targeted Humboldt as an apologist for European colonialism and even for slavery in the Americas.

The unabridged retranslation in this edition turns over a new leaf. It is a fundamental departure from the image of Humboldt that Thrasher's *The Island of Cuba* created, and that has gone unchallenged for more than 150 years. *Ad fontes*—it is time to read what Alexander von Humboldt really wrote.

Nashville / Potsdam, April 2009

Note on the Text

The translation in this edition is based upon the full text of the freestanding French edition of Alexander von Humboldt's two-volume *Essai politique sur l'île de Cuba* from 1826, including all footnotes and tables. The numbers in the outer margins of each page refer back to that edition. Prior English and Spanish translations of this text have used its earlier, significantly shorter versions from 1825/26 and have often included others' materials in both the text and the footnotes, making it difficult at times to distinguish between Humboldt's own words and the writings of his translators and commentators. To avoid such confusion, we have placed our own additions and corrections in [square brackets], reserving the use of parentheses entirely for Humboldt himself.

We firmly believe that maintaining the original structure of this intensely nonlinear text keeps the reader engaged more directly and more fully with Humboldt's conceptual and narrative process. Throughout our work, we have retained his paragraphing and have also respected his decision neither to break his narrative into chapters nor to provide a table of contents. In place of a more conventional table of contents, we offer the following inventory of Humboldt's subsections to give readers a preliminary sense of how his narrative moves and what larger topics it covers. We have added several bracketed sections here to identify sections that other translations erroneously turned into chapters.

Because Humboldt typically supplied indexes for his work, even though he did not do so in this case, we felt quite at liberty to create a detailed index of our own to provide readers with additional navigational assistance to a text that has an often baffling amount of very specific information on many topics. Our inventory only suggests them in the broadest of strokes.

Throughout, we have also kept intact Humboldt's seemingly capricious use of capitalization and italics and all words, phrases, and quotations originally written in Spanish and other languages (such as Latin and Italian). We have not corrected occasional math errors in the tables. Only where meanings are not evident from context do we add our own translations of foreign words and phrases in [square brackets]. Humboldt himself rarely converts currencies and other units of measure, illustrating their variety across different parts of the world. We have followed his practice, offering explanations at the beginning of our annotations of currencies and units of measure no longer in use. To avoid encumbering this already densely textured work with endnote numbers, we have used ▾ to signal the presence of one or more annotations, which follow the text of the translation. The annotations are ordered in accordance with the pagination of the original text. To help readers trace at least part of the vast and intricate scientific network that Humboldt created, we provide a briefly annotated bibliography of the sources he acknowledges in this work.

Fold-outs of Humboldt's two Cuba maps accompany this edition to assist the interested reader in locating the many places on that island to which Humboldt refers in this work. The editorial note affords more information about the thinking behind this translation and about the historical and scholarly contexts for *The Political Essay on the Island of Cuba*. Select materials from these contexts can be found at www.press.uchicago.edu/books/humboldt/.

The HiE Team

POLITICAL ESSAY
ON THE
ISLAND OF CUBA

REASONED ANALYSIS OF THE MAP OF THE ISLAND OF CUBA

The map that accompanies the Political Essay on the Island of Cuba is part of the *Atlas géographique et physique des régions équinoxiales du Nouveau-Continent* [Geographical and physical Atlas of the equinoctial regions of the New Continent], of which twenty-two plates have already appeared. In this Atlas, as in my *Mexican* atlas [*Atlas géographique et physique sur la Nouvelle-Espagne*], I endeavored to correct our knowledge of America's interior on the basis of results from the astronomical observations I made and largely analyzed[1] during my voyages to the north of Lima and on the Amazon River. I sketched some maps partly in the field and partly after my return to Europe. Others were drawn—based either on my own sketches or on the [geographical] measurements I took—by expert geographers who were kind enough to aid in the publication of my works. In both cases, any errors in the Atlas of equinoctial America are mine alone. I trust that readers, in passing judgment on these attempts at gradually perfecting our knowledge of the Geography of Spanish America, will take into account exactly when each map was published. People will determine whether the author used all materials available to him at the time and of whose existence he was aware, whether he combined them with skill, and whether he enriched them with his own observations.

I.VIII

It is very easy to draw and revise maps in countries that have already been sites of important geodesic work: When the relations between distance and location have been determined precisely through triangulation, there are no longer any variations. The study of America's Geography, however, is far from such a state of perfection that rules out trial and error and careful choices

I.IX

1. See the results of these initial analyses, many copies of which circulate in America. I compared my data to Mr. Oltmanns's conclusive results in the ▾*Recueil d'observations astronomiques et de mesures barométriques* [Collection of astronomical observations and barometric measurements], Vol. I, p. XX, on which I collaborated with this meticulous, humble scholar between 1807 and 1811.

between quite variable data. Most of the coastline (in the north of Cuba, in Chocó, in Guatemala and Mexico from Tehuantepec to San Blas) has not been surveyed with care. Inland, only a few scattered astronomical observations can guide the Geographer. Accuracy increases when these points are sufficiently close together, grouped systematically, and linked by ▾*chronometric lines*. But to avoid the danger of altering mutually dependent coordinates over time, it is indispensable to outline the nature of the data that form the basis for each map. That is how, in my work in south America, the steppes of Venezuela (llanos), the Orinoco, the Casiquiare, and the Río Negro came to form a single *system of coordinates* that is linked by time lines to Cumaná and Caracas, I.X whose positions are based on conclusive astronomical observations.[1] Farther to the west, I connected the Río Magdalena, the Bogotá plateau, Popayán, Pasto, Quito, the Amazon, and Lower Peru in a second *system* from 10° 25′ northern latitude to 12° 2′ southern latitude. This latter group of coordinates, with Cartagena de Indias on one side and Callao de Lima on the other, has recently been connected to the former grouping by way of a *chronometric line* from west to east. In March of 1824, ▾Mr. Roulin, Mr. Rivero, and Mr. Boussingault extended Bogotá's time zone to the mouth of the Meta River which is situated about 6′ to the east of the Indian village Cariben. They discovered the difference between the meridian of the mouth of this river and Bogotá's meridian—0^h 26′ 7″—, while my own observations[2] from April 1800, made on a rock (*Piedra de la Paciencia*) that rises in the midst of the *Boca del Meta* I.XI and in Santa Fé de Bogotá from July and September 1801, show a difference in longitude of 0^h 25′ 58″. A series of measurements conducted inland connect Cumaná or the Orinoco Delta to the shores of the South Sea near Callao in Peru.

I offer this example, which has a *chronometric line* of 640 leagues and contains many intermediate coordinates determined during conclusive observations, to illustrate how the free governments of America might procure maps of their vast territories quickly and at little cost through astronomical methods. I also mention this example to emphasize the need for carefully analyzing existing work. Without apprising other Geographers of the level of accuracy that one believes one has attained, there is no way to know either how to perfect, through the correction of intermediate points, what has already been mapped or how to draw attention to places that are as yet insufficiently charted. The dissemination of these analyses is indispensible for the advancement of astronomical geography, particularly at a time when dramatic changes

1. Solar eclipses, Jupiter's moons, and lunar distances.

2. *Recueil d'observations astronomiques*, Vol. I, pp. 222[f]; Vol. II, pp. 23[5f].

in our knowledge of coordinates and configurations are resulting in new maps being drawn. If the relationship or the relative dependence of certain numbers of positions were to remain inadequately known, then future changes in knowledge would expose serious errors. I.XII

For my map of the island of Cuba, I used the astronomical observations of the most experienced Spanish seafarers and observations that I myself had occasion to make to the west of the port of Trinidad, on Cape San Antonio, in Havana, between that city and Batabanó, and in the *Jardines y Jardinillos* from Punta Matahambre to the Mouth of the Río Guaurabo. My collected data were published in great detail in the *Recueil d'observations astronomiques*, Vol. II, pp. 13–147, 567. On the map of the island of Cuba, first drawn up in 1819 and published in 1820, the port of Batabanó, the Flamenco, Piedras, and Diego Pérez Keys, the port of Trinidad, and Cabo Cruz are situated toward the south, in their true positions. But on this early map, the latitude of the northern coast[1] of the Isle of Pines and the entire configuration of Cuba's southern coast, from Cape San Antonio to the eastern edge of the Cayos [Laberinto] de las Doce Leguas were as erroneous as they were in the otherwise generally praiseworthy maps that the ▾*Depósito hidrográfico de Madrid* had published up to that point. The important corrections of Cuba's southern coastline that the naval Lieutenant ▾Don Ventura de Barcaíztegui and the frigate Captain Don José del Río made in 1795 and 1804, respectively, were not published until 1821. I adopted their corrections of the coastline between Punta de la Llana and Cape San Antonio, and between the Cabeza del Este de los Jardinillos and Cabo Cruz (excepting Trinidad's coordinates), in the second edition of my map of the island of Cuba (1826). The middle part—from long. 83° 30′ to 86° 20′—between the Cortés Lagoon, the Isle of Pines, and the Bay of Pigs, is copied from a sketch that my learned friend, ▾Don Felipe Bauzá, former director of the Depósito Hidrográfico in Madrid, kindly drew for me during my stay in London in May of 1825. When he entrusted me with this draft, this tireless member of the ▾Malaspina expedition told me that he had merged my determinations with Mr. del Río's plottings, and that he was busy completing a large map of the island of Cuba, on four sheets, for which purpose he was subjecting his materials to renewed scrutiny. Mr. Bauzá's name is enough to ensure the excellent quality of this map. I.XIII I.XIV

The history of Cuba's Geography has gone through the same phases as did that of the other Antillean islands and the east coasts of the New Continent. Initially, all points were placed too far to the west. Following what he called *las reglas de la Astronomía* [the rules of Astronomy], Christopher Co-

1. Compare Purdy, *Columbian Navigator*, p. 175.

lumbus[1] concluded that Cape San Antonio was located 75° to the west of the meridian of Cádiz. This 3½° mistake was compounded by another 4° in I.XV ▾First Mate Pedro de Medina's famous world map[2] from 1576. The *Quarterón* by ▾Bartolomé de la Rosa, preserved in the Depósito de Mapas in Madrid, I.XVI still places Havana 79° 14′ west of the Cádiz meridian in 1755, an error of 3° 9′, despite the fact that ▾Cassini[3] had already deduced the capital's actual longitude with an error of less than 45″ from the observations of eclipses of the moon and of Jupiter's moons that ▾Don Marco Antonio de Gamboa had made in Havana between 1715 and 1725. ▾Mr. Oltmanns used Bürg and Triesnecker's tables to good effect in analyzing[4] and recalculating Gamboa's observations; he arrived at an average result of 5^h 38′ 57″. The actual longitude of Havana's Morro is 5^h 38′ 49″, a surprising concurrence for these kinds of observations. If Don Bartolomé de la Rosa's *Quarterón* errs in its ▾absolute longitudes and once more places Havana 3¼° too far to the west, he still, as ▾Mr. Espinosa notes, provides relative longitudes with rare precision. The differences I.XVII between meridians for Havana's Morro, Punta de Guanos, and Key

1. In June of 1494, the Admiral also observed a lunar eclipse on Saint-Domingue's southern coast, another in September of 1494 near Adamana (today Saona Islet), a little to the west of Cabo Engaño. He found the difference with the Cádiz meridian to be 5h 23′, an error in longitude of 8° 45″ (Herrera, *Historia de las Indias occidentales*, Dec. I, pp. 56 and 58).

2. See the French translation by Nicolás de Nicolai, geographer to King Henri the Second, p. 64. This world map situates London at 58° latitude, an 18° difference with the meridians of Cape San Antonio and Temixtitlan (Mexico), with an error of 4°. Mexico's real longitude, as it was recognized (in 1778) by ▾Velázquez and [León y] Gama and confirmed by Don Dionisio Galiano (in 1791) and by myself (in 1803), is 6^h 45′ 42″. If ▾Mr. de Navarrete, whose literary talents and vast erudition I esteem greatly, had read the Reasoned Analysis for my Atlas of New Spain (*Essai politique sur le royaume de la Nouvelle-Espagne*, Vol. I, p. XV), he would not have reproached "a foreign traveler" in Mr. Zach's *Correspondance astronomique*, Vol. XIII, p. 56. He would not have taken recourse to the lunar eclipses that the Jesuit Sánchez had observed in 1584, and he would have been convinced that, when I published my own observations of satellites, lunar distances, azimuth, and time differences, I hastened to point out that my late friend, Don Dionisio Galiano, had arrived at 6^h 45′ 49″ for the longitude of Mexico City *before I did*, although the Gulf of Mexico map published in 1799 by Madrid's *Depósito hidrográfico* and a note that Mr. Espinosa sent me before my departure for Cumaná, indicated 6^h 52′ 8″. I was even the first (*Recueil d'observations astronomiques*, Vol. II, p. 496) to publish the Mexican observations from the Malaspina expedition. (To express more concisely the meridians on which longitudes are based in this report, I will henceforth use the simple initials Gr., Cz., and P. to designate the Greenwich, Cádiz, and Paris meridians, just as in thermometric observations.)

3. *Mémoires de l'Académie Royale*, 1729, p. 412.

4. *Recueil d'observations astronomique*, Vol. II, pp. 20–31.

Largo at the mouth of the Bahama Channel are exactly right. But such precise coordinates, so important for ships wishing to avoid Florida's sand bars in leaving the Channel and the Placer de los Roques (Salt Keys), can already be found in ▾Captain Francisco de Seixas y Lobera's old unpublished maps from 1692.[1]

Returning from his voyage to California where he and the ▾Abbé Chappe had observed Venus's passage, ▾Don Vicente Doz stopped over on the island of Cuba. He put Havana's longitude at 85° 7′, an error greater than half a degree. A similar longitude (85° 10′) was adopted in the famous *Mapa* of the *Gulf of Mexico*, the *Seno Mexicano*, which ▾Don José de San Martín Suárez drafted in 1787 after he had consulted with the ship captains and mates whom he had met in Havana. Too widely used for too long, this map has caused many a shipwreck.

I.XVIII

Since 1792 and 1795, a new era for Geography has begun on the island of Cuba and all the shores of the Antilles basin. Works by Barcaíztegui, ▾Rigada, Churruca, Ferrer, del Río, Cevallos, and Robredo came one after the other, correcting the contours of the coastline. And, thanks to the calculations and learned expositions of Mr. Ferrer[2] and Mr. Oltmanns,[3] Havana became one of the ports with the most precisely determined astronomical position in America. Between 1790 and 1794, Don Ventura de Barcaíztegui plotted the coastline between Santiago de Cuba and Punta Maternillos at the eastern entrance to the *Old Bahama Channel.* Don José del Río's works (1802–1804) include the southern coast between Cape San Antonio and Cape Cruz. What little we have known (since 1792) even about the *Old Channel* is due to the zealous efforts of *Capitán de Correos* Don Juan Henrique de la Rigada.[4] But in this

1. [Espinosa,] *Memorias de los navegantes españoles*, Vol. I., p. 93; Vol. II, p. 45.

2. *Connaissance des temps*, 1817, pp. 318–37. *Transactions of the American Philosophical Society*, Vol. VI, p. 107.

3. *Recueil d'observations astronomiques*, Vol. II, pp. 47–54 and 81, where Mr. Oltmanns's *État de la Géographie de l'île de Cuba* from 1809 can be found, p. 81.

4. [Moreno,] *Nueva Carta del Canal de Bahama*, 1805, according to observations by Don Dionisio Galiano (on the vessel San Fulgencio in 1799), by ▾Don Mariano Isasbirivil (on the vessel Goleta Elisabet in 1798), by Don Francisco Montes (on the Navío Ángel in 1799), and by Don Tomás Ugarte (on the Navío San Lorenzo in 1794). Coordinates and longitude differences between Matanzas, Cayo de Sal (at the western edge of the Placer de los Roques), Bajo Nicolao, Cayo de Piedras, Cruz del Padre, and eastern Megano are of premier importance for navigational safety. I also took into account the *Depósito*'s old works, especially for the first edition of my map: [Bauzá,] *Seno Mexicano*, 1799 (corrected in 1805); [Bauzá, Lángara, and Moreno,] *Carta de una parte de las Islas Antillas*, 1799 (corrected 1805); [Ferrer,] *Carta de la Isla de Santo-Domingo y parte oriental del Canal Viejo de Bahama*, 1802.

I.XIX region between Punta Maternillos and the port of Matanzas, much remains to be done through the study of astronomy, as is the case farther west between Bahía Honda and Cape San Antonio. Longitude positions there are entirely unclear, and unfortunately, these inaccuracies extend over a length of 135 nautical leagues.

As for the interior of the island of Cuba, it is a *terra incognita*, except for the triangle between Bahía Honda, Matanzas, and the Batabanó Surgidero [Outlet]. It is within this triangle that I astronomically determined the posi- I.XX tions of Fondadero, near the Town of San Antonio de los Baños, Río Blanco, Almirante, Antonio de Beita, the village of Managua, and San Antonio de Barreto. To chart the island's interior to the east of Güines, I used two rough sketches that I drew in Havana between 1803 and 1805. But these two sketches contradict each other too often. Cuba's general shape depends on determining exact coordinates for Cape San Antonio, Havana, Batabanó, Cape Cruz, and Maisí Point. Havana and Batabanó determine the *minimum* width of the island, which is 8⅓ nautical leagues, while old maps (even the ones that the Depósito published in 1799) give us 16 leagues. ▾However great the flaws in my map of the interior of Cuba, it is at least the first to offer contours traced according to the astronomical coordinates that we owe to the efforts of Spanish mariners. The names of all *ciudades* and *villas* are marked but without a guarantee as to their precise respective distances. These indications are im- I.XXI portant for those involved in statistical research on unequal population distribution. On old maps, the length, composition, and similarity of names (San Felipe y Santiago del Bejucal, Santiago de las Vegas or Compostela, San Antonio Abad or de los Baños) caused much confusion. Since I have typically referenced the sources that I consulted, I will restrict myself to a small number of partial examples.

Havana.—Because of the time difference of Nueva Barcelona, the chronometer, after 26 days of sailing on very choppy seas, measured $5^h\,38'\,40''$ for Havana's Morro, assuming that Nueva Barcelona is situated at $4^h\,28'\,19.2''$. The eight eclipses of Jupiter's moons that I observed with Don Dionisio Galiano, together with Mr. Robredo's far more numerous observations,[1] gave Mr. Oltmanns $5^h\,38'\,52.5''$ or $84^\circ\,43'\,7.5''$ as a conclusive result. Since my return to Europe, especially between 1806 and 1812, ▾Don José Joaquín de Ferrer and Don Antonio Robredo have observed many more stellar occultations I.XXII in Havana than have, up to now, been noted in any other part of America. In notes that Mr. Ferrer entrusted to Mr. Arago on his visit to Paris (in June 1814) and which were published in *Connaissance des Temps* in 1817, the Spanish

1. *Recueil d'observations astronomiques*, Vol. II, p. 89.

navigator, whose premature death all friends of science mourn, placed the Morro at 84° 42′ 44″. But in another more recent unpublished account he entrusted to Mr. Bauzá, Arago settled on 84° 42′ 19″, assuming Cádiz's location at 8° 37′ 45″ to the west of Paris. In the *Recueil d'observations astronomiques*, Mr. Oltmanns and I gave 13° 45′ 52″ as the difference between the meridians of the Morros of Havana and of Veracruz. Mr. Bauzá, who reopened the debate[1] over the coordinates of Havana, Veracruz, and Puerto Rico, arrived at 13° 45′ 40.5″, which differs less than a second from our result. According to Mr. Givry, the difference in meridians between Havana's Morro and Martinique's Fort Royal during the *Bayadère* expedition is 21° 21′ 26″.

Bahía Honda.—According to Ferrer,[2] the Potrero de Madrazo, the bay's southernmost point, is lat. 22° 56′ 7″ and long. 0° 49′ 26″ to the west of Havana's Morro. Basing his conclusions on this observation, Mr. Bauzá places the mouth of the bay at 85° 31′ 11″, between Morillo and Punta de Pescadores, assuming Havana's Morro to be at 84° 42′ 19″. I.XXIII

Cabo San Antonio.—My chronometer read 87° 17′ 22″ at the anchorage, and I situated the cape at 2° 34′ 15″ to the west of Havana's Morro. In *Memorias del Depósito Hidrográfico de Madrid*, Mr. Espinosa settled on 87° 8′ 41″. But since he situates Havana's Morro a little farther west[3] than I do, one has to adhere to the *Memorias* in determining the resulting discrepancy between the meridians at 2° 24′ 27″. Mr. Del Río,[4] however, had also arrived at 78° 39′ 0″ Cz, or 87° 16′ 45″ P., which differs from my result by only 37″. Captain Monteath measures 87° 19′ 23″, but this result seems to derive from a longitude between Port-Royal and Jamaica that British navigators have not uniformly established.[5] I.XXIV

Batabanó.—The Spanish original of Don José del Río's map[6] proposes lat. 22° 42′ 30″ and long. 84° 43′ 15″. Mr. Espinosa had indicated lat. 22° 43′ 10″ in the Table of positions. Mr. Oltmanns derived lat. 22° 43′ 19″, long. 84° 45′

1. [Bauzá,] *Sobre la situación geográfica de La Habana, Veracruz y Puerto Rico*, 1826 (unpublished).

2. *Connaissance des temps*, 1817, pp. 301–35.

3. The *Memorias* place the Morro first at 76° 0′ Cz; then, more precisely, at 76° 6′ 29″ Cz (Vol. II, pp. 67 and 91).

4. Findings from original observations that I received from Mr. Bauzá, who placed Cabo San Antonio at 87° 17′ 22″.

5. Mr. Oltmanns arrived at 79° 5′ 30″ by measuring Mercury's passage and the moons height; Mr. Bauzá found 79° 13′ 30″; De Mayne and Sabine arrived at 79° 13′ 30″ by means of lunar distances.

6. The French edition published by the royal naval Depot proposes lat. 22° 44′, long. 84° 42′.

56″ from Mr. Lemaur's geodesic calculations. Using different combinations, Mr. Bauzá settled on lat. 22° 43′ 34″ and long. 84° 46′ 23″.

Tetas de Managua.—Having made observations to the north and south of the Tetas in Managua village and in San Antonio de Barreto,[1] I assumed that the *eastern Teta* was at 22° 57′ 38″. ▾Don Pedro de Silva's trigonometric calculations, which Mr. Robredo sent me, suggest a more northerly latitude. These calculations, however, depend on absolute positions for the clock tower of Guanabacoa and the Mirador [overlook][2] of the Marquis of Real Socorro.

I.XXV

Trinidad.—I discussed this city's latitude during my second trip to Havana,[3] and I did not adopt the 21° 42′ 40″ coordinates from the new Spanish map drawn up on the basis of Mr. Del Río's observations. Three stars observed under less favorable conditions during the one night when I could make observations in Trinidad gave me 21° 48′ 20″. Gamboa and ▾Mr. de Puységur had already come up with 21° 46′ 35″ and 21° 47′ 15″, respectively. Coming from the *Jardinillos* near the Isle of Pines, I measured a longitude difference of 2° 22′ between Havana's Morro and the Pueblo de Trinidad at the stern, due to Havana's time difference. This longitude coincides[4] with that on Mr. Del Río's special map, which gives 82° 23′ 45″. Puerto Casilda is at 3′ 30″ to the south of the city but along the same meridian. According to his unpublished notes, Del Río places Boca de Guaurabo (South Point) at lat. 21° 42′ 24″ and long. 73° 49′ 45″ Cz.

I.XXVI

Cabo de Cruz.—I adopted Mr. Ferrer's coordinates: lat. 19° 47′ 16″ and long. 4° 38′ 29″ to the east of Havana's Morro. Del Río[5]: lat. 19° 49′ 27″ and long. 80° 3′ 27″.

The Morro of Santiago de Cuba.—Reporting ▾Don Ciriaco Cevallos's observations about Puerto Rico's coordinates, Mr. Oltmanns suggests 78° 21′ 42″. Mr. Bauzá adopts 78° 16′ 41″ for the Morro of Santiago and 77° 35′ 36″ for the Puerto of Guantánamo, and my map situates this port at 77° 38′.

Punta de Maisí.—Here is another place whose coordinates are chronometrically dependent on those of Puerto Rico. Doubts have once again been raised about Puerto Rico's longitude, which was once thought to have been

1. *Relation historique*, Vol. III [Q], p. 365. [Q refers to the quarto, or folio, edition.]

2. *Recueil d'observations astronomiques*, Vol. II, 567. According to Ferrer, the eastern Teta [teat or mound] is at lat. 22° 58′ 18.5″ and long. 0° 2′ 48″ to the Morro's west; according to Del Río, at lat. 22° 0′; the French Depot's map shows lat. 22° 1″.

3. *Recueil d'observations astronomiques*, Vol. II, p. 72.

4. *Memorias del Depósito* [that is, Espinosa, *Memorias sobre las observaciones astronómicas*] (Vol. II, p. 64): Trinidad, Pueblo, long. 82° 23′ 31″; my own chronometer: 82° 21′ 7″.

5. I continue to refer to this officer's original observations, which Mr. Bauzá gave me.

determined with extreme precision. Mr. de Zach[1] even finds that it is inaccurate by as much as 5′ or 6′. Results differ by this much depending on whether one confuses the observations with an entirely different value or keeps them separate. By assuming the Morro of Puerto Rico to be at 59° 50′ 44.5″ Cz, Mr. Bauzá arrives at 76° 26′ P. for Punta de Maisí. I.XXVII

ˇDon José Luyando's excellent chronometers measured lat. 21° 39′ 40″ and long. 70° 46′ 23″ to the west of Cádiz for Punta de Maternillos and for the three following points: Punta de Mangles 19° 52′ 33″, Cayo de Moa 21° 17′ 10″, and Cayo de Guinchos, 18° 2′ 9″ to the east of the castle of San Juan Ulúa, which we situate at long. 98° 29′. Following the original summary of Don José del Río's observations, I add: the mouth of Río San Juan,[2] NW Point, lat. 21° 48′ 18″, long. 74° 3′ 5″ Cz; the mouth of the Jagua, lat. 22° 1′ 7″, long. 74° 18′; Punta Matahambre, NW edge, lat. 22° 21′ 34″, long. 75° 53′ 29″; Cayo Flamenco, lat. 22° 1′ 0″, long. 75° 20′ 8″; Cayo de Don Cristóbal, the southernmost[3] point, lat. 22° 50′ 3″, long. 75° 35′ 30″; Piedras de Diego Pérez, lat. 22° 1′ 39″, long. 75° 18′ 15″; Cayo Piedras[4] (not to be confused with the other Key of this name near Boca Grande to the east of Cayo Bretón), lat. 21° 57′ 39″, long. 74° 49′ 48″. I.XXVIII I.XXIX

ˇCaptain De Mayne, who has profoundly enriched our knowledge of Antillean Geography, situated the SE cape of Anguilla island at lat. 23° 29′ 30″ and long. 79° 27′ 0″ Gr., or 81° 47′ 15″ P. Mr. Bauzá prefers 81° 45′ 19″.

1. *Correspondance astronomique*, Vol. XIII, p. 128. Don José Sánchez Cerquero (today the Director of the Observatorio de la ciudad de San Fernando) situated the Morro of Puerto Rico at 68° 27′ 15″ in 1816, according to the calculations of Aldebaran's occultation of October 21, 1793. According to Mr. Ferrer (*Connaissance des temps*, 1817, p. 322), 68° 28′ 3″; according to Mr. Bauzá, 68° 28′ 29″; Mr. de Zach, 68° 31′ 3″. Calculations from Aldebaran's only occultation had given Mr. Oltmanns 68° 35′ 15″ (*Recueil d'observations astronomiques*, Vol. II, p. 125). The average occultations of lunar distances and chronometrical readings is 68° 32′ 30″, but Mr. Oltmanns prefers 68° 33′ 30″. Puerto Rico therefore fluctuates between 68° 28′ and 68° 34′, and its position is far less certain than those of Havana, Veracruz, Cumaná, and Cartagena. By assuming Puerto Rico to be at 59° 50′ 44.5″ Cz, Mr. Bauzá, through laborious research, arrives at a longitude difference of 16° 12′ 16.5″ between Havana's Morro and Puerto Rico's, and 30° 0′ between Veracruz and Puerto Rico.

2. *Relation historique*, Vol. III [Q], p. 478. I gave a list of all anchorages on the island of Cuba on pp. 384 and 385.

3. Certainly not the same *Cayo* whose latitude I approximated at 22° 10′ (*Observations astronomiques*, Vol. II, p. 110).

4. I arrived at lat. 21° 56′ 40″ but long. 1° 8′ 44″ to the west of Batabanó. One should not forget that absolute longitudes are all based on that of Batabanó, which I situate at 84° 45′ 56″ and Mr. del Río places at 84° 43′ 15″.

I remain quite uncertain about the true coordinates of the Town of [Puerto] Príncipe [the city of Camagüey]. Gamboa observed there several stars' upper meridians and (on August 15, 1714) the immersion of Jupiter's first moon. Mr. Oltmanns determined the latitude very precisely at 21° 26′ 34″; this seems right. But if one adopted a longitude of 80° 39′ 30″, the Town of Príncipe would almost coincide with the meridian of Sabana la Mar near Punta de Judas to the east of the point where I placed Morón, following unpublished maps that had been sent to me from Havana. Given the current state of Geographical knowledge on *The Old Bahama Channel*, this way of attaching Príncipe to the north coast strikes me as unfounded. Quite likely, there are huge errors in longitude to the west of Punta Maternillos. But could they be as much as one degree? To this day, we do not know. Mr. Ferrer and Mr. Luyando have already acknowledged a 28′ error for the Cayo de Guinchos. Mr. Bauzá informs me that the map drawn up on Count Jaruco's orders (a map in which distances and the coastline's configuration are highly flawed) shows the Villa (today, Ciudad) de Santa María del Puerto Príncipe situated at S 36° W of the Silla de Cayo Romano, which is 54 miles away. But how can one reconcile such a westward position with Don Francisco María Celi's unpublished map on which the Villa de Puerto Príncipe is situated barely 0° 16′ to the west of the mouth of the Río Máximo and, simultaneously, in the meridian of Cayo Confites[1]? In the second edition of my Cuba map, I removed the name of Puerto Príncipe, which I had borrowed from Jefferys's map. It is nevertheless certain (and Celi's manuscript chart indicates this) that there was once a settlement called *Embarcadero del Príncipe* to the east of Punta Curiana, between the mouths of the Caonao and Jiguey Rivers.

I.XXX

I.XXXI

According to Gamboa's fine observations of latitude, the Town of Sancti Spíritus is situated at 21° 57′ 37″. A single lunar eclipse caused the longitude to fluctuate between the meridians 81° 47′ and 82° 9′.

The Caymans.—I discussed elsewhere[2] the position of these islets, which have wandered on our hydrographic maps for quite some time now. At various times, the fine maps of the Depósito de Madrid have situated Grand Cayman's NE cape at 82° 58′ (from 1799 to 1804), at 83° 40′ (in 1809), and once more at 82° 59′ (in 1821). The latter position from Barcaíztegui and Del Río's map is identical to the one that I believed I could deduce by measuring the

1. Celi's extremely detailed chart, for which he used a compass, shows a *Serranía de piedra imán* [mountain of magnetic stone] 17 leagues to the west of the town of Príncipe. Magnetic attractions could have wreaked havoc with readings.

2. Compare my *Recueil d'observations astronomiques,* Introduction, p. XLIII, Vol. II, p. 114; *Relation historique* [Vol. III, Q], p. 329. [Espinosa,] *Memorias del Depósito hidrográfico*, Vol. II, p. 66.

sun's height a few times at a 12-mile distance on heavy seas while the first mates said that, according to the compass readings, we were on the meridian of the island's center. The horizon was hard to make out and foggy, but the ▾hour angle coincided enough to remove uncertainty as to the vessel's position within 12″. Should we conclude that Louis Berthoud's chronometer malfunctioned to such a degree that, 6 days later, the same instrument showed exactly 87° 17′ 22″ for the longitude of Cape San Antonio? More likely, I was not across from the Center of Grand Cayman, and the play of magnetic attractions was confusing the compass readings. Here are some more data: ▾Purdy's map, based on Captain Livingston's observations (1823), has 83° 52′ for Grand Cayman's SW cape and 83° 24′ for its NE cape. A map of Cuba's south coast, French naval Depot edition from 1824 corrected by ▾Captain Roussin who, in collaboration with the knowledgeable hydrographer Mr. Givry, has added so much to our knowledge of Brazil's geography, shows: 83° 46′ (lat. 19° 24′) for the NW cape. Captain De Mayne's map has 83° 49′ 15″ (lat. 19° 22′ 30″) for the NW cape and 85° 47″ (lat. 19° 14′) for the SW cape. It is the latter position that I adopted for my second edition of the map of the island of Cuba. Mr. Sabine reports the location of his observations on magnetic intensity[1] at lat. 19° 25′ (?) and long. 83° 25′ 15″.

I.XXXII I.XXXIII

Del Río's map situates *Little Cayman* (*Caimán Chico occidental*, to Spanish sailors) at 82° 25′ for its NW longitude. Mr. Bauzá, however, adopts 82° 2′ (lat. 19° 44′). I situated the eastern cape of *Cayman Brac* (*Caimán Chico oriental*, to Spanish sailors) at 82° 7′ 37″ by chronometrically[2] linking this point to Trinidad de Cuba after 36 hours of sailing. The time difference with Puerto Rico had given Mr. Cevallos 81° 59′ 36″, assuming that Aguadilla was at 0° 59′ 54″ to the west of Puerto Rico's Morro and that the latter, following Mr. Oltmanns, was at 68° 33′ 80″. Such doubts as to the location of Grand Cayman and the two Little Caymans islands, which sailors sometimes confuse with each other, will not be cleared up until a single observer, equipped with several chronometers, will have examined the three islets in turn and determined their lengths and their respective distances[3] by connecting them to the meridian of Cabo San Antonio.

I.XXXIV

By taking this cape as the basis for all studies of the southern coast of the island of Cuba, one can measure the extent to which the results of different ob-

1. *Pendulum Experiments*, 1826, p. 401.

2. *Recueil d'observations astronomiques*, Vol. II, p. 113.

3. William Dampier had already judged the space between the *Cayman Chico occidental* and *Cayman Grande* at no more than 15 nautical leagues (*Voyages and Descriptions*, ed. from 1696, Vol. II, Part I, p. 30).

servers actually disagree with one another. Frigate Captain Don José del Río, for instance, does not provide a longitude for Havana's Morro in his unpublished notes. But, by tying the *Jardinillos* to Cape San Antonio, which he situates no more than 37″ farther east than I do, one can see that this sailor assumes that the *Cayos* are generally 4′, sometimes even 6′ to 9′, farther east than I do.

Meridian Differences for Cabo		
San Antonio and Cayo Flamenco	3° 18′ 52″.	Del Río.
	3° 13′ 50″.	Humboldt.
Piedras de Diego Pérez	3° 20′ 45″.	Del Río.
	3° 14′ 20″.	Humboldt.
Cayo de Piedras	3° 49′ 12″.	Del Río.
	3° 40′ 10″.	Humboldt.

I.XXXV Farther east, the differences suddenly become much smaller when one considers differences in longitude between Cape San Antonio and the following places:

	Del Río.	Humboldt.
Río San Juan	4° 35′ 55″	4° 36′ 33″
Boca de Jagua	4° 21′ 0″	3° 23′ 0″
Trinidad[1] (town	4° 53′ 0″	4° 56′ 15″

I doubt that Cape San Antonio has ever been connected to Cabo Cruz by continuous triangulation. And when using chronometers, the imprecision of the hour angles taken above the sea's horizon can compound the inaccuracies that arise from discrepancies between clocks. What makes me believe that the error is not so much on my part is that my longitudal readings in the *Jardinillos* are broadly in agreement with Mr. Espinosa's (see the Introduction to my *Recueil d'observations astronomiques*, Vol. I, p. XLVI). The average difference I.XXXVI is no more than 12″ to 15″.

	Northern Latitude		Longitude East of Batabanó	
Place Names	Espinosa	Del Río	Espinosa	Humboldt
Cayo Flamenco	22° 2′ 30″	22° 1′ 0″	0° 46′ 11″	0° 42′ 24″
Cayo de Don Cristóbal	22° 12′ 4″	22° 5′ 30″	0° 25′ 11″	0° 24′ 56″
Piedras de Diego Pérez	22° 0′ 40″	22° 1′ 39″	0° 46′ 41″	0° 42′ 54″
Cayo de Piedras	21° 56′ 40″	21° 57′ 39″	1° 8′ 46″	1° 8′ 44″
Punta Matahambre	22° 18′ 5″	22° 21′ 34″	0° 11″	0° 6[′] 56″

1. Map of the Río Guaurabo, created by frigate captain José del Río in 1803.

As for the *Jardinillos*'s latitudes, which are not the same in Mr. del Río's manuscripts and in Mr. Espinosa's table, I should mention here that I did not determine a single one on dry land, but that they are only approximate and derived from earlier meridian altitudes.

The map of the island of Cuba was compiled by ▾Mr. Lapie, squadron chief of the royal corps of geographer-engineers of France, who has recently acquired further title to geographers' esteem thanks to his excellent work on Greece and the Archipelago.

TABLE
OF GEOGRAPHICAL POSITIONS FOR THE ISLAND OF CUBA,
DETERMINED BY ASTRONOMICAL OBSERVATIONS

Names of Places.	Northern Latitude.	Longitude West of Paris.	Names of Observers, and Comments.
HAVANA, signal light of the Morro	23° 9′ 24.3″	84° 43′ 7.5″	Robredo, Ferrer, Galiano, Humboldt (definite result from Mr. Oltmanns in 1808). In 1817, Ferrer decided on 84° 42′ 44″; later by 21 stellar occultation on 84° 42′ 19″.
TETA ORIENTAL DE MANAGUA	22 58 3	84 40 0	Lemaur, Ferrer, Humboldt.
MANAGUA, village	22 58 48	84 37 34	Humboldt; long. uncertain, lat. certain at around 10″ or 12″.
SAN ANTONIO DE BARRETO	22 56 34	....	Humboldt.
RÍO BLANCO	22 51 24	84 31 15	*Id.*
EL ALMIRANTE	22 57 36	84 36 7	*Id.*
SAN ANTONIO DE BEITIA	22 53 25	84 39 13	*Id.*
EL FONDEADERO	22 51 34	84 54 30	(near the town of San Antonio de los Baños), Humboldt.
LOS GÜINES	22 50 27	....	Lemaur.
INGENIO DE SEIBABO	22 52 15	....	*Id.*
SAN ANTONIO DE LOS BAÑOS	22 53 31	....	*Id.*
MADRUGA, village	22 55 0	84 12 23	Ferrer.
CAFETAL DE SAN RAFAEL	22 57 16	84 9 28	Ferrer.
MESA DEL MARIEL	22 57 24	85 0 20	Ferrer (median of Guanajay).
TORREÓN DEL MARIEL	....	85 3 14	Ferrer.
MATANZAS, town	23 2 28	83 57 59	*Id.*
PAN DE MATANZAS	23 1 55	84 2 49	*Id.*
PUNTA DE GUANOS	23 9 27	84 1 7	Ferrer.
MADRAZO	22 56 7	85 32 33	Ferrer (southernmost corner of the coast of Bahía Honda).
MORILLO DE BAHÍA HONDA	22 59 0	85 31 15	*Id.*
PAN DE GUAIJABÓN	22 47 31	85 44 36	*Id.*
CABO SAN ANTONIO	21 49 54	87 17 22	Humboldt.
BATABANÓ	22 43 19	84 45 56	Lemaur.

I.XXXVII

Names of Places.	Northern Latitude.	Longitude West of Paris.	Names of Observers, and Comments.
CAYO DE DON CRISTÓBAL	22° 10′ 0″	84° 21′ 0″	Humboldt.
CAYO FLAMENCO	22 0 0	84 3 32	*Id.*
LAS PIEDRAS DE DIEGO PÉREZ	21 58 10	84 3 2	Humboldt. The latitudes of the Jardines y Jardinillos are not based on actual observations but have been inferred at a distance from the Keys' meridian.
CAYO DE PIEDRAS	21 56 40	83 37 12	
BOCA DE JAGUA, WESTERN POINT	22 1 7	85 4 22	
BOCA DEL RÍO SAN JUAN, NORTHERN POINT	21 48 18	82 40 50	Del Río, Humboldt.
TRINIDAD, town	21 47 20	82 21 7	Gamboa, Puységur, Humboldt (lat. debated).
CABO DE CRUZ	19 47 16	80 3 52	
SANTIAGO DE CUBA (MORRO)	19 57 29	78 16 41	Cevallos, Bauzá.
PUERTO DE GUANTÁNAMO	....	77 35 36	Bauzá.
CABO BUENO	20 6 10	76 33 32	Ferrer.
CABO MAYSÍ	20 16 40	76 30 25	Ferrer (Bauzá, long. 76° 26′).
CAYO DE MOA	....	77 12 0	Luyando.
PUNTA DE MULAS	21 4 35	77 56 32	Ferrer.
PUNTA MATERNILLOS	21 39 40	79 24 15	Luyando.
CAYO DE GUINCHOS	....	80 27 0	Luyando; in the Old Bahama canal.
CAYO VERDE	22 5 6	79 59 32	Ferrer.
CAYO DE LOBOS	22 24 50	79 55 43	*Id.*
CAYO CONFITES	21 11 44	80 3 45	*Id.*
CAYO SANTA MARÍA	22 39 24	81 16 50	*Id.*
XXXVIII STA. MARÍA DE PUERTO PRÍNCIPE, town	21 26 34	....	Gamboa, Oltmanns.
SANCTI SPÍRITUS, town	21 57 36	....	Oltmanns.
ISLA ANGUILLA, SE cape	23 29 30	81 45 19	De Mayne.

I.XXXIX The table of positions for the island of Cuba depended on a very small number of places, the most important having been discussed in the preceding pages. As almost all these coordinates depend on the exact determination of Havana's meridian (specifically, that of the Morro), one has to consider that Mr. Ferrer (in his Journal from 1814) and Mr. Bauzá (according to an Account that Mr. Ferrer compiled a short time before his death) situate this meridian 23″ and 48″ farther east than Mr. Oltmanns does. When I have indicated Mr. Oltmanns's old results in my table of coordinates, I have done so only to remain consistent with other points in the tables in my *Recueil d'observations astronomiques*. Besides, I have only recorded differences in longitude between the Morro and other points (capes, keys, etc.), and for these, a margin of error of 3″ is dwarfed by *varying interpretations* of the data. By excluding solar eclipses—the February 21, 1803 and the June 16, 1806 eclipses gave longitude readings too far to the west—and only considering occultations (Mr. Ferrer had published 16 of these by 1814), I arrived at 84° 42′ 18.5″ for the Morro of I.XL Havana. Of these 16 occultations, 10 do not deviate from the average result by more than 1″.

Tables of coordinates would be more useful to sailors and geographers if they typically showed the limits within which each longitude oscillates, as far as we know now. It is not easy to derive results from different kinds of calculations. In this process, which would require the calculation of probabilities, Geographers proceed by trial and error. For example, from the same number of stellar occultations oscillating around an average longitude of 2″ to 8″, one can come to very different conclusions depending on whether one takes the average of all observations or whether one excludes some of them. The problem is even more difficult to resolve when one oscillates between the margin of error from a small number of occultations, solar eclipses, or planetary passages, and the margin of error from a large number of satellites, lunar heights, and lunar distances. The extremes in longitude between which each loca- I.XLI tion swings should be taken in the same way that yearly *high* and *low* temperatures are considered. These limits should recall that it is quite likely for a given place (for example, the port of Cartagena) to be situated no farther east than 77° 47′ 50″ and no farther west than 77° 51′ 15″, according to our current state of knowledge in astronomical Geography. As observations whose results are closer to the extreme limits do not offer an equal degree of certitude, the longitude that we can currently consider the most likely is certainly not the average between extreme longitudes. The following table offers, in a limited space, an attempt at compiling all more or less definitive results for 20 coordinates founded on observations of celestial phenomena. The common expression "chronometric longitude" is exceedingly vague, if one does not know what position was adopted as a point of departure. I always added this element to meridian differences that were obtained through chronometers.

Names of Positions.	Farthest Extensions.	Comments.
CUMANÁ (Castillo de San Antonio)	66° 29′ 15″ and 66° 31′ 10″	Probably 66° 30′ 0″. *Solar eclipse. Saturn. Lunar distance* (Solar eclipse 4h 25′ 45″. Saturn 4h 25′ 37.5″. Lunar distance 4h 25′ 32.5″. Chronometric difference of meridians of Cumaná and Sta. Cruz de Tenerife 3h 11′ 52″; hence chronometrical longitude 4h 26′ 4″. Humboldt, Oltmanns).
LA GUAIRA (jetty)	69 23 10 and 69 29 00	Probably 69° 27′ 0″. – *Saturn. Lunar distance* (Saturn 69° 30′, Ferrer, Oltmanns. Lunar distance 69° 18′, Ferrer, but tables from ▼Mason).
CARTAGENA DE INDIAS (cathedral)	77 47 50 and 77 51 15	Probably 77° 50′. – *Passage of Mercury. Occultation. Saturn.* (Passage of Mercury 77° 46′, Fidalgo, Robredo, ▼Tiscar. Occultation 77° 47′ 54″, Fidalgo, Tiscar. Occultation 77° 48′ 15″, ▼Noguera, Oltmanns. Occultation 77° 51′ 45″, Ferrer. Solar eclipse 77° 49′ 55″, Tiscar, Robredo. Saturn 77° 51′ 15″, Noguera, Oltmanns. Chronometric difference of meridians of Cartagena and of Havana's Morro 6° 54′ 15″; hence longitude 77° 48′ 4″, Humboldt).
HAVANA (Morro)	84 42 19 and 84 43 10	Probably 84° 42′ 19″. – *Occultation. Solar eclipse. Saturn* (21. Occultation 84° 42′ 19″, Ferrer, Robredo. Solar eclipse 84° 44′ 24″, Robredo, Ferrer; but according to the most recent tables, Oltmanns 84° 43′ 4″. Saturn 84° 42′ 54″, Humboldt, Galiano, Robredo, Oltmanns. Chronometric difference of meridians of the Morro and of Puerto Rico 16° 12′ 16.5″, Bauzá).
PUERTO RICO (Morro)	68 27 45 and 68 34 00	Probably 68° 33′ 30″. – *Occultation. Lunar distance* (Aldebaran's occultation under unfavorable circumstances, 4h 33′ 22″, Churruca, ▼Lalande; 4h 33′ 36″, ▼Méchain; 4h 33′ 58.6″, Triesnecker; 4h 34′ 7.6″, ▼Wurm; 4h 33′ 38″, Ferrer; 4h 34′ 22.9″, Oltmanns; 4h 33′ 46″, Cerquero; 4h 34′ 4″, Zach. Lunar distance 68° 24′ 41″, Ferrer, but in more recent tables Oltmanns 68° 27′ 45″. Chronometric longitude through Havana 68° 30′ 3″; through Veracruz, 68° 29′, Bauzá, Oltmanns).
I.LXIII FORT-ROYAL (Martinique)	63 25 40 and 63 28 6	Probably 63° 26′ 0″. – *Passage of the moon. Saturn. Chronometric* (Lunar passage 63° 26′ 0″, ▼Pingré, Oltmanns. Chronometric difference of meridians of Fort-Royal and of Cap Français, 11° 10′ 36″, hence chronometric longitude 63° 27′ 34″; of Fort-Royal and Falmouth on the isle of Antigua 0° 44′ 0″; hence chronometric longitude 63° 28′ 6″, Borda).

Names of Positions.	Farthest Extensions.	Comments.
PORT ROYAL (Jamaica)	79° 3′ 45″ and 79° 13′ 30″	Probably 79° 5′ 30″. – *Passage of Mercury. Right ascension of the moon* (Passage of Mercury 79° 3′ 45″, ▼Macfarlane, ▼Candler, Oltmanns. Right ascension of the moon 79° 7′ 15″, Macfarlane Oltmanns. Chronometric longitude, 79° 13′ 30″, Sabine; 79° 12′ 45″, De Mayne).
FORT WILLOUGHBY (Barbados)	61 55 45 and 61 57 30	Probably 6° 56′ 48″. – *Occultation. Saturn* (5 Occultations 4h 7′ 43.7″, ▼Maskelyne, Oltmanns; 12 sat. 4h 7′ 50″, Maskelyne, Oltmanns).
ILE ANHATOMIRIM (Brazil)	50 58 12 and 31 1 15	Probably 51° 1′ 14″. – *Lunar distance. Chronometric* (Lunar distance 51° 1′ 17″, ▼Duperrey. Chronometric difference of meridians of the [Ilha de] Anhatomirim and Santa Cruz de Tenerife, 32° 27′ 48″; hence chronometric 51° 0′ 53″, Roussin, Givry; of Anhatomirim and the Ratos Island 5° 25′ 32″, Givry, ▼Fouque, ▼Lartigue; hence chronometric longitude 51° 0′ 46″).
RIO DE JANEIRO (Ratos island [that is, Rat Island; also Snake Island; today known as Fiscal Island])	43 32 33 and 43 36 55	Probably 45° 35′ 14″. – *Saturn* (to the number of 285 lm. and Em.). *Lunar distance. Chronometric* (70 sat. 45° 36′ 55″, Dorta First sat. only 45° 36′ 40″. Chronometric longitude 45° 35′ 14″, Givry; 45° 32′ 33″, Fouque; 45° 36′ 22″, ▼Freycinet).
MONTEVIDEO	58 30 22 and 58 37 10	Probably 58° 34′ 20″. – *Passage of Mercury. Occultation. Saturn* (Passage of Mercury 58° 30′ 22″, Malaspina; Occultation 58° 37′ 11″, Malaspina; Saturn 58° 30′ 55″, Varela).
VALPARAISO (Rosario castle).	74 00 00 and 74 11 00	Probably...... – *Occultation. Solar eclipse. Saturn. Lunar distance* (Occultation 73° 51′ 15″, ▼Hall, ▼Foster; but according to Oltmanns, 74° 11′ 19″. Solar eclipse, 74° 8′ 15″, ▼Feuillée and Méchain; 74° 7′ 21″, Feuillée and Triesnecker. Saturn 74° 0′ 25″, Malaspina, Méchain; 74° 14′ 15″, Oltmanns. Lunar distance 73° 59′, Lartigue. Chronometric difference of meridians of Valparaiso and Callao 5h 30′ 40″, Malaspina; 5h 31′ 47″, Hall; 5h 30′ 43″, Lartigue; hence average chronometric longitude, 74° 3′ 27″. Chronometric difference of meridians of Valparaiso and Quilca, 0° 49′ 2″).
COQUIMBO	73 38 00 and 73 47 45	Probably...... – *Occultation. Saturn* (2 occultations 73° 47′ 45″, Malaspina, Tiscar; 2 sat. 73° 38′ 0″, Malaspina. Chronometric difference of meridians of Coquimbo and Valparaiso 0° 16′ 16″, average of Malaspina and Hall; of Coquimbo and Callao, average of ▼Atrevida, of Descubierta, and of Basil Hall, 5° 47′ 19″; hence, chronometric longitude, 73° 46′ 44″. Bauzá prefers for Valparaiso 74° 3′ 18.5″; for Coquimbo 73° 43′ 34″).

I.LXIV

	Names of Positions.	Farthest Extensions.	Comments.
I.LXV	CALLAO (San Felipe Fortress.)	79° 33′ 00″ and 79° 35′ 10″	Probably 79° 34′ 30″. – *Passage of Mercury. Saturn. Lunar distance* (Passage of Mercury 79° 34′ 30″, Humboldt and Oltmanns. Six sat. 79° 31′ 55″, observed at Lima, Oltmanns. One sat. 79° 35′ 54″, Malaspina, Oltmanns. Lunar distance 79° 29′ 41″, Lartigue; 79° 34′ 5″, Duperrey).
	GUYAQUIL (city jetty)	82 14 00 and 82 18 25	Probably 82° 18′ 10″. – *Occultation. Lunar eclipse. Chronometric* (Occultation 82° 18′ 11″, Malaspina, Oltmanns. Lunar eclipse compared to 6 observations corresp. 82° 18′ 25″, Malaspina and Oltmanns. Chronometric difference of meridians of Guayaquil and of Callao 2° 43′ 40″, Humboldt; hence, chronometric longitude, 82° 18′ 10″; of Guayaquil and of Callao 2° 39′ 52″, Malaspina; 2° 33′ 36″, Hall).
	QUITO (central square)	81 4 15 and 81 6 30	Probably 81° 4′ 38″. – *Saturn. Lunar eclipse. Lunar distance* (Saturn, 5^{h} 24′ 17″, Ulloa, ▼Godin, Oltmanns. Lunar eclipse 5^{h} 24′ 19″, Ulloa, Oltm. Lunar distance 5^{h} 24′ 26″, Humboldt. Chronometric difference of meridians of Quito and of Popayán 0^{h} 8′ 20.3″; hence chronometric longitude 5^{h} 24′ 21″, Humboldt).
	PANAMA (cathedral)	81 38 45 and 81 44 50	Probably...... – *Occultation. Saturn* (2 occultations 81° 38′ 17″, Malaspina, Tiscar; 2 sat. 81° 47′ 15″, Malaspina. Chronometric difference of meridians of Panama and of Acapulco 20° 33′ 5″, Malaspina; hence chronometric longitude 81° 36′ 28″. Several other chronometric combinations, through Portobelo and Cartagena de Indias, Mr. Bauzá, longitude 81° 43′ 33″).
	ACAPULCO (jetty)	102 9 30 and 102 13 00	Probably 102° 9′ 33″. – *Occultation. Saturn. Lunar distance* (Occultation 6^{h} 48′ 50.5″, Malaspina, Oltmanns. Sat. 6^{h} 48′ 58, Malaspina, Oltmanns. Lunar distance 6^{h} 48′ 26″, Humboldt. Chronometric difference of meridians of Acapulco and of San Blas 0^{h} 21′ 22″, Malaspina; 0^{h} 21′ 38″, Hall; hence average chronometric longitude 6^{h} 48′ 58″; of Acapulco and of Guayaquil 1^{h} 19′ 27″, Humboldt; hence chronometric longitude 6^{h} 48′ 39.8″).
I.LXVI	SAN BLAS (Contaduria)	107 35 40 and 107 38 50	Probably 107° 35′ 48″. – *Occultation. Saturn. Lunar distance* (Occultation 107° 38′ 42″, Hall and Foster one sat.; 107° 34′ 35″, Malaspina and Oltmanns; lunar eclipse 107° 36′ 45″, Malaspina, Oltmanns; lunar distance 107° 37′ 24″, Hall; Mr. Bauzá decides on 102° 12′ 41″ for Acapulco; for San Blas 107° 37′ 4″).

Names of Positions.	Farthest Extensions.	Comments.
VERACRUZ (jetty)	98° 28′ 00″ and 98° 30′ 15″	Probably 98° 29′ 0. – *Occultation. Saturn. Lunar distance. Chronometric* (Occultation 6h 33′ 57″, Ferrer, Oltmanns. Saturn 6h 33′ 52″, Ferrer and Oltmanns. ▾Hypsometer operations 6h 34′ 1″, Humboldt. Through a solar eclipse observed at Tabasco 6h 33′ 54″, Ferrer. Chronometric difference of meridians of Veracruz and Morro of Puerto Rico, 2h 0[′] 0″ Bauzá; of Veracruz and of Havana's Morro 13° 45′ 44″, Montes, Ferrer, Isasbirivil; hence chronometric longitude 98° 28′ 3″; of Veracruz and Cap François 23° 50′ 8″, Borda, Ferrer, Churruca; hence, chronometric longitude 98° 28′ 18″).

(When examining the limits within which the longitudes oscillate, one can get quite an accurate idea of the current state of our knowledge of American astronomical Geography. All in all, these coordinates fluctuate on average within a little less than 15″; for half of the longitudes noted here, the range of extremes does not even exceed 7.7″.)

POLITICAL ESSAY ON THE ISLAND OF CUBA

The political importance of the island of Cuba stems from the size of its surface area, which is fifty percent greater than that of the island of Haiti, the admirable fertility of its soil, its naval bases, the nature of its population, three-fifths of which are free men, and, above all, the advantages of Havana's geographical position. The northern Caribbean Sea, known as the Gulf of Mexico, forms a circular basin of more than 250 leagues in diameter. It is a Mediterranean with two outlets, of which the coastline, from the tip of Florida to ▾Cabo Catoche on the Yucatán, currently belongs exclusively to the United States of Mexico and the United States of North America. The island of Cuba, or, rather, its coastline between Cape San Antonio and the city of Matanzas situated at the mouth of the Old Channel, blocks the Gulf of Mexico to the southeast. Other than the outlet between Bahía Honda [in Pinar del Río Province] and the Florida lowlands, Cuba does not allow the ocean current, known as the *Gulf-stream*,[1] any other openings besides a channel between the Capes of San Antonio and Catoche toward the south and the Old Bahama Channel toward the north. Near the northern outlet, where the trade routes of many peoples cross, lies the beautiful port of Havana, fortified both by nature and by many works of manmade artistry. The fleets that leave from this port, built in part with Cuban Cedrela and Mahogany, can fight at the opening of the Mexican Mediterranean and threaten the opposite coasts, much as the fleets that leave from Cádiz can dominate the ocean near the ▾Pillars of Hercules. The Gulf of Mexico, the Old Channel, and the Old Bahama Channel meet at Havana. The opposing direction of the currents and the atmospheric disturbances—quite violent at the beginning of winter—lend these waters at the edge of the tropics a peculiar character.

I.2

I.3

1. Vol. I, pp. 122–41. Note [from the front page]: All references in the notes that do not specify a work are to the octavo edition of the *Relation historique*. [References to the quarto edition of the French text are marked here as Q.]

The island of Cuba is not only the largest of the Antilles (its surface area differs little from that of England proper, without Wales); thanks to its narrow, elongated shape, it also has an extensive coastline. It neighbors on Haiti, Jamaica, the southernmost province of the United States (Florida), and the easternmost province of Mexico (Yucatán). This circumstance merits the most serious attention, because the countries that are accessible by sea in ten to twelve days—Jamaica, Haiti, Cuba, and the southern portion of the United States (from Louisiana to Virginia)—are home to more than 2.8 million Africans. ▾Ever since Santo Domingo, Florida, and New Spain broke away from the Spanish metropole, the island of Cuba has been linked only by religion, language, and mores to the countries that surround it, countries that had been subject to the same laws for centuries.

I.4

Florida forms the last link of the long chain of federal states whose northern reaches touch the basin of the Saint Lawrence Stream and which extends from the region of palm trees to a region of the most severe winters. The inhabitants of New England regard as public dangers the increasing growth of the black population, the larger number of *slave states* and their appetite for cultivating colonial commodities. They vow that the ▾Straits of Florida—the current border of the great American Federation—shall not be crossed save in the name of free trade founded upon equal rights. They fear that Havana might fall under the domination of a European power even more menacing than Spain. They are no less desirous that the political ties that once bound Louisiana, Pensacola, and Saint Augustine in Florida to the island of Cuba should remain severed forever.

I.5

The extreme barrenness of Florida's soil and the lack of inhabitants and cultivation have made its proximity of negligible importance to Havana's commerce. This is not at all the case with Mexico's coast which, in a semicircle extending from the heavily frequented ports of Tampico, Veracruz, and Alvarado to Cape Catoche, almost touches the western part of the island of Cuba at the Yucatán peninsula. Commerce between Havana and the port of Campeche is very active. It grows despite the new government in Mexico, because the equally illicit trade with the more distant coasts of Caracas or Colombia involves only a small number of ships. In these unsettled times, provisions of salted meats for the slaves,[1] called *tasajo*, are obtained more easily and with less danger from Buenos Aires and the plains of Mérida than from the plains of Cumaná, Barcelona, and Caracas. For centuries, the island of Cuba and the archipelago of the Philippines drew from the coffers of New Spain the funds necessary for domestic administration and the maintenance of fortifi-

I.6

1. Vol. IV, p. 72; VI, pp. 96–97.

cations, arsenals, and shipyards (*situados de atención marítima*). As I have detailed elsewhere,[1] Havana was the military port of New Spain, and, until 1808, it received more than 1.8 million piasters annually from the Mexican treasury. In Madrid, people had long been used to viewing the island of Cuba and the archipelago of the Philippines as Mexican dependencies situated at rather unequal distances to the east and to the west of Veracruz and Acapulco but tied nonetheless to the Mexican metropole. In those days, that metropole was itself a European colony, bound to Europe by all the ties of trade, mutual assistance, and the most ancient of affections. Little by little, the growth of its own wealth made unnecessary the financial assistance Cuba had customarily drawn from the Mexican treasury. Of all of Spain's possessions, this island 1.7 prospered the most. After the ▾troubles in Saint-Domingue, the port of Havana ascended to the first rank of commercial centers. A happy concurrence of political circumstances, the moderation of royal officials, and the clever, prudent, and no doubt self-interested conduct of the inhabitants has made it possible for Havana to enjoy uninterrupted free trade with foreign nations. Its revenue from customs duties has grown so prodigiously that the island of Cuba supports its own needs and, during the wars between the metropole and the continental Spanish colonies, could also furnish considerable sums to the army that had fought in Venezuela, to the garrison of San Juan de Ulúa castle, and to very costly, and very often useless, naval provisioning.

I visited the island twice, once for three months, and again for a month and a half. ▾I was privileged to enjoy the trust of people who, by their talents and their positions as administrators, property owners, or businessmen, 1.8 were all well placed to furnish me with information about Cuba's growing public wealth. The particular protection with which the Ministry of Spain honored me made their confidence quite legitimate. I dare flatter myself that I also earned it through the moderation of my principles, my circumspect demeanor, and the peaceable nature of my endeavors. Not in thirty years has the Spanish government hindered the publication of the most valuable statistical documents on the state of commerce, agriculture, and finance, not even in Havana. I consulted these documents, and the relationships that I have kept up with America since my return to Europe have moved me to complete the materials that I had gathered on the region. In the company of ▾Monsieur Bonpland, I traveled through the environs of Havana, the beautiful Güines Valley, and the coast between Batabanó and the port of Trinidad. After succinctly describing the appearance of the locales and the singular changes of a climate so different from that of the other Antilles, I will examine the island's

1. *Essai politique sur le royaume de la Nouvelle-Espagne*, Vol. II, pp. 823–24.

1.9 population, its *area* calculated from a scrupulous survey of the coastline, the objects of trade, and the state of public revenue.

Havana's appearance from the entrance of the port is one of the most pleasant and picturesque on the coastline of tropical America north of the equator. Celebrated by travelers of all nations, this site has neither the luxurious vegetation that lines the banks of the Guayaquil, nor the wild majesty of Rio de Janeiro's rocky shoreline, two ports in the southern hemisphere. But it has the grace that, in our climates, adorns scenes of cultivated nature, blending the majesty of plant life with an organic vigor typical of the Torrid Zone. In this mixture of gentle impressions, the European forgets the dangers that threaten him at the heart of the Antilles' populous cities. He tries to take in the diverse elements of a vast landscape: the fortified castles that crown the rocks to the left of the port, which is an interior basin surrounded by villages and farms; the palm trees that grow to a prodigious height; the city half-hidden behind a forest of masts and sails. When entering the port of Havana, one passes 1.10 between the Morro (*Castillo de los Santos Reyes*) and the small fort, *San Salvador de la Punta*. The opening is only between 170 and 200 toises wide for three-fifths of a mile. Having exited the bottleneck and passed the beautiful castle of *San Carlos de la Cabaña* and the *Casa Blanca* to the north, one arrives at a basin shaped like a cloverleaf whose main axis, extending from SSW to NNE, is 2⅕ miles long. This basin connects three coves—Regla, Guanabacoa, and Atarés—the latter of which has sources of freshwater. Surrounded by walls, the city of Havana forms a promontory bounded by the Arsenal to the south and by the Punta fort to the north. There are 5 to 6, rather than 8 to 10, fathoms of water above the remains of some *submerged vessels* and the shoals of Luz. The castles of *Santo Domingo, Atarés*, and *Carlos del Príncipe* guard the city to the west; they are 660 and 1,240 toises from the interior wall by land. The suburbs (*arrabales* or *barrios extra muros*) of Horcón, Jesús María, Guadalupe, and Señor de la Salud occupy the middle terrain which, each 1.11 year, narrows the *Campo de Marte*. Havana's great buildings—the cathedral, the *Casa del Gobierno*, the admiralty, the arsenal, the post office or *Correo*, and the Tobacco Factory—are remarkable more for their solid construction than for their beauty. Most roads are narrow and unpaved. The cobblestones come from Veracruz, an expensive distance for things so heavy. Shortly before my arrival, someone had the strange idea of compensating for the scarcity of cobblestones by bundling together large tree trunks, as they do in Germany and Russia when building embankments across swampland. This plan was soon abandoned, and newly arrived travelers found, to their surprise, the most beautiful logs of *Cahoba*, a mahogany, submerged in Havana's mud. For lack of an effective police force, few cities in Spanish America appeared more hideous at the time of my voyage. One marched knee-deep in mud. The

1.12 multitude of *volantes*, Havana's typical carriages, the dray carts loaded with cases of sugar, and the porters elbowing passersby made the pedestrians' lives positively stressful and humiliating. The smell of *tasajo* (poorly cured meat) infested the houses and the tortuous roads. The police have now alleviated these inconveniences and have recently made very sensible improvements to the overall cleanliness of the roads. The houses are airier, and the *Calle de los Mercadores* is quite lovely. Here, as in our European cities, the network of poorly laid out streets can only slowly be corrected.

There are two beautiful promenades. One, the *Alameda,* is located between the Paula hospice and the theater, which an Italian artist, ▾Mr. Peruani, decorated very tastefully in 1803. The other lies between the Castillo de la Punta and the *Puerta de la Muralla*. The latter, also called the *paseo extra muros*, has deliciously fresh air: after sunset, it bustles with carriages. ▾The Marquis de la Torre built the paseo. Of all the island's governors, he was the one to initiate the first and most successful drive to improve the police and the municipal 1.13 government. ▾Don Luis de las Casas, whose name remains equally dear to Havana's residents, and the Count of Santa Clara expanded and improved these promenades. Near the *Campo de Marte* is the botanical garden, an object worthy of the government's attention, and something else altogether, whose appearance at once aggrieves and appalls: ▾the barracks [barracones] in front of which the pitiable slaves are exposed for sale. Since my return to Europe, a marble statue of King Charles III has been installed on the *extra muros promenade* in a place that had initially been designated for a monument to Christopher Columbus. Columbus's ashes were brought to the island of Cuba after the end of Spanish rule in Saint-Domingue. Since ▾Fernando Cortés' ashes were transferred from one church to another in Mexico City in the same year [1796], the two men who embodied the conquest of America both received new tombs in the same period at the end of the eighteenth century.

The most majestic of palm trees, the *Palma real*, lends a special character to the landscape around Havana. This is the *Oreodoxa regia* from our de- 1.14 scription of American palm trees.[1] Its trunk, 60 or 80 feet in height, is slender though a bit bulbous in the middle. Its glistening upper part, a delicate green shade newly formed by the contraction and expansion of the leafstalk, contrasts with the rest, which is off-white and cracked. It is as if two columns have been set on top of one another. The Cuban *Palma real* has plumed fronds that reach straight up into the sky and are curved only at the top. The bearing of this plant recalls the *Vadgiai* palm tree that covers rocks in the waterfalls of the Orinoco and balances its long branches above a fog of spume. As

1. [Bonpland, ▾Kunth, and Humboldt,] *Nova genera et species plantarum*, Vol. I, p. 305.

in any densely populated place, the vegetation thins out here. Around Havana, in the Regla amphitheater, the palm trees that so delight me are vanishing year after year. Wetlands that I once saw covered by bamboo forests I.15 are now cultivated and drained. Civilization advances; the earth, more and more stripped of its vegetation, barely offers any trace of its wild abundance. Houses cover everything from the Punta to San Lázaro, the Cabaña to Regla, and from Regla to Atarés. Those that circle the bay are plain and elegant in their construction. One sketches the floor plans and then orders them from the United States, as one might order a piece of furniture. When yellow fever reigns in Havana, people retire to these country homes and the hills between Regla and Guanabacoa, where the air is purer. In the coolness of the night, when boats cross the bay and thanks to the water's phosphorescence leave behind trails of light, these rustic sites offer charming and peaceful retreats to the inhabitants who flee the populous city's tumult. To judge accurately the progress of agriculture, visitors should see the little *chacaras* of corn and other subsistence plants, the pineapples (*ananas*) on the fields of Cruz de Piedra, and the Bishop's Garden (*Quinta del Obispo*), which has recently become an exquisite place.

I.16 Surrounded by walls, Havana proper is no more than 900 toises long and 500 toises wide. But more than 44,000 souls, of whom 26,000 are blacks and persons of mixed race, are crammed into this narrow area. A population of almost equal size has found refuge in the two large suburbs of *Jesús María* and *Salud*. Salud [health] does not deserve its pretty name: the air temperature there is probably lower than the city's, but the roads could have been wider and more clearly defined. For thirty years, Spanish engineers have been making war on the inhabitants of the *arrabales* [suburbs]. These engineers insist to the government that the houses are too close to the fortifications and that the enemy could entrench itself there with impunity. They do not have the courage to demolish these areas and displace a population of 28,000 in *Salud* alone. Since the great fire of 1802, *Salud* has been expanded considerably: first, shacks were built, and then, little by little, the shacks became houses. The inhabitants of the *arrabales* presented many plans to the King I.17 which called for their enclosure within Havana's fortifications and for legalizing their title to the land which, to this very moment, rests on nothing more than tacit consent. They wanted to make Havana an island by digging a long moat from the Puente de Chaves, near the Matadero, to San Lázaro. The distance, by way of a natural channel lined with mangroves and sea grape, or cocolloba, is nearly 1,200 toises between the arsenal and the Castillo de Atarés, which lies already beyond the bay. In this way, the city would have had three layers of fortification from the landward side in the west. First, on the outside, the Atarés and Príncipe ramparts on the hills, then the projected moat, and finally the city wall and the former path that the Count of Santa Clara paved for 700,000 piasters. To defend Havana from the west is of paramount importance. As long as control of Havana proper and of the middle part of the bay is maintained, the *Morro* and the *Cabaña*—the former requiring 800 men, the latter 2,000—are impregnable: provisions can be brought in from Havana to maintain the garrison even as the city absorbs considerable losses. Prominent French engineers have ensured that the enemy would need to begin the capture of the city by bombarding the *Cabaña*, which is a beautiful fortress. But the enemy's garrison, shut up in the casements, could not resist the insalubrious climate for long. ▾The British took the *Morro* without gaining control of Havana. But back then, the *Cabaña* and the *Fort* number 4, which dominate the *Morro*, did not yet exist. The most important fortifications to the south and the west are the *Castillos de Atarés y del Príncipe* and the battery of *Santa Clara*. I.18

OFFICIAL CENSUS (PADRON) OF HAVANA (THE CITY PROPER), ACCORDING TO DIFFERENCES IN COLOR, AGE, AND SEX, IN 1810 I.19

	Men			Women			
Colors.	*a.* from one day to 15 years	*b.* from 15 to 60 years	*c.* from 60 to 100 years	*d.* from one day to 15 years	*e.* from 15 to60 years	*f.* from 60 to 100 years	Total of Men and Women *g.*
Whites	3,146	6,057	348	2,860	5,478	476	18,365
Free Pardos	804	1,103	116	725	1,515	141	4,414
Free Blacks	833	1,149	133	819	2,308	284	5,886
Enslaved Pardos	227	153	194	197	119	183	1,073
Enslaved Blacks	1,781	4,699	78	1,561	5,224	94	13,437
TOTAL	6,791	13,161	869	6,162	14,644	1,178	43,175

OFFICIAL CENSUS OF THE OUTLYING AREA (ARRABAL) OF SALUD, IN 1810 I.20

Colors.	*a.*	*b.*	*c.*	*d.*	*e.*	*f.*	*g.*
Whites	3,261	1,312	874	3,687	1,812	744	11,690
Free Pardos	460	779	40	190	1,000	8	2,477
Free Blacks	500	2,489	17	587	3,026	113	6,732
Enslaved Pardos	100	220	8	77	189	11	605
Enslaved Blacks	448	3,552	15	558	2,300	42	6,915
TOTAL	4,769	8,352	954	5,099	8,327	918	28,419

Official census of the arrabal of Jesús María, in 1810

Colors.	*a.*	*b.*	*c.*	*d.*	*e.*	*f.*	*g.*
Whites	658	720	274	480	974	257	3,363
Free Pardos	326	399	169	268	551	174	1,887
Free Blacks	499	628	304	370	838	314	2,953
Enslaved Pardos	83	32	58	74	77	56	400
Enslaved Blacks	508	719	241	347	976	231	3,022
Total	2,047	2,518	2,046	1,530	3,416	1,032	11,625

I.21

Official census of the arrabal of Horcón, in 1810

Colors.	*a.*	*b.*	*c.*	*d.*	*e.*	*f.*	*g.*
Whites	132	329	49	218	287	31	1,046
Free Pardos	72	62	17	64	91	18	324
Free Blacks	44	30	11	41	60	16	202
Enslaved Pardos	37	17	10	34	17	10	125
Enslaved Blacks	56	544	16	71	96	10	593
Total	341	782	103	428	551	85	2,290

Official census of the arrabal of Cerro, in 1810

Colors	*a.*	*b.*	*c.*	*d.*	*e.*	*f.*	*g.*
Whites	259	302	8	258	252	4	1,083
Free Pardos	27	31	1	35	34	2	130
Free Blacks	15	33	2	10	40	2	102
Enslaved Pardos	0	0	0	0	0	0	0
Enslaved Blacks	144	343	7	72	118	1	685
Total	445	709	18	375	444	9	2,000

I.22

Official census of the arrabal of San Lázaro, in 1810

Colors.	*a.*	*b.*	*c.*	*d.*	*e.*	*f.*	*g.*
Whites	221	414	82	223	396	59	1,385
Free Pardos	34	44	5	55	66	11	215
Free Blacks	22	34	18	26	63	18	181
Enslaved Pardos	22	27	1	23	19	2	94
Enslaved Blacks	71	294	30	77	223	18	713
Total	360	813	136	404	767	108	2,588

Official census of the arrabal of Jesús del Monte, in 1810

Colors.	*a.*	*b.*	*c.*	*d.*	*e.*	*f.*	*g.*
Whites	868	390	187	565	486	223	2,719
Free Pardos	22	16	24	32	21	11	126
Free Blacks	45	51	112	82	94	62	446
Enslaved Pardos	0	0	0	0	0	0	0
Enslaved Blacks	181	204	60	52	111	90	698
Total	1,116	661	383	731	712	386	3,989

Official census of Regla, in 1810

I.23

Colors.	*a.*	*b.*	*c.*	*d.*	*e.*	*f.*	*g.*
Whites	353	430	22	331	415	25	1,576
Free Pardos	20	45	0	41	64	0	170
Free Blacks	14	30	2	13	42	3	104
Enslaved Pardos	0	0	0	0	0	0	0
Enslaved Blacks	37	105	5	132	86	3	368
Total	424	610	29	517	607	31	2,218

General summary of Havana's population (the city proper plus the outlying areas La Salud or Guadalupe, Jesus María, Horcón, Cerro, San Lázaro, Jesus del Monte, and Regla), in 1810. I. According to color, age, and sex

I.24

	Men				Women				
Colors.	from one day to 15 years	from 15 to 60 years	from 60 to 100 years	total of men	from one day to 15 years	from 15 to 60 years	from 60 to 100 years	total of women	Total of Men and of Women
Whites	8,888	9,914	1,844	20,646	8,624	11,100	1,819	21,543	41,189
Free Pardos	1,775	2,479	380	4,626	1,410	3,342	365	5,117	9,743
Free Blacks.	2,032	4,744	599	7,375	1,948	6,471	810	9,229	16,604
Enslaved Pardos	469	469	271	1,209	405	421	262	1,088	2,297
Enslaved Blacks	3,226	10,260	452	13,938	2,870	9,134	489	12,493	26,471
Total	16,390	27,906	3,538	47,834	15,255	29,468	3,747	48,470	96,304

1.25

II. According to Barrios

Names of Arrabales.	White.	Free Pardos.	Free Blacks.	Enslaved Pardos.	Enslaved Blacks.	Total.
Havana	18,361	4,414	5,880	1,073	13,437	43,175
La Salud	11,690	2,477	6,732	605	6,915	28,419
Jesus María	3,363	1,887	2,953	400	3,022	11,625
Horcón	1,046	324	202	125	593	2,290
Cerro	1,083	130	102	0	685	2,000
San Lázaro	1,385	215	181	94	713	2,588
Jesús del Monte	2,719	126	446		698	3,989
Regla	1,576	170	104	0	368	2,218
Total	41,227	9,743	16,606	2,297	26,431	96,304
		28,349		28,728		

1.26

Recapitulation

Whites		41,227
Free Pardos	9,743	26,349
Free Blacks	16,606	
Enslaved Pardos	2,297	28,728
Enslaved Blacks	26,431	
		96,304

In these tables, the term *pardos*, persons of color, describes everyone who is not *moreno*, that is, pure black. The land army, sailors, royal marines, monks, nuns, and nonresident foreigners, the *transient population,* are not included in the 1810 census, whose results many highly esteemed, recently published works accidentally attributed to 1817. The Havana garrison generally comprises up to 6,000 men; the number of foreigners typically reaches 20,000. Thus, the population of Havana and its seven outlying areas no doubt currently (1825) surpasses 130,000. The following table presents the population growth in Havana and its outlying areas from the 1791 census, ordered by the Captain-general Don Luis de las Casas, until 1810.

1.27

Census years	Whites	Free People of Color.	Slaves	Total	Proportion of the Three Classes
1791	23,737	9,751	10,849	44,337	53..22..25..
1810	41,227	26,349	28,720	96,296	43..27..30..
Increase	17,490	16,598	17,871	51,967	

Increase	whites	73	percent
	free people of color	171	
	slaves	165	
	all classes	117	

We add here the population growth during half of this period, from 1800 to 1810, but only for the *barrio extramuros de Guadalupe*: 1.28

Year.	Whites.	Free People of Color.		Total of Free People of Color.	Slaves.		Total of Slaves.	Total.
		Pardos.	Blacks.		Pardos.	Blacks.		
1800	3,323	1,087	1,243	2,330	92	1,766	1,858	7,511
1810	11,690	2,477	6,732	9,209	605	6,915	7,520	28,419
Increase	8,367	1,087	5,489	6,879	513	5,149	5,762	20,908

Increase	whites	251	percent
	freedmen	295	
	slaves	310	
	the three classes	278	

We can see that the population has more than doubled in twenty years, from 1791 to 1810. During the same period, the population of New York, the largest city of the United States, grew from 33,200 to 96,400. Its population is now about 140,000, slightly larger than Havana's and almost equal to that of Lyon. Mexico City, which had 170,000 habitants in 1820, still seems to me the New Continent's foremost city. It is perhaps a relief to the free states of this part of the world that the Americas contain only six cities of 100,000 souls or more: Mexico City, New York, Philadelphia, Havana, Rio de Janeiro and Salvador da Bahía. In Rio de Janeiro, there are 105,000 blacks in a population of 135,000; in Havana, whites make up two-fifths of the entire population. Women are as numerous in Havana as they are in the main cities of the United States and Mexico.[1] 1.29

1. The censuses of Boston, New York, Philadelphia, Baltimore, Charleston, and New Orleans give 109 to 100 as the ratio of women to men. In Mexico City, there are 92,838 women and 76,008 men; this creates an even stranger ratio: 122:100. I have already addressed this observation in another place (*Essai politique sur le royaume de la Nouvelle-Espagne*, Book II, Chap. VII, Vol. I, pp. 140–41), where I noted that, in Mexico and in the United States, the number of living men exceeds that of women, provided that one applies the same criteria both to villages and cities, whereas in Europe the relationship is inverted. In the United States (the entire country), the number of living men in relation to living women is 100 to 97. If one consults the official 1820 census, in which the partial sums are hardly exact, one finds that in the vast territory of the United States, there

1.30–31 The congregation of poorly acclimated foreigners in a cramped and populous city surely increases mortality. Yellow fever has had far less of an impact on mortality rates than is popularly believed. When the number of imported blacks is not very large, and when commercial activity does not attract many poorly acclimated sailors either from Europe or the United States, then births almost equal deaths.[1] Here are tables for five years for Havana and its outlying areas (*barrios extramurales*).

1.32

Years	Marriages	Births	Deaths
1813	386	3,525	2,948
1814	390	3,470	3,622
1820	525	4,495	4,833
1821	549	4,326	4,466
1824	397	3,566	3,697

This table, which shows extreme fluctuations due to the unequal influx of foreigners, gives 1 to 33.5 as the average birth rate and 1 to 33.2 as the average death rate in relation to the general population, assuming a total population of 130,000 for greater Havana. According to the latest and most precise studies of France's population, the corresponding rates, for all of France, are 31⅔ to 1 and 39⅔ to 1. For Paris, between 1819 and 1823, they are 1 to 28 and 1 to 31.6.
1.33 The circumstances that influence these figures in large cities are so complicated and so variable that one can hardly judge the number of inhabitants by the numbers of births and deaths. In 1806, a period during which the population of Mexico City was barely above 150,000, the number of deaths and births was 5,166 and 6,155, respectively, while in Havana, which had 130,000 inhabitants, the corresponding numbers were about 3,900 and 3,880. Havana has two hospitals with a sizeable number of patients: the general hospital (*Caridad* or *San Felipe y Santiago*) and the military hospital (*San Ambrosio*).[2]

are 3,993,206 white men and 3,864,017 white women, for a total of 7,857,223. By contrast, there were 7,137,014 men and 7,254,613 women in Great Britain in 1821; 1,478,900 men and 1,512,030 women in Portugal in 1801; 2,432,431 men and 2,574,452 women in the Kingdom of Naples in 1818; 1,599,487 men and 1,721,160 women in Sweden in 1805; 2,268,180 men and 2,347,090 women in Java in 1815. In Sweden, the ratio of living women to men appears to be 100 to 94; in the Kingdom of Naples, 100 to 95; in France, Portugal, and Java, 100 to 97; in England and Prussia, 100 to 99. This is how different occupations and ways of life affect the mortality of men!

1. See the *Guía de Forasteros de la Isla de Cuba* for 1815, p. 245; for 1825, p. 363; this statistical manual is far better prepared than the majority of those published in Europe. 5,696 people were vaccinated in Havana in 1814; in 1824, almost 8,100.

2. On the average mortality rates in hospitals in Veracruz and Paris, see my *Essai politique sur le royaume de la Nouvelle-Espagne*, Vol. II, pp. 777 and 784.

1.34

Annual Movement	Military Hospital of San Ambrosio			Central Hospital of San Felipe y Santiago		
	1814.	1821.	1824.	1814.	1821.	1824.
Bed-ridden since the previous year	226	307	264	153	251	127
Admitted during the course of the year	4352	4829	4160	1484	2596	2196
Sum	4578	5136	4424	1637	2847	2323
Deceased	164	225	194	283	743	533
Released cured	4208	4623	3966	1224	1948	1651
Remained ill and bed-ridden	206	283	264	130	156	139

1.35 In the *general hospital*, more than 24 percent die in an average year; in the military hospital, barely 4 percent. It would be unjust to attribute this enormous difference to the treatments used by the religious order of San Juan de Dios which runs the former establishment. Without a doubt, patients afflicted with *vómito*, yellow fever in its worst form, also enter the hospital of San Ambrosio, but the majority of its patients have minor, even insignificant, ailments. The general hospital, on the other hand, receives the old, the incurable, and blacks with only months to live, of whom planters or masters (*los amos*) want to rid themselves so that they will not have to take care of them. It is fair to say that, as a general matter, public safety improvements have also bettered the state of public health in Havana. But these changes have a beneficial effect only for the natives. Foreigners from northern Europe and America suffer under the general influence of the climate. They will continue to do so even when the roads are as clean as one could possibly wish. The influence of the coastline is such that inhabitants of the island's interior contract *vómito* 1.36 as soon as they arrive in Havana. The city markets are well provisioned. Having carefully documented the price of commodities and foodstuffs that 2,000 beasts of burden bring to the markets of Havana daily, studies have found that the consumption of meat, corn, manioc, legumes, liquor, milk, eggs, feed, and smoking tobacco reaches 4,480,000 piasters per year.

We spent December, January, and February making observations in the vicinity of Havana and on the beautiful plains of Güines. We found the noblest hospitality in the family of Mr. Cuesta and in the house of Count O'Reilly. Along with Mr. Santa María, Mr. Cuesta owned one of the most important mercantile houses in the Americas. We stayed at Mr. Cuesta's home and placed our collections and instruments in Count O'Reilly's vast mansion, whose terraces were especially favorable for astronomical observations. At that time, Havana's longitude was uncertain by more than one-fifth of a degree.[1] 1.37

1. Humboldt, *Recueil d'observations astronomiques*, Vol. II, pp. 53, 8[2].

Mr. Espinosa, the knowledgeable director of the *Depósito hidrográfico de Madrid*, suggested 5h 38′ 11″ in a table of positions that he gave me upon my departure from Madrid. Mr. de Churruca placed the Morro at 5h 39′ 1″. In Havana, I had the pleasure of meeting one of the ablest officers in the Spanish navy, ▾Don Dionisio Galiano, the captain of the vessel that had rounded the Straits of Magellan. Together, we observed a series of eclipses of Jupiter's moons, whose average result gave us 5h 38′ 50″. In 1805, Mr. Oltmanns deduced from the observations I had reported that the Morro was 5h 38′ 50.5″ = 84° 43′ 7.5″ to the west of the Paris meridian. This longitude was confirmed by the 15 stellar occultations that Mr. Ferrer observed and measured between 1809 and 1811.[1] This excellent observer proposes 5h 38′ 50.9″ as the definitive result. As for the magnetic dip, I found it with ▾Borda's compass (Dec. 1800) 53° 22′ from the former sexagesimal division. Twenty-two years later, this dip was no more than 51° 55′, according to the extremely precise 1.38 observations that ▾Captain Sabine made during his memorable voyage to the coasts of Africa, America, and Spitzbergen [Norway]; it had decreased by 1° 27′. Farther to the east but also in the northern hemisphere, in Paris,[2] the decrease was 1° 11′ over nineteen years (1798 to 1817). In Paris (in October 1796), my gradient needle had made 245 oscillations in ten minutes. As I approached the magnetic equator, the number of oscillations decreased. In San Carlos de Río Negro (north. lat. 1° 53′ 42″), this number[3] was no more than 216. On the basis of these readings, I had predicted that magnetic forces decrease from the pole to the equator. My surprise was even greater when these 1.39 oft-repeated observations gave me 246 oscillations in Havana, proving that the intensity of forces was greater in the western hemisphere at 23° 8′ of latitude than in Paris at 48° 50′. I have already demonstrated elsewhere that the *isodynamic lines* must not be confused with the *lines of equal magnetic dip*, and Captain Sabine[4] confirmed through observations probably more precise

1. *Connaissance des temps*, 1817, p. 330.

2. In 1798, I had found 69° 51′ in Paris in collaboration with the Chevalier de Borda by changing the poles many times; in 1806 Mr. Gay-Lussac obtained 69° 12′; Mr. Arago found 68° 40′ in 1817, and 68° 7′ in 1824. All these experiments were undertaken with instruments of the same construction.

3. *Relation historique*, Vol. VIII, pp. [26,] 27, 28, 346. These results need to be corrected relative to temperature.

4. Sabine, *Account of Experiments to Determine the Figure of the Earth by Pendulum*, 1825, p. 483, 494. The intensity of magnetic forces is weaker below the magnetic equator near the coast of West Africa than near the west coast of southern America. Through the decrease of forces from the magnetic equator that passes between Micuipampa and Cajamarca (about lat. 7° 1′ south, long. 80° 40′, height 1,500 toises) to Paris, I obtained the ratio of 1.0000 to 1.3482. Mr. Sabine finds the decrease, from a point on the magnetic equator

than my own the rapid augmentation of force in tropical America. This able naturalist posited the intensity of forces in Havana and London to be 1.72 to 1.62 (designating 1 as the force below the magnetic equator near St. Thomas in the Gulf of Guinea). The position of the northern magnetic pole (lat. 60° long. 82° 20′ west) is such that the polar distance is smaller in Havana than the polar distances in London and in Paris. I discovered (on January 4, 1801) the magnetic dip at Havana at 6° 22′ 15″ to the east. In 1732, Harris had put it at 4° 40′. How can we say that it does not change in Jamaica when it has undergone so many variations on the island of Cuba? 1.40

SURFACE AREA, TERRITORIAL DIVISION, CLIMATE.—The island of Cuba is surrounded by shallows and reefs along two-thirds of its coastline. Because navigators avoided these *dangers*, the island's true configuration remained unknown for a long time. Above all, the distance between Havana and the port of Batabanó was exaggerated. It was only after the *Depósito hidrográfico de Madrid*, the most wonderful establishment of its kind in all of Europe, had published the writings of the frigate Captain Don José del Río and the lieutenant Don Ventura de Barcaíztegui that the *area* of Cuba was calculated with any precision. 1.41. With their data, the shape of the Isle of Pines and the southern coast between Puerto Casilda and Cabo Cruz (behind the *Cayos de las Doce Leguas*) took on a very different aspect on our maps. On the basis of studies published at the *Depósito* through 1807, ▾Mr. Lindenau[1] had found that the island of Cuba, not counting the small neighboring islands, was 2,255 sq. lg. (15 to a degree) and with them 2,318 sq. lg. The latter result is equal to 4,102 square nautical leagues (20 to a degree). Using slightly different data, Mr. Ferrer arrived at 3,848 square nautical leagues.[2] In order to present the most precise result possible according to the current state of astronomical knowledge, I engaged my esteemed friend Mr. Bauzá, who became famous due to his important and rigorous work, to calculate the *area* based on the four-page map of the island of Cuba that he is about to finish. 1.42 This knowledgeable geographer was the answer to my prayers. *He found* (in June 1825) *that the surface area of the island of Cuba, not including the Isle of Pines, is* 3,520 *square nautical leagues and* 3,615 *square nautical leagues with this island.* He deduced from these twice-repeated calculations that the island of

---

near St. Thomas (lat. 0° 5″ north, long. 4° 24′ east, height 3 t.) to London, in the ratio 1 to 1.62. In comparing my experiments with oscillation to those of ▾Mr. de Rossel, Mr. Biot and Mr. Hansteen had already noticed that the magnetic force was smaller in the meridian from Surabaya to the island of Java than in Peru ([Hansteen and Hanson,] *Untersuchungen über den Magnetismus der Erde*, Part I, 70).

1. Zach, *Monatliche Correspondenz*, Dec. 1807, p. 312.

2. Unpublished notes.

Cuba is one-seventh smaller than had previously been believed: it is 33 percent larger than Haiti or Saint-Domingue. Its surface area is roughly equal to that of Portugal and one-eighth less than that of England without Wales. The Antilles occupy an *area* half as large as that of Spain, and the island of Cuba alone almost equals the other Greater and Lesser Antilles in its surface area. Its greatest width, from Cape San Antonio to Point Maisí (in the direction WSW-ENE and then WNW-ESE) is 227 leagues.[1] Its greatest length (in the N-S direction), from Point Maternillos to the mouth of the Magdalena River near Tarquino Peak, is 37 leagues. The island's average width at four-fifths of its length, between Havana and Puerto Príncipe, is 15 leagues. In the best cultivated part, between Havana (lat. from the center of the city 23° 8′ 35″) and

I.43 Batabanó (lat. 22° 43′ 24″), the isthmus is a mere 8⅓ nautical leagues. We will soon see that this proximity of the northern and southern coastlines makes the port of Batabanó very important from a commercial and military standpoint. Among all the great islands of the globe, it is Java (4,170 sq. lg.) that most resembles the island of Cuba in area and shape. Cuba boasts a coastal rim of 520 leagues, of which 280 belong to the southern coast between Cape San Antonio and Point Maisí.

In his calculation of the *area*, Don Felipe Bauzá posits the longitude of Cape San Antonio to be at 87° 17′ 22″, the Morro of Havana at 84° 42′ 20″, Batabanó at 84° 46′ 23″, and Punta Maisí at 76° 26′ 28″. He follows ▾Don José Sánchez Cerquero in situating Puerto Rico at 68° 28′ 29″. The first two of these longitudes fall within 3″ or 4″ of my own observations. *Observations astronomiques*, Vol. I, pp. 90f, and *Relation historique*, Vol. III [Q], p. 360. The geodesic calculations of ▾Don Francisco Lemaur, the skilled engineer and recent commander of the castle of San Juan de Ulúa, gave me 84° 45′

I.44 56″ for Batabanó, which I verified in Havana (at Count O'Reilly's mansion). Mr. Ferrer posits 76° 30′ 25″ for Cape Maisí, while persisting in situating Puerto Rico at 68° 28′ 3″. *Connaissance des Temps*, 1817, p. 323. I will not insist on this longitude for Puerto Rico, which has already provoked lively discussions and for which three correspondent observations of ▾Aldebaran's occultation (Oct. 21, 1793) gave Mr. Oltmanns 68° 35′ 43.5″ and 68° 33′ 30″. These figures are based on the overall observations of occultations, distances, and time differences. *Observations astronomique*, Vol. II, p. 125 and 139. Old and slightly imprecise calculations put the island of Cuba at either 6,764 *leguas planas* or *leguas legales españolas* (5,000 varas or 26 1/6 to a degree), equal to 906,458 *caballerías* (432 square varas or 35 English acres) (according

1. Still in nautical leagues 2,854 toises (20 to a degree), if the opposite is not explicitly stated.

to the *Patriota Americana*, 1812, Vol. II, p. 292, and the *Documentos sobre el tráfico de negros*, 1814, p. 136) or 52,000 square miles (with 640 *acres* or 1/11.97 square nautical leagues). Melish, *A Geographical Description of the United States*, p. 444. Morse, *New System of Modern Geography*, p. 238. The following table gives a better sense of the territorial significance of the island of Cuba in relation to the rest of the Antilles:

1.45

Islands.	Surface in Square Nautical Leagues.	Total Population.	Population per Square League.
Cuba, according to Mr. Bauzá	3615	715,000	197
Haiti, according to Mr. de Lindenau	2450	820,000	334
Jamaica	460	402,000	874
Puerto Rico	322	225,000	691
Greater Antilles	6847	2,147,000	313
Lesser Antilles	940	696,000	740
Antillean Archipelago	7787	2,843,000	365

The terrain of more than four-fifths of the island of Cuba is very low. Its soil is covered with secondary and tertiary formations permeated by granite gneiss, syenite, and euphotide. To this day, we know little more about the country's geognostic configuration than the relative age and composition of its terrain. We know only that the highest mountain range is located in the island's southeast, between Cabo Cruz, Punta Maisí, and Holguín. This mountainous region, situated to the northwest of Santiago de Cuba, is called the *Sierra* or the *Montañas de Cobre* [Copper Mountains]; it seems to be more than 1,200 toises[1] above sea-level. According to this supposition, the peaks of the *Sierra* would tower above those of Jamaica's Blue Mountains and above those of the Pico de la Selle and the Massif de la Hotte in Saint-Domingue. The *Sierra de Tarquino*,[2] fifty miles to the west of Santiago de Cuba, belongs to the same group as the *Copper Mountains*. A chain of hills crosses the island from ESE to WNW; it approaches the southern coast between the meridians of Puerto Príncipe and Villa Clara. Farther to the west, toward Álvarez and Matanzas in the *Sierras de Gavilán*, *Camarioca*, and *Madruga*, the hills approach the

1.46

1.47

1. Are the *Montañas de Cobre*, as some pilots claim, visible from the coasts of Jamaica, or, what is more likely, only from the north slope of the Blue Mountains? In the first case, their height would exceed 1,600 toises, supposing a refraction of one-twelfth. It is certain that the mountains of Jamaica are visible from the summit of the *Cuchillas* or *Lomas* of Tarquino (*Patriota Americano*, Vol. II, p. 282).

2. Lat. 19° 52′ 57″; long. 79° 11′ 45″, according to Mr. Ferrer.

northern coastlines. Near the mouth of the Guaurabo River at the town of Trinidad, I saw to the NW the ▾*Lomas de San Juan*[1] [San Juan Hills] which form needles or horns of more than 300 toises in height[2] and whose slopes incline more or less regularly to the south. This limestone group is an imposing sight when one is at anchor near Cayo de Piedras. The coastlines of Jagua and Batabanó are very low, and I believe that to the west of the meridian of Matanzas, there is not a single elevation higher than 200 toises, except for the Pan de Guaijabón. In the island's interior, the land, gently rolling as in England, is no more than 45 to 60 toises above sea level.[3] The objects most visible from afar and most famous among sailors are the *Pan de Matanzas*,[4] a truncated cone shaped like a little monument; the *Arcos de Canasí*, which look like small segments of a circle from Puerto Escondido and Jaruco; the *Mesa de Mariel*,[5] the *Tetas de Managua*,[6] and the *Pan de Guaijabón*.[7] This descending layer of limestone formations toward the north and the west points to underwater connections between these rock formations and the equally low terrain of the Bahamas, Florida, and the Yucatán.

I.48

I.49

With intellectual culture and formal education having long been restricted to Havana and the surrounding districts, the profound ignorance of the Geognosy of the *Montañas de Cobre* is unsurprising. ▾Don Francisco Ramírez, a

1. Lat. 21° 58′; long. 82° 40′.

2. This estimate is based on elevation angles that I took at sea at approximately known distances.

3. The village of Wajay, situated 15 nautical miles from Havana, S 25° W, 38 toises above sea level: the ridge line from the village of Bejucal to the Taverna del Rey, 48 toises.

4. Height 197 toises. Lat. 23° 1′ 55″; long. 84° 3′ 36″, assuming with Mr. Oltmanns that Havana's Morro is at long. 84° 43′ 8″. While sailing, I determined that the Arcos de Canasí were 115 toises high.

5. Middle of Guanajay, in the Mesa, lat. 22° 57′ 24″; long. 85° 0′ 20″. Torreón del Mariel, 85° 3′ 14″.

6. The astronomical position of the two limestone hills situated E-W and known as the Tetas de Managua is of great importance for determining the location of Havana. I observed the latitudes not at the foot of the eastern Teta [teat or mound], but in the village of Managua and at San Antonio de Barreto, and I connected the *Teta oriental* to these two places. I place the *Teta oriental de Managua* at lat. 22° 58′ 48″. Mr. Ferrer puts it at 22° 58′ 19″; long. 84° 40′ 19″, while Captain Don José del Río has 84° 37′. Mr. Ferrer's longitude seems preferable to me; in the French copy of the Map of the River, the Tetas are placed at 84° 34′! Trigonometric equations by Mr. Francisco Lemaur assign them to 84° 39′ 52″. Mr. Silva finds the differences of latitude between the Mirador of the Marquis of Real Socorro, Havana, and the Teta oriental de Managua to be 8,666.85 toises.

7. Lat. 22° 47′ 31″; long. 85° 44′ 37″; height 390 toises. The Sierra de los Órganos and the Sierra del Rosario is farther to the west on the northern coast; the Sierra of the Río Puerco is farther to the south.

traveler and a student of Mr. Proust who is well versed in chemistry and mineralogy, told me that the island's western part is granite, and he recognized gneiss and primitive schist there. It is probably in these granite formations that the alluvial deposits of *gold dust* originated. They were ardently[1] exploited at the beginning of the conquest, much to the detriment of the na- I.50–51

1. In *Cubanacán*, that is, in the island's interior, near Jagua and Trinidad where the water carried the gold dust right into the limestone soil. (Manuscripts by ▾Don Félix de Arrate of 1750, and Don Antonio López, 1802.) Martyr d'Anghiera, the wittiest of the *conquest*'s authors, said (Dec. III, Book IX, p. 24 D. and p. 63 D., 1533): "Cuba is richer in gold than Hispaniola (Haiti); and, at the very moment of writing, we have assembled 180,000 gold castellanos." If this estimate is not exaggerated—and I am much inclined to believe that it is—it would prove a degree of extraction and plunder equal to 3,600 gold marks. Herrera estimates the *quinto del Rey* in Cuba at 6,000 pesos, which would indicate an annual yield of 2,000 gold marks at 22 carats, purer than the gold of Cibao on Saint-Domingue (see my *Essai politique sur le royaume de la Nouvelle-Espagne*, Vol. II, p. 648 on the value of *castellanos de oro* and the sixteenth-century *peso ensayado*). In 1804, all the mines in Mexico produced 7,000 gold marks, in Peru 3,400. It is difficult to distinguish how much of the gold that the first *Conquistadors* sent to Spain came from panning and how much had accumulated on the surface for centuries and been gathered by natives, who were robbed indiscriminately. Assuming a yield of 3,000 gold marks from panning in the two islands of Cuba and Haiti (in Cubanacán and Cibao), we arrive at an amount three times less than the gold that the tiny province of Chocó produced annually (from 1790–1805). To assume such a former abundance is hardly implausible. If one is surprised that panning in Cuba and Saint-Domingue yields such poor results these days, compared to the vast quantities that were once extracted there, one must recall that yields from panning for gold also declined in Brazil between 1760 and 1820, from 6,600 kg gold to less than 595 (*Relation historique,* Vol. X, pp. 317f). Gold nuggets weighing many pounds, as they are currently found in Florida and the Carolinas, prove the original abundance of the entire Antillean basin from the island of Cuba to the Appalachian Mountains. Furthermore, it is entirely natural that we see a sharper decline in the yield from panning for gold than in the yield from the subterranean extraction of veins. Without a doubt, metals no longer renew themselves in the cracks of veins (by sublimation), in alluvial plains, or near the course of rivers where plateaus are higher than the level of neighboring waterways. But in rocks with metal-bearing veins, the miner does not know the entire field of extraction. He might *follow* the works to deepen them and then cross *neighboring veins*. Alluvial plains are generally rich in metals only to a shallow depth; they typically rest on rocks that are entirely barren. Their superficial position and the uniformity of their composition make it easier to know the limits of their mining resources, which can be depleted in a short time wherever workers congregate and panning water abounds. I think that these considerations, taken from the history of the *conquest* and the science of the miner's art, can throw some light on the problem of Haiti's mineral wealth, which is now the object of heated debates. On this island, as in Brazil, it would be more profitable to attempt subterranean extractions (in veins) in primitive and intermediary terrains rather than resume the panning abandoned during centuries of barbarity, pillage, and carnage.

1.52 tives. Traces of gold dust are still found in the Holguín and Escambray Rivers, as is widely known in the vicinity of Villa Clara, Sancti Spíritus, Puerto del Príncipe [Camagüey], Bayamo [Granma], and the Bahía de Nipe. Is it possible that in the sixteenth century[1]—a time when the Spanish had been more interested in America's natural resources than they were later—the *Conquis-* 1.53 *tadores* may have spoken of abundant copper because of these formations of amphibolic schist, in which transition thonschiefer is mixed with diorite and euphotide and whose counterparts I have found in the hills of Guanabacoa?

The island's central and western parts contain two *formations of compact limestone, one of clayish sandstone and another of gypsum*. The first somewhat resembles the formation of the Jura, not by virtue of its deposits or its superposition, which are unknown to me, but because of its appearance and composition. It is white or yellow with a light ochre tint, with even cracks that are sometimes conchoidal, sometimes smooth. It has somewhat narrow layers with some nodules, often hollow, of pyrogenous silex (Río Canímar, two leagues to the west of Matanzas), and pectinite, cardite, terebratula, and madrepore fossils[2] not so much dispersed throughout the mass as concentrated 1.54 in clusters. Between the Potrero [cattle farm] of the Count of Mopox and the harbor of Batabanó, I did not come across any oolitic layers but found plenty of porous, almost bubblelike strata that resembled the spongy layer of Jurassic limestone in Franconia, near Donzdorf, Pegnitz, and Kirchenthumbach [Germany]. Yellowish and cavernous strata with cavities of 3–4 inches in diameter alternate with strata that are completely compressed[3] and have fewer fossils. The chain of hills that borders the Güines plain to the north and connects to the Lomas de Camoa and the Tetas de Managua belongs to this last variety, reddish white and almost *lithographic*, like the Jurassic limestone in Pappenheim [Germany]. Embedded in the compact and cavernous layers are veins of brown and ochreous iron. Perhaps the red earth, or *tierra colorada*, so sought after by planters (*hacendados*) of coffee, comes from the decomposition of some superficial layers of oxidized iron mixed with silica and clay 1.55 or with a marly reddish sandstone[4] layered on top of the limestone. I call this formation *Güines Limestone* to distinguish it from another near Trinidad in

1. "Hay buen cobre in Cuba" [there is good copper in Cuba] (in the eastern part that I visited at the time). Gómara, *Historia de las Indias,* XXVII [Vol. I, fol. LI, p. 113].

2. I saw neither the gryphites nor the ammonites of Jurassic limestone, neither the nummulites nor cerites of coarse sandstone.

3. Since the western part of the island lacks deep ravines, the alternation is recognizable while traveling from Havana to Batabanó, the deepest layers (at a 30° to 40° incline to the NE) appearing as one advances.

4. Ferrous sand and sandstone; *Iron-sand*?

the *Lomas de San Juan*, which formed much more recently from steep peaks that recall the *Caripe Limestone* Mountains in the vicinity of Cumaná.[1] This formation also encompasses large caverns near Matanzas and Jaruco. I have never learned of any fossil bones found here. The numerous caverns that accumulate rainwater and swallow little streams cause occasional landslides.[2] I believe that Cuba's gypsum belongs not to the tertiary but to the secondary terrain. It is mined in many places to the east of Matanzas, at San Antonio de los Baños, where it contains sulfur, and in the Keys off San Juan de los Remedios. One must not confuse this sometimes porous, sometimes compact (Jurassic?) *Güines Limestone* with another formation so recent that it is likely still evolving today. I refer to the *limestone agglomeration* that I saw in the *keys* that line the coast between Batabanó and the bay of Jagua, mainly to the south of the Zapata swamps at Cayo Buenito, Cayo Flamenco, and Cayo de Piedras. Sounding has proven that these are rocks that rise up abruptly from a depth of 20 to 30 fathoms. Some are at the very surface of the water, while others rise one-fourth to one-third of a toise above the surface. There are sharp fragments of madrepore corals and shells of 2 to 3 cubic inches cemented by grains of quartz sand. Landfill, in which we can distinguish only the *detritus* of shells and coral, covers these rocks' entire uneven surface. This tertiary formation is no doubt related to formations on the coasts of Cumaná, Cartagena, and Guadeloupe's Grande Terre which I mentioned in my Geognostic Table of South America.[3] Recently, ▾Mr. de Chamisso and Mr. Gaimard have shed much light on the *formation of coral islands of the* South Seas. At the foot of the Castillo de la Punta, which is situated near Havana on a ridge of cavernous rocks[4]

I.56

I.57

1. *Relation historique*, Vol. X, pp. 286 and 287.

2. For example, the ruins of the tobacco mills at the former royal plantation.

3. See [*Essai géognostique,*] Vol. X, pp. [302]ff. In his *Historique physique des Antilles françoises* (Vol. I, pp. 136, 138 and 543), ▾Mr. Moreau de Jonnès also clearly distinguishes the *Roche à Ravois* in Martinique and in Haiti, which is porous, filled with terebratulites, anemones, and other debris from pelagic shells somewhat analogous to *Güines Limestone* on the Island of Cuba, from pelagic limestone sediment called *Platine* or *Maçonne bon Dieu* on Guadeloupe. In the *Cayos* of the island of Cuba or in the *Jardinillos del Rey y de la Reyna*, all the coral rock that rises above the surface of the water seems to me fragmentary, that is to say composed of shattered blocks. It is, however, probable that, in the depths, it sits upon masses of still living polypary lithophytes.

4. The surface of these banks, blackened and weathered by the water, exhibits cauliflower patterns like those observed on lava currents. Is the change in color produced by the water due to the manganese whose presence is evident from dendrites? (Vol. VIII, pp. 24ff). The sea, which enters through fissures in the rock and through a cave at the base of the *Castillo de Morro*, compresses the air there and forces it out with an extraordinary noise. This noise explains the phenomenon of the *baxos roncadores* (snoring rocks), well

1.58 covered with vibrant green seaweed and living coral, there are a great many madrepores and other lithophyte corals. At first, one might think that this limestone rock, which comprises most of Cuba, resulted from the uninterrupted work of nature: from creative and partially destructive organic forces still active, even now, deep in the heart of the Ocean. But the impression that these limestone formations are new fades as soon as one leaves the coast or recalls the series of *coral rocks* enclosed by formations of different ages, such as muschelkalk, Jura limestone, and *common limestone*.[1] One can find the same 1.59 coral rocks of the Castillo de la Punta in the high mountains of the interior, where they are accompanied by fossilized bivalves far different from those that currently live on the coasts of the Antilles. Without wanting to assign a definitive place in the table of formations to the *Güines Limestone* at the Castillo de la Punta, I have no doubt about the relative antiquity of this rock in relation to the *conglomerate limestone of the Keys* situated to the south of Batabanó and to the east of the Isle of Pines. The earth underwent great changes between the periods when these two *terrains* were formed, one encompassing the great caverns of Matanzas, the other growing day by day through the agglutination of coral fragments and quartz dust. The latter of these formations seems to sit on the southern side of the island of Cuba, sometimes on the (Jurassic) limestone of Güines, sometimes (toward Cape Cruz) directly on primitive rock.[2] In the Lesser Antilles, coral has even covered the residual products of volca- 1.60 nic activity. Many of Cuba's *Cayos* [Keys] have freshwater; I have found very good water right in the middle of the *Cayo de Piedras*.[3] When one considers the very small size of these islets, it is hard to believe that these freshwater deposits are simply nonevaporated rainwater. Does this prove an underwater connection between the coastal limestone and the limestone that forms a base for the lithophytic coral, such that hydrostatic pressure pushes Cuba's freshwater through the coral rock of the Keys? This is the case in Jagua bay where, in the middle of the sea, there are freshwater springs frequented by manatees.

---

known to sailors who travel from Jamaica to the mouth of the San Juan de Nicaragua River or the island of San Andrés.

1. See Cuvier and Brongniart, *Description géologique des environs de Paris*, p. 269, on the accumulation of coral in the Calcaire Grossier of Paris (cerite and nummulite coral). ▾Maraschini, *Sulle formazioni delle rocce del Vicentino saggio geologico*, p. 177.

2. I have already noted this *indifference to superposition* in Vol. X, pp. 301ff.

3. According to my observations: lat. 21° 56′ 40″; long. 83° 37′ 12″ (*Observations astronomiques*, Vol. II, pp. 111[f]).

To the east of Havana, a group of syenite and euphotide rocks[1] pierces the secondary formations in a remarkable fashion. The southern side of the 1.61 bay and the hills of the Morro and the Cabaña to the north consist of Jurassic limestone. But on the eastern side of the two Ensenadas [inlets] of Regla and Guanabacoa, the whole formation is *transitional*. Moving north to south, one can see, in the clear light of day, syenite composed of partially decayed amphibole, a little quartz, and reddish-white, occasionally crystallized, feldspar. This beautiful syenite, whose strata incline to the northwest, alternates twice with serpentine rock. The intercalated serpentine layers are three toises thick. Farther south, toward Regla and Guanabacoa, the syenite disappears, and the entire surface is covered with serpentine rock, rising in hills of 30 to 40 toises from east to west. The serpentine rock has abundant cracks. It is bluish-grey on the outside, leek- and asparagus-green on the inside, covered with manganese dendrite and crisscrossed by tiny veins of asbestos. It contains neither granite nor amphibole but a metallic diallage spread throughout the rock mass. Serpentine can break like bone or wood, sometimes in a conchoidal, or shell-shaped, fashion. This was the first time that I found a 1.62 metallic diallage in the tropics. Many serpentine rocks have magnetic poles; others are of such a homogeneous consistency and have such an oily sheen that, from a distance, one would be tempted to mistake them for pitchstone. If only these beautiful substances were used in the arts in Cuba as they are in many parts of Germany. When one approaches Guanabacoa, the serpentine rock shows 12–14 inch thick veins filled with fibrous quartz, amethyst, and superb stalactiform chalcedony covered with small protuberances. Perhaps one day, chrysophrase will also be found there. Copperish pyrites, said to be mixed with silver-gray copper, appear in the middle of these veins, but I did not find any trace of such gray copper ore. It is probably the metallic diallage that gives the Cerros of Guanabacoa their centuries-old reputation for being rich in gold and silver. Petroleum[2] bubbles up here and there from some 1.63

1. A succinct description of this group that I had composed in Spanish was published in Havana in 1804 under the title ▼*Noticia mineralógica del Cerro de Guanabacoa comunicada al Ex. Sr. Marqués de Someruelos, Capitán General de la Isla de Cuba*.

2. Are there sources of oil other than Guanabacoa in Havana Bay, or should we admit that the source of betún *líquido*, that is, pitch, which ▼Sebastián de Ocampo used in 1508 to caulk his ships, is exhausted? It is nevertheless this source that drew Ocampo's attention to the port of Havana when he gave it the name *Puerto de Carenas*. It is said that there are abundant sources of petroleum in the island's eastern part (*manantiales de betún y chapopote* [wells of tar and bitumen]) between Holguín and Mayarí and on the coastline of Santiago de Cuba. An islet (Siguapa) was recently discovered near Punta Hicacos which has a surface of solid, clayey bitumen. This mass recalls the asphalt of Vallorbe in

cracks in the serpentine rock. Freshwater springs are frequent there, and they I.64 contain a bit of sulfurous hydrogen and iron oxide. The baths of Barreto are very agreeable, but their temperature is almost the same as the air. The geognostic constitution of this group of serpentine rocks deserves particular attention because of its isolation, its veins, its connections with the syenite, and its *elevation* across formations of fossil shells. Feldspar with a soda base (compact feldspar), together with diallage, forms the euphotide and serpentine rocks; together with hypersthene, it forms hypersthenite; with amphibole, it forms diorite; with pyroxene [augite], dolerite and basalt; with granite, eclogite.[1] These five variations of feldspar shot through with oxidized iron and titanium sphene are dispersed over the entire earth and probably have similar origins. It is easy to distinguish two formations of euphotides: one lacks amphibole, even when it alternates with amphibolic rocks (Joria in Piedmont, Regla on the island of Cuba). It is rich in pure serpentine and metallic dial- I.65 lage and sometimes in jasper (Tuscany, Saxony). The other is permeated with amphibole that sometimes gives way to diorite;[2] there are no layers of jasper but sometimes rich veins of copper (Silesia, Mussinet in Piedmont, the Pyrenees, Parapara in Venezuela, and the Copper Mountains in the American west). In Scotland and Norway, the latter euphotide formation, blended with hypersthenite thanks to an admixture of diorite, sometimes develops genuine serpentine strata. To this day, no volcanic rocks from more recent epochs (such as trachytes, dolerite, or basalt) have been found on the island of Cuba. I do not even know if they have been found elsewhere in the Greater Antilles, whose geognostic composition differs fundamentally from that of the chain I.66 of volcanic and limestone islands between the island of Trinidad and the Virgin Islands. Earthquakes, typically less calamitous in Cuba than in Puerto

the Jura Limestone. Is the Guanabacoa serpentine formation repeated near Bahía Honda in the Cerro del Rubí? The hills of Regla and Guanabacoa offer botanists, at the feet of a few sparse palm trees, Jatropha pandurifolia, Jatropha integerrima Jacq, Jatropha fragrans, Petiveria alliacea; Pisonia loranthoides, Lantana involucrate, Russelia sarmentosa, Ehretia havanensis, Cordia globosa, Convolvulus pinnatifidus, C. calycinus, Bignonia lepidota, Lagascea mollis Cav., Malpighia cubensis, Triopteris lucida, Zanthoxylum Pterota, Myrtus tuberculata, Mariscus havanensis, Andropogon avenaceus Schrad., Olyra latifolia, Chloris cruciata, and a large number of Banisteria, whose golden flowers adorn the landscape (see our *Florula Cubæ Insulæ* in the *Nova genera et species plantarum*, Vol. VII, p. 470).

1. Reuthberg, near Dolau (Bayreuth); Saualpe (Styrie).

2. On a serpentine that follows veins of grünstein (diorite) like a shadow near Loch Clunie in Perthshire, see ▾MacCulloch in *Edinburgh Journal of Science*, 1824, July, pp. 3–16. For a vein of serpentine and the alternations that it produces on the banks of the Carity near West Balloch in Forfarshire, see ▾Charles Lyell [in] *Transactions of the Geological Society of London*, Vol. III, p. 43.

Rico or Haiti, occur more often in the eastern region between Cape Maisí, Santiago de Cuba, and the Ciudad de Puerto Príncipe. Perhaps the fault that is thought to cross the granite plate between Port-au-Prince and Cape Tiburón [in Santo Domingo], which caused whole mountains to tumble into the sea in 1770, extends laterally across this region.[1]

The cavernous texture of the limestone formations (*soboruco*) that I have just described, the steep angle of their banks, the narrowness of the island, the ubiquity of *deforested* plains, the proximity of mountains forming an elevated range on the southern coast—all these factors can be considered the principal causes of the lack of rivers and the island's aridity, particularly in Cuba's western part. By contrast, Jamaica and many of the Lesser Antilles, which have forest-covered volcanic peaks, are more favored by nature.[2] The terrain most famous for its fertility is found in the districts of Jagua, Trinidad, Matanzas, and Mariel. The Güines Valley owes its reputation for fertility solely to artificial irrigation (*zanjas de riego*). Despite the absence of large rivers and the soil's uneven fertility, the island of Cuba offers the most diverse and pleasing landscape at every step, thanks to its hilliness, its evergreens, and the distribution of its vegetation. The Mammea and the Calophyllum Calaba, two trees with large, tough, and glossy leaves; five species of palms (the Royal Palm, or Oreodoxa regia, the common coconut palm, the Cocos crispa, the Corypha Miraguama and the C. maritima); and little bushes always laden with flowers all adorn the hills and the savannas. The Cecropia peltata indicates humid areas. One might be tempted to believe that the island had originally been a forest of palms, lemon trees, and wild oranges. The latter, which rarely exceed 10 to 15 feet in height and have minuscule fruits, probably predate the arrival of the Europeans[3] who brought the domesticated *agrumi* [an Italian citrus fruit]. Most often, the lemon tree and the orange tree do not grow in the same place. When new colonists clear the land by burning it, they determine the quality of the soil based on whether it is covered by one or the other grouping of *social plants*, preferring the soil of the orange tree, or

I.67

I.68

1. ▾Dupuget, in *Journal des mines*, VI, p. 58, and Leopold von Buch, *Physicalische Beschreibung der Canarischen Inseln*, 1825, p. 403.

2. [Moreau de Jonnès,] *Histoire physique des Antilles*, Vol. I, pp. 44, 118, 287, 295, 300.

3. See my *Essai politique sur le royaume de la Nouvelle-Espagne*, Vol. II, p. 415. The most enlightened inhabitants of the island correctly recall that the domesticated orange trees that came from Asia maintain the size and all the properties of their fruit when they become wild (this is also the opinion of ▾Mr. Gallesio in his *Traité du Citrus*, p. 32). Brazilians do not doubt that the *small bitter orange* called *laranja da terra*, which grows wild far from human settlements, is American in origin (Caldcleugh, *Travels in South America*, Vol. I, p. 25).

naranja, to that of the small lemon tree (*limón*). Deforestation, *desmonte,* is a real problem in a country where sugarmills have yet to be widely retooled so 1.69 that they use only *bagazo*, or pressed-out sugarcane pulp, as fuel. The soil's aridity increases as it is stripped of the trees that serve as shelters against the sun and whose leaves produce a ▾caloric updraft against a perpetually serene sky, causing water vapor to precipitate in the cooled air.

There are few rivers worthy of attention, among them the Güines, which, in 1798, was intended to be connected to a *small canal* that would cross the island along the meridian of Batabanó; the Armendaris or Chorrera River whose waters are channeled to Havana by the ▾*Zanja de Antonelli*; the Río Cauto to the north of the town of Bayamo; the Máximo, which springs east of Puerto Príncipe; the Sagua la Grande near Villa Clara; the Palmas, which empties into the sea near Cayo Galindo; the little Jaruco and Santa Cruz rivers between Guanabo and Matanzas, which are navigable for some miles from their mouths and are favorable for transporting loads of sugar crates; the San 1.70 Antonio, which, like many others, disappears into limestone caves; the Guaurabo to the west of the port of Trinidad; and the Galafre River in the fertile Filipinas district, which empties into the Laguna de Cortés. The most abundant springs gush forth on the southern coast from Jagua to Punta de Sabina, where the soil is extremely swampy for 46 leagues. The abundance of water seeping through cracks in the stratified rock creates hydrostatic pressure that causes freshwater to bubble up far from the shore in the midst of salt water. The district of Havana is not among the most fertile, and, due to demand from the capital, the few sugar plantations that border on the capital itself have given way to livestock farms (*potreros*) and fairly sizable fields of corn and feed. Cuban agriculturists distinguish between two sorts of soil, often intermixed like the squares of a chessboard: black soil (*negra* or *prieta*), which is clayey and filled with natural compost, and red soil (*bermeja*), which is more siliceous and mixed with oxidized iron. Although the *tierra negra* is typically 1.71 preferred for sugarcane cultivation because of its capacity for retaining water, and *tierra bermeja* for coffee cultivation, many sugar plantations are nevertheless on red soil.

Havana's climate is typical for the edge of the Torrid Zone: it is a tropical climate whose temperature variations over the course of the year are similar to those of the temperate zone. The positions of Calcutta (lat. 22° 34′ N), Canton (lat. 23° 8′ N), Macao (lat. 22° 12′ N), Havana (lat. 23° 9′ N), and Rio de Janeiro (lat. 22° 54′ S) in relation to the Ocean and the Tropics of Cancer and Capricorn—that is, equidistant from the equator—are of the utmost importance for meteorology. This field can only advance by determining certain *numerical elements* that form the indispensable basis for the laws we seek to uncover. Because the vegetation seems identical near the borders of the Torrid Zone and at the equator itself, it is common to confuse the zone between 0° and 10° with that between 15° and 23° lat. The region of palm trees, bananas, and arborescent grasses even extends beyond the two tropics. But it would be foolhardy to apply what one observes at the edge of the tropics to what might take place in the plains bordering the equator (as recently happened after ▾Dr. Oudney's death in discussions concerning the soil elevation at which ice can form in the kingdom of Bornu [Nigeria]). To correct these errors, one has to be very familiar with the average annual and monthly temperatures and with the temperature range at Havana's latitude during different seasons. Only in this way can one prove, through exact comparison with other points equidistant from the equator (such as Rio de Janeiro and Macao), that the large decreases in temperature observed on the island of Cuba are due to the influx of cold air masses from the temperate zones toward the Tropics of Cancer and Capricorn. Four years of solid observations suggest that the average temperature of Havana is 25.7° C (20.6° R), only 2° C higher than it is in those parts of America closest to the equator.[1] The proximity of the sea raises the average yearly temperature on the coasts. But in the interior of the island, where the north wind blows with the same force and the land rises a mere 40 toises[2] above sea level, the average temperature reaches only 23° C (18.4° R), far from surpassing Cairo and Lower Egypt. The differences between the hottest and coldest average temperatures reach 12° C in the interior and 8° C in Havana; by contrast, they barely reach 3° C on the coast, in Cumaná. July and August, the warmest months on the island of Cuba, have an average temperature of 28.8° C, perhaps even 29.5° C, which is nearly as hot as the equator itself. The coldest months are December and January: their mean temperature is 17° C in the island's interior and 21° C in Havana, that is, 5° C to 8° C below the mean temperature at the equator during the same months but still 3° C warmer than the hottest month in Paris. As for extreme temperatures[3] measured with a centigrade thermometer in the shade, the conditions at the border of the Torrid Zone approach those of the equator (lat. 0° to 10°

I.72

I.73

I.74

1. The mean temperature of Cumaná (lat. 10° 27′) is 27.7° C. It is even asserted that, in the Lesser Antilles between 13° and 16° lat., Guadeloupe's is 27.5° C, Martinique's 27.2° C, and Barbados's 26.3° C. Moreau de Jonnès, *Histoire physique des Antilles*, Vol. I, p. 186.

2. Barely 6 toises more above sea level than Paris (second floor of the Royal Observatory).

3. Mr. Lachenaie holds that he saw the centesimal thermometer at 39.3° C in the shade in 1800 (on Saint-Rose on the island of Guadeloupe). But we do not know if his instrument was precise and free from the effects of radiant heat. In Martinique, the records are 20° and 35°.

north or south). A thermometer at 38.4° C (30.7° R) in Paris will only reach 33° C in Cumaná. In Veracruz, over the course of thirteen years, the temperature reached 32° C (25.6° R) only once; in Havana, Mr. Ferrer saw it oscillate only between 16° and 30° C over the course of three years (1810–1812). In the manuscripts that I have at hand, Mr. Robredo cites a temperature of 34.4° C (27.5° R) as remarkable for 1801. According to Mr. Arago's intriguing studies, Paris reached extreme temperatures between 36.7° and 38° C (29.4° and 1.75 30.7° R) four times in ten years (between 1793 and 1803). The convergence of the two periods when the sun passes through its zenith often causes intense heat on Cuba's coastline, as it does in all places near the border of the Torrid Zone between the parallels of 20° and 23.5°. This heat wave lasts not so much for entire months but for a few consecutive days. In an average year, the temperature does not rise above 28° C to 30° C in August. I have seen people complain of excessive heat when the temperature reaches 31° C (24.8 R). A temperature of 10° C or 12° C is already somewhat rare during the winter. But when the north wind blows for several weeks, carrying cold air from Canada, ice can sometimes form at night on the plains near Havana.[1] Following the observations of Mr. Wells and Mr. Wilson, we can assert that 1.76 the caloric updraft produces this effect when the thermometer is around 5° C and even 9° C above freezing. Mr. Robredo, however, assures me that he has seen the thermometer drop as low as 0° C. That a thick layer of ice could form near sea level in the Torrid Zone should surprise the naturalist even more, for in Caracas (lat. 10° 31′), at an altitude of 477 toises, nothing freezes above 11° and, closer to the equator, it is necessary to go to 1,400 toises to see ice.[2] What is more, there is no more than a 4° to 5° difference in latitude between Havana and Saint-Domingue, and between Batabanó and Jamaica. Yet, in Saint-Domingue, Jamaica, Martinique, and Guadeloupe, the lowest temperature in the plains[3] is between 18.5° C and 20.5° C.

1.77 It would be interesting to compare Havana's climate with the climates of Macao and Rio de Janeiro. One is situated in the same place in the *northern* tropics but on the *east* coast of Asia, the other on the *east* coast of America but in the *southern* tropics. Rio de Janeiro's mean temperatures are based on

1. This random cold had already struck the first voyagers. "In Cuba," said Gómara, "sometimes one feels the chill." *Historia de las Indias*, fol. XVI [fol. LI, "De la isla de Cuba"].

2. It is not even seen at Quito (1,490 toises), situated in a narrow valley where a frequently foggy sky diminishes the force of radiation.

3. The observation of 18.5° C is by Mr. Hapel Lachenaie. Mr. Le Dru asserts that he never had seen the thermometer descend below 18.7° C. But he believes that snow falls on the mountains of Loquillo on that same island!

3,500 observations by ▾Mr. Bento Sánchez Dorta; those of Macao are based on 1,200 observations that the Abbé Richenet was kind enough to communicate to me.[1]

	Havana lat. 23° 9′ N	Macao lat. 22° 12′ N	Rio de Janeiro lat. 22° 54′ S
Average annual temp	25.7°	23.3°	23.5°
of the hottest month.	28.8°	28.4°	27.2°
of the coldest month.	21.1°	16.6°	20.0°

Despite the frequent north and northwesterly winds, Havana's climate is warmer than that of Macao and Rio de Janeiro. Macao gets cold because of the west winds during the winter which are frequent on the eastern coastline of any large continent. The proximity of an extremely broad landmass covered with mountains and plateaus renders the variation of temperature over the course of the different months of the year more unequal in Macao and Canton than on an island caressed by the warm waters of the *Gulf-stream* to the west and the north. In addition, the winters are far colder in Canton and in Macao than in Havana. In 1801, the mean temperatures in December, January, February, and March were between 15° and 17.3° C in Canton and between 16.6° and 20° C in Macao, while in Havana, they are generally between 21° and 24.3° C. Nevertheless, Macao's latitude is 1° farther south than Havana's, and the latter city and Canton are on the same parallel within a single minute. Yet, although the isothermal lines, or lines of equal heat, are *concave* toward the pole in the *system of climates of east Asia* and in the *system of climates of eastern America*, the cold on the same geographical parallel is nevertheless more considerable in Asia.[2] During nine years (1806–1814), the Abbé Richenet, who made use of Six's excellent *maximum* and *minimum* thermometer, saw the instrument descend to 3.3° and 5° C (38° and 41° F). In Canton, the thermometer at times almost reaches zero, and, due to caloric updraft, ice forms on the terraces of the houses. Although this very low temperature never lasts for more than a day, the British merchants who reside in Canton enjoy using their fireplaces from November to January, whereas in Havana, one never really

I.78

I.79

1. Once I compare all the notations of this respectable and hardworking clergyman, the partial results from Macao should undergo some slight changes. See Vol. X, [p. 407].

2. Such is the difference in *climate* between the east and west coasts of the Old Continent that in Canton (lat. 23° 8′), the average yearly temperature is 22.9° C, while in Santa Cruz on Tenerife (lat. 28° 28′) it is 23.8° C, according to ▾Mr. von Buch and Mr. Escolar. Canton, situated on an east coast, takes part in a *continental climate*; Tenerife is an island near the west coast of Africa.

feels the need to warm oneself before the *brazero* [brazier], a metal box filled with coal. Hail is frequent and unusually large in Canton and Macao, while it has barely been observed in Havana over the course of fifteen years. In the three locales, the temperature sometimes hovers between 0° and 4° C for several hours. But snow—and this strikes me as remarkable—has never once been noted. Large temperature drops notwithstanding, the banana and palm trees around Canton, Macao, and Havana make the landscape every bit as beautiful as it is in plains far closer to the equator.

1.80

It is a happy circumstance for advancing the study of meteorology in our current state of civilization that we can gather so much numerical data on the climates of places situated almost directly in the two tropics. Five of the largest commercial cities—Canton, Macao, Calcutta, Havana, and Rio de Janeiro—are in this position. Moreover, in the northern hemisphere, Europeans frequent Muscat, Syene, New Santander, Durango, and the northernmost of the Sandwich Islands, and in the southern, Bourbon [Réunion], Île de France [Mauritius], and the port of Cobija between Copiapo and Arica. These places offer the naturalist the same advantages of location as Rio de Janeiro and Havana. Climatology advances slowly because data is collected randomly in places where civilization is just beginning to develop. These points form little clusters separated from each other by immense spaces of *lands unknown* to meteorologists. To deduce the laws of nature governing the distribution of heat across the globe, we must make observations in a spirit that conforms to the needs of an emerging science and determine which numerical constants are the most important. New Santander, on the east coast of the Gulf of Mexico, probably has a lower mean temperature than the island of Cuba. The atmosphere in Mexico likely experiences the winter cold that comes from the large continent to the northwest. By contrast, when we leave the *climatic system of eastern America* and cross the Atlantic basin, or, rather, its submerged valley, and fix our attention on the African coast, we find that the isothermal lines stretch convexly up toward the pole in the *system of cis-Atlantic climates* on the *west* coast of the old continent. There, the Tropic of Cancer passes between Cape Bojador and Cape Blanco near Rio do Ouro, which lies within the inhospitable borders of the Sahara desert. The mean temperature of these places should be much higher than that of Havana both because of their position on an *east coast* and because of the proximity of a desert that radiates heat and scatters molecules of sand in the atmosphere.

1.81

1.82

We have seen that temperature drops on the island of Cuba are of such short duration that they do not usually affect banana trees, sugarcane, or other tropical vegetation. It is well known how easily plants with a vigorous constitution resist a passing cold spell and that the orange trees of Genoa survive snowfall and cold temperatures up to 6° to 7° C below freezing.[1] Because Cuba's plant life has all the characteristics of the vegetation in regions closer to the equator, it is surprising to find plants from the temperate zones and the mountains of equatorial Mexico, even in the plains. In other works, I have often drawn botanists' attention to this extraordinary phenomenon in plant Geography. Pines (Pinus occidentalis) are not found in the Lesser Antilles. According to ▾Mr. Robert Brown, they are absent even from Jamaica (between 17¾° and 18½° lat.), despite the elevation of the Blue Mountains. One only begins to encounter them farther north in the mountains of Saint-Domingue and throughout the island of Cuba,[2] which extends between the 20° and 23° parallels. There, they grow to 60 to 70 feet in height; and it is truly remarkable that *Cahoba*[3] mahogany and Pines flourish in the same plain on the Isle of Pines. In the southeastern regions of the island of Cuba, pines can also grow on the slopes of the Copper Mountains, where the soil is arid and sandy. This Conifer covers the inland plateau of Mexico. At least the samples that Mr. Bonpland and I gathered from Acaguisotla, Nevado de Toluca, and the

I.83

I.84–85

1. Gallesio, [*Traité du citrus*,] p. 55.

2. Mr. Barataro, the knowledgeable student of Professor Balbi's, whom I have consulted on the Pinus occidentalis positions on Saint-Domingue, assured me that he saw this tree near Cape Samana (lat. 19° 18′) in the plains among other plants of the warm region, and that it is found only on mountains of medium height in Saint-Domingue and Puerto Rico, not on the highest mountains. According to what travelers report, the pines on the island of Cuba and on the Isle of Pines to the south of Batabanó are genuine pine trees with imbricated pinecones similar to the Swartz Pinus occidentalis; they are not (as I had suspected for some time) Podocarps. Moreover, the first Spaniards who visited the Antilles sometimes confused Pines with Podocarps, and a passage by Herrera ([*Historia de las Indias occidentales*,] I, p. 52) proves beyond the shadow of a doubt that the *Pinos del Cibao* of which Christopher Columbus spoke after his second voyage were Conifers with monocarpic fruit—that is, true Podocarps. "These very high Pines," said the Admiral, "that have no pinecones are composed in such a fashion by nature that they resemble the olive trees of Seville" [Herrera y Tordesillas, *Historia*, vol. 1, chap. 12; Las Casas, *Historia de las Indias*, book 1, chaps. 91 and 92]. In my first description of Bertholletia, I already noted, following ▾Laet ([*Novus orbis*,] Vol. VIII, pp. 178ff.), how naïve and typical were the descriptions of the first voyagers who were unaccustomed to use technical terms whose value they did not suspect. Are the pines of the Guanajay and Rattan Islands (by 16.5° lat.), useful for mast construction, Podocarps or a variety of *Pinus*? (Herrera, [*Historia de las Indias occidentales*,] Dec. I, p. 131; Laet, *Novus orbis*, p. 341; Juarros, *Compendio de la historia de la ciudad de Guatemala*, Vol. II, p. 169; Tuckey, *Maritime Geography*, Vol. IV, p. 294). We do not know if the Isle of Pines, situated at 8° 57′ lat. to the east of Portobelo, received its name by error from the first explorers. In equinoctial America, between the parallels of 0° and 10°, I have not even seen Podocarps below 1,100 toises above sea level.

3. Swietenia Mahogany L.

Cofre de Perote do not seem to differ much from the Antilles' Pinus occidentalis described by ▾Swartz. Yet, the pines that we saw at sea level on the island of Cuba at 20° and 22° latitude, and only on the island's south side, do not grow lower than 500 toises on the Mexican continent between 17.5° and 19.5°. On the path from Perote to Jalapa on the eastern mountains facing the island of Cuba, I noticed that 935 toises is the pine tree line. Yet, two degrees to the south, on the western mountains between Chilpanzingo and Acapulco

I.86 near Quasiniquilapa, the pine tree line is 580 toises and perhaps 450 toises at some other points. These anomalies of position are very rare in the Torrid Zone and probably due less to temperature[1] than to soil quality. In the system of plant migration, we should suppose that the Cuban Pinus occidentalis came from the Yucatán before the channel opened between Cape Catoche and Cape San Antonio, and not from the United States, which is, after all, rich in Conifers. In Florida, the species whose botanical geography we are tracing here does not exist.

I.87 The following table details temperature observations on the island of Cuba:

Observations on Wajay

Month	1796. F.	1797. F.	1798. F.	1799. F.	Average in degrees centigrade
January	65°	64°	68°	61°	18°
February	72	66	69	63	19.5
March	71	64	68.5	64	19.3
April	74	68	70	68	21.1
May	78.5	77	73	76	24.7
June	80	81	83	85	27.8
July	82.5	80	85	87	28.6
August	83	84	82	84	28.4
September	81	81.5	80	76	26.4
October	78	75.5	79.5	73	24.5
November	75	70	71	61	20.6
December	63	67.5	60	59	16.7
Annual average	75.2°	73.2°	74.2°	71.4°	23.0°

As I noted above, the village of Wajay is situated 5 nautical leagues from Havana on a plateau 38 toises above sea level. The partial mean temperature

1. See a table that gives the position of Conifers and fruitless and flowerless trees with indications of their required temperatures in [Humboldt, Bonpland, and Kunth,] *Nova genera et species plantarum*, Vol. II, p. 27. Pines are no longer found around Jalapa on the eastern slope of the Mexican plateau at 700 toises, although the temperature falls below 12° C.

in December 1795 was 18.8° C; in January and February of 1800, temperatures rose from 13.8° to 18.9° (according to a ˺Nairne thermometer).

OBSERVATIONS ON HAVANA

1.88

Month	1800. Degree cent.	Average of 1810–1812
January	...	21.1°
February	...	22.2
March	21.1	24.3
April	22.7	26.1
May	25.5	28.1
June	30.0	28.4
July	30.3	28.5
August	28.3	28.8
September	26.1	27.8
October	26.6	26.4
November	22.2	24.2
December	23.8	22.1
Average	25.7	25.7

Wajay, interior of the isle of Cuba			Havana, coasts	Cumaná, lat. 10° 27′
Dec–Feb	18.0°	C.	21.8°	26.9°
March–May	21.7		26.2	28.7
July–August	28.2		28.5	27.8
Sept.–Nov	23.8		26.1	26.8
Aver. temp.	22.9		25.7	27.6
Coldest month	16.7		21.1	26.2
Warmest	28.6		28.8	29.1
Rome, lat. 41°53″ aver. t. 15.8°. Warmest month				25.0
				Coldest 5.7°

These are genuine averages deduced from each day's high and low temperatures. However, the results that Don Antonio Robredo obtained in the village of Wajay and in Havana (1800) are perhaps a few tenths of a percent too high, because he used three daily observations simultaneously (at 7 in the morning, noon, and 10 at night). Mr. Ferrer's averages, derived from his observations over three years—1810, 1811, and 1812 (Vol. X, p. 449)—are the most precise figures we have on Havana's climate. The tools that this deft seafarer used during six months in 1800 were better calibrated than Mr. Robredo's

1.89

instruments. Mr. Robredo himself admits that his Havana apartment did not allow the air to flow freely ("*no muy ventilada*" [not well ventilated]), while his situation at Wajay was ideal: it was "*un lugar abierto de todos vientos pero cubierto contra el sol y la lluvia*" [a place open to the wind but protected from the sun and the rain]. In the second half of December 1800, I almost always saw the centigrade thermometer between 10° and 15° C. In January, it dropped to 7.5° C at the ▾Hacienda del Río Blanco. In the countryside near Havana, at

I.90 50 toises above sea level, the water sometimes froze to some depth. Mr. Robredo, a superb observer, communicated this observation to me in 1801. He repeated the experiment in December of 1812, after violent north winds had blown for almost a month. Because in Europe it snows in the plains when the temperature is a little above freezing, it is doubly surprising that no snowfall has ever been reported anywhere on the island, not even in the Lomas de San Juan (San Juan Hills) or on the high mountains of Trinidad. Only white frost (*escarcha*) has been seen on the summits of these mountains and on the *Cobre*. It would seem that conditions beyond a rapid temperature drop in the upper atmosphere are necessary to produce snowfall and ice. We have already noted (Vol.VI, pp. [17]9ff; Vol. X, pp. 334ff) that ice is never seen in Cumaná and so rarely in Havana that it seems to occur only once every fifteen to twenty

I.91 years during lightning storms accompanied by SSW gales. On the coastline of Jamaica in Kingston, a temperature drop down to 20.5° C (69° F) during sunrise is regarded as an extraordinary phenomenon.[1] On that island, one must ascend to 1,150 toises in the Blue Mountains to see 8.3° (in August); similarly, in Cumaná, at 10° lat., I never saw the thermometer below 20.8° C (see Vol. XI, pp. 10ff). Temperature changes are abrupt in Havana. In April 1804, the temperature in the shade went from 32.2° to 23.4° C over the course of three hours, a difference of 9° C. This is a considerable change for the Torrid Zone, and twice the drop that one would find farther south, on the coast of Colombia. In Havana (lat. 22° 8′), people complain about the cold when the temperature drops rapidly to 21° C; the same is true in Cumaná (lat. 10° 28′) when the temperature drops to 23° (see Vol. XI, pp. 10ff). In Havana, in April 1804, the temperature of water that had been exposed to thorough evaporation and was thus regarded as very fresh was 24.4° C (19.5° R); at the same time, the mean

I.92 daily temperature rose to 29.3° C (see XI p. 18). During the three years of Mr. Ferrer's observations (1810–1812), the thermometer was never below 16.4° C (February 20, 1812) nor above 30° C (August 4 of the same year). I have seen it as high as 31.2° C as early as April (1801). But years passed without the tem-

1. ▾Edwards, *The History, Civil and Commercial, of the British Colonies in the West Indies*, 1793, Vol. I, p. 183.

perature rising to 34° C (27.2° R) even once. It would be very interesting to bring together all the observations about the heat of the earth's core at the far reaches of the tropical zone. I have found it to be between 22° and 23° C in the limestone caverns near San Antonio de Beita and at the sources of the Río Chorrera (*Recueil d'observations astronomiques*, Vol. I, p. [134]). Mr. Ferrer found it to be 24.4° C in a well that was 100 feet deep. These observations, perhaps made under less than favorable circumstances, would indicate a temperature of the earth below the mean temperature of the one in Havana. On the coastline, it is about 25.7° C, compared to the island's interior, where it is 23° C at 40 toises of elevation. These results hardly conform to what has been observed in the temperate and arctic zones. Do the deep currents that carry polar water to the equatorial regions decrease the temperature of the earth's core on smaller islands? We have already treated this tricky subject while reporting experiments undertaken in the Guacharo cavern near Caripe (*Relation historique*, Vol. III, p. 144, 145, 194, and 195). In the wells of Kingston and the low country of Guadeloupe, however, the thermometer is known to have reached 27.7°, 28.6°, and 27.2° C, that is, a temperature more or less equal to the mean air temperature in the same locales. 1.93

The large temperature drops in countries situated at the far reaches of the Torrid Zone are connected to oscillations in barometric pressure not observed closer to the equator. In Havana, as in Veracruz, strong north winds interrupt the regular patterns of atmospheric pressure. On the island of Cuba, I have observed that, while the barometer is steady at 0.765 mm during a northern breeze, it will typically drop to 0.756 millimeters or even lower with the south wind. We have already remarked elsewhere that barometric averages during the months when the mercury is highest (December and January) differ by 7 or 8 millimeters from the monthly averages when the pressure is lowest (August and September). This difference is almost as much as it is in Paris and 5 to 6 times more than it is between the equator and 10° lat. north and south of the equator. 1.94

Averages of	December	0.76656^{m}	per	22.1°	cent. of T.
	January	0.76809		21.2	
	July	0.76453		28.5	
	Aug	0.76123		28.8	

Over the course of the three years (1810–1812) during which Mr. Ferrer took these readings,[1] the extreme variations on days when the mercury rose or fell the most exceeded at most 30 millimeters. To give an impression of each

1. *Connaissance des temps*; Vol. IX, p. 300.

1.95 month's accidental oscillations, I add here a table[1] based on Don Antonio Robredo's unpublished notes, which provide the observations for 1801 within a hundredth of an inch:

	Maxima	Minima	Averages	Aver. Temp.
January	30.35^{po}	29.96^{po}	30.24^{po}	14.5^{po} R.
February	30.38	30.01	30.26	15.6
March	30.41	30.20	30.32	15.5
April	30.39	30.32	30.35	17.2
May	30.44	30.38	30.39	19.4
June	30.36	30.33	30.34	22.2
July	30.38	29.52	30.22	22.4
August	30.26	30.12	30.16	22.8
September	30.18	29.82	30.12	21.0
October	30.16	30.04	30.08	18.6
November	30.18	30.09	30.12	16.5
December	30.26	30.02	30.08	12.1

1.96 Hurricanes are far rarer on the island of Cuba than on Saint-Domingue, Jamaica, and the Lesser Antilles situated to the southeast of Cape Cruz. We must not confuse the extremely violent northeasterly winds (*los nortes*) with the *hurricanes* that blow more often from the south-south-east and the south-south-west. When I visited Cuba, there had not been a proper hurricane since August of 1794; the storm of November 2, 1796 was fairly weak. In Cuba, the season for these sudden and terrifying atmospheric movements, when the wind, often accompanied by lightning and hail, blows from all the points of the compass, is from the end of August through September and especially in October. Sailors particularly fear the months of July, August, and September through mid-October in Saint-Domingue and the Caribbean islands. Hurricanes are most frequent there in August, meaning that they occur later as one 1.97 moves to the west. In March, there are sometimes fierce gales from the southeast, even in Havana. No one in the Antilles believes anymore that hurricanes come in predicable patterns.[2] From 1770 to 1795, there were 17 hurricanes in the Caribbean islands, while from 1788 to 1804, not a single storm hit Marti-

1. In this table, the monthly *averages* are the true averages based on daily *highs* and *lows*. The monthly *extremes* indicate the barometric pressure on the two days when the barometer was highest and lowest. Because the table should give only the differences between highs each month and not the absolute average highs, the highs are not reduced to zero temperature, and the level of the basin was not adjusted.

2. See the discussion of this important phenomenon in [Moreau de Jonnès,] *Histoire physique des Antilles*, Vol. I, pp. 325, 350, 355, 376, 387.

nique, which had experienced three in 1642. Hurricanes are rarer at the two ends of the long chain of the Antilles (at the SE and NW extremities). The islands of Trinidad and Tobago have the advantage of never having experienced their effects, and in Cuba, violent ruptures of the atmospheric equilibrium are very rare. When hurricanes do occur, their destructive forces hit the sea rather than human settlements, and more on the south and southeast coasts than toward the north and northwest.[1] As early as 1527, a hurricane partly destroyed the famous ▾Pánfilo [de] Narváez expedition in the port of *Trinidad de Cuba*. 1.98

Below are the barometric movements according to unpublished notes by Captain ▾Don Tomás de Ugarte during that hurricane of August 27 and 28, 1794, which caused the loss of many ships in Havana bay.

August 25	16^{h}	30.po 04
	20	03
	noon	02
Average temp.	4	02
85.8° Fahr.)	8	01
	midnight.	01
August 26	16^{h}	30.00
	20	00
(Aver. temp. 88°)	noon	00
	4	29.99
	midnight.	98
August 27	16^{h}	29.95
	18	94
	20	90
(Aver. temp. 81°)	22	89
	noon	86
	2	84
	4	82
	6	80
	7	80
	8	79
	10	77
	10.5	76
	11	73
	11.5	69
	midnight.	63
August 28	12.5^{h}	29.59
	13	58
August 28.	13.5^{h}	29.57
	14.	56
	14.5	54
	15	52
	15.5	50
	16	51
	18	52
(Aver. temp. 83°)	18.5	54
	19	59
	19.5	63
	20	67
	20.5	70
	21	72
	21.5	74
	22	75
	22.5	76
	noon	78
	2	79
	2.5	82
	3.5	83
	6	84
	7	87
	8	89
	9	90
	10	93
	11	96
	midnight.	30.01

1. This difference between the two coasts is also visible in Jamaica.

The hurricane began the morning of the 27th; its force increased as the barometer dropped; and it finished on the evening of the 28th. We have already indicated above that, on October 25, 1810, Mr. Ferrer saw his barometer drop to 744.72 millimeters at a temperature of 24° C, due to a furious SSW gust. Previously, his barometer had indicated a yearly average of 763.71 millimeters at 26° C.

I.99

Among the causes for the temperature drop during the winter, I might have cited the large number of shoals that surround Cuba. They reduce the temperature by a few degrees because the water molecules cool at the surface and then drop to the bottom. Other reasons might be the polar currents that move into the tropical Ocean's depths or the mingling of deep and shallow water at the point where the ocean's bottom drops off sharply.[1] The stream of warm water (*gulf stream*) that abuts the northwest coastline and whose speed is often diminished by winds from the north and northeast compensates in part for this lowering of the temperature. The chain of shoals that surrounds the island and looks like a shadow on our maps is fortunately broken at several points, which permits commercial access to the coast. In general, the parts of the island that are most exempt from *dangers* (reef, sandbanks, rocks) are on the southeast between Cabo Cruz and Punta Maisí (72 nautical leagues) and in the northwest between Matanzas and Cabañas (28 nautical leagues). In the southeast, the tall Ur-mountains turn the coastline into more of a cliff: this is where we find the ports of Santiago de Cuba, Guantánamo, Baitiquiri, and Baracoa (around Punta Maisí). ʼBaracoa was the first place settled by Europeans. The entrance to the Old Bahama Channel from Punta de Mulas to the WNW of Baracoa to the new settlement called Puerto de las Nuevitas del Príncipe is also free of sandbanks and breakers. Here, sailors find excellent anchorages a little to the east of Punta de Mulas in the three coves of Tánamo, Cabonico, and Nipe, and to the west in the ports of Sama, Naranja, Padre, and Nuevas Grandes. Near this last port and, rather astonishingly, at roughly the same meridian of the island as the beginning of the shoals of *Buena Esperanza* and *Las doce leguas*, the Keys of the Old Bahama Channel begin their uninterrupted line running right to the Isle of Pines. This line extends for 94 leagues from Nuevitas to Punta Hicacos. The Old Bahama Channel is at its narrowest just opposite the islands of Cayo Cruz and Cayo Romano, where it is barely 5 to 6 leagues wide. Here, the Great Bahama Bank is most pronounced. The keys closest to the island of Cuba and the parts of the Bank that are not covered by water (Long Island, Eleuthera) have, like Cuba, a very elongated shape. If the sea level were twenty or thirty feet lower,

I.100

I.101

1. See Vol. I, pp. 100; II, pp. 72, 73, 74; V, pp. 190–191.

an island larger than Haiti would appear on the water's surface. The chain of reefs and keys that borders the navigable part of the Old Bahama Channel toward the south leaves little basins without shoals between the Channel and Cuba's coastline. These basins connect with many ports that have good anchorages, such as Guanajay, Morón, and Remedios. I.102

After exiting through the Old Bahama Channel, or, rather, the San Nicholas Channel, between *Cruz del Padre* and a bank of the Salt Cliffs, or Cayos de Sal, of which the lowest offer sources of freshwater,[1] one finds once more, from Punta de Hicacos to Cabañas, a coastline free of *dangers*. In this interval, the coastline furnishes the anchorages of Matanzas, Puerto Escondido, Havana, and Mariel. Farther away, to the west of Bahía Honda—a tempting prize for any maritime power that happens to be an enemy of Spain—begins a chain of shoals, the *Bajos de Santa Isabel* and the *Bajos de los Colorados*, which extends to Cape San Antonio without interruption. From this cape all the way I.103 to Punta de Piedras and the Bahía de Cortés, nearly the entire coastline is a *sheer* drop and does not allow for any soundings. But between Punta de Piedras and Cabo Cruz, almost the whole southern part of Cuba is surrounded by shoals, of which the Isle of Pines is merely a small portion peeking above the waves. To the west, this formation is known as *Royal Gardens* (*Jardines y Jardinillos*) and, to the east, as *Cayo Bretón*, *Cayos de las Doce Leguas*, and *Bancos de Buena Esperanza*. Along the entire south coast, the shoreline is not free of *dangers*, except between the Bay of Pigs and the mouth of the Río Guaurabo. These waters make for difficult navigation. I had occasion to determine several points of latitude and longitude there en route from Batabanó to Trinidad de Cuba and Cartagena de Indias. The resistance to currents that the Isle of Pines highlands and the extraordinary elongation of Cape Cruz offer might be said to favor both the accumulation of sand deposits and the work of the stony corals that thrive in the tranquil and shallow waters. Of this I.104 145-league stretch of coastline in the south, only one-seventeenth offers free access between Cayo de Piedras and Cayo Blanco a little to the east of Puerto Casilda. Here, small vessels frequent anchorages such as El Surgidero del Batabanó, La Bahía de Jagua, Puerto Casilda, or Trinidad de Cuba. Beyond the latter port, near the mouth of the Río Cauto and Cabo Cruz inside the *Cayos de las Doce Leguas* [or "Laberinto de las Doce Leguas," the "twelve-league

1. Cayos del Agua (lat. 23° 58′, long. 82° 36′), in the Placer de los Roques or the Cayo de Sal. I place the Cayo del Agua a little more to the west than Captain Steetz did on the interesting maps that accompany his *Instruction nautique sur les Passages à l'île de Cuba*, 1825, p. 55, where he placed the Havana Morro at 84° 39′ and the Pan de Matanzas at 83° 58′. Mr. Ferrer, meanwhile, places them at 84° 42′ 44″ and 84° 3′ 12″ by means of averages that merit complete confidence.

labyrinth"], the lagoon-rich coast is almost entirely inaccessible and equally deserted.

Here are the most precise figures I could gather on the position of Cuba's harbors:

To the east of Cape Cruz (lat. 19° 47′ 16″, long. 80° 4′ 15″):

Santiago de Cuba	lat. 19° 57′ 29″, long. 78° 18′
Guantánamo Bay	lat. 19° 54′, long. 77° 36′
Puerto Escondido	lat. 19° 54′ 55″, long. 77′ 24″
Baitiquiri	lat. 20° 2′, long. 77° 12′

To the northwest of Cape Maisí	lat. 20° 16′ 40″, long. 76° 30′ 25″:
Puerto de Malta	lat. 20° 17′ 10″, long. 76° 43′
Baracoa	lat. 20° 20′ 50″, long. 76° 50′
Maravi	lat. 20° 24′ 11″, long. 77° 17′
Puerto de Navas	lat. 20° 29′ 44″, long. 77° 20′
I.105 Cayaguaneque	lat. 20° 30′, long. 76° 56′
Taco	lat. 20° 31′ 17″, long. 77° 0′
Jaragua	lat. 20° 32′ 44″, long. 77° 3′
Puerto de Cayo Moa	lat. 20° 42′ 18″, long. 77° 14′
Yaguaneque	lat. 20° 42′, long. 77° 22′
Cananova	lat. 20° 41′ 30″, long. 77° 24′
Cebollas	lat. 20° 41′ 52″, long. 77° 28′
Tánamo	lat. 20° 42′ 41″, long. 77° 37′
Puertos de Cabonica y Livisa	lat. 20° 42′ 11″, long. 77° 46′
Nipe	lat. 20° 44′ 40″, long. 77° 51′
Banes	lat. 20° 52′ 50″, long. 78° 1′

To the northwest of Punta de Mulas	lat. 21° 5′, long. 77° 57′:
Sama	lat. 21° 5′ 50″, long. 78° 11′

In the Old Bahama Channel:

Naranjo	lat. 21° 5′ 23″, long. 78° 19′
Vita	lat. 21° 6′, long. 78° 25′
Barrirai	lat. 21° 4′ 9″, long. 78° 27′
Jururu	lat. 21° 3′ 39″, long. 78° 28′
Gibara	lat. 21° 6′ 12″, long. 78° 33′
Puerto del Padre	lat. 21° 15′ 40″, long. 78° 49′
Puerto del Malagueta	lat. 21° 16′, long. 78° 58′
Puerto del Manatí	lat. 21° 23′ 44″, long. 79° 7′
Puerto de Nuevas Grandes	lat. 21° 26′ 50″, long. 79° 13′
Puerto de las Nuevitas del Príncipe	lat. 21° 38′ 40″, long. 79° 2′
Guanajay	lat. 21° 42′, long 80° 11′

Embarcadero del Príncipe lat. 21° 44′, long. 80° 23′
Between Río Jiguey and Punta Carina to the NNE of Hato de Guanamacar
Morón lat. 22° 4′, long. 80° 56′ I.106
Puerto de Remedios lat. 22° 32′, long. 81° 56′
Puerto de Sierra Morena lat. 23° 3′, long. 82° 54′

To the west and to the southwest of Punta Hicacos (lat. 23° 10′, long. 83° 32′):
Matanzas lat. 23° 3′, long. 83° 54′
Puerto Escondido lat. 23° 8′, long. 84° 12′
Mouth of the Río Santa Cruz lat. 23° 7′, long. 84° 18′
Jaruco lat. 23° 9′, long. 84° 25′
Havana lat. 23° 9′, long. 84° 43′
Mariel lat. 23° 5′ 58″, long. 85° 2′
Puerto de Cabañas lat. 23° 3′, long. 85° 13′
Bahía Honda (the southernmost shore of the bay near Potrero de Madrazo) lat. 20° 56′ 7″, long. 85° 32′ 10″

To the east of Cape San Antonio (lat. 21° 50′, long. 87° 17′ 22″):
Surgidero del Batabanó lat. 22° 43′ 19″, long. 84° 45′ 56″
Bahía de Jagua lat. 22° 4′, long. 82° 54′
Trinidad de Cuba's two ports, that is:
Puerto Casilda lat. 21° 45′ 26″, long. 82° 21′ 7″
Mouth of the Río Guaurabo lat. 21° 45′ 46″ and long. 82° 23′ 37″

There are many lagoons—Vertientes, Santa María, Curajaya, Yaguabo, Junco, etc.—but no proper ports between Trinidad de Cuba and Cabo Cruz.

I.107 The positions of the 50 ports and anchorages on island of Cuba were the result of studies that (in 1826) led me to correct the island's 1820 map. The latitudes are, in large part, taken from *Portulano de la América septentrionale construído en el Depósito hidrográfico de Madrid* (1818), but my longitudes differ from it considerably. The *Portulano* places Havana's Morro at 84° 37′ 45″ or 5′ too far to the east. Consult Bauzá, *Derrotero de las Islas Antillas*, 1820, p. 487, and Purdy, *Columbian Navigator*, p. 175. I preferred the positions that Mr. Ferrer assigned to Capes Cruz and Maisí and the Punta de Mulas, and it is to these same capes that I reduced many points that Don José del Río and Don Ventura Barcaíztegui had determined. I base my conclusions on my own observations, distancing myself from the position these skilled seafarers assigned to Puerto Casilda. Mr. Bauzá, who adopts my positions for Batabanó and Punta Matahambre, prefers nevertheless long. 76° 26′ 28″ for Punta Maisí, who follows Don José Sánchez Cerquero in placing Puerto Rico at 68° 28′ 29″. Mr. Cerquero's assembly of rather heterogeneous observations nevertheless gives him 68° 26′ 30″, while ‸Mr. de Zach sees 68° 31′ 0″ as a more I.108

probable result. [Zach,] *Correspondance astronomique*, Vol. XIII, p. 125, 128. After a consideration of all the data, Mr. Oltmanns determined an average of 68° 33′ 30″. See my *Recueil d'observations astronomique*, Vol. II, p. 139.

On the island of Cuba, as was once the case in all of Spain's American possessions, one must distinguish between *ecclesiastical*, *military-political*, and *financial* districts. We will omit the *judicial* hierarchy, which has created much confusion among modern geographers,[1] since the island has only had a single ▾*Audiencia*, in Puerto Príncipe [Camagüey], since 1797; its jurisdiction extends from Baracoa to Cape San Antonio. The division into two bishoprics dates from 1788 when Pope Pius VI named the first bishop of Havana. Having once been under the Archbishop of Saint-Domingue, along with Louisiana and Florida, the island of Cuba had had only one bishopric since its discovery. Pope Leo X founded it in 1518 at Baracoa [Guantánamo province], in the island's westernmost corner. Four years later, this bishopric was moved to Santiago de Cuba. But the first bishop, ▾Friar Juan de Ubite, arrived only in 1528. At the beginning of the nineteenth century (in 1804, to be precise), Santiago de Cuba was elevated to an archbishopric. The border between the dioceses of Havana and of Cuba [Oriente] follows the meridian of Cayo Romano, about 80.75° long. west of Paris, between the town of Sancti Spíritus and the city of *Puerto Príncipe* [Camagüey]. For the purposes of military-political government, the island is divided into two *gobiernos* [administrative units] under the same Captain: In addition to the capital itself, the *gobierno* of Havana comprises the districts of Puerto Príncipe [Camagüey] and of *Cuatro Villas* [Las Villas]; Trinidad, today a city; Sancti Spíritus; Villa Clara; and San Juan de los Remedios [now Remedios, part of Villa Clara province]. The Captain-general, who is also the Governor of Havana, appoints a *Teniente Gobernador* [deputy] in Puerto Príncipe, as he does in Trinidad [Sancti Spíritus] and Nueva Filipina [Pinar del Río]. Like the magistrate's jurisdiction, the territorial jurisdiction of the Captain-general extends to eight places with their own *pueblos de Ayuntamiento* [town halls]: the cities of Matanzas, Jaruco, San Felipe y Santiago, Santa María del Rosario; the towns of Guanabacoa, Santiago de las Vegas, Güines, and San Antonio de los Baños. The *gobierno* of *Cuba* consists of Santiago de Cuba, Baracoa, ▾Holguín, and Bayamo [Granma]. The current borders of the *gobiernos* are therefore not the same as those of the bishoprics. For example, until 1814, the district of Puerto Príncipe, with its 7 parishes, was under both the *gobierno* of Havana and the archbishopric of Cuba [Oriente].[2] In the 1817 and 1820 censuses, Puerto Príncipe

I.109
I.110

1. Vol. IV, pp. 70 and 71.

2. [Arango,] *Documentos sobre el tráfico de los negros*, 1814, pp. 127, 130.

[Camagüey] was combined with Baracoa [Guantánamo] and Bayamo under the administrative jurisdiction of the district of *Cuba* [Oriente]. It remains for me only to describe a third, financial division. In the March 23, 1812 *assessment*, the island was repartitioned into three provinces, known as *Intendencias* or *Provincias*—Havana, Puerto Príncipe [Camagüey], and Santiago de Cuba—whose lengths from east to west were about 90, 70, and 65 nautical leagues, respectively. The administrator of Havana exercises the prerogatives of a Superintendent General, who is a delegate of the Royal Treasury of the Island of Cuba, the *Superintendente general subdelegado de Real Hacienda de la Isla de Cuba*. Following this division, the province of *Cuba* [Oriente] comprises Santiago de Cuba, Baracoa, Holguín, Bayamo, Gibara, Manzanillo, Jiguaní, Cobre, and Tiguaros. The Province of *Puerto Príncipe* [Camagüey] includes the city of Puerto Príncipe, Nuevitas, Jagua, Sancti Spíritus, San Juan de los Remedios, Villa de Santa Clara, and Trinidad. The district farthest to the west, *Havana* Province, consists of everything to the west of *Cuatro Villas*, an area over which the capital's financial manager had lost control. When the cultivation of the land is more uniformly advanced, the division of the island into 5 provinces will likely seem most convenient and most linked with the historical memory of the *conquest's* first days: into *Vuelta Abajo* (from Cape San Antonio to the beautiful village of Guanajay and Mariel); *Havana* (from Mariel to Álvarez); *Cuatro Villas* (from Álvarez to Morón); *Puerto Príncipe* (from Morón to Río Cauto), and *Cuba* (from the Río Cauto to Punta Maisí).

I.111

My map of the island of Cuba, however imperfect when it comes to the interior, is still the only map on which one can find the 13 cities, or *ciudades*, and 7 towns, or *villas*, that are the objects of the divisions I have just explained. The border between the bishopric of Havana and *Santiago de Cuba* runs from the mouth of the little Santa María River (long. 80° 49″) along the southern coast through the parish of San Eugenio de la Palma, through the *haciendas* of Santa Ana, Dos Hermanos, Copey, and Ciénaga toward the Punta de Judas (long. 80° 46′), along the northern coast opposite Cayo Romano. During the reign of the Cortes, the Royal Spanish Courts, it was agreed that the ecclesiastical boundary would duplicate the provincial border of Havana and Santiago de Cuba (*Guía de Forasteros de la Isla de Cuba*, 1822, p. 79). The diocese of Havana consists of 40 parishes; Cuba's diocese has 22. Established at a point in time when cattle ranches (*haciendas de ganado*) took up most the island, these *parroquias* were too vast and too poorly adapted to the needs of contemporary civilization. The bishopric of Santiago de Cuba consists of the 5 *cities* of Baracoa, Cuba, Holguín, Guisa, and Puerto Príncipe and the town of Bayamo. In the bishopric of San Cristóbal de la Habana, there are 8 *cities*—Havana, Santa María del Rosario, San Antonio Abad or de

I.112

I.113

los Baños, San Felipe y Santiago del Bejucal, Matanzas, Jaruco, La Paz, and Trinidad—and 6 towns—Guanabacoa, Santiago de las Vegas o Compostela, Santa Clara, San Juan de los Remedios, Sancti Spíritus, and San Julián de los Güines. Among Havana's inhabitants, the most popular territorial division is that of the ▾*vuelta arriba* and *vuelta abajo*, that is, to the east and to the west of the Havana meridian. The island's first governor to take the title of Captain-general (1601) was ▾Don Pedro Valdés. There had been 16 other governors before him, beginning with the famous Settler and *Conquistador* ▾Diego Velázquez, a native of Cuéllar, whom Admiral [Diego] Columbus himself had appointed in 1511.

POPULATION—I have just examined the territory, the climate, and the geological makeup of a country that offers a broad field for human civilization. In what follows, I will assess how the wealthiest of the Antilles, with such a vital natural environment, will one day affect the balance of power in insular I.114 America. I will do so by comparing Cuba's current population to the numbers of inhabitants that a largely unexploited territory of 3,600 square nautical leagues fertilized by tropical rains might hypothetically have been able to sustain. Three successive censuses of highly variable precision produced these results:

1775 a population of	170,862
1791	272,140
1817	630,980

According to the last assessment, whose details I will outline below, Cuba had 290,021 whites, 115,691 free people of color, and 225,268 slaves. These results are more or less consistent with the interesting work that the Municipality of Havana submitted to the Royal Spanish Courts in 1811, which counted approximately 600,000, of which 274,000 were whites, 114,000 freedmen, and 212,000 slaves. If one takes into account the various omissions in the 1817 census, the introduction of slaves to Cuba—Havana customs registered more than 41,000 slaves in only three years: 1818, 1819, and 1820—, and the increase I.115 in the number of free people of color as well as whites, which a comparison between the 1810 and 1817 censuses in the island's eastern part revealed, one would arrive at the following figures for the end of 1825:

Frees		455,000
whites	325,000	
free people of color	130,000	
Slaves		260,000
	Total	715,000

Consequently, the population of the island of Cuba today is barely different from that of the British Antilles, and it is practically double that of Jamaica. In Cuba, where slavery has put down its roots very deeply, there are striking contrasts among its inhabitants who are classified according to origin and state of freedom. The table below illustrates these contrasts, which prompt the gravest of reflections. I.116–17

Antilles compared with one another and with the Continental States	Total Population	Whites	Free people of color, Mixed, and Blacks	Slaves	Distribution among the classes	
Isle of Cuba	715,000	325,000	130,000	260,000	Whites	0.46
					Free people of color	0.18
					Slaves	0.36
						1.00
Jamaica	402,000	25,000	35,000	342,000	Whites	0.06
					Free people of color	0.09
					Slaves	0.85
						1.00
All of the British Antilles	776,500	71,350	78,350	626,800	Whites	0.09
					Free people of color	0.10
					Slaves	0.81
						1.00
All the Antilles	2,843,000	482,600	1,212,900	1,147,500	Whites	0.17
					Free people of color	0.43
					Slaves	0.40
						1.00
United States of North America	10,525,000	8,575,000	285,000	1,665,000	Whites	0.81
					Free people of color	0.03
					Slaves	0.16
						1.00
Brazil	4,000,000	920,000	1,020,000	2,060,000	Whites	0.23
					Free people of color	0.26
					Slaves	0.51
						1.00

This table[1] shows that, on the island of Cuba,[2] free men encompass 64 percent of the entire population; in the British Antilles, barely 19 percent. In all I.118

1. This table presents data to the end of 1823; only Cuba's population dates are from 1825. If we allow 936,000 for the population of Haiti (see Vol. XI, pp. 158 and 159 [also this edition, p. 315]) rather than 820,000, then we have 2,959,000 for the whole of the Antilles, of which 1,329,000, or 45 instead of 43 percent, would be free people of color.

2. In 1788, free men constituted 13 percent of the population in the French part of St. Domingue (whites: 8 percent; free people of color: 5 percent), and slaves, 87 percent.

of the Antilles, men of color (black and of mixed race, free and enslaved) add up to 2,360,000 or 83 percent of the total population. If the laws in the Antilles and the legal status of people of color do not change for the better soon and if we continue to talk without acting, political supremacy will pass into the hands of those who have the power of labor, the will to emancipate themselves, and the courage to endure long privations. This bloody catastrophe will occur as a necessary result of circumstances without any involvement on

I.119 the part of Haiti's free blacks, who will not have to abandon the isolation in which they have lived up to now. Who would dare predict the impact that an *African Confederation of the Free States of the Antilles*, situated between Colombia, North America, and Guatemala, would have on the politics of the New World? The fear of this eventuality no doubt affects the spirit more than do the principles of humanity and justice. But the whites on each island believe their power to be unassailable. Any coordinated action on the part of blacks seems to them impossible; any change, any concession accorded to the enslaved population a sign of cowardice. There is no pressure: the horrible catastrophe of St. Domingue was nothing more than the result of the government's incompetence. Such are the illusions that reign among the vast majority of colonists in the Antilles and that also hinder improvements in the status of black people in Georgia and the Carolinas. More than any other island in the Antilles, the island of Cuba has the ability to prevent this shipwreck. This island has 455,000 free men and 260,000 slaves. By measures that are at once

I.120 humane and prudent, it should be possible to prepare for the gradual abolition of slavery. We must not forget that free blacks and mixed-race people have outnumbered slaves throughout the Antilles ever since Haiti's emancipation. On the island of Cuba, the numbers of whites, and especially the numbers of freedmen who could easily join up with them, are growing rapidly. Without the illegal continuation of the slave trade after 1820, the slave population would have diminished quickly. If the progress of human civilization and the firm will of the new states of free America were to put a stop to this infamous commerce, the enslaved population would dwindle more considerably for some time due to the existing imbalance between the sexes and the continuing manumissions. This decline will end only when the death–birth ratio among slaves offsets the effects of manumissions. Whites and freedmen already make up nearly two-thirds of the island's total population, and the current growth in their numbers marks, at least in part, the decrease of slaves

I.121 in the total population. Among the slaves, the proportion of women to men is 1 to 4 on sugar plantations (excluding mixed-race slaves); for the entire island, it is 1 to 1.7. In the cities and on farms where slaves serve either as domestics or work during the day both for themselves and for their masters, the ratio is 1 to 1.4, even (as is the case in Havana[1]) 1 to 1.2. My observations below show that these calculations are based on numbers that can be regarded as the *upper limit.*

Some worry that the island's total population will shrink once the slave trade is abolished in fact and not merely in law (as it has been since 1820); that it will be impossible to continue the large-scale cultivation of sugar cane; and that, in the future, Cuba's agricultural sector will be restricted to coffee and tobacco plantations and the raising of livestock. These tempting prognoses are founded on arguments that do not strike me as sufficiently sound. One forgets that ▾sugar plantations, many of which are shorthanded and wear out the blacks with countless *night shifts*, employ no more than 1/5 of the total slave population. Calculating the total population growth *rate* on the island of Cuba for a time when no more black Africans are imported also requires one to take into account and counterbalance complicated variables that affect each population group differently: that is, growth factors vary greatly among whites, freedmen, and slave laborers. For slaves, such factors additionally depend on whether they work on sugar, coffee, and tobacco plantations, on cattle ranches, or as domestics, artisans, and day laborers in the towns and cities. Given these variables, one should not rush to any gloomy conclusions but, rather, wait until the government has obtained reliable statistical data. The spirit in which even the earliest censuses (such as the one of 1775) were undertaken, using rubrics of age, sex, race, and states of civil liberty, deserves the highest praise. What was lacking were only the means for doing them properly. One sensed that the inhabitants' peace of mind depended in part on knowing the occupations of black people, that is, their numerical distribution on sugar plantations, on the farms, and in the cities. In order to remedy the evil of slavery, to avert public dangers, and to console the unfortunate members of a race that suffers and is feared more than anyone cares to admit, it is necessary to clean out the wound. There are restorative forces in any intelligently lead social body—as there are within an organic body—, and with those one can eradicate even the most ingrained of evils.

I.122

I.123

In 1811, the Municipality and the Chamber of Commerce of Havana assumed the total population of the island of Cuba to be 600,000, of which

1. It seems to me likely enough that, at the end of 1825, the total population of people of color—mixed-race and free and enslaved blacks—consisted of 160,000 in the cities and 230,000 in the countryside. In an 1811 text presented to the Spanish Cortes, the *Consulado* [merchant guild] estimated 141,000 people of color in the cities and 185,000 in the countryside. [Arango,] *Documentos sobre los negros*, p. 121. The large congregation in the cities of persons of mixed race, along with free and enslaved blacks, is characteristic of the island of Cuba.

326,000 were men of color (free and enslaved, mixed-race and black). We see below how this total population was distributed across the different parts I.124 of the island, in the cities and in the rural areas. These figures show the numbers of free blacks and slaves relative to the total number of people of color.

Cuba's Territorial Divisions	Free People of color	Slaves	People of color, both free and enslaved
I. Western REGION (Jurisdiction of Havana)			
in the Towns	0.11	0.11 ½	0.22 ½
in Rural Areas	0.01 ½	0.34	0.35 ½
II. Eastern REGION (Cuatro Villas, Puerto Príncipe, Cuba)			
in the Towns	0.11	0.09 ½	0.20 ½
in Rural Areas	0.11	0.10 ½	0.21 ½
Total	0.34 ½	0.65 ½	1.00

It follows from this table, which subsequent research could certainly improve, that in 1811, nearly three-fifths of the people of color lived in the Ju- I.125 risdiction of Havana between Cape San Antonio and Álvarez. In this region, the cities had as many mixed-race persons and free blacks as they did slaves. But the urban *population of color* was a third smaller than its rural counterpart. By contrast, on the island's eastern part, from Álvarez to Santiago de Cuba and Cape Maisí, people of color in the cities almost equaled in number those on plantations and farms. We will see that, between 1811 and the end of 1825, 185,000 African blacks arrived on Cuba's shores by both legal and illegal means. Havana customs alone registered nearly 116,000 between 1811 and 1820. Without a doubt, this newly introduced population more often found its way into rural areas than into the cities, changing the relationships between the island's eastern and western parts and between the cities and the country that were believed to have existed in 1811. Black slaves proliferated on the eastern plantations. We know that, between 1811 and 1825, the numbers of both I.126 free and enslaved people of color, be they of mixed race or black, did not grow by more than 64,000, or one-fifth, despite the fact that 185,000 *negros bozales* ["unseasoned" slaves] were imported directly from Africa during that time. This awful truth shows the *relative distributional* changes to be within limits far narrower than would have been believed earlier.

We saw above that, if one assumed a population of 715,000 (which I believe to be about *the lower limit*), the *relative population* of the island of Cuba would amount to 197 individuals per square nautical league by the end of 1825. This is almost two times less than St. Domingue's population and four times less than that of Jamaica. If Cuba were as well cultivated as Jamaica is, or, better put, if their population *density* were the same, Cuba would have 3,615×874, or 3,159,000, inhabitants,[1] more people than live today in the entire Republic of Colombia or in all of the Antilles. Yet, Jamaica still has 1,914,000 uncultivated *acres*. I.127

The oldest official *padrones* and *censos* [surveys and censuses] with which I could familiarize myself during my sojourn in Havana are those that the Marquis de la Torre and Don Luis de las Casas[2] commissioned in 1774 and 1775 and in 1791, respectively. It is well known that both censuses were executed with extreme sloppiness, and that a large portion of the population might not have been included. The 1775 *Padrón*, with which the ▾Abbé Raynal was familiar, furnished the following results: I.128

Men	whites	54,555
	free mixed-race.	10,021
	free blacks	5,959
	enslaved mixed-race	3,518
	enslaved blacks.	25,256
		99,309

1. Supposing Haiti's population to be 820,000, we have 334 inhabitants per square nautical league. Supposing 936,000, the relative population is 382. Cuban authors think that the island of Cuba can support 7.5 million inhabitants (see [Zayas y Chacón and Benítez,] *Reclamación de los representantes de Cuba contra la ley de arancéles*, 1821, p. 9). Even according to this hypothesis, the relative population would still not equal Ireland's. Some British geographers put Jamaica at 4,090,000 acres, or 534 square nautical leagues.

2. This governor founded the *Patriotic Society*, the Agriculture and Trade Board, or *Junta de agricultura y comercio*, a public library, the *Consulado*, the House for poor girls (*Casa de beneficiencia de niñas indigentes*), the botanical Garden, a chair of mathematics and free primary schools (*escuelas de primeras letras*). He tried to alleviate the barbaric forms of criminal justice and created the noble post of *defender of the poor*. From the same period dates Havana's embellishment, the opening of the Güines trail, the construction of ports and dikes, and, what is much more important, the protection extended to periodicals that edify the public spirit. Don Luis de las Casas y Aragorri, Captain-General of the island of Cuba (1790–1796), was born in the village of Sopuerta in Biscay. He fought with the highest distinction in Portugal, Pensacola, the Crimea, Algiers, Mahon, and on Gibraltar. In July of 1800, he died in Puerto Santa María at the age of 55. See the précis of his life by Friar Juan Gonzáles (del Orden de Predicadores [of the Preachers' Order]) and Don Tomás Romay.

Women	whites	40,864
	free mixed-race	9,006
	free blacks	5,629
	enslaved mixed-race	2,206
	enslaved blacks	13,356
		71,061

Of a total of 170,370, the Jurisdiction of Havana alone encompasses 75,617. I.129 I have not had any occasion to verify these figures against official documents. The 1791 *Padrón* mentioned—and this number is consistent with the registries—272,141 inhabitants, 137,800 in the Jurisdiction of Havana: 44,337 in the capital; 27,715 in the Jurisdiction's other *ciudades* and *villas*; and 65,748 in rural districts, the *partidos del campo*. The simplest reflections reveal what is contradictory in the results[1] of this undertaking. The total of 137,800 inhabitants in the Jurisdiction of Havana appears to consist of 73,000 whites, 27,600 free people of color, and 37,200 slaves. So whites exist in a 1 to 0.5 ratio to slaves, instead of the ratio of 1 to 0.83 that has been observed in the city and the country for quite some time. In 1804, I discussed Don Luis de las Casas's census with those who had an intimate knowledge of the place. While determining the value of the omitted quantities through partial comparisons, I.130 we surmised that the island's population in 1791 could not have been any less than 362,700. From 1791 to 1804, this population grew from the numbers of *bozales* that, according to the customs' registries, increased to 60,393 during this period, from immigration from Europe and St. Domingue (5,000), and, finally, from the slightly higher number of births in relation to deaths in a country where one-fourth to one-fifth of the entire population is condemned to a celibate life. The combined effect of these three causes of growth—assuming an annual loss of only seven percent among the *negros bozales*—was estimated at 60,000; this is the source of the *lower limit*[2] of about 432,080 for

1. ▾Andrés Cavo, *De vita Josephi Juliani Parenni Havanensis* (Rome, 1792), p. 10. Some copies read 151,150 instead of 137,800.

2. In the figure of 432,000, I counted, for 1804, 234,000 whites, 90,000 free people of color, and 108,000 slaves. (The 1817 census counted 290,000 whites, 115,000 free people of color, and 225,000 slaves.) I estimated the black slave population by assuming that a plantation produces 80 to 100 arrobas of sugar per black and that the average population of a sugarmill, or *ingenio*, was 82 slaves. At the time, there were more than 350 sugar plantations. In the seven parishes of Guanajay, Managua, Batabanó, Güines, Cano, Bejucal, and Guanabacoa, we found, through an exact census, 15,130 slaves in 183 *ingenios* ([Arango,] *Expediente*, p. 134. *Representación del Consulado de la Habana del 10 Julio* 1799, manuscript). The relationship between sugar production and the number of blacks employed on sugar plantations is very difficult to determine. On some plantations, 300 blacks barely produce 30,000 arrobas of sugar; on others, 150 blacks produce nearly 27,000 arrobas

1804. The 1817 census sets the population at 572,363, which should be considered only as the lower limit. It substantiates my findings from 1804, which have since been used in numerous statistical works. According to customs' registries alone, 78,500 blacks were imported between 1804 and 1816. I.131–32

To this day, the most important documents we have on the island's population were published on the occasion of ▾a famous proposition that Mr. [Guridi y] Alcocer and Mr. Argüelles made in the assembly of the Spanish Courts on March 26, 1811 against the slave trade in general and against the practice of enslaving blacks born in the colonies. As pieces of supporting evidence, these precious documents accompany the presentations[1] that ▾Don Francisco de Arango, one of the most progressive and best informed statesmen, made before the Courts in the name of the *Consulado* [merchant guild] and the Patriotic Society of Havana. These documents state that "no general census exists besides the one attempted in 1791 under the wise administration of Don Luis de las Casas and that, since this time, we have been limited to partial censuses in the most populous districts." The results published in 1811 are based on incomplete data and estimates of population growth from 1791 to 1811. In the following table, we have adopted the island's division into four parts: 1) the *Jurisdiction of Havana*, or the *western part* between Cape San Antonio and Álvarez; 2) the *Jurisdicción de Cuatro Villas*, with 8 parishes, situated I.133

per year. The number of whites can be pinned down through militia enrollments. In 1804, there were 2,680 regulars, or *disciplinados*, and 21,831 *militiamen* known as *rurales*, despite the extreme ease with which one could dodge service and the countless exemptions granted to *Abogados*, *Escribanos*, *Médicos*, *Boticarios*, *Notarios*, *Sacristanes y Sirvientes de Iglesia*, *Ministros de Escuela*, *Mayorales*, *Mercadores* [lawyers, clerks, physicians, apothecaries, notaries, clergy and servants of the Church, schoolmasters, overseers, merchants], and all those who were considered *nobility*. Compare [Arango,] *Reflexiones de un Habanero sobre la independencia de esta isla*, 1823, p. 17. In 1817, men between the ages of 15 and 60 who were considered capable of bearing arms broke down as follows: (1) the free class: 71,047 whites, 17,862 free persons of mixed race, 17,246 free blacks (a total of 106,155 free men); (2) slaves: 10,506 persons of mixed race and 75,393 blacks (a total of 85,899 slaves; the total of free men and slaves between the ages of 15 and 60 is 192,054). Taking as a basis the ratio between draftees and the general population in France (Peuchet, *Statistique*, pp. 243, 247), we find that this estimate of 192,054 would presuppose a population of less than 600,000. The *contingents* of the three classes of whites, freedmen, and slaves are 0.37, 0.18, and 0.45, respectively, while the proportions of these classes in the general population are more probably 0.46, 0.18, and 0.36.

1. *Representación del 16 de Agosto 1811, que por encargo del Ayuntamiento, Consulado y Sociedad patriótica de la Habana, hizo el Alférez mayor de aquella ciudad, y se elevó a las Cortes por los espresados cuerpos*. This document is printed among the *Documentos sobre el tráfico y esclavitud de negros*, [Arango,] 1814, pp. 1–86, which I cited above. Some of the general results of Mr. Arango's work had already been published in 1812 in the *Patriota de la Habana*, Vol. II, p. 291.

to the east of Álvarez; 3) the *Jurisdicción de Puerto Príncipe*, with 7 parishes; 4) the *Jurisdicción de Santiago de Cuba*, with 15 parishes. The latter three dis-
I.134 tricts comprise the island's eastern part.

Population in 1811

Territorial Divisions	Whites	Free People of Color	Slaves	Total
I. Eastern part	113,000	72,000	65,000	250,000
Jur. of Cuba	40,000	38,000	32,000	110,000
Jur. of Puerto Príncipe	38,000	14,000	18,000	70,000
Jur. of Cuatro Villas	35,000	20,000	15,000	70,000
II. Western part	161,000	42,000	147,000	350,000
Havana and outlying areas	43,000	27,000	28,000	98,000
Rural Areas	118,000	15,000	119,000	252,000
Isle of Cuba	274,000	114,000	212,000	600,000

The relationship between the castes will remain a political problem of
I.135 great importance until such a time that wise legislation calms inveterate hatreds by granting greater equality of rights to the oppressed classes. In 1811, the number of whites on the island of Cuba surpassed the number of slaves by 62,000 and came within one-fifth of the number of free people of color and slaves. During that same period, *whites*, who made up 9 percent of the total population in the British and French Antilles, comprised 45 percent of Cuba's population. The number of the *free people of color* rose to 19 percent, that is, double what one finds in Jamaica and Martinique. The 1817 census, modified by the Provincial Delegation, or *Deputación Provincial*, counted no more than 115,700 freedmen and 225,300 slaves. This comparison proves that (1) the numbers of freedmen were estimated with little precision both in 1811 and in 1817 and (2) the mortality rate for blacks was so high that, from 1811 to 1817, the numbers of slaves increased by no more than 13,300, despite the fact that customs *registered* more than 67,700 African blacks for that period.

The decrees of the Courts (on March 3 and July 26, 1813) and the necessity of knowing the population numbers so as to convene *provincial, county, and parish electoral councils*, the *juntas electorales de provincia, de partido* and *de*
I.136 *parroquias,* led the administration in 1817 to replace the *estimates* attempted in 1811 with a new census. My information here is based on an unpublished note that the American deputies to the *Courts* officially conveyed to me. To date, only portions of these results have been printed, both in the *Guías de Forasteros de la Isla de Cuba* (1822, p. 48, and 1825, p. 104) and in the *Reclamación hecha contra la ley de Arancéles* [by ▾Zayas y Chacón] (1821, p. 7).

1817 CENSUS (EXCLUDING 58,617 TRANSEUNTES AND BLACKS IMPORTED IN THE SAME YEAR)

Principal territorial divisions (Provincias y Gobiernos.)	Partidos	Parroquias	Civil status White people's military and religious status		Whites	Free people of color	Slaves	Total
I. PROVINCE OF HAVANA	12	94	...	...	197,658	58,506	136,213	392,377
a) Gobierno político de la *Havana*	10	69	Civil.	123,566				
			Rel.	644	135,177	40,419	112,122	
			Milit.	10,967				
b) Gobierno de *Matanzas*	1	12	Civil.	9,501				
			Rel.	10	10,617	1,676	9,594	
			Milit.	1,106				
c) Gobierno de *Trinidad* with the 3 towns of S. Spíritus,	1	13	Civil.	50,332				
Remedios, and Villa Clara			Rel.	80	51,864	16,411	14,497	
			Milit.	1,452				
II. PROVINCE OF CUBA	5	34			59,722	57,185	63,079	179,986
a) Gobierno político de *Cuba* with the 3 Tenencias de	4	28	civil.	30,587				
Bayazo, Holguín, and Baracoa			Rel.	171	33,733	50,230	46,500	
			Milit.	2,975				
b) Ten. Gobern. de *Puerto Príncipe*	1	6	Civil.	24,830				
			Rel.	129	25,989	6,955	16,579	
			Milit.	1.030				
POPULATION OF THE ISLE OF CUBA, according to the 1817 census	17	128	...	...	257,380	115,691	199,292	572,363

I.138 It might seem surprising that the estimate submitted to the Courts in 1811 arrives at a total greater than that of the *actual* census of 1817 by 28,000; but this only seems to be a contradiction. The last census was likely less imperfect than the 1791 census. However, it fell short in counting the existing population because of the fear that censuses usually inspire in people who tend to see them as precursors of new taxes. Moreover, the Deputación *Provincial* [Provincial Delegation] believed that it should make two adjustments when it conveyed the census results to Madrid. First, they added the 32,641 whites (*transeúntes del comercio y los buques entrados* ["transients," that is, "traveling merchants and people on merchant vessels"]) whom commerce brings to the island of Cuba and who, according to harbor masters' logs, are part of the crews. Second, they added the 25,976 *negros bozales* who were imported in 1817. These additions, according to the *Deputación Provincial*, resulted in a total of 630,980 for 1817, which breaks down into 290,021 whites, 115,691 free people of color, and 225,261 slaves. It is erroneous, I think, that the *Guías* I.139 [Almanacs] published in Havana and the many unpublished tables sent to me recently provide a total of 630,980 not for the end of 1817 but for the beginning of 1820. The *Guías*, for example, add 25,976 to the 199,292 slaves of the 1817 *censo* as an "*aumento que se considera de 1817 a 1819*" [estimated increase from 1817 to 1819]. Yet, if one looks at the customs registries, it appears[1]

1. [Poinsett,] *Notes on Mexico*, p. 217. In this work, the 1817 census arrives at 671,079 instead of 630,980. This difference comes from an error in the figures for *free men of color*. Mr. Poinsett's table shows 28,373 free black men and 26,002 women; 70,512 free mixed-race men and 29,170 women; for a total of 154,057 free people of color. Yet, the *censo*, based on the *Guías*, and my unpublished table count only 115,699, a difference of 38,358. By substituting 32,154 for the 70,512 free men, we arrive at a less shocking ratio between the two sexes, which, moreover, harmonizes this figure with the ratio that we see among free blacks. Also, if there were 70,000 free mixed-race men and 28,000 free black men on the island of Cuba, how would one find, even following Mr. Poinsett, the number of individuals permitted to bear arms to consist of an almost equal number of mixed-race men and free blacks (17,862 and 17,246)? How could there have been only 9,700 free persons of mixed race of both sexes and 16,600 free black men and black women in Havana, as the 1810 census claims (see Vol. XI, p. 201 [this edition, p. 32])? The *Notes on Mexico*, which is known not to be particularly precise, indicates for 1817 on the entire island a) 32,302 mixed-race slaves and 166,843 black slaves in the ratio 1 to 5; b) 74,821 female slaves of all colors and 124,324 male slaves in the ratio 1 to 1.7. In Havana, however, where mixed-race slaves are far more numerous than they are in the countryside, their ratio to black slaves is no more than 1 to 11; and in the Filipinas jurisdiction (*Memorias de la Sociedad económica de la Habana*, 1819, no. 31, p. 232), one finds in 1819 out of 3,634 slaves, 1,049 women (52 mixed race women, 437 creole black women [nègresses creoles], and 560 bozales, or recently imported, black women [negresses bozales]) and 2,585 men (91 mixed-race, 548 creole blacks [nègres creoles], and 1,946 bozales).

that the numbers of blacks imported in these 3 years was 62,947. To wit: 25,851 in 1817, 19,902 in 1818, and 17,194 in 1819. The judicious author of the *Letters from the Havana* dedicated to ▼Mr. Croker, First Secretary of the Admiralty, believes the population of people of color, both free and enslaved, to be 370,000 in 1820. But he regards[1] the added total of 32,641 that the *Junta provisional* [Provisional Council] proposed as too high. He posits that, in 1820, the total white population was no more than 250,000, and he concedes no more than 238,796 whites (129,656 men and 109,140 women) as the result of the 1817 *censo*. The real figure published for many successive years in the *Guía* is 257,380. I.140 I.141

Why be surprised by partial contradictions in population tables drawn up in America when we recall the difficulties that one had to overcome, in the center of European civilization in England and France, each time that one undertook the huge operation of a general census? One *knows*, for example, that Paris's population in 1820 was 714,000; given the number of deaths and the estimated birthrate, it is *believed* to have been 530,000 at the beginning of the eighteenth century (▼Chabrol, *Recherches statistique sur la ville de Paris*, 1823, p. XVIII). But during ▼Minister Necker's time, nearly one-sixth of this population was unknown. One knows that the population in England and Wales grew by 3,104,683 from 1801 to 1821. Yet, birth and death registries furnish no evidence of a growth higher than 2,173,416, and it is impossible to attribute 931,267 to immigration from Ireland to England alone (Powell, *Statistical Illustrations on the British Empire* 1825, p. XIV *and* XV). These examples do not prove that one must be wary of all calculations from the field of political economy. But they do confirm that one should not use numerical variables until they have been discussed and their margin of error has been determined. Otherwise, one might be tempted to compare the different degrees of probability that statistical results provide for geographical positions, lunar eclipses, distances between the moon and the sun, and stellar occultations in the Ottoman Empire to statistics for Spanish and Portuguese America, France, or Prussia. I.142 I.143

1. There are also several numerical errors in the *Letters from the Havana* [Jameson], pp. 16–18 and 36. For 1817, slaves are estimated at 124,324 instead of 199,292; for 1819, at 181,968, "forming an advantage of 143,050 over the white population." However, the white population was then already above 290,000. I believe it to be at least 325,000 in 1825, and a *Habanero* who is among the best informed about the localities had even posited 340,000 for 1823. [Arango, *Reflexiones de un habanero*] *Sobre la independencia de Cuba*, p. 17. On some parts of the island—for example, in San Juan de los Remedios and in Filipinas for the year 1819—the tables were drawn up with extreme care by ▼Don Joaquin Vigil de Quiñones and Don José de Aguilar.

To adapt a census from twenty years earlier to a different period, one must know the growth *rate*. Yet, this *rate* is based only on the 1791, 1810, and 1817 censuses, which were undertaken only in the eastern and least populous region of the island. When comparisons rely on sets that are too small and that are derived from particular circumstances (for example, in seaports or in cantons where sugar plantations are heavily concentrated), they will not produce numerical results that are valid for the whole country. One generally believes that the number of whites grows more in the countryside than in the cities; that free people of color, who prefer farming to exercising a skill in the cities, multiply with greater speed than all other classes; and that the numbers of black slaves, among whom unfortunately only one-third more women

I.144

would be needed to balance out the number of men, would shrink more than 8 percent annually.

We saw above that the white population of Havana and its environs grew by 73 percent in twenty years, while the population of free people of color increased by 171 percent. In the eastern part, the numbers of whites and freedmen doubled almost everywhere during the same period. The ranks of free people of color grow through the movement from one caste to another, and the ranks of slaves increase notably through the activity of the slave trade. The white population grows insignificantly through migration from Europe,[1] the Canary Islands, the Antilles, and the Mainland: it grows by itself. Rare are examples of *official whitening* in the form of *white letters* that the *Audiencia* accorded to high-yellow families [familles d'un jaune pale].

In 1775, the province of Cuba [Oriente] had a population of 171,626; in

I.145

1806, an official census in the *Jurisdiction of Havana* ascertained 277,364. Included in this dominion were 6 *ciudades* (the capital and its surroundings, Trinidad, San Felipe y Santiago, S. María del Rosario, Jaruco, and Matanzas), 6 *villas* (Guanabacoa, Sancti Spíritus, Villa Clara, San Antonio, San Juan de los Remedios, and Santiago), and 31 pueblos (*Patriota americano,* Vol. II, p. 300). The growth rate would thus have been no more than 61 percent in 31 years. It would seem much higher if one could compare half of this period. In fact, the 1817 Padrón counts a population of 392,377 for the same area of the country then called the *Provincia de la Habana*, encompassing the *Administrations,* or *Gobiernos,* of the capital, Matanzas, and Trinidad, or *Cuatro Villas*. This proves a growth rate of more than 41 percent in 11 years. In comparing the populations of the capital and the province of Cuba for 1791 and 1810,

1. In 1819, for example, only 1,702 individuals arrived, among whom 416 were from Spain, 384 from France, 201 from Ireland and England. Illness took one-sixth to one-seventh of nonacclimated whites.

one must not forget that the findings on growth are a bit too high, because the first of these censuses provided the occasion for more omissions than the second. I think that we get closer to the truth by comparing the most recent *censos* (1810 and 1817) for the province of Cuba: in 1810, 35,513 whites, 32,884 free people of color, 38,834 slaves, for a total of 107,231; and in 1817, 33,733 whites, 50,230 free people of color, and 46,500 slaves, for a total of 130,463. The 6-year growth is greater than 23,200, or 21 percent, for there is probably an error for whites in the second census. The number of whites and the number of free men in general is so considerable in the *Cuatro Villas* district that, in the 6 *partidos* [counties] of San Juan de los Remedios, San Agustín, San Anastasio del Cupey, San Felipe, Santa-Fé, and Sagua la Chica, there was, in an *area* of 24,651 *caballerías*, a total population of 13,722 in 1819, composed of 9,572 whites, 2,010 free people of color, and 2,140 slaves. By contrast, in the 10 *partidos* of the Jurisdiction of Filipinas, there were nearly 9,400 free men in a total population of 13,026; 5,871 whites, 3,521 free people of color (including 203 free *negros bozales*); and 3,634 slaves. The ratio of freedmen to whites was therefore 1 to 1.7. 1.146

Nowhere in the world where slavery reigns are manumissions as frequent as on the island of Cuba. Far from hindering them or making them onerous, as do the French and British laws, Spanish law favors freedom. ▾The right that every slave has either to *buscar amo* (change masters) or to emancipate himself if he can pay the purchase price; the religious sentiment that inspires many a lenient master to grant freedom to a certain number of slaves in his will; the practice of engaging a multitude of blacks in domestic service and the affections that are born from this proximity to whites; the ease with which slaves can work for themselves and pay only a certain portion of their earnings to their masters—all these are the principal causes that allow so many slaves in the cities to attain their freedom. I would include the lottery and games of chance among the ways in which a slave can obtain the funds to free himself, if it were not for the fact that undue confidence in these hazardous venues often led to the most disastrous of consequences. The position of free people of color is more felicitous in Havana than in the nations that have boasted a highly advanced culture for centuries. Unknown in Havana are the barbaric laws[1] that are still being invoked these days in other places, laws under which freedmen, who cannot receive assistance from whites, can be deprived of their freedom and *sold for the profit of the treasury* if they are convicted of harboring runaway blacks! 1.147 1.148

1. Decision of the Sovereign Council of Martinique, June 4, 1720. March 1st, 1766 Ordinance, §7.

▾Since the original population of the Antilles has entirely disappeared (the Caribbean *Zambos*, a mix of indigenous peoples and blacks, having been transported from the island of Saint Vincent to that of Roatán [Rattan, Ruatan] in 1796), the current population of the Antilles (2,850,000) should be considered a mixture of European and African blood. Racially pure blacks make up almost two thirds of this mixture, whites one-sixth, and mixed-race people one-seventh. In Spain's continental colonies, descendents of the vanished Indians can be found among the *mestizos* and *zambos*, mixtures of Indians with whites and blacks. This comfort is absent in the Antilles. The I.149 state of society in the Antilles at the beginning of the sixteenth century was such that, with rare exceptions, the new colonists did not mix with the indigenous peoples any more than the British of Canada do today. Cuba's Indians have disappeared much like the ▾Guanches of the Canary Island did. Nonetheless, 40 years ago, there were families in both Guanabacoa and Tenerife that tried to extract a small pension from the government by claiming a few drops of Indian or Guanche blood. It is no longer possible to determine Cuba's or Haiti's populations at the time of Christopher Columbus. How can one claim, following otherwise very judicious historians, that the island of Cuba had a million inhabitants[1] at the time of the conquest in 1511 and that only 14,000 remained in 1517! The statistics that one finds in the writings of the Bishop of Chiapa [Bartolomé de las Casas] are full of contradictions. I.150 And if it is true that the devout Dominican, Friar Luis Bertrán, whom the ▾*encomenderos* persecuted[2] much like the British planters today do the Methodists, predicted upon his return that "the 200,000 Indians on the island of Cuba would die as victims of the Europeans' cruelty," one must at least conclude that the indigenous race was far from being extinguished between the years 1555 and 1569.[3] According to ▾Gómara, however (such is the confusion among historians of the period), there were already no more Indians on the island of Cuba by 1553.[4] To conceive of how vague the first Spanish voyagers' estimates must have been at a time when one knew nothing of the population of any province on the Spanish Peninsula, one has only to remember I.151 that the number of inhabitants that Captain Cook and other sailors attributed

1. Albert Hüne, *Historisch-philosophische Darstellung des Negersclavenhandels*, 1820, Vol. I, p. 137.

2. See some of the intriguing revelations in Juan de Marieta, *Historia de todos los Santos de España*, Book VII, p. 174.

3. One does not know anything with any precision until the period of Friar Luis Bertrán's return to San Lucar in 1569. He was consecrated priest in 1547. Marieta, pp. 167 and 175 (compare also *Patriota*, Vol. II, p. 51).

4. [Gómara,] *Historia de las Indias*, fol. XXVII.

to Tahiti and the Sandwich Islands[1] varied from 1 to 5 at a time when statistics were already capable of precise comparisons. Given the immense fertility of its soil and its being surrounded by seas with an abundance of fish, one imagines that the island of Cuba could have fed several millions of these Indians who neither drink alcohol nor eat meat and who have cultivated corn, man- I.152 ioc, and many other nutritious roots. But if this population concentration had taken place, would it not have manifested itself in a civilization more advanced than the one Columbus's diaries proclaimed? Would the peoples of Cuba have remained below the level[2] of the Lucaya Islands? Whatever activities one wishes to posit as the causes of their destruction—be it the tyranny of the *conquistadores*, the unreason of the governors, the excruciating work of panning for gold, smallpox or the frequency[3] of suicide—it would be difficult to I.153

1. On the rapid decline of the population of the archipelago of the Sandwich Islands since Captain Cook's voyage, see ▼Gilbert Farquhar Mathison, *Narrative of a Visit to Brazil, Peru, and the Sandwich Islands*, 1825, p. 439. We know with some certitude from the reports of the missionaries who changed the face of things in Tahiti by profiting from internal dissension that, in 1818, the entire archipelago of the Society Islands [French Polynesia] had no more than 13,900 inhabitants, 8,000 of whom lived in Tahiti. Should we believe in the 100,000 estimated for Tahiti alone during the time of Cook? The Bishop of Chiapa was no more vague in his estimates for the indigenous population of the Antilles than modern writers on the population of the group of Sandwich Islands, which they fix sometimes at 740,000 (▼Hassel, *Historisch-statistischer Almanach für 1824*, p. 384), and sometimes at 400,000 (Hassel, *Statistischer Umriss*, 1824, Issue 3, p. 90). According to Mr. de Freycinet, this group contains no more than 264,000.

2. Gómara, *De menor politica*, p. XXI ["Las islas Lucayos," *Historia*, Vol. I, fol. XLI, p. 88 in *Historia*]. That the indigenous peoples of equinoctial America generally avoid an animal- and milk-based diet is already noted in ▼Pope Alexander VI's famous bull of 1493: "making diligent search in the ocean sea, [they] have found certain remote islands and mainlands that were not heretofore found by any other. In which, as is said, many nations inhabit, living peaceably and going naked, not accustomed to eat flesh. And as far as your messengers can conjecture, the nations inhabiting the aforesaid lands and islands, believe that there is one God creature in heaven" (*Car. Coquel. Bull. amp. Coll.* [*Inter Caetera*], Vol. III, Part III, p. 234.) In these very Antilles where the people fear the influence of ▼*zemís*, small cotton fetishes (Petrus Martyr, [Anghiera,] *Opus Epistolarum*, fol. XLVI), monotheism (belief in a *Great Spirit* superior to the *zemís*) was presumably widespread!

3. This mania of entire families hanging themselves in huts and caves, of which ▼Garcilaso speaks, was without a doubt caused by despair: however, rather than bewail the savagery of the sixteenth century, one wanted to exculpate the *conquistadores* by attributing the disappearance of the indigenous people to their *penchant for suicide*. See *Patriota*, Vol. II, p. 50. All the sophistry of this genre is found in the work that ▼Mr. Nuix published *on the humanity of the Spanish during the conquest of the Americas* (*Reflexiones imparciales sobre la humanidad de los Españoles contra los pretendidos filósofos y políticos, para illustrar las historias de Raynal y Robertson, escrito en Italiano por el Abate Don Juan Nuix,*

conceive how at least three or four hundred thousand Indians—I would not
I.154 go so far as to say one million—could have completely vanished in 30 or 40 years. The war against the ▼Cacique Hatüey was very short and restricted to the island's easternmost corner. Few complaints were lodged against the administration of the first two Spanish governors, Diego Velázquez and ▼Pedro de Barba. The oppression of the indigenous people only dates to the arrival of the cruel ▼Hernando de Soto around 1539. If we postulate, following Gómara, that there were no more Indians fifteen years later under the government of ▼Diego de Majariegos (1554–1564), we should be compelled to admit that considerable remnants of this population saved themselves in dugouts bound for Florida, believing, in keeping with ancient traditions, that they were returning
I.155 to their ancestral homelands. Only the mortality rate among black slaves that one observes in the Antilles in our day can throw some light on these numerous contradictions. To Christopher Columbus and to Velázquez, the island of
I.156 Cuba would have appeared heavily populated,[1] assuming it had, for instance,

---

y traducido al castellano por ▼*Don Pedro Varela y Ulloa, del Consejo de S. M.*, 1782). The author considers (p. 186) the expulsion of the Moors under ▼Philip III a religious and meritorious act and ends his work congratulating (p. 293) American Indians "on having fallen into the hands of the Spanish whose behavior has always been the most humane and whose government has been the wisest." Many pages of this book recall "*the salutary rigors of the dragonnade*," and the odious passage in which a man, known by his talent and his virtues as the ▼Count de Maistre (*Soirées de Saint-Pétersbourg*, Vol. II, p. 121), justifies Portugal's inquisition, "because it only made a few drops of guilty blood flow." To what sophistry one has recourse when one wants to defend religion, national honor, or the stability of governments while exonerating everything that outrages humanity in the actions of the clergy, of peoples, and of laws! It is in vain that one attempts to destroy the power most solidly established on earth, that of witnessing history.

1. Columbus recounts that the island of Haiti was attacked a few times by a race of black people, *gente negra*, who lived farther to the south and to the southwest. He hoped to visit them during his third voyage, ▼because the blacks possessed the *guanín* metal [copper alloy], several samples of which the Admiral had procured during his second voyage. These samples, tested in Spain, had been found to be composed of 0.63 gold, 0.14 silver, and 0.19 copper (Herrera, *Historia de las Indias occidentales*, Dec. I, book 3, chapter 9, p. 79). In fact, Balboa discovered this black tribe on the Isthmus of Darien. "This conquistador," Gómara said (*Historia de las Indias*, fol. XXXIV), "entered the province of Quareca: he did not find any gold there but some black slaves belonging to the lord of the place. He asked the lord from where he had obtained them; the response was that people of this color lived somewhat nearby, and that one was constantly at war with them. These blacks," Gómara added, "resembled the blacks of Guinea, and no one saw any others in America" (en las Indias yo pienso que no se han visto negros despues). This passage is remarkable. Hypotheses were made in the sixteenth century much as we make them today, and Petrus Martyr (*Oceanica*, Dec. III, book 1, p. 43) imagined that these men whom Bal-

been peopled to the same degree that it was when the British arrived in 1762. The crowds that the appearance of European ships at various points drew along the coastline easily fooled the first travelers. In 1792, however, the island of Cuba had no more than 200,000 inhabitants, with the same *Ciudades* and *Villas* that it has today. Forty-two years suffice to leave only the memory of a people treated like slaves, exposed to the folly and brutality of masters, excessive labor, starvation, and the ravages of smallpox. In many of the Lesser Antilles under British domination, the population shrinks 5 to 6 percent a year; in Cuba, it shrinks by more than 8 percent. But the annihilation of 200,000 people in 42 years presupposes an annual loss of 26 percent, hardly a believable number. One can, however, aver that the mortality among Cuba's indigenous population was much greater than that among the blacks who were bought at very high prices.[1]

I.157

I.158

When studying the island's history, one observes that its colonization went from east to west and that today, the initially settled places are the most sparsely populated—a pattern that holds throughout the Spanish colonies. Cuba's first white settlements were established in 1511 when the *conquistador* and colonizer, or *poblador,* Velázquez, on orders from Don Diego Columbus, disembarked at Puerto de Palmas, near Cape Maisí, known then as *Alfa y Omega*. Velázquez subjugated the cacique Hatüey, a fugitive from Haiti, who had retired to the eastern part of the island of Cuba and had become the

boa saw, the ▾Quarecas, were Ethiopian blacks who (*latrocinii causa*) had overrun the seas and shipwrecked on the coastlines of America. But the blacks of Sudan are hardly pirates, and it would be easier to conceive that some Eskimos in their hidebound canoes would be able to come to Europe than that some Africans came to Darien. Those who hypothesized a mix of Polynesians with Americans preferred to consider the Quarecas a race of Papuans similar to the *negritos* of the Philippines. These tropical migrations from west to east, from the westernmost part of Polynesia to the Isthmus of Darien, are very difficult, despite the fact that the winds blow from the west for weeks. Above all, it would be good to know if the Quarecas were really similar to the blacks of the Sudan, as Gómara said, or if they were nothing but a very dark-skinned [très-basanés] race of Indians (with smooth and straight hair) who, from time to time (before 1492), overran the same island of Haiti that today is the domain of Ethiopians. On the passage of the ▾Caribs from the Lucaya Islands to the Lesser Antilles without touching any of the Greater Antilles, see Vol. IX, pp. 35 and 36.

1. The number of *registered slaves* was 17,959 in Dominica in 1817; 28,024 in Grenada; 15,893 in Saint Lucia; 25,941 in Trinidad. In 1820, these same islands counted no more than 16,554, 25,677, 13,050, and 23,537 slaves. Losses were therefore (according to the status of the registries) one-twelfth, one-eleventh, one-fifth, and one-eleventh *in three years* (unpublished *documents* graciously conveyed to me by Mr. Wilmot, Undersecretary of State at the Department of Colonies of Great Britain). We saw above that, before the abolition of the trade, Jamaica's slave population shrank by 7,000 each year.

head of a confederation of minor indigenous nobility. In 1512, they began to build the city of Baracoa, which was later followed by Puerto Príncipe, Trinidad, the Villa de Sancti Spíritus, Santiago[1] de Cuba (1514), San Salvador de Bayamo, and San Cristóbal de la Habana. At first, Havana was founded (1515) on the island's southern coast in the *Partido* of Güines; four years later, it was transferred to Puerto de Carenas whose position, at the entrance of the two Bahamian straits—*el Viejo y el Nuevo* [the Old and the New]—seemed more favorable for commerce than the southwest coast of Batabanó.[2] Since the sixteenth century, civilization's progress has powerfully influenced the relationship between the castes. These relationships vary between the districts with nothing but cattle ranches and those where the soil had been worked for a long time: in the seaports and interior cities, in the places where colonial crops are grown, and in places that produce corn, vegetables, and feed.

I.159 I.160

I. The *Jurisdiction of Havana* is experiencing a decline in the *relative population* of whites. This is not the case in the cities of the interior and in all of the *vuelta abajo*, the western regions used for tobacco plantations that employ free labor. In 1791, Don Luis de las Casas' census counted 137,800 souls in the Jurisdiction of Havana, among whom *whites*, *free people of color*, and *slaves* existed in the ratio of 53 to 20 to 27 percent. In 1811, after the heavy importation of slaves, the ratio was believed to be 46 to 12 to 42 percent. In districts with large sugar and coffee plantations (*partidos de grandes labranzas*), whites form barely a third of the population. The *ratio of castes* (taking this expression to mean the relationship of each caste to the total population) oscillated between 30 and 36 percent for whites, between 3 and 6 percent for free people of color, and between 58 and 67 percent for slaves. Meanwhile, in districts dedicated to the cultivation of tobacco and in the *vuelta abajo*, the ratio was 62 to 24 to 14 percent and even 66 to 20 to 14 percent in districts with *ganadería* [cattle breeding].

I.161

1. *Patriota*, Vol. II, p. 280. *Manuscrits de Don Felix de Arrate y Acosta*, compiled in 1750 on the basis of the official documents saved from the great fire of Havana in 1538. I am surprised to see (*Guía*, 1815, p. 73) that the ▾Franciscan friars of Santiago de Cuba date their convent's founding to 1505, the complete reconnaissance of the coastline by Sebastián de Ocampo dating only to 1508.

2. See Vol. XI, pp. 236 ff [this edition, pp. 45, 168.]. [Arango,] *Documentos*, p. 116. The tree beneath which the Spanish celebrated the first mass (at Puerto de Carenas) is still there. After its discovery, the island that was officially called the *siempre fiel Isla de Cuba* [always faithful Island of Cuba], was successively named *Juana*, *Fernandina*, *Isla de Santiago*, and *Isla del Ave María*. Its crest dates from 1516.

II. About the population growth in the *Jurisdiction of Cuatro Villas* and Puerto Príncipe as well as Cuba, we have more precise information than for the western regions. *Cuatro Villas* has felt the effects that come from the inhabitants' different occupations. In the districts of Sancti Spíritus, where cattle farms prosper, and in San Juan de los Remedios, where commerce in contraband with the Bahamas is quite frequent, the whites have grown in number from 1791 to 1811. By contrast, whites have decreased in number in the eminently fertile Trinidad district where sugar plantations are now developing at an extraordinary pace. In Villa Clara, free people of color are gaining on other classes. I.162

III. In the *Jurisdiction of Puerto Príncipe* [Camagüey], the entire population has almost doubled in twenty years. It has grown by 89 percent, as in the most beautiful parts of the United States. However, Puerto Príncipe's surrounds are no more than immense plains dotted with half-wild cattle herds. A recent visitor to the island[1] commented that the landowners have no other care than to bury in their coffers the money that their overseers make and to exhume it for the card games and court cases that they pass on from one generation to the next.

IV. In the entire *Jurisdiction of Cuba* [Oriente], the ratio between the three classes has changed little in 20 years. The Partido [county] de Bayamo is still distinguished by the large number of free people of color (44 percent), a number that increases from year to year, as it does in Holguín and in Baracoa. In the environs of Cuba province, coffee plantations flourish, which leads to a considerable increase in the number of slaves.[2] I.163–64

1. Masse, *L'île de Cuba*, 1825, p. 302.

2. In the table published by the Secretary of the Consulado, ▾Mr. del Valle Hernández ([Arango,] *Documentos*, p. 149, and *Patriota*, Vol. II, p. 283), the slaves of Bayamo are estimated at 16,733; this figure tallies neither with the sum total of 47,984 nor with the quotient of 0.26. As it is more likely that typographical errors affected one rather than two figures, I substituted the number of slaves (12,633) that one derives by the quotient and the sum total at the same time. The table of the four districts of the province of Cuba [Oriente] is the *unmodified* result of the censuses; it sets the Province of Cuba's population at 106,331. In the *general table of the island of Cuba* (see Vol. XI, p. 310 [this edition, p. 74]), the *censo* results are altered either by reducing them to round numbers or by augmenting them, as is explicitly avowed in the *Documentos*, p. 137. Thus, the contradictions are only *seeming*. I do not know why only the number of slaves in the Jurisdiction of Cuba was decreased in the general table, but this change only concerns one-tenth of the enslaved population in the island's eastern part. As there are *variantes lecciones* [different outcomes] in every result from the censuses, I will add that the other *Padrones* counted 98,780 for the four dis-

I.164

I. Four districts in the province of Cuba

Districts	Whites	Free people of color	Slaves	Total	Relations of the three classes to the total population			
Cuba	1791	7,926	6,698	5,213	19,837	0.40	0.33	0.27
	1810	9,421	6,170	8,836	24,427	0.38	0.25	0.37
Baracoa	1791	850	1,381	169	2,400	0.35	0.57	0.08
	1810	2,060	1,319	664	4,043	0.51	0.33	0.16
Holguín	1791	4,116	1,001	5,862	10,979	0.37	0.09	0.54
	1810	8,534	4,542	16,850	29,926	0.28	0.13	0.59
Bayamo	1791	6,584	9,132	7,287	23,003	0.29	0.40	0.31
	1810	14,498	20,853	12,633	47,984	0.30	0.44	0.26
Total	1791	19,476	18,212	18,521	56,219	0.34	0.33	0.33
	1810	34,513	32,984	38,834	106,331	0.32	0.31	0.37

Until the final years of the eighteenth century, the number of female slaves on sugar *plantations* was extremely small. And, what should be quite surprising, this state of affairs resulted from a prejudice founded on "religious I.165 scruples" about the introduction of women, whose price was generally a third lower than that of men.[1] Slaves were forced into celibacy under the pretext of avoiding moral disorder! Only the Jesuits and the Bethlemite monks renounced this grave prejudice; they alone tolerated female blacks on their plantations. If the doubtlessly quite imperfect 1775 census already counted 15,562 slave women and 29,366 male slaves, one must not forget that this census covered the entire island and that sugar plantations, even today, employ no more than a quarter of the enslaved population. Since 1795, the Havana *Consulado* began seriously to occupy itself with the project of ascertaining the enslaved population's growth independently of variations caused by the slave trade. Don Francisco Arango, whose views were always insightful, proposed to impose a tax on plantations that did not have at least one-third females among its slaves. He also wanted to levy a 6-piaster tariff for every black slave introduced to the island; female slaves (*negras bozales*) would be exempt from I.166 this tariff. Although these measures were not adopted because the *colonial assemblies* always resisted coercive means, the desire to multiply marriages and better care for the children of slaves awakened during this period, and a *royal decree* (from April 22, 1804) recommended these ends "to the conscience and

tricts of Cuba in 1810 and 48,033 for the district of Puerto Príncipe (*Documentos*, pp. 137 and 150). An 1800 census listed 53,267 for Cuatro Villas.

1. [Arango,] *Documentos*, p. 34.

the humanity of the colonists." According to ▼Mr. Poinsett, the 1817 census counts 60,322 slave female blacks and 106,521 black slaves. In 1777, the ratio of black women slaves to men was 1 to 1.9; 40 years later, it had barely changed in a sensible fashion[1]: it was 1 to 1.7. One should attribute this insignificant change to the enormous quantity of *negros bozales* introduced since 1791. The importation of female blacks was not considerable, except between 1817 and 1820, such that the male black slaves working in the cities became a smaller 1.167 fraction of the total. In the *partido* [county] of Batabanó, which, in 1818, had a population of 2,078 on 13 ▼*ingenios* [sugar plantations] and 7 *cafetales* [coffee plantations] there were 2,226 male blacks and only 257 slave female blacks (ratio = 8 to 1). In the Jurisdiction of San Juan de los Remedios (which, in 1817, counted a population of 13,700 on 17 sugar plantations and 73 *cafetales*), there were 1,200 male blacks and 660 slave female blacks (ratio = 1.9 to 1). In the Jurisdiction of Filipinas (which, in 1819, had a population of 13,026), there were 2,494 male blacks and 997 slave female blacks (ratio = 2.4 to 1). And if on the entire island of Cuba black male slaves exist in a 1.7 to 1 ratio to female slaves, there is barely a 4 to 1 ratio on sugar plantations.

The first importation of blacks to the island's eastern portion occurred in 1521 and did not exceed 300. The Spanish had then much less of an appetite for slaves than the Portuguese did: 12,000 blacks were sold in Lisbon[2] in 1539, as in our day (to the eternal shame of Christian Europe), the Greek slave trade 1.168 occurs in Constantinople and Smyrna. In Spain, the commerce in slaves was regulated during the sixteenth century. The Court bestowed the privilege of trading slaves, which Gaspar de Peralta purchased for all of Spanish America in 1586. In 1595, ▼Gómez Reynel bought it, and, in 1615, Antonio Rodríguez de Elvas did so. The total importation was then no more than 3,500 slaves per year, and Cuba's inhabitants, who primarily raised livestock, barely received any. ▼During the War of Succession, the French relinquished Havana to trade slaves for tobacco. The British *asiento* [contractual monopoly] revived the importation of blacks for a short time. In 1763, however, even though the occupation and the presence of foreigners had created new needs, the number of slaves still did not amount to 25,000 in the Jurisdiction of Havana and to 32,000 on the entire island. The total number of African blacks imported

1. In the British Antilles, out of a slave population of 627,777 in 1823, there were 308,467 males, and 319,310 females: a 3⅕ percent preponderance of females. Only Trinidad and Antigua, like Demerary, have more males than females among slaves. See [Powell,] *Statistical Illustrations of the British Empire*, 1825, p. 54.

2. Bryan Edwards, *The History, Civil and Commercial, of the British Colonies in the West Indies*, Vol. III, p. 202. See also Vol. I, pp. 422ff.

between 1521 and 1763 was probably[1] 60,000. Among their descendents are
1.169 the free mixed-race persons, the majority of whom inhabit the island's eastern part. From 1763 until 1790, when the commerce in slaves was unfettered by the monarchy, Havana received 24,875 slaves (4,957 through the *Compañía de Tabacos*, from 1763 to 1766; 14,132 through a contract with the ▾Marquis de Casa Enrile, from 1773 to 1779; and 5,786 through a contract with Baker and Dawson, from 1786 to 1789). If one estimates at 6,000 the importation of slaves to the island's eastern part for the same 27 years (from 1763 to 1790), one arrives at a total of 90,875 since the discovery of the island of Cuba or, rather, between 1521 and 1790. We will soon see that, due to the ever-increasing ▾slave trade, the 15 years after 1790 furnished more slaves than the two and a half centuries that had preceded the era of free trade. The slave trade quickly doubled when England and Spain agreed that the transatlantic trade would end north of the Equator after November 22, 1817 and that it would be abolished everywhere on May 30, 1820. Posterity will one day find it hard to believe that the King of Spain accepted from England a sum of 400,000 pounds
1.170 sterling as compensation for the damages that might result from the cessation of this barbaric trade. Below is the number of African blacks imported solely through the port of Havana based on its customs registries:

1790	2,534	1806	4,395
1791	8,498	1807	2,565
1792	8,528	1808	1,607
1793	3,777	1809	1,162
1794	4,164	1810	6,672
1795	5,832	1811	6,349
1796	5,711	1812	6,081
1797	4,552	1813	4,770
1798	2,001	1814	4,321
1799	4,919	1815	9,111
1800	4,145	1816	17,737
1801	1,659	1817	25,841
1802	13,832	1818	19,902
1803	9,671	1819	17,194
1804	8,923	1820	4,122
1805	4,999	31-year total	225,574

The yearly average for this period[2] was 7,470 and, for the last 10 years, 11,542.
1.171 The latter number should be at least a quarter higher, as much because of the

1. [Arango,] *Documentos*, pp. 39 and 118.
2. Other unpublished notes in my possession show 23,560 slaves for 1817.

illegal trade and the customs oversights as because of the legal imports from Trinidad and Santiago de Cuba. The findings are:

For the entire island,	from 1521 to 1763	60,000
	from 1764 to 1790	33,409
For Havana alone,	from 1791 to 1805	91,211
	from 1806 to 1820	131,829
		316,449
Increase in illegal trade from 1791 and 1820,		56,000
including the island's eastern region		372,449

We saw *above* that, during the same 300 years, Jamaica received 850,000 blacks from Africa,[1] or, more precisely estimated, more than 677,000 in 108 years (from 1700 to 1808). But today, Jamaica does not have a population of 380,000 blacks and persons of mixed race, free and enslaved! The island of Cuba offers a more comforting picture, with 130,000 free people of color, while Jamaica, out of a total population half as large, counts only 35,000. The island of Cuba received from Africa:

I.172

before 1791	93,500
from 1791 to 1825 at least	320,000
	413,500
Due to the small number of female blacks imported in 1825 there were	
free and enslaved blacks	320,000
Mixed-race	70,000
men of color	390,000

A similar calculation, based on slightly different numerical variables, was submitted to the Spanish Courts on July 20, 1811. It attempted to prove that, until 1810, the island of Cuba had received fewer than 229,000 African blacks.[2] In 1811, the island was *represented* by a population of 326,000 blacks and mixed-race persons, both enslaved and free, which is 97,000 more than the

I.173

1. See Vol. XI, p. 145 [below p. 379]. I add here that all the British colonies of the Antilles, which today have no more than 700,000 blacks and mixed-race persons, free and enslaved, received, according to customs registries, 2,130,000 blacks from the coasts of Africa in 106 years (from 1680 to 1786)!

2. According to a note published by the Havana Consulado (*Papel periódico*, 1801, p. 12), the average price of 15,647 *bozales* imported between 1797 and 1800 was put at 375 piasters per head. At the same rate, the 307,000 blacks from Africa imported between 1790 and 1823 would have cost the island's inhabitants 115,125,000 piasters.

African imports.[1] Leaving aside the fact that the whites had a part in the existence of 70,000 persons of mixed race;[2] leaving aside the natural offspring I.174 that so many thousands of blacks imported over the years should have produced, one exclaims: "What other nation or human society can render such an advantageous account of the effects of this disgraceful trade in blacks, this *desgraciado tráfico*!" I respect the feelings that prompt these lines. I repeat that, in comparing the island of Cuba to Jamaica, the result of the comparison seems to place Spanish law and the mores of Cuba's inhabitants in a more positive light. The comparisons demonstrate a state of affairs more favorable to the physical preservation and emancipation of blacks in Cuba. But how sad a spectacle it is to see Christian and civilized peoples discussing among themselves I.175 who, over the course of three centuries, has made fewer Africans perish by reducing them to slavery! I will not extol the treatment of blacks in the southern parts of the United States,[3] but there are degrees in humanity's suffering. The slave who has a cabin and a family is less miserable than he who is penned as if part of a herd. The greater the number of slaves who live with their families in shacks that they take to be their property, the greater their proliferation. For the United States, one counts:

1. My calculation ends with 1825 and estimates 413,500 blacks imported since the *conquest*. If the figures that go up to 1810 and were sent to the Cortes show 229,000 ([Arango,] *Documentos*, p. 119), then there is a difference of 184,500. Yet, according to the registries of the Havana customs house, the number of *negros bozales* brought in through this port between 1811 and 1820 was greater than 109,000. We must increase this figure, (1) according to principles admitted by the *Consulado* itself, by one-fourth (or 27,000) for the legal imports in the island's eastern part; and (2) by the results of illegal trade from 1811 to 1825.

2. The work that the *Consulado* undertook in 1811 on the probable distribution of 326,000 people of color, both free and enslaved, encompasses some very remarkable materials that only great familiarity with local conditions could have furnished to the administration. A) *Cities* (western part): in Havana, 27,000 free people of color and 28,000 slaves; the 7 pueblos of the *Ayuntamiento*, 18,000; and so, in the entire Jurisdiction of Havana, there are 36,000 free people of color and 37,000 slaves. *Cities* (eastern part): 36,000 free people of color and 32,000 slaves. Total for cities: 72,000 free people of color and 69,000 slaves, or 141,000. B) *Rural areas*: Jurisdiction of Havana, 6,000 free people of color and 110,000 slaves. Eastern part: 36,000 free people of color and 33,000 slaves. Total for the rural areas (*campos*), 185,000. [Arango,] *Documentos sobre los negros*, p. 121.

3. On the comparative state of misery among the slaves of the Antilles and the United States, see [Macaulay], *Negro Slavery in the United States of America and Jamaica*, 1823, p. 31. In 1823, Jamaica counted 170,466 male slaves; 171,916 females. It was determined that, in 1820, the United States had 788,028 male slaves and 750,100 females. It is therefore not the disproportion between the sexes that is causing the lack of natural population growth in the Antilles!

1790	480,000	slaves
1791	676,696	
1800	894,444	
1810	1,191,364	
1820	1,541,568	

The annual increase[1] over the last ten years (not including the emancipation of 100,000 slaves) was 26 out of a thousand, which will cause a doubling in 27 years. Yet, I say with ▾Mr. Cropper[2] that, if the slaves in Jamaica and Cuba had multiplied in the same proportion,[3] these two islands would have nearly reached their current population—the one after 1795, the other after 1800—without 400,000 blacks having been weighed down with irons on the coasts of Africa and shipped to Port-Royal and Havana. I.176–77

The mortality of blacks varies widely on the island of Cuba, as it does throughout the Antilles, depending on the type of work, the humanity of masters and *overseers*, and the number of female blacks who can care for the sick. There are plantations where 15 to 18 percent perish annually. I have heard coldhearted discussions about whether it would be better for the owner not to wear out slaves with excessive work and have to replace them less often as a result, or to take from them everything possible in a few years and be forced to buy *negros bozales* more frequently. Such are the rationalizations of greed when humans use other humans as beasts of burden! It would be unjust to doubt that the mortality of blacks has diminished greatly on the island of Cuba over the last 15 years. To their credit, many owners are improving the I.178

1. The increase in the number of black slaves from 1790 to 1810 (from 514,668) is due (1) to natural family growth; (2) to 30,000 blacks imported in the 4 years (1804 to 1808) during which South Carolina's law unfortunately, once again, allowed import by trade; (3) the acquisition of Louisiana, where there were 30,000 blacks at the time. The additions resulting from these two last causes account for only one-eighth of total growth and are offset by the manumission of more than 100,000 blacks who disappeared from the registries in 1810. The slave population grows a little less rapidly (in the exact proportion of 0.02611 to 0.02915) than the total population of the United States. But, where slaves form a considerable proportion of the population, as they do in the southern states, the numbers grow faster than those of the whites (Morse, *New System of Modern Geography*, 1822, p. 608).

2. *Letter addressed to the Liverpool Society*, 1823, p. 18.

3. The number 480,000 for the year 1770 is not based on an actual census. It is no more than an approximation. Mr. Albert Gallatin thinks that the United States, which, in late 1823, had a population of 1,665,000 slaves and 250,000 free people of color, and in consequence a total of 1,915,000 blacks and mixed-race persons, never received more than 300,000 blacks from the coasts of Africa, or 1,830,000 fewer than the British Antilles imported between 1680 and 1786. Today, the population of blacks and persons of mixed race in the Antilles barely surpasses a third that of the United States.

plantation regime. The average mortality of recently imported blacks is still between 10 and 12 percent.[1] On many well-run sugar plantations, it could be reduced to 6 or 8 percent. The loss of *negros bozales* differs according to the I.179 time of their introduction. The most favorable time is from October to January when the season is salubrious and the food on the plantations is plentiful. In the hottest months, 4 percent die already *during the auctions*, as was the case in 1802. Here are the most promising means of preventing the destruction of blacks: increasing the number of female slaves—so useful for caring for their husbands and their ill compatriots; exempting them from work during pregnancy and for taking care of their children; housing family units in separate cabins; more provisions; more days of rest and not overworking them. People familiar with the internal workings of plantations think that, under the current state of affairs, the number of black slaves would diminish by 5 percent annually, if the illegal slave trade ended altogether. It is a reduction nearly equal to that of the British Lesser Antilles, except for St. Lucia and Grenada. These islands were warned by parliamentary discussions fifteen years before I.180 the slave trade's definitive abolition, and they had time to import more female blacks. On the island of Cuba, abolition was more sudden and more unexpected.

In the official documents published in Havana, a comparison was attempted between Cuba's *population density* (the relation of population to the island's *area*) and the population density of the least populated parts of France and Spain. As no one knew the true *area* of the island then, these calculations could not be exact. We saw above that the entire island contains roughly 200 individuals per square nautical league (20 to a degree). This is one third less than Cuenca, Spain's most thinly populated province, and four times less than Hautes Alpes, France's least populous département. The inhabitants of the island of Cuba are so unequally distributed that one can consider practically five-sixth of the island unpopulated.[2] In some parishes (Consolación, Macuriges, Hanábana), there are no more than 15 inhabitants per square league in the midst of pasture land. By contrast, in the Bahía Honda, I.181 Batabanó, and Matanzas triangle (more precisely, between Batabanó, the Pan de Guaijabón, and Guamacaro), there are more than 300,000 inhabitants

1. We have received assurances that in Martinique, where there are 78,000 slaves, the average mortality is 6,000. Births among slaves still do not rise above 1,200 annually. On losses in the British Antilles, see Vol. III [Q], p. 336 [this edition, pp. 314ff.]. Before the abolition of the slave trade, Jamaica lost 7,000 individuals annually, or 2.5 percent. Since this period, the population decrease has practically been nil. [African Institution,] *Review of the Registry Laws by the Committee of the African Institute*, 1820, p. 43.

2. [Arango,] *Documentos*, p. 136. See also Vol. IX, pp. 251 and 257.

on 410 sq. lg.—one-ninth of the island's total *area* and three-sevenths of the island's population—and more than six-sevenths of its commercial and agricultural wealth. This triangle still has no more than 732 inhabitants per square league. It is not quite as large as two *medium size* départements in France, and the *population density* is half. But one must not forget that even in this little triangle between Guaijabón, Guamacaro, and Batabanó, the southern part is somewhat underpopulated. The *Parroquias* [parishes] with the most sugar plantations are: Matanzas with Naranjal, or Cuba Mocha and Yumurí; Río Blanco del Norte with Madruga, Jibacoa, and Tapaste; Jaruco, Güines, and Managua with Río Blanco del Sur, San Gerónimo, and Canoa; Guanabacoa with Bajurayabo and Sibarimón; Batabanó with Guara and Buenaventura; San Antonio with Govea; Guanajay with Bahía Honda and Guaijabón; Cano with Bauta and Guatao; Santiago with Wajay and Trinidad. The Parroquias that are the least populated and service only the *cría de Ganado* [cattle raising] are: Santa Cruz de los Pinos, Guanacapé, Cacaragícaras, Piñar del Río, Guane, and Baja in the *Vuelta abajo* [the west]; Macuriges, Hanábana, Guamacaro, and Álvarez in the *Vuelta arriba* [the east]. The *hatos* [cattle ranches] that occupy uncultivated land of 1,600 to 1,800 *caballerías* are disappearing little by little. And if the settlements attempted at Guantánamo and Nuevitas did not enjoy the rapid successes that had been expected, other settlements, for example those in the Jurisdiction of Guanajay, have done quite well (Francisco de Arango, *Expediente*, 1798, manuscript). I.182

We already noted above how much the population of the island of Cuba has grown over the course of centuries. A native of a northern country much less favored by nature, I recall that under an administration sympathetic to the progress of agricultural industry, the March of Brandenburg [Germany], whose soil is mostly sandy, feeds almost double the population of the island of Cuba, using a third of the land area. The population's extremely uneven distribution, the absence of inhabitants on a large portion of the coastline, and the enormous extent of coastal areas make the military defense of the entire island impossible. Neither enemy invasion nor illegal commerce can be prevented. Havana is without doubt a well-defended locale that rivals the most important places of Europe in its fortifications. The *Torreones* [towers] and the fortifications of Cogimar, Jaruco, Matanzas, Mariel, Bahía Honda, Batabanó, Jagua, and Trinidad can muster fairly lengthy resistance. But two-thirds of the island is nearly defenseless, and even vessels that can maneuver quickly could barely protect these parts. I.183

The spread of intellectual culture, which is almost entirely limited to the white class, is as uneven as the population's. In the ease and politesse of its manners, Havana high society resembles that of Cádiz and the richest commercial cities of Europe. But once one leaves the capital or the neighboring I.184 plantations where the wealthy landowners live, one is struck by the contrast between this state of partial and local civilization and the simplicity of mores that reigns on isolated farms and in small towns. Habaneros were the first among the wealthy inhabitants of the Spanish colonies who visited Spain, France, and Italy. In Havana, one has always been very well informed about European politics and about the schemes at the Courts that could either make or break a government minister. Their knowledge of current events and their foresight have immensely helped the residents of the island of Cuba steer clear, in part, of obstacles to the development of colonial wealth. In the interval between the ▾Peace of Versailles and the beginning of the Saint-Domingue Revolution, Havana seemed ten times closer to Spain than to Mexico, Caracas, and New Granada. Fifteen years later, when I visited the colonies, this illusion of unequal distance had already diminished considerably. Today, the independence of the continental colonies, the importation of foreign industry, I.185 and the financial needs of new states have multiplied the ties between Europe and America. With distances shortening thanks to improvements in transportation, and Mexicans and Guatemalans[1] competing to visit Europe, the majority of former Spanish colonies, at least those on the Atlantic Ocean, seems equally close to our continent. Such are the changes within a short period of time, and they are accelerating at an ever-faster pace. They are the effect of enlightenment and long held-back activity. They make less striking the contrasting customs and levels of civilization that I had observed at the beginning of this century in Caracas, Bogotá, Quito, Lima, Mexico City, and Havana. Basque, Catalán, Galician, and Andalucian[2] influences become less noticeable I.186 each day. And perhaps, as I write these lines, it might no longer be fair to characterize the diverse nuances of national culture in the six capitals I just named in the same way that I had attempted to do elsewhere.[3]

The island of Cuba does not have the impressive and abundant institutions that date from long ago in Mexico. But in Havana, the patriotism of its inhabitants, energized by the productive competition between the different centers of American civilization, will lead to the expansion and perfection of the island's institutions at a time when the political circumstances and the confidence that ensure domestic tranquility will permit it. The Patriotic Society of Havana (established in 1793); those of Sancti Spíritus, Puerto Príncipe,

1. *Los Centro-Americanos,* as the Constitution of the Federal Republic of Central America decreed on November 22, 1824, calls them.

2. Vol. IV, pp. 150, 151, and 152.

3. Vol. IV, pp. 206 and 207.

and Trinidad that depend on it; the University with its chairs in theology, jurisprudence, medicine,[1] and mathematics established in 1728 in the convent of the *Padres Predicadores*;[2] the chair in political economy founded in 1818; the chair in agricultural botany; the Museum and School of descriptive anatomy, owing to the enlightened efforts of ▾Don Alejandro Ramírez; the public library; the Free School of Drawing and Painting; the Nautical School; the Lancaster Schools; and the Botanical Garden: some of these institutions are relatively young, others are quite old. Some are awaiting progressive improvements; others await total reform designed to harmonize them with the spirit of the century and the needs of Society. I.187

AGRICULTURE—When the Spanish first settled the islands and the American continent, their main crops were subsistence plants, much as they still are in old Europe. This stage in people's agricultural life, which is the most natural and comforting, continues to this day in Mexico, Peru, as well as in the cold and temperate regions of Cundinamarca—wherever whites control wide expanses of territory. Edible plants, like bananas, manioc, corn, European grains, potatoes, and quinoa remained the basis of continental agriculture at various sea levels in the tropics. Indigo, cotton, coffee, and sugarcane do not appear in the region save in isolated clusters. For two and a half centuries, Cuba and the other islands of the Antillean archipelago all cultivated the same plants that had fed half-wild natives and filled the large island's vast savannas with cattle herds. Around 1520, Piedro de Atienza planted the ▾first sugarcane in Saint-Domingue, and people even built sugarcane presses powered by hydraulic cylinders.[3] But the island of Cuba barely took part in these industrial stirrings. More remarkably, the *Conquest*'s historians,[4] in 1553, still report no sugar ex- I.188 I.189

1. In Havana alone, there were, in 1825, more than 500 practicing doctors, surgeons and pharmacists; to wit: 61 medicos [doctors], 333 cirujanos latinos or romancistas [surgeons], and 100 farmaceúticos [pharmacists]! On the entire island, one counted in the same year: 312 lawyers (198 of them in Havana) and 94 escribaños [clerks]. The increase in lawyers alone was such that, in 1814, there were still only 84 in Havana and 130 on the entire island.

2. Cuba's clergy is neither numerous nor very wealthy, if one disregards the Bishop of Havana and the Archbishop of Cuba, the former of whom has 110,000 piasters, the latter 40,000 piasters in annual rents. The canons receive 3,000 piasters. The number of churchmen does not exceed 1,100, according to the official censuses that I have at hand.

3. On the *trapiches* [animal-driven mills] or *molinos de agua* [water-driving mills] of the sixteenth century see ▾[Fernández de] Oviedo, *L'Histoire naturelle et generalle des Indes*, book 4, chapter 8.

4. López de Gómara, *Conquista de México* (Medina del Campo 1553), fol. CXXIX.

ports other than *Mexican sugar* destined for Spain and Peru. Far from contributing to trade what we call today *colonial products*, Havana exported only hides and leather until the eighteenth century. After livestock rearing came tobacco cultivation and bee keeping, with the first hives (*colmenares*) having come from Florida. Soon, *wax* and *tobacco* became more important than *animal hides*, which, in their turn, were replaced by *sugarcane* and *coffee*. The cultivation of each of these products did not supersede older crops. During these different phases of agricultural development and to this day, sugar has I.190 consistently been Cuba's most profitable commodity per annum, despite the widely observed tendency to allow coffee plantations to predominate. Today, tobacco, coffee, sugar, and wax exports, by both legal and illegal means, are reaching 14 or 15 million piasters, based on current commodity prices.

SUGAR. Customs registries for 64 years show the following exports from Havana:

From	1760 to 1763, annual average, at the most	13,000	cases
From	1770 to 1778	50,000	
In	1786	63,274	
	1787	61,245	
	1788	69,221	
	1789	69,125	
	1790	77,896	
	1791	85,014	
	1792	72,854	
	1793	87,970	
	1794	103,629	
	1795	70,437	
	1796	120,374	
	1797	118,066	
	1798	134,872	
	1799	165,602	
	1800	142,097	
	1801	159,841	
	1802	204,404	
I.191	1803	158,073	
	1804	193,955	
	1805	174,544	
	1806	156,510	
	1807	181,272	
	1808	125,875	
	1809	238,842	
	1810	186,672	

From	1811 to 1814, annual average	206,487
In	1815	214,111
	1816	200,487
	1817	217,076
	1818	207,378
	1819	192,743
	1820	215,593
	1821	236,669
	1822	261,795
	1823	300,211
	1824, not a very fertile year	245,329

This table, the most extensive published to date, was compiled from a large number of official manuscripts that people sent me, including the *Aurora* and the *Papel periódico de la Havana*, the *Patriota Americano*, the *Guías de Forasteros de la Isla de Cuba*, the *Sucinta Noticia de la situación presente de la Havana*, 1800 (manuscript), the *Reclamación contra la ley de Arancéles* [Zayas y Chacón and Benítez], 1821, and the *Redactor general de Guatemala*, 1825, July, p. 25. According to customs registries, which are a less reliable source, 183,960 cases of sugar were loaded in Havana between January 1 and November 5, 1825. The two months of November and December are missing for 1823, but we know that 23,600 cases of sugar were shipped between January and November of that year. I.192

To calculate the full volume of Cuba's sugar exports, one must supplement Havana's export numbers with (1) exports from other *authorized* ports, especially Matanzas, Santiago de Cuba, Trinidad, Baracoa, and Mariel, and (2) the probable volume of illegal trade. During my stay on the island, exports from Trinidad de Cuba were still estimated at no more than 25,000 cases. When examining Matanzas's customs registries, it is necessary to avoid *double counting* and to distinguish[1] carefully between sugar exported directly to Europe and sugar destined for Havana. In 1819, actual transatlantic exports from Matanzas were no more than one-thirteenth of Havana's. By 1823, I found them to be already one-tenth. According to two customs tables, one showing exports from Havana and the other exports from both Havana and Matanzas, the former shipped 300,211 cases of sugar and 895,924 arrobas of coffee, the latter 328,418 cases of sugar and 979,864 arrobas of coffee. According to this data, we can add at least 70,000 cases shipped from other ports to the 235,000 cases that is the average for the port of Havana for the last eight years. Thus, estimating customs fraud at one-fourth, we get more than 380,000 cases (nearly 70 million kg) of sugar for the island's total exports, legal and illegal. In I.193

1. [Jameson,] *Letters from the Havana*, pp. 91, 95.

I.194 1794, individuals with considerable knowledge of the place already estimated[1] Havana's consumption at 298,000 arrobas, or 18,600 cases, of sugar, and the entire island's consumption at 730,000 arrobas, or 45,600 cases. If we recall that the island's population during this period[2] was nearly 362,000, 230,000 of whom were free men, and that it is now 715,000, 455,000 of whom are free men, we must assume a total consumption of 88,000 cases for 1825. If the figure for consumption were 60,000, that would suggest at least 440,000 cases, or 81 million kg, for the total production of the sugar plantations. This is a *numerical limit* that would shrink no more than one-fifteenth, even if we estimated that domestic consumption was half that between 1794 and 1825.

To get a better sense of Cuba's agricultural wealth, we must compare the island's production during years of mediocre fertility with the production and I.195 export of sugar in the rest of the Antilles, Louisiana, Brazil, and the Guianas.[3]

ISLAND OF CUBA, according to estimates previously discussed: production, I.196 at least 440,000 cases; legal exports, 305,000 cases or 56 million kg; with contraband, 380,000 cases (70 million kg). This is nearly one-seventh less than the average exports from Jamaica.

1. ▾Don Antonio López Gómez, *Historia natural y política de la Isla de Cuba*, 1794 (unpublished), chapter I, p. 22. I do not know what sort of research was the basis for estimating consumption for the entire island at 25,000 to 30,000 cases. It was given to me as a definitive figure in 1804, before I knew of Mr. López Gómez's manuscript. Maybe they based consumption for the entire island on that of Havana, which can be measured more easily. The amount of sugar used in this city, whether in the manufacture of chocolate and preserves, or in people's diets, is far beyond what one could imagine in Europe, even in southern Spain.

2. See Vol. XI, p. 306 [this edition, p. 217].

3. In the following calculations, we adhere to results given by *customs registries* without increasing the figures to conform to always vague hypotheses of the effects of illegal trade. In the weight reductions, we assumed 1 *quintal* or 4 *arrobas* = 100 Spanish pounds = 45.976 kg; 1 *arroba* = 25 Spanish pounds = 11.494 kg; 1 *caja de azúcar* from Havana = 16 arrobas = 183.904 kg; 1 cwt = 112 British pounds = 50.796 kg. The latter estimate is based on ▾Mr. Kelly's work, which assumes that 453.544 grams = 1 pound of weight. Basing his calculations on the weight of an inch cube of distilled water under the conditions required by the new British law, ▾Mr. Francœur came up with only 453.296 grams for one pound, which means that 1 cwt = 50.769 kg, which is within 5/1,000nds [or one two-hundredth] of ▾Mr. Riffault's result in the second edition of Thomson's *Chimie*, Vol. I, p. XVII. Following Mr. Kelly, I presume that 1 cwt = 50.79 kg, but I had to mention the doubts that remain concerning such an important factor. In the *Prices-Current* that was printed in Havana, the Spanish quintal is estimated to be 46 kg, while the conversion of *Hundred-Weight* that is used commercially in Paris is also 50.792 kg.

JAMAICA. Production[1] (that is domestic consumption + exports) in 1812, according to ▼Mr. Colquhoun's slightly high estimate, 135,592 *hogsheads* equal to 14 cwt, or 96,413,648 kg. Exports in 1722, when the island had no more than 60,000 slaves, were 11,008 hds; 35,000 hds in 1744; 55,761 hds, or 780,654 cwt, in 1768 (with 166,914 slaves);[2] 1,417,758 cwt,[3] or 72,007,928 kg, in 1823 (with 342,382 slaves). It follows from the data that Jamaica's exports in the very fertile year of 1823 were no more than one-eighteenth greater[4] than those of the island of Cuba, which rose in the same year by legal means to 370,000 cases, or 68,080,000 kg. Calculating the average from 1816 to 1824, we find, according to the documents that I owe to the generosity of ▼Mr. Charles Ellis, that exports from Jamaica to the ports of Great Britain and Ireland were 1,597,000 cwt (81,127,000 kg).

I.197

BARBADOS (with 79,000 slaves), GRENADA (with 25,000 slaves), and SAINT VINCENT (with 24,000 slaves) are the three islands that produce the most sugar of all the British Antilles. In 1812, their exports to Great Britain were 174,218 cwt, 211,134 cwt, and 220,514 cwt, respectively. In 1823, they were 314,630 cwt, 247,360 cwt, and 232,577 cwt. Together, Barbados, Grenada, and Saint Vincent export an amount of sugar that still does not equal what Guadeloupe and Martinique ship to France each year. The three British islands have 128,000 slaves on 43 square nautical leagues; the two French islands have 178,000 slaves on 81 sq. lg. The island of Trinidad, which, after Cuba, Haiti, Jamaica, and Puerto Rico, is the largest of the Antilles, has, according to Mr. Lindenau and Mr. Bauzá, an *area* of 133 sq. lg. However, in 1823, it exported only 186,891 cwt (9,494,000 kg), the result of the labor of 23,500 slaves. On this island conquered from the Spanish, agriculture has grown rapidly; in 1812, production did not yet exceed 59,000 cwt.

I.198

BRITISH ANTILLES. The cultivation of sugarcane began as a branch of colonial industry in Jamaica in 1673. In an average year, exports from the entire British Antilles to Great Britain's ports totaled: 400,000 cwt between 1698 and 1712; a million cwt between 1727 and 1733; 1,485,377 cwt between 1761 and

1. Colquhoun, *Wealth of the British Empire*, p. 378.

2. Stewart, *A View of the Past and Present State of the Island of Jamaica*, 1823, p. 17.

3. [Powell,] *Statistical Illustrations*, p. 57. See Note A, at the end of the tenth Book.

4. 1812 sugar exports from Jamaica to ports in Great Britain and Ireland were, according to Colquhoun, 1,832,208 cwt, or 93,076,166 kg; in 1817, 1,717,259 cwt for Great Britain alone.

1765; 2,021,325 cwt between 1791 and 1795 (with 460,000 slaves); 3,112,734 cwt in the very fertile year of 1812; and 3,005,366 cwt in 1823 (with 627,000 slaves).[1] From 1816 to 1824, the average was 3,053,373 cwt. Today, Jamaica exports to Great Britain's ports more than half the sugar of the British Antilles, where there is one slave for every 1.8 people. The British Antilles' exports to Ireland are 185,000 cwt.

I.199–201

FRENCH ANTILLES. Exports to France: 42 million kg. In 1810, Guadeloupe exported 5,104,878 pounds of refined sugar and 37,791,300 pounds of unrefined sugar. Martinique exported 53,057 barrels (one thousand pounds per barrel) of sugar and 2,699,588 gallons (four pints of Paris per gallon) of molasses, for a total of 95,955,238 pounds for the two islands.[2] Between 1820 and

1. For 1812, see Colquhoun's work; information for 1823 is based on a work recently published under the title of *Statistical Illustrations of the British Empire* [by Powell]. Having access to partial data only, I was able to convince myself that exports between 1812 and 1823 came from nearly the same islands that England had possessed since the ▾Treaty of Paris [1763]. Only the islands of Tobago and St. Lucia were added by 1823, which accounts for an additional 175,000 cwt of sugar. Estimates before 1812 are from Mr. Edwards (in *The History, Civil and Commercial, of the British Colonies in the West Indies*, Vol. I, p. 19) and concern some nearby islands in the same part of the Antilles with insignificant production at that time. We can observe that, since 1812, sugar exports to England have not increased any further, while the number of slaves appears not to have changed significantly—if we can assume that omissions in the *records* were the same in 1812 as in 1823. In 1812, we counted (with St. Lucia, the Bahamas, and the Bermuda Islands) 634,100 slaves; in 1823, 630,800. Research prior to the publication of the *Statistical Illustrations* had given me 626,800 slaves (see Vol. XI, pp. 145 and passim [this edition, pp. 378 and passim]). I did not want to make use of the tables published for the years 1807–1822, in which exports from islands temporarily conquered and from the Dutch Guianas (Demerary, Berbice and Suriname, before the Peace of Paris) were included under the name "sugar from the British West Indies." This geographical confusion has given rise to the notion of a growth in production larger than it was in reality. Average exports from 1809–1811 and 1815–1818, for example, were ([Powell,] *Statistical Illustrations*, p. 56) 3,570,803 and 3,540,993 cwt. But, after eliminating Demerary and Berbice's 370,000 cwt of sugar from Anglo-America's total, we arrive at no more than 3,185,000 cwt for the 15 Antilles currently under British rule. Taking the same corrections into account, 1822 alone gives 2,933,700 cwt, and this result conforms within one forty-secondth to the figure I gave in the text for the year 1823 (3,005,366 cwt). Mr. Edwards, according to the last edition of his excellent work on the West Indies, believes that average exports from the British Antilles, during the period between 1809 and 1811, was 4,210,276 cwt. In this estimate, too high by a third, Antilles sugar was no doubt confused with sugar from the Guianas, Brazil, and all other parts of the world, for *total sugar imports* to Great Britain were no greater than an average of 4,242,468 cwt between 1809 and 1811.

2. *Official Notes.*

1823, the French Antilles exported to France 142,427,968 kg unrefined sugar and 19,041,840 kg refined sugar, a total of 161,469,808 kg, which amounts to an annual average of 40,367,452 kg.

THE ARCHIPELAGO OF THE ANTILLES. By estimating at 18 million kg the exports of the Dutch, Danish and Swedish Lesser Antilles, which have only 61,000 slaves, we arrive at almost 287 million kg unrefined and refined sugar exports for the entire Antillean Archipelago: I.202

165 millions or	58 percent from the	British Antilles (626,800 slaves)
62	22 percent	Spanish Antilles (281,400 slaves)
42	14 percent	French Antilles (178,000 slaves)
18	6 percent	Dutch, Danish, and Swedish Antilles (61,300 slaves)

▾Sugar exports from Saint-Domingue are nearly nonexistent at this time. In 1788, they were 80,360,000 kg; in 1799, they were still believed to be as high as 20 million. If they had remained at levels typical of the island's most prosperous period, they would have increased the Antilles' total sugar exports by 28 percent, almost 18 percent for all of America. Currently, Brazil, the Guianas, and Cuba together, with their 2,526,000 slaves, provide nearly 230 million kg, that is, (excluding contraband) three times as much sugar as Saint-Domingue exported during its period of greatest abundance. Since 1789, enormous agricultural growth in Brazil, Demerary, and Cuba has made up for Haiti's shortfall and has rendered imperceptible the abandonment of this republic's sugarmills.

French, British, and Dutch GUIANA [Suriname]. Total exports at least 40 million kg. Between 1816 and 1824, British Guiana exported an average of 557,000 cwt, or 28 million kg. In 1823, exports from Demerary and Essequibo (with 77,370 slaves) to Great Britain's ports were 607,870 cwt and from Berbice (with 23,400 slaves) 56,000 cwt, for a total of 33,717,757 kg. We can hazard 9 to 10 million kg for Dutch Guiana,[1] or Suriname. Exports from Surinam were 15,882,000 pounds in 1823; 18,555,000 in 1824; and 20,266,000 I.203 I.204

1. In his very informative work, *Nederlandsche Bezittingen in Azia, Amerika en Afrika* (1818, Vol. II, pp. 188, 202, 204, 214), a Dutch author, Mr. Van den Bosch, estimates no more than 32,408,293 pounds of sugar for the three colonies of Demerary, Essequibo, and Berbice (with 85,442 slaves) in 1814. According to the same author, Surinam, with barely 60,000 slaves, exported nearly 20,477,000 pounds of sugar in 1801. These exports have barely varied since then: they are generally 17,000 barrels (at 550 kg per barrel). Cayenne

in 1825. ˚Mr. Thuret, the Consul General for the King of the Netherlands in Paris, collected these notes.

BRAZIL. 1,960,000 slaves were exported from this vast country, where sugarcane is cultivated from the *Capitanía general* of Río Grande to the parallel[1] of Porto Alegre (lat. 30° 2′); such exports are far higher than is commonly believed.[2] According to precise information, exports were 200,000 cases (650 kg per case) or 130 million kg in 1816, one-third of which was sent to Belgium by way of Hamburg, Bremen, Trieste, Livorno, and Genoa; the rest went to Portugal, France, and England. In 1823, England received no more than 71,438 cwt or 3,628,335 kg. Sugar is typically very expensive on the coast of Brazil. Production of Brazilian sugar has declined since 1816 because of national unrest: during years of massive drought, exports barely reached 140,000 cases. Those who are particularly familiar with this branch of American commerce think that as soon as political stability returns to Brazil, sugar exports will reach 192,000 cases, or 125 million kg, of which 150,000 cases will be refined and 42,000 unrefined. People believe that Rio de Janeiro will furnish 40,000 cases, Bahía 100,000, and Pernambuco 52,000, estimates that do not even account for years of exceptional fertility.

EQUINOCTIAL AMERICA and LOUISIANA today provide a total of 460 million kg sugar for European and the U.S. trade (such are the results of meticulous analysis of all partial data):

is beginning to produce 1 million kg. Estimates of the three Guianas' black population (see Vol. XI, p. 167 [this edition, pp. 318ff]) are probably one-seventh too high.

1. On limits of cultivation in the southern hemisphere, see Auguste de Saint-Hilaire, *Aperçu d'un voyage au Brésil*, p. 57. To the north of the Tropic of Cancer, we put sugar production in Louisiana for 1815 at 15 million pounds or 7,350,000 kg (Pitkin, [*Statistical View*,] p. 249).

2. In the statistical work that appeared under the title *Commerce du dix-neuvième siècle* [by Moreau de Jonnès], Vol. II, p. 238, sugar exports from Brazil to Europe were estimated at no more than 50,000 cases. But according to customs registries in Hamburg, this port alone received 44,800 cases of Brazilian sugar in 1824 and more than 31,900 cases (at 680 kg per case) in 1825. England and Belgium imported more than 10,000 cases during the same period. Mr. Auguste de Saint-Hilaire believes that exports from Bahia have risen to only 60,000 cases in recent years. According to the official documents that Mr. Adrien Balbi assembled, we find that Portugal imported 34,692,000 kg Brazilian sugar in 1796, 36,018,000 kg in 1806, and 45 million kg in 1812.

287 millions or	62 percent from the Antilles	(1,147,500 slaves)
125	27 percent from Brazil	(2,060,000 slaves)
40	9 percent from the Guyanas	(206,000 slaves)

We will shortly see that Great Britain alone, with its population of 14.4 million, consumes more than one-third of the 460 million kg sugar that the New Continent produces in countries where the slave trade has brought together 3,314,000 wretched slaves to satisfy this appetite! These days, sugarcane cultivation is so widespread in the different parts of the globe that material or political causes that would suspend or destroy the efforts of industry in one of the Greater Antilles would not have the same impact on the price of sugar and on the general trade activities of Europe and the United States as they did before, when intensive agriculture was confined to a smaller space. Spanish writers have often compared the wealth from sugar production on the island of Cuba to the riches from the mines of Guanajuato in Mexico. In effect, at the beginning of the nineteenth century, Guanajuato furnished a quarter of Mexico's silver and a sixth of all American silver. Today, the island of Cuba exports (by legal means) one-fifth of the Antillean Archipelago's sugar and one-eighth of all the sugar that flows from equinoctial America to Europe and the United States. 1.207

On the island of Cuba, people distinguish three grades of sugar, with each grade reflecting the level of purity attained through *refining* (*grados de purga*). The upper part of each loaf, an upside-down cone, yields *white* sugar, the highest grade; the loaf's middle part consists of yellow sugar known as *quebrado;* and the lower part, which is the cone's tip, consists of *cucurucho*, the lowest grade. All of Cuba's sugars are thus refined. There is only a very small amount of unrefined sugar, or moscovado—a corruption of *azúcar mascabado*. Because each part of the cone has a different shape and thus a different volume, the loaves (*panes*) also have different weights. Generally, they weigh one *arroba* after refining. Ideally, the *maestros de azúcar* [sugar masters or master technicians] want each sugar loaf to yield five-ninths white, three-ninths *quebrado*, and one-ninth *cucurucho*. White sugar is more expensive when it is sold alone and not as part of a *surtido* [variety] sale, in which three-fifths white and two-fifths of yellow sugar are mixed together in the same lot. For mixed lots, prices generally differ by 4 Spanish reales (*reales de plata*); for white sugar, the difference rises to 6 or 7 reales. Many factors have caused significant fluctuations in the price of sugar: the Saint-Domingue revolution; prohibitions imposed by the *continental system*; the enormous consumption of sugar in England and the United States; and the progress of agriculture in Cuba, Brazil, Demerary, the Island of Bourbon [Réunion], and Java. During 1.208

a period of twelve years, from 1807 to 1818, prices fluctuated between 3 and I.209 7 reales[1] and between 24 and 28 reales, respectively, which proves proportional fluctuations on a scale from 1 to 5. During these same years, the price of sugar in England varied[2] no more than 33 to 73 shillings per quintal, that is, on a scale from 1 to 2.2. By using the price of Havana sugar in Liverpool over the course of a few months rather than taking the average price for the whole year, we can see a variation between 30 shillings (in 1811) and 134 shillings (in 1814), that is, an increase at a rate of 1 to 4.4. High prices between 16 and 20 reales per arroba remained steady in Havana for five years almost without interruption, from 1810 to 1815, while prices have dropped by a third since 1822, to 10 and 14 and, more recently, in 1826, even to 9 and 13 reales. I offer these details to give a more precise idea of a sugar plantation's net profit and thus of the sacrifices that a landowner inclined to content himself with more modest profits might make to improve the lives of his slaves. Sugar cultivation I.210 is still profitable even at the current price of 24 piasters per case (taking the average between the price of *blanco* and *quebrado*). Yet, a proprietor whose medium-sized sugarmill has an output of 800 cases now sells his harvest for only 19,200 piasters, compared to the 28,800 piasters it was worth twelve years ago (at 36 piasters per case).[3]

During my stay in the Güines Plains in 1804, I tried to gather some precise information on the *numbers* pertaining to sugar production: a large *ingenio* that produces 32,000 to 40,000 *arrobas* (367,000 to 460,000 kg) of sugar generally has 50 caballerías,[4] or 650 hectares. Sugarcane is grown on one-half of

1. In Havana sugar prices, the two figures always indicate the price of *quebrado* and *blanco* per *arroba*. The strong piaster equals 8 reales and is worth 5 francs and 43 centimes, according to the official exchange rate. On the market, it is worth 13 centimes less.

2. See the tables for prices between 1807 and 1820 in [Powell,] *Statistical Illustrations of the British Empire*, p. 56, and from 1782 to 1822 in Tooke, *Thoughts and Details on the High and Low Prices*, 1824, Appendix to Part II, pp. 46–53.

3. The agrarian unit of measure, called the *caballería*, comprises 18 *cordelas* (a *cordel* has 24 *varas*) or 432 *varas* per square cordel. Consequently, each vara = 0.835 meters; according to ▾Rodríguez, a caballería comprises 186,624 square varas, or 130,118 square meters, or 32 2/10 English acres.

4. There are only a few *plantations* on the entire island of Cuba that can produce 40,000 arrobas: these are the Río Blanco *ingenios*—or those of the ▾Marquis de Arcos, and Don Rafael O'Farrill, and Doña Felicia Jáurregui. Sugarmills with an annual production of 2,000 cases or 32,000 arrobas (about 368,000 kg) are already considered very large. In the French colonies, generally only a third or a quarter of land is dedicated to *food plantation* (bananas, yams, sweet potatoes); in the Spanish colonies, a greater area is lost to pasturage. Such is the natural consequence of the former habits of having *haciendas de Ganado* [cattle ranches].

these (less than 10 percent of a square nautical league) (*cañaveral*), while the other half is used for cultivating foodstuffs and for raising cattle (*potrero*). Land prices naturally vary with soil quality and proximity to the ports of Havana, Matanzas, and Mariel. In a radius of 25 leagues around Havana, we can estimate one caballería at two or three thousand piasters. For a yield of 32,000 arrobas (2,000 cases of sugar), the sugar plantation, or *ingenio*, must have at least 300 blacks. A seasoned adult slave sells for 450 to 500 piasters; an unseasoned bozal black goes for 370 to 400 piasters. A black slave probably costs 45 to 50 piasters a year in food, clothing, and medicine; taking into account capital interest and subtracting holidays, that is more than 22 soles per day. Slaves are given *tasajo* (sun-cured meat) from Buenos Aires and Caracas. When tasajo is too expensive, they are fed salted cod (*bacalao*) and *comestibles* (*vianda*) such as calabash, yams, sweet potatoes, and corn. In 1804, an arroba of tasajo was worth 10 to 12 reales in Güines. Today (1825), it costs 14 to 16. On an *ingenio* such as the one we are considering here (with an output of 32,000 to 40,000 *arrobas*), it is necessary to have (1) three pieces of equipment with cylinders either pulled by oxen (*trapiches*) or moved by two water wheels; (2) following the old Spanish method, 18 wood-fired boilers (*piezas*), whose slow burning causes enormous wood consumption; or, following the French method, a *reverberatory furnace* (introduced in 1801 by Mr. Bailly of Saint-Domingue under the auspices of Don Nicolás Calvo), 3 *clarificadoras* [clarifiers], 3 *pailas* [large boiling pans], and 2 *traines de tachos* [sugar pan trains] (each train has 3 *piezas*)—in total, 12 *fondos* [basics]. It is popularly said that 3 *arrobas* of refined sugar yield 1 barrel of *molasses* and that molasses alone are enough to defray the costs of running the plantation. This is so where rum is distilled in abundance. Thirty-two thousand *arrobas* of sugar yield 15,000 *barrels of molasses* (at 2 *arrobas*), from which one can make 500 *pipas of aguardiente* at 25 piasters. A balance sheet based on these numbers would show the following for 1825:

I.211 I.212 I.213

Value of 32,000 *arr.* of sugar (blanco and quebrado), at 24 piasters per case or 16 arrobas	48,000	Piasters
Value of 500 *pipas de aguardiente*	12,500	
	60,500	Piasters

Estimated annual costs of an *ingenio* are 30,000 piasters.

Moreover, the used capital consists of 50 caballerías of land, at 2,500 piasters	125,000	Piasters
300 blacks, at 450 piasters each	135,000	
buildings, mills	80,000	
cisterns, cylinders, animals, and general inventory	130,000	
	470,000	Piasters

It follows from this calculation that to establish an *ingenio* that can yield 2,000 *cajas* [cases] a year, an investor would get 6 1/2 percent interest, assuming the current price of sugar and assuming that he follows the old Spanish method. I.214 This interest rate is not considerable for an establishment that is not exclusively agricultural and whose costs remain constant but whose output sometimes decreases by more than a third. It is rare that one of the large *ingenios* can produce 32,000 cases of sugar for several consecutive years. It should not be surprising, then, that people preferred to cultivate rice over sugarcane when the price of sugar was very low in the island of Cuba (4 or 5 piasters per quintal). The profits of long established estate owners (*hacendados*) derives from (1) the fact that start-up costs were much less 20 or 30 years ago when a caballería of good land cost no more than 1,200 to 1,600 piasters, instead of 2,500 to 3,000 piasters, and an adult black sold for 300 instead of 450 or 500 piasters; and (2) the fact that very high sugar prices balanced out very low ones. Sugar prices differ so widely over the course of 10 years that capital interest varies from 5 to 15 percent. In 1804, invested capital was no more than 400,000 piasters, and, based on the value of sugar and liquor, gross I.215 output would have risen to 94,000 piasters. Yet, from 1797 to 1800, the average[1] price of a case of sugar was sometimes 40 instead of the 24 piasters I had to assume in calculations for 1825. When a sugar plantation, a large mill, or a mine is run by the individual who founded the establishment, interest-rate estimates based on the landowner's capital investment should not guide those who, buying it secondhand, attempt to assess the benefits of different types of industry.

The calculations I made on the island of Cuba suggest that, on average, a hectare yields 12 cubic meters of ▾*magma*, from which at most 10 to 12 percent raw sugar can be extracted by means of the processes still used today. In Bengal, one needs 6 pounds of magma, according to ▾Mr. Beckford; according to Mr. Roxburgh, one needs 5.6 pounds, because 28 deciliters of sugarcane magma yield 450 grams of unrefined sugar. It follows from this that if one compares magma to a salty liquid, this liquid contains 12 to 16 percent crystallizable I.216 sugar, depending on the soil's fertility. On good land in the United States, sugar maple (Acer saccharinum) yields 450 grams of sugar per 18 kg sap, or 2½ percent. This is the same amount of sugar that sugar beets yield, if one compares the quantity of extracted sugar to the tuberous roots' entire mass. 500 kg unrefined sugar is extracted from 20,000 kg beets cultivated in good soil. As sugarcane loses half its weight when the juice is pressed out, it yields a weight

1. *Papel periódico de la Havana*, 1801, no. 12.

equal to the plant mass, or six times as much raw sugar as the beet—provided one compares not the juice but the volume of the tubers of the Beta vulgaris to that of the stalks of the Saccharum officinarum. Depending on the nature of the soil, the amount of rain, the distribution of heat from season to season, and the more or less premature tendency of the plant to flower, sugarcane magma will vary in its constituent parts. It is not only, as the sugar *maestros* claim, the greater or lesser amount of sugar in the compound that accounts for this difference. Rather, the difference derives from the ratios between the crystallizable and uncrystallizable sugar (Mr. Proust's liquid sugar), the albumin, the gum, the green starch, and the malic acid. The amount of crystallized sugar might be the same, and yet, depending on the methods that one uses, the amount of light brown sugar, or cassonade, extracted from the identical volume of sugarcane magma may differ considerably, because the proportion of the substances that accompany crystallizable sugar varies. When combined with some of these substances, the crystallizable sugar forms syrup that does not crystallize and remains in the form of molasses. Too abrupt a rise in temperature seems to accelerate and increase the loss. These factors explain why the *maestros de azúcar* view themselves as *bewitched* from time to time during a given season, because, even with the same care, they cannot *make the same amount of sugar* every single time. These factors also explain why either more or less light brown sugar, called cassonade, is extracted from the same sugar magma when one alters one's method—for example, by changing either the heat or the speed of heating. It cannot be repeated often enough: no one should expect large savings in the manufacture of sugar simply by altering the construction and arrangement of the boilers and the furnaces. One can expect such savings only from improving the chemical processes, from a more intimate familiarity with how lime, alkaline, and animal carbon react, and from the exact determination of the *maximum* temperatures to which the sugar magma should be successively exposed in the different boilers. The inspiring analyses of sugar, starch, gum, and woody substances that ʻMr. Gay-Lussac and Mr. Thénard produced, together with the works in Europe on grape and beet sugar, and the research of Mr. Dutrône, Mr. Proust, Mr. Clarke, Mr. Higgins, Mr. Daniell, Mr. Howard, Mr. Braconnot, and Mr. Derosne have laid the groundwork for these improvements. But everything still has to be put into practice, even in the Antilles. One certainly would not be able to improve amalgamation processes in Mexico significantly without having first examined the nature of minerals put in contact with mercury, sodium chloride, *magistral* [powdered copper pyrites], and lime during a long stay in Guanajuato or in Real del Monte. Similarly, improving technical processes in sugarmills

I.217

I.218

requires a chemist on the cutting edge of plant chemistry who can analyze small amounts of *sugar magma* extracted from plants grown on different soil at different times of the year on different *ingenios*. For this, ▾one can use either common or *creole* sugarcane, Tahitian sugarcane [which is a light color], or *Guinea* cane, which is red. Without this preliminary work, which should be done by someone from one of Europe's famous laboratories, with solid grounding in beet sugar production, one can only achieve partial improvements. Otherwise, the sugarcane production process will remain what it is today: the result of a more or less felicitous fumbling.

1.219

In soil that can be irrigated or in which plants with tuberous roots preceded the cultivation of sugarcane, a caballería of fertile land yields three or four thousand *arrobas*, instead of 1,500 *arrobas*, which, in turn, yield 2,660 to 3,540 kg sugar (both *blanco* and *quebrado*) per hectare. If we posited 1,500 *arrobas* and, on the basis of Havana's prices, 24 piasters per case of sugar, we would find that the same hectare might produce 870 francs worth of sugar, or 288 francs worth of wheat, assuming an eightfold increase in crop yield and a price of 18 francs per one hundred kg wheat. I have commented elsewhere that, in comparing two branches of agriculture, we must not forget that sugarcane cultivation requires enormous capital outlay. Currently, for instance, one would need to invest 400,000 piasters for an annual output of 32,000 arrobas, or 368,000 kg, on a single sugar plantation. According to Mr. Beckford[1] and Mr. Roxburgh, an acre of irrigated terrain (4,044 square meters) in Bengal yields 2,300 kg raw sugar, or 5,700 kg per hectare. Where there are vast stretches of land with this level of fertility, as is the case in the East Indies, the price of sugar is, unsurprisingly, low. The output of a single hectare in the East Indies is double that of the best soil in the Antilles, and the cost of hiring a free Indian is nearly three times less than the cost of black slave labor on the island of Cuba.

1.220

Calculations for 1825 show that a Jamaican plantation of 500 *acres* (or 15½ *caballerías*), on which 200 acres are dedicated to sugarcane, would yield 2,800 cwt, or 142,200 kg, sugar through the labor of 200 slaves, 100 oxen, and 50 mules. This plantation, including the slaves, would be worth 43,000 pounds sterling. According to this estimate from ▾Mr. Stewart, one hectare would yield 1,760 kg unrefined sugar, given the high quality of Jamaican sugarcane. We saw above that, by assuming a large sugar plantation in Havana to encompass 25 caballerías, or 325 hectares, which yield 32,000

1.221

1. ▾[Tennant,] *Indian Recreations* ([Fleming, *A Catalogue of Indian Medicinal Plants*?] Calcutta, 1810, p. 73), Roxburgh [in] *Oriental Repertory*, Vol. II, p. 425.

to 40,000 cases, we arrive at either 1,130 or 1,400 kg refined sugar (*blanco* and *quebrado*) per hectare. This figure more or less tallies with that for Jamaica, considering that sugar loses weight during *refining* when raw sugar is converted into refined sugar, that is, *azúcar blanco y quebrado*. On Saint-Domingue, a carreau (3,403 square toises = $1\frac{29}{100}$ hectares) is estimated to be 40, sometimes 60, quintals. For 5,000 pounds, we still get 1,900 kg unrefined sugar per hectare. If one assumes—as one should when speaking of the output of the entire island of Cuba—that one caballería (13 hectares) of average fertility yields 1,500 arrobas of refined sugar (a mixture of *blanco* and *quebrado*), or 1,330 kg per hectare, then it follows that 60,872 hectares, or 1.222 19.75 square nautical leagues (roughly one-ninth the size of an average département in France), suffice to produce the 430,000 cases of refined sugar that the island of Cuba uses for its own consumption and for legal and illegal exports. Surprisingly, fewer than twenty square nautical leagues can yield an annual output whose value (calculating one case in Havana at a rate of 24 piasters) is more than 52 million francs. To provide all the raw sugar that 30 million Frenchmen require, which currently amounts to 56 to 60 million kg, 1.223 one needs[1] only $9\frac{5}{6}$ square nautical leagues for sugarcane in the tropics and $37\frac{1}{2}$ square nautical leagues for beets in the temperate zone! In France, a hectare of *good* soil seeded or planted with beets produces between ten thousand and thirty thousand kg beets. Soil of average quality produces 20,000 kg, which would yield $2\frac{1}{2}$ percent, or 500 kg, of unrefined sugar. 100 kg of this raw sugar, however, yields 50 kg refined sugar, 30 kg brown sugar (cassonade), and 20 kg moscovade. A hectare of beets, therefore, produces 250 kg refined sugar.

Just before my arrival in Havana, some samples of beet sugar had been ordered from Germany. It was said that ▾beet sugar "threatened the existence of the *sugar islands* in America." The planters were stunned to realize that beets were, in fact, quite similar to sugarcane. But they deceived themselves into believing that the high price of manual labor in Europe and the difficulty 1.224 of separating the crystallizable sugar from such a large mass of vegetal pulp

1. ▾Mr. Barruel calculates 67,567 arpents of water and forests (11 square nautical leagues) for 15 million kg unrefined beet sugar (*Moniteur* from March 22, 1811). In tropical agriculture, I get 1,900 kg raw sugar per hectare. I owe this very specific data on the manufacture of beet sugar to the friendship and generous communications of Baron Delessert, my associate at the Academy of Sciences. Through his botanical publications, his immense herbariums, and a library equally rich in works on science and political economy, he has facilitated for many years now the editions of different parts of my *Voyage aux régions équinoxiales*.

would render the whole operation largely unprofitable. Chemistry has meanwhile overcome these difficulties. In 1812, France had more than 200 beet sugar factories working with very varied success and producing one million kg unrefined sugar, that is, *five-eighths* of France's current sugar consumption. Today, there are even fewer factories that, run with intelligence, still produce more than a half million kilograms.[1] The inhabitants of the Antilles, who

I.225 are well versed in Europe's affairs, fear neither the sugar from beet, from rags, grapes, chestnuts, mushrooms, or coffee from Naples nor indigo from the Riviera. Thankfully, the hope of seeing the condition of slaves in the Antilles improve does not depend on the success of these minor European agricultural endeavors.

I mentioned on several occasions that, until 1762, the island of Cuba contributed to trade no more agricultural products than today do Veragua and the Isthmuses of Panama and Darien, the three least industrious and most neglected of the Central American provinces. The British occupation of Havana, a seemingly unfortunate political event, woke everyone up. The city was evacuated on July 6, 1764, and the first stirrings of local industry date from this

I.226 memorable time. The construction of new forts in accordance with a grand plan[2] suddenly put a great deal of silver into circulation. Later on, the slave trade, once it had become free,[3] furnished hands for the sugarmills. Bureaucratic tangles notwithstanding,[4] Cuba's growing prosperity was due to the following successive causes: the opening to trade of all Spanish ports and, intermittently, even to those of the neutral powers; Don Luis de las Casas's sensible administration; the founding of the *Consulado* [merchant guild] and the *Pa-*

1. Although the current price of unrefined sugar is 1.50 francs per kg in the ports, beet sugar still offers an advantage in certain locales, for example in the environs of Arras. Beet sugar would be established in other parts of France, if the price of Antilles sugar were to rise to 2 or 2.25 francs per kg and if the government did not impose a tax on beet sugar to compensate for the loss of duties attendant on the reduction of consumption of sugar from the Antilles. The manufacture of beet sugar is particularly profitable where it is integrated in the rural economy, the improvement of the soil, and the feeding of livestock. It is not a type of agriculture independent of local circumstances, as the cultivation of sugarcane is in the tropics.

2. It is said that the construction of the *Cabaña* fortress alone cost 14 million piasters.

3. *Real cédula de 28 de Febrero de 1789*.

4. The complication of *autoridades y jurisdicciones* is such that the Report [by Valle Hernández] on the *present situation of the island of Cuba*, p. 40, lists 25 types of civil and ecclesiastical courts (*Juzgados*). This fragmentation of supreme authority explains what was said above (pp. 162ff. [this edition, pp. 85ff.; Vol. XI, pp. 363ff.]) about the ever increasing number of lawyers.

triotic Society; the destruction of the French colony of Saint-Domingue[1] and the ballooning of the price of sugar that inevitably ensued; the improvement of machinery and boilers largely by refugees from ▾Cap-Français; the closer ties between the sugarmill owners and the Havana merchants, who had heavily invested in agricultural enterprises (sugarmills and coffee plantations). I.227

Sugarmilling experienced the greatest changes between 1796 and 1800. First, mules (*trapiches de mulas*) began to replace oxen (*trapiches de bueyes*). Then, in Güines, hydraulic wheels (*trapiches de agua*) were introduced, which the first *conquistadores* had already used on Saint-Domingue. Finally (in Ceibabo), the ▾Count Jaruco y Mopox funded experiments with steam engines (*bombas de vapor*). Today, 25 steam engines are used in various sugarmills on the island of Cuba. At the same time, the cultivation of Otahití [Tahitian] sugarcane became more common. Preparatory boilers (*clarificadoras*) were introduced and the reverberatory furnaces were set up more efficiently. I.228 On a large number of plantations, prosperous landowners (to their credit) showed a noble concern for sick slaves, for importing more female blacks, and for educating children.

In 1775, the number of sugarmills (*ingenios*) on the whole island was 473. In 1817, there were more than 780. Not a single plantation among the initial group produced even a fourth of the sugar that even a *second-rate ingenio* produces today. Progress of this branch of agricultural industry is not measured by the number of sugarmills alone. In Havana province, there were:

In	1763	70	sugarmills
	1796	305	
	1806	480	
	1817	625	

1. At three points: in August, 1791, in June, 1793, and in October, 1803. It is particularly the unfortunate and bloody expedition of ▾Generals Leclerc and Rochambeau that accomplished the destruction of Saint-Domingue's sugar plantations.

Table of Agricultural Wealth in the Province of Havana, 1817

Partidos	Sugarmills (*Ingenios de azúcar*)	Coffee plantations (*cafetales*)	Potreros[1]	Haciendas de Cria	Tobacco plantations (*Vegas*)	Churches	Houses
Havana	1		12			31	16,613
Villa de Santiago	43	17	190		30	52	3,327
Bejucal	49	14	62			6	872
Villa de San Antonio	4	124	51	51	76	10	1,684
Guanajay	122	295	96			30	1,139
Guanabacoa	9	1	1			36	3,654
Filipinas		16	48	196	883	13	1,822
Jaruco	133	81	148		5	8	1,793
Güines	78	35	124	1	10	17	2,055
Matanzas	95	83	200	12		10	1,954
Santa Clara	14	78	220	267	100	7	3,441
Trinidad	77	35	45	403	150	24	3,914
Total	625	779	1,197	930	1,601	224	42,268

1. So as not to mischaracterize agriculture in the Spanish colonies, I am abstaining from using French words in place of well-worn Spanish terms. The *Hatos* or *Haciendas de cria* and *Potreros* are types of livestock ranches. But the former, which are generally 2 to 3 leagues in diameter and are not enclosed, have animals that are almost entirely wild. They require no more than 3 or 4 men on horseback (*peones*) roaming the countryside in search of cows and mares that have just given birth so as to *brand* the young animals. *Potreros* consist of enclosed pastures, a small part of which is often turned over to corn, banana, and manioc cultivation. Potreros are used to fatten the animals born on the *Hatos;* livestock breeding occurs as well, but it is of secondary concern (*de pequeñas crias*).

This table separates the districts (Trinidad and Santa Clara) that still preserve an earlier fondness for rural life and for *Hatos* [cattle ranches] both from the tobacco districts (Filipinas, Trinidad) and from those districts with the most sugar plantations (Jaruco, Guanajay, Matanzas, and San Antonio Abad). Partial growth is quite significant. In 1796, there were only 73 sugarmills in Jaruco and Río Blanco del Norte; in Güines and of Matanzas, there were no more than 73, 25, and 27, respectively. By comparison, in 1817, there were 133, 78, and 95. I.230 I.231

The increase in tithes is one of the clearest signs of an increase in agricultural wealth. We record them here for the last 15 years. Tithes (*rentas decimales arrendadas*) were due the bishopric of Havana[1] every four years:

from	1789 to 1792 for	792,386	piasters
	1793 to 1796 for	1,044,005	
	1797 to 1800 for	1,595,340	
	1801 to 1804 for	1,864,464	

In the last period, tithes on average rose to 2,330,000 francs annually, although sugar only accounted for half a tithe, or five percent. To show the ratios between not so much the production but the export of rum and molasses (*miel de purga*) and the export of refined sugars over a span of several years, I record below the figures from 1815–1824, basing them on Havana customs registries: I.232

Years	Pipas of rum	Bocoyes of molasses	Cases of refined sugar
1815	3,000	17,874	214,111
1816	1,860	26,793	200,487
1817	...	30,759	217,076
1818	3,219	34,990	207,378
1819	2,830	30,845	92,743
1822	4,633	34,604	261,795
1823	5,780	30,145	300,211
1824	3,691	27,046	245,329

Based on the average of the last five years, the export of 1,000 cases of refined sugar (183,904 kg) corresponds to the export of 17 *pipas* of sugar rum and 130 *bocoyes* of molasses.[2] I.233

1. *Official documents* that separate for each period of time the output of 40 *Parroquias* [parishes] and *Casas excusadas* [exempt buildings], that is, habitations whose tithes are reserved for the construction of churches and hospitals.

2. *One pipa of aguardiente* = 180 *frascos* or 67.5 gallons; 1 *bocoy* = 6 *barriles*. The pipa of *aguardiente de caña* that today is worth 25 piasters in Havana was worth more than 35 between 1815 and 1819. The *bocoy de miel de purga* was worth 7 reales de plata. It is widely

The enormous costs that the large *ingenios* and the frequent domestic turmoil that excess and disorganization generate all too often place landholders in a state of absolute dependence on merchants.[1] The most common loans I.234 are those in which capital is advanced to the *hacendado*, who agrees to pay two piasters above the market rate at harvest time for every quintal of coffee and two reales de plata for every arroba of sugar. This is how a harvest of one thousand cases of sugar is sold in advance (*refacción*) at a 4,000-piaster loss. The high volume of business and the lack of currency are so pronounced in Havana that the government itself is often forced[2] to take out loans at 10 percent and that individuals loan out at 12 or 16 percent. The enormous profits that the slave trade generates—on the island of Cuba sometimes as much as 100 and 125 percent per voyage—have considerably contributed to higher interest rates. Many speculators borrow money at 18 and 20 percent, thereby invigorating this shameful and deplorable business.

The first sugarcane planted with care on virgin soil will yield crops for 20 to 23 years; after that, it must be replanted every three years. The Matamoros I.235 Hacienda has a cane field (*cañaveral*) that has been harvested continuously for 45 years. Currently, the most fertile land for sugar production is around Mariel and Guanajay. Tahitian Cane, known as *Caña de Otahití* and recognizable from afar by its fresher green tint, has the advantage of yielding, on identical soil, one-fourth more juice and a more fibrous and thicker *bagazo* higher in combustible fiber. Filled with the pride of the semieducated, the refiners (*maestros de azúcar*) insist that Otahití sugarcane juice (▾*guarapo*) is easier to work with and that it also requires less lime or potash for crystallization.[3] Although the *South Sea Cane* probably produces the thinnest stalks after 5 or 6 years of cultivation, the knots are still farther apart than they are with *Caña criolla* or *Caña de la tierra*. The initial fear that South Sea Cane would grad-

accepted that three loaves of sugar yield a *barril de miel de purga* [molasses], about 2 arrobas. During the *refining process*, one adds to the first layer of wet clay (*barro*), pounded by the feet of animals in a shed (*piza*), another layer of clay (*barrillo*). After removing these layers, the refined sugar is left in the cone (*horma*) for another eight days so that the small remainder of the molasses can drain away entirely (*para escurrir y limpiar*).

1. Contracts between capitalist merchants and the *hacendados* have caused the latter losses of 30 to 40 percent particularly in 1798, a period when so many new sugarmills were being built. Laws prohibit all loans that exceed a 5 percent interest rate, but people know how to circumvent these regulations by means of fictive contracts (▾Sedano, *Sobre la decadencia del ramo de azúcar*, 1812, p. 17).

2. I recall the *Emprestito de la Intendencia de la Havana* of November 5, 1804 [loan by the colonial government in Cuba].

3. The moment that lime is added, the *foam* blackens. The tallow and other fatty bodies push the foam (*cachasa*) to the bottom and reduce it.

ually degenerate into ordinary[1] sugarcane has fortunately proved groundless. 1.236 On the island of Cuba, sugarcane is planted during the July-October rainy season and harvested from February to May.

Because clearing the land has been too fast, the island has become deforested, and the sugarmills are beginning to lack fuel; the fires in the old boilers (*tachos*) always required a little *bagazo* (sugarcane drained of its juice). Only since refugees from Saint-Domingue introduced reverberatory furnaces has there been an attempt at not using any wood at all and to burn only *bagazo*. The older furnace and boiler design used one *tarea* of wood—160 cubic feet—to produce 5 arrobas of sugar; or, for 100 kg unrefined sugar, 278 cubic feet of lemon tree and orange tree wood. In Saint-Domingue's reverberatory furnaces, a cart of bagazo, which is about 495 cubic feet, produces 640 1.237 pounds of raw sugar, which means that 158 cubic feet of bagazo yield 100 kg sugar. During my stay in Güines, and particularly in Río Blanco at the Count of Mopox's residence, I tried out many new things to reduce the waste of fuel, to line the firebox with heat-retaining substances, and to help slaves suffer less while tending the fire. A long stay in Europe's salt mines and my exposure to applied ▾halurgy during my early youth had given me the idea for these devices that I managed to adapt with some success. Wooden lids placed on the *clarificadoras* [clarifiers] accelerated the evaporation process, which led me to believe that a system of lids and portable frames equipped with counterweights could be used on other boilers. This project merits further study. But the amount of *sugarcane juice* (*guarapo*), of crystallized sugar extracted and lost, of fuel, time, and costs needs to be evaluated with care.

Many imprecise claims have been made about the price of sugarcane in 1.238 discussions about the possibility of replacing sugar from the colonies with *beet sugar* from Europe. Here are some data that can serve as the basis for more exact comparisons. In Europe, the price of sugar from the colonies is based (1) on the initial purchase price; (2) on the cost of shipping and insurance; and (3) on customs duties. Currently, the purchase price in the Antilles is no more than one-third the sale price in Europe. When an equal mix of white and brown sugar (*blanco y quebrado*) costs 12 reales de plata[2] per ar- 1.239

1. On these varieties and the history of their introduction, see Vol. V, pp. 102, 103, 104, 218, and 219. The cases of sugar that come from the *Mississippi* in barges that carry 3,000 cases are made of pine and cypress. In 1804, they cost 14 to 18 reales a piece.

2. Without a doubt current profits for Havana planters (*hacendados*) are far lower than is generally believed in Europe. However, ▾Don José Ignacio Echegoyen's very old calculation on the *cost of sugar production* seem to me somewhat exaggerated. This man, with his extensive experience in the technical domain, figured that the production of 10,000 arrobas of sugar necessitated an expense of 12,767 piasters and a capital investment of 60,000

roba in Havana, a *caxa* of 184 kg is worth 126.48 francs. Consequently, the price of 100 kg refined sugar is 68.69 francs, assuming that one piaster equals 5.27 francs. In the French colonies, the initial purchase price is 50 francs for 100 kg raw sugar, or 50 centimes per kg. Shipping and insurance are also 50 centimes. Duties are 49.50 francs for every 100 kg, or 49½ centimes per kg. Hence the figure of 1.50 franc for the total price of unrefined sugar in European ports (such as Le Havre). Beet juice cultivated in temperate climes yields I.240 no more than a third or a fourth of the crystallized[1] sugar than the tropical *sugarcane juice* does. But beet factories have the advantage in terms of shipping, insurance, and duties, which amount to 10 soles, or two-thirds, of the total price of a pound of unrefined sugar from the colonies. If colonial sugar were entirely replaced by native sugar, French customs would lose more than I.241 29 million francs yearly, assuming the current status quo.

Europeans widely and erroneously believe that most slaves in the part of the Antilles known as the *sugar colonies* are put to work in sugar factories. Such beliefs influence the way in which people imagine what would happen if the *slave trade* were to cease. Sugarcane cultivation is clearly one of the most important reasons for keeping it alive. But a simple calculation proves that the entire slave population in the Antilles is almost three times larger than the

---

piasters on the part of the owner. The cost would therefore be 55 francs for every 100 kg. And, supposing their value to be 65 francs (about 24 piasters per *caxa*), the 60,000 piaster capital investment would yield no more than a 3⅘ percent return, assuming such unfavorable circumstances. This calculation, given to me in Havana, dates back to 1798, a period when manufacturing costs and the price of land and blacks were far lower than they are today. But we should not forget that (1) that molasses and rum production, which are worth 25 piasters per *pipa*, can raise the value of manufactured sugar by one-fourth and is left off the ledgers; (2) that Mr. Echegoyen composed his memorandum to prove the degree to which tithing on sugar production was onerous and believed he could get away with exaggerating *hacendados'* expenses (see Vol. XI, p. 389 [see this edition, p. 115]; *Patriota*, Vol. II, p. 65, and the already cited account by Don Diego José de Sedano, *Sobre la decadencia del ramo de azúcar*, 1812, p. 5).

1. Count Chaptal also assumes no more than 210 kg unrefined sugar for 10,000 kg beet root, or 2.1 percent of the entire weight (*Chimie appliquée à l'agriculture*, Vol. II, p. 452). As thoroughly stripped roots give 70 percent juice, we can assume that one extracts 3.5 percent unrefined sugar from beet juice in an average year. In some locales, among them Touraine [France], this juice contains up to 5 percent crystallizable sugar, while in Java, the magma contains 25 to 30 percent sugar! The yield per hectare in soil of average fertility does not, however, differ greatly on Java from yields that we have already had occasion to consider for the island of Cuba (pp. 396 and 397). Mr. Crawfurd estimates that 1,285 pounds of refined sugar comes from an English acre on Java, which means 1,445 kg per hectare (*History of the Indian Archipelago*, Vol. I, p. 476).

number that works in sugarmills. Already seven years ago, I demonstrated[1] that, if the 200,000 cases of sugar that the island of Cuba exported in 1812 were produced by large plantations, fewer than 30,000 slaves would have sufficed for this type of industry. In order to defeat prejudices founded upon false numerical estimates and in the name of humanity, it is worth recalling at this point that slavery's evils affect far more individuals than would be needed for agriculture. This would be so even if one were to admit—which I am far from doing—that sugar, coffee, indigo, or cotton could only be cultivated by [1.242] slave labor. On the island of Cuba, it is generally held that it takes 150 blacks to produce 1,000 cases (184,000 kg) of refined sugar, or, in round numbers, one adult slave produces a little more than 1,200 kg.[2] To produce 440,000 cases [1.243] therefore takes no more than 66,000 slaves. If we add to this number 36,000 for coffee and tobacco cultivation, we find that, of the 260,000 slaves who exist on the island of Cuba today, about 100,000 would suffice for the three great branches of colonial industry on which commerce is based. Besides, tobacco is almost entirely cultivated by whites and free men. I have shown (in Vol. XI, p. 300), basing my assertion on the highly respectable authority of the *Consulado de la Havana*, that one-third (32 percent) of the slave population lives in the cities and therefore remains distant from any form of agricultural work. Yet, if we take into consideration (1) the number of children on the *haciendas*, who cannot work, and (2) that small plantations or *dispersed estates* require a far greater number of blacks to produce the same quantities of sugar [1.244] than do large *consolidated estates*, we find that, out of 187,000 slaves who live

1. *Relation historique*, Vol. V, pp. 28[9] and 2[90].

2. On Saint-Domingue, it was assumed that it took four-fifths of a field slave for one carreau of a large, beautiful estate. But on farms dispersed throughout the entire island, it took 3 slaves per carreau according to the ▼Marquis of Gallifet. Yet, if a carreau ($1\frac{29}{100}$ hectares) yields 2,500 kg unrefined sugar, then we get 833 kg per slave. Mr. Moreau de Jonnès has even demonstrated that calculations for all of cultivated land in the French colonies yield no more than $33\frac{1}{3}$ quintals, or 1,640 kg per carreau (*Commerce au* XIXe *siècle*, Vol. II, pp. 308, 311). In Jamaica, ▼Mr. Whitmore estimates that one needs no more than one black slave to produce one hogshead of sugar (or 711 kg). Already the compiler of the *Representación del Consulado de La Habana* to the Courts appears impressed by the large amount of sugar produced in Cuba with fewer slaves than in Jamaica ([Arango,] *Documentos*, p. 36). In the *Sucinta Noticia de la situación de la Isla de Cuba, en Agosto 1800*, compiled by one of Havana's wealthy property owners [Antonio del Valle Hernández], I find the following claim: "Our lands are so fertile that we get 160 to 180 arrobas in very advantageous locations, and one hundred arrobas of white and brown sugar per black on the whole island. On Saint-Domingue, one figures 60, on Jamaica, 70 arrobas of unrefined sugar." Converted into kg, we get 1,194 kg refined sugar for Cuba and 804 kg unrefined sugar for Jamaica.

in rural areas, at least one-fourth, or 46,000, produce neither sugar, nor coffee, nor tobacco. The slave trade is not only barbaric; it is also unreasonable, because it fails to achieve its end. It is like a current of water that one has channeled across a great distance to the colonies, where the water does not reach the soil for which it had been intended. Those who keep repeating that only black slaves can cultivate sugarcane seem unaware that there are 1,148,000 slaves in the Antilles Archipelago, and that only five or six hundred thousand of them produce all the colonial crops in the entire Antilles.[1] Just examine the I.245 current state of Brazil's industry! Calculate how many hands it takes to provide Europe with the sugar, coffee, and tobacco that leave Brazil's ports! Visit Brazil's gold mines, which are barely worked these days! And then ask yourself whether *Brazil's industry* really requires the enslavement of 1,960,000 blacks and mixed-race people. More than three-quarters of these Brazilian slaves do not pan for gold; nor do they produce any colonial crops—the very crops that, one assures us with such seriousness, render the slave trade *a necessary evil* and an *inevitable political crime*.[2] I.246

COFFEE.—Like the improvements in boiler construction in the sugar factories, coffee cultivation dates back to the arrival of emigrants from Saint-Domingue, especially between the years 1796 and 1798. A hectare yields 860 kg, the product of 3,500 bushes. Havana province has

in	1800	60	*cafetales*
in	1817	779	

Because the coffee tree is a shrub that produces good harvests only every four years, coffee exports from the port of Havana were no more than 50,000 arrobas, even in 1804. They reached

1. To prove how far this calculation is from being an exaggeration, we recall that 287 million kg sugar and 38 million kg coffee were exported from the Antillean Archipelago, and that on large estates and under conditions of only average fertility, 800 kg sugar and 500 kg coffee (the product of 2,000 bushes) take one black to produce, then one finds that 435,000 field hands are needed for export sugar and coffee production. If we augment this number with children and the larger numbers required for small holdings that lag behind in productivity by a third to a half, we arrive at no more than 652,000 slaves out of 1,148,000 slaves of all ages and both sexes in the Antilles (see Vol. XI, pp. 160 and 161). In Cuba, the Consulado assumed 69,000 slaves in the cities and 143,000 in the fields for 1811.

2. A very knowledgeable traveler, Mr. Caldcleugh (*Travels in South America*, Vol. I, p.79), also estimates the Brazilian slave population at 1.8 million, although he presumes that the whole population is no more than 3 million (see Vol. IX, pp. 177 and 178).

in	1809 per	320,000	*arrobas*
	1815	918,263	
	1816	370,229	
	1817	709,351	
	1818	779,618	
	1819	642,716	
	1820	686,046	
	1822	501,429	
	1823	895,924	
	1824	661,674	

These figures reflect large variations in customs fraud and in the abundance of harvests in light of the fact that results from 1815, 1816, and 1823, which one might assume to be less precise, were recently checked against customs registries. In 1815, when the price of coffee was 15 piasters per quintal, the value of Havana's exports exceeded 3,443,000 piasters. In 1823, exports from the port of Matanzas reached 84,440 arrobas. It is likely that the island's total exports, both legal and illegal, are more than 14 million kg during years of average fertility. I.247

I. Registered exports, annual average, from 1818 to 1824:		
a) in Havana	694,000	arrobas.
b) in Matanzas, Trinidad, Santiago de Cuba, etc	220,000	
II. Customs fraud[1]	304,000	
Total	1,218,000	

It follows from this calculation that coffee exports for the island of Cuba are higher than Java's, which ▾Mr. Crawfurd[2] estimated at 190,000 *piculs*, or 11.8 million kg, and higher than Jamaica's, which, according to customs regis- I.248

1. According to data gathered *in situ,* customs fraud is far more considerable for coffee than for sugar exports. I estimated the former at one third and the latter at one fourth of *registered quantities.* Coffee bags that should contain 5 arrobas often have 7 to 9, so that lately, proprietors have been asked to make *declaraciones juradas* [sworn statements].

2. It was only because of a mistaken conversion from tons to *pounds* (assuming 54,260 tons = 486,158,960 lbs) that this esteemed author was led to believe that Java's exports (25,840,000 lbs or 11,628,000 kg) amounted to two-sevenths of the coffee exports from the British Antilles and to one-nineteenth of Europe's consumption ([Crawfurd,] *History of the Indian Archipelago*, Vol. III, p. 374). The 54,260 tons (20 cwt or 1,016 kg) that Mr. Crawfurd takes to be Europe's demand for coffee does not equal 218 million kg but instead 55,128,000 kg, an estimate even lower than the one on which I settled in 1818 (*Relation historique*, Vol. V, pp. 87, 88 and 296). It is believed that the whole of Arabia contributes no more than 7 to 8 million kg coffee to the Persian, Indian, and European market (Page, *Traité d'économie politique et de commerce des colonies*, Vol. I, p. 30).

I.249 tries, reached[1] no more than 169,734 cwt, or 8,622,478 kg, in 1823. In the same year, Great Britain received[2] 194,820 cwt, or 9,896,856 kg, from the entire British Antilles, proving that Jamaica alone produced six-sevenths of it. Guadeloupe contributed 1,017,190 kg to the French metropole in 1810; Martinique 671,336 kg. In Haiti, where coffee production before the French Revolution was 37,240,000 kg, Port-au-Prince exported no more than 91,544,000 kg in 1824. It seems that *today, total legal coffee exports from the Antillean Archipelago reach more than 38 million kg.* This is nearly five times France's level of consumption which, in an average year between 1820 and 1823, was 8,198,000 kg.[3]
I.250 Great Britain's consumption is still[4] no more than 3½ million kg. However, trade and production of this crop have increased so much in both hemispheres that Great Britain has exported coffee at different stages of its trade:

in	1788	30,862	cwt (per 50.8 kg)
	1793	96,167	
	1803	268,392	
	1812	641,131	
	1814	1,193,361	
	1818	456,615	
	1821	373,251	
	1822	321,140	
	1823	296,942	

I.251 Exports for 1814 were 60.5 million kg, what one could take to be almost the entirety of Europe's consumption at that time. Great Britain—always taking this name in its true sense, as designating only England and Scotland—consumes today *nearly two and a half times less coffee than and three times as much sugar* as France.

1. In 1812, Mr. Colquhoun put Jamaica's exports to the ports of the United Kingdom at 28,385,395 British pounds or 12,773,427 kg, and imports from the entire British Antilles (excluding islands temporarily conquered) at 31,871,612 British pounds, or 14,342,225 kg (*Wealth of the British Empire*, p. 378; *Relation historique*, Vol. V, p. 81 and passim).

2. [Powell,] *Statistical Illustrations*, p. 54. Exports from British Guyana in 1823 were 72,644 cwt or 3,690,315 kg.

3. Rodet, *Du Commerce exterieur*, p. 153. Of these 8 million kg coffee, Paris alone appears to consume more than 2.5 million.▼Chateauneuf, *Recherches sur les consommations de Paris*, 1821, p. 107.

4. Prior to the year 1807, when coffee duties were reduced, Great Britain's consumption was not even 8,000 cwt (less than half a million kg). In 1809, it increased to 45,071 cwt, in 1810 to 49,147 cwt, in 1823 to 71,000 cwt, and in 1824 to 66,000 cwt (or 3,552,800 kg). [Liverpool East India Association,] *Report of the Commission of the Liverpool East-India Association*, 1822, p. 38, and Nichols, *London Price Current*, 1825, p. 63 [citation listed under *Prince's London Price Current* in the bibliography].

Much like sugar prices in Havana are calculated per *arroba*, 25 Spanish pounds (or 11.49 kg), coffee prices are always per quintal (or 45.97 kg). We have seen the price of coffee fluctuate from 4 to 30 piasters. In 1808, it even fell below 24 *reales*. Between 1815 and 1819, the price was between 13 and 17 piasters per quintal; today, it is at 12 piasters. It is likely that coffee cultivation on the whole island of Cuba employs no more than 28,000 slaves who, in an average year, produce 305,000 Spanish quintals (14 million kg) or, based on current market value, 3,660,000 piasters. Meanwhile, 66,000 blacks produce 440,000 cases (81 million kg) of sugar, which, at a price of 24 piasters, are worth 10,560,000 piasters. It follows from this calculation that a slave currently produces 130 piasters worth of coffee and 160 piasters worth of sugar. It is almost needless to point out here that these figures vary with the price of these two crops, whose fluctuations often go in opposite directions. To throw some light on agriculture in the tropical region, I also consider domestic consumption together with legal and illegal exports. I.252

TOBACCO.— Cuban tobacco is famous throughout Europe, where the habit of smoking adopted from the natives of Haiti was introduced toward the end of the sixteenth and the beginning of the seventeenth century. It was generally hoped that tobacco cultivation would become an important object for trade unfettered by an odious monopoly. The good intentions that the government has displayed since it abolished the ▾*Factoría de tabacos* six years ago have not improved this branch of industry as much as one might have expected. Tobacco planters lack capital. Leasing land is becoming very expensive, and the preference for growing coffee impedes the cultivation of tobacco. I.253

The earliest data we have on the amount of tobacco that the island of Cuba contributed to the stores of the metropole go back to 1748. According to Raynal, a far more precise writer than is generally believed, this quantity was 75,000 arrobas in an average year between 1748 and 1753. Between 1789 and 1794, the island's tobacco crop rose yearly to 250,000 arrobas. But from then until 1803, with the increasing price of land and the attention turned exclusively to coffee and sugar plantations, the royal monopoly's (*estanco*) petty annoyances and obstacles to foreign trade have gradually decreased production by more than one half. The belief is, however, that the island's total tobacco production once again reached three to four thousand arrobas between 1822 and 1825.

Cuba's domestic consumption of tobacco exceeds 200,000 arrobas. Until 1761, the *Havana Trade Company* delivered Cuban tobacco to the Peninsula's royal factories in accordance with contracts with the *Real Hacienda* [treasury] that were renewed periodically. The government's *Factoría de tabacos* I.254

replaced this company and exploited the monopoly in its turn. Tobacco planters were paid according to three categories: *suprema*, *mediana*, *e infima* [supreme, medium, and very poor]. In 1804, the respective prices were 6, 3, and 2½ piasters per *arroba*. If we collate the various prices with the quantities produced, we find that the Royal Factory paid an average price of 16 piasters per quintal for tobacco leaves. Because of production costs, a pound of *cigarros* costs the administration 6 *reales* (or ¾ piasters), even in Havana. A pound of *polvos delgados con color* [fine snuff] costs 3½ *reales*; a pound of *polvos suaves* [common soft snuff], also known as Seville *cucaracheros*, costs 1½ *reales*.

In good years, when the harvest brought in 350,000 *arrobas* of leaves (thanks to the advances that the *Factory* had made to struggling tobacco planters), 128,000 *arrobas* were destined for the Peninsula, 80,000 for Havana, I.255 9,200 for Peru, 6,000 for Panama, 3,000 for Buenos Aires, 2,240 for Mexico, and 1,000 for Caracas and Campeche.[1] To arrive at the sum of 315 million—considering that the harvest loses 10 percent of its weight in *merma y averia* [damage and breakage] during manufacture and transport—we should assume that 80,000 *arrobas* were consumed in the island's interior (*en los campos*) where the government has no authority. The maintenance of 120 slaves and production costs reached no more than 12,000 piasters a year, but the employees of the *Factoría* cost 541,000 piasters.[2] In good years, the 128,000 arrobas sent to Spain as cigars and snuff (*rama y polvos*) often exceeded 5 mil- I.256 lion piasters, based on Spain's fixed rate. It is surprising to see that export records for 1816 (documents published by the *Consulado*) indicate no more than 3,400 *arrobas* of exports. For 1823, records show no more than 13,900 arrobas of *leaf tobacco* (*en rama*) and 71,000 pounds of *bundled tobacco* (*torcido*), at a combined estimated value of 281,000 piasters. For 1825, we find no more than 70,302 pounds of cigars and 167,100 pounds of loose leaf-tobacco and strips. But it must be remembered that no branch of contraband is more active than that in cigars. Although the tobacco of the *Vuelta abajo* is the most renowned, considerable exports come from the island's eastern region. I am somewhat skeptical about a total export of 200,000 boxes of cigars (a value of 2 million piasters) that many travelers assume for these past few years. If har-

1. *De la Situación actual de la Real Factoría de Tabacos de la Habana en Abril 1804* (official unpublished document). In Seville, sometimes 10 to 12 million pounds of tobacco amassed, and revenue from the Peninsula's *Renta del Tabaco* reached 6 million piasters in good years.

2. According to figures from the *Royal treasury* published in 1822, housing and pensions for retired employees still came to 18,600 and 24,800 piasters a year after the *Factoría de tabacos* in Havana had closed.

vests were that plentiful, why would Cuba import tobacco for lower-class consumption from the United States?

After sugar, coffee, and tobacco, three products of great importance, I will not speak of *cotton*, *indigo*, or *wheat* on the island of Cuba. These two branches of colonial industry are of little consequence, and the proximity of the United States and of Guatemala renders competition nearly impossible. The state of El Salvador, part of the *Central-American* Federation, contributes 12,000 *tercios*, or 1,800,000 pounds, to the indigo trade, exports worth more than two million piasters. To the great surprise of those who have traveled in Mexico, wheat cultivation thrives near Cuatro Villas at elevations slightly above sea level, but it has not taken root more broadly. The flour is exquisite, but colonial crops are more attractive to workers, and the fields of the United States, this Crimea of the New-World, yield harvests far too lavish for the trade in domestic grains to be sufficiently protected by the prohibitive tariff system on this island right across from the deltas of the Mississippi and the Delaware. Similar obstacles hinder flax, hemp, and wine production. Cuba's inhabitants themselves perhaps do not even know that wine making began on their island with wild grapes during the first years of the *conquest*.[1] The types of grapes native to America have given rise to the widespread misconception that the true Vitis vinifiera is native to both continents. The *parras monteses* [mountain grapes] that yield "the slightly sour wine of the island of Cuba" were probably identical to the Vitis tiliaefolia that ▾Mr. Willdenow described from our plant collection. To this date, grapes have not been cultivated[2] with an eye toward producing wine anywhere in the northern hemisphere south of 27° 48′ or the latitude of Ferro Island, one of the Canaries, and 29° 2′, or the latitude of Bushire in Persia.

I.257

I.258

I.259

1. "Wine has been made from many mountain grapes, though it is somewhat sour" (Herrera, [*Historia de las Indias occidentales*,] Dec. I, *p.* 233). ▾Gabriel de Cabrera has discovered a tradition in Cuba that appears to be quite similar to the Semitic peoples' story of Noah experiencing the effects of fermented grapes for the first time. He adds that the idea of two races of men—one *naked*, the other *dressed*—is connected to this American tradition. Did Cabrera, preoccupied with the myths of the Hebrews, misinterpret the natives' words, or (what seems more likely) did he add to the analogies that teach us irrefutably that there were ancient traditions shared among the peoples of both worlds? Examples of those are *the woman as serpent*, the *fight between two brothers*, the *Great Flood*, ▾*Coxcox's raft*, the *exploratory bird*, and so many other myths? See my *Vues des Cordillères et Monumens des peuples indigènes de l'Amérique*, Pl. XIII and XXVI; Vol. I, pp. 114, 235, 237, 376; Vol. II, pp. 14, 128, 17[6], 177, 19[8], 392 (octavo ed.).

2. Leopold von Buch, *Physicalische Beschreibung der Canarischen Inseln*, 1825, p. 124.

WAX—It is not produced by native bees (Mr. Latreille's Melipones) but by bees introduced from Europe via Florida. This trade became important only after 1772. The entire island's exports, which were no more than 2,700 arrobas[1] in an average year between 1774 and 1779, were estimated (taking into consideration customs fraud) at 42,700 arrobas in 1803, 25,000 of which were destined for Veracruz. Mexico's churches consume an enormous amount of Cuba's wax. Prices vary from 16 to 20 piasters per arroba. According to customs registries, Havana's only exports were:

1.260

in	1815	23,398	*arrobas*
	1816	22,365	
	1817	20,076	
	1818	24,156	
	1819	19,373	
	1820	16,939	
	1822	14,450	
	1823	15,692	
	1824	16,058	
	1825	16,505	

Trinidad and Baracoa's small port also have a brisk trade in wax from the rather uncultivated areas in the island's east. Near sugar factories, many bees die by *intoxicating* themselves on the molasses of which they are extremely fond. Generally, wax production decreases as agriculture spreads. Given wax's current price, exports of this material by both legal and illegal means are worth half a million piasters.

TRADE—We have already noted elsewhere that the importance of the island 1.261 of Cuba is founded not only on the wealth of its products and its population's demand for commodities and European merchandise but also on the port of Havana's advantageous location at the entrance to the Gulf of Mexico where the great trade routes of two worlds intersect. Even when agriculture and industry were in their infancy and contributed to trade barely 2 million piasters in sugar and tobacco, Abbé Raynal[2] said that "the island of Cuba alone is worth a kingdom to Spain." These memorable words were something of a prophecy. Ever since the Spanish metropole lost Mexico, Peru, and so many other states that declared independence, statesmen discussing the Peninsula's political interests must have given serious thought to these words.

1. Raynal, [*Histoire philosophique*,] Vol. III, p. 257.
2. *Histoire philosophique*, Vol. III, p. 257.

The island of Cuba, to which the court in Madrid has wisely accorded a great deal of commercial freedom for a long time now, exports, by legal and illegal means, more than 14 million piasters worth of its only native products: sugar, coffee, tobacco, wax, and hides.[1] This is about a third of what Mexico provided in precious metals during the period[2] of its mines' greatest productivity. One could say that Havana and Veracruz[3] are for the rest of America what New York is for the United States. The capacity of the 1,000 to 1,200 merchant vessels that annually enter the port of Havana (excluding inland navigation) is growing to 150,000 or 170,000 tons.[4] Additionally, even in times of peace, one often sees 120 to 150 warships at anchor in Havana. From 1815 to 1819, products registered at this port's only customs house (sugar, rum, molasses, coffee, wax, and leather) reached 11,245,000 piasters in an average year. In 1823, exports registered at less than two-thirds their market value amounted to over 12½ million piasters (not counting 1,179,000 piasters in specie). It is very likely that the entire island's imports, both legal and illegal and estimated at the actual price of commodities, merchandise, and slaves, are worth about 15 to 16 million piasters today; barely 3 or 4 million are re-exported. Havana imports far beyond its own needs, trading its colonial commodities for manufactured goods from Europe in order to resell some of those in Veracruz, Trujillo, Guaira, and Cartagena.

I.262

I.263

I.264

Fifteen years ago, I discussed in another work[5] parts of the tables that were

1. Taking the lowest prices in recent years, we can count among these products: 380,000 cases of sugar (at 24 piasters) = 9,120,000 piasters; 305,000 quintals of coffee (at 12 piasters) = 3,660,000 piasters (Vol. XI, pp. 369, 370, 384, 385; Vol. XII, pp. 7ff [this edition, pp. 118ff]). Given commodity prices in 1810 to 1815, the exports of the island of Cuba will rise to a value of 18 or 19 million piasters. Fortunately, the sugar production rose with the decline in prices: these prices, in 1826, are barely 22 piasters a case, while, in 1801, they rose to 40 piasters.

2. In 1805, 27,165,888 piasters of gold and silver coins were minted in Mexico. But, taking the average during ten years of political calm between 1800 and 1810, we arrive at barely 24.5 million piasters.

3. In 1803: Veracruz's imports were 15 million piasters; exports (not including precious metals) amounted to 5 million piasters. In Havana, re-exports will grow after the *depot*'s establishment.

4. In 1816, the tonnage of New York's trade was 299,617, that of Boston 143,420. Ship capacity is not, however, an exact measure of commercial wealth. Countries that export rice, flour, lumber, and cotton need more vessel capacity than tropical regions, whose products (cochineal, indigo, sugar, and coffee), though they are of considerable value, take up little space.

5. *Essai politique sur le royaume de la Nouvelle Espagne*, Vol. II, p. 746; and *Relation historique*, Vol. IX, pp. 307 and 308.

published "under the misleading term *trade balance*." I emphasized the unreliability of these supposedly transparent trade accounts, which people, tempted by false principles of political economy, believe should be assessed I.265 on total cash sales only. The tables below show two years (1816 and 1823) of *Balanzas y Estados de Comercio* [trade balances and statements] compiled on order of the government. I have not changed a single figure, because they show *numbers limited to a minimum value* (and this is already an enormous advantage in the assessment of hard-to-know quantities). Prices indicated in these *statements* are neither those at the point of origin nor at the destination ports. These are fictive estimates, *official values*, as they are known in Great Britain's customs houses.[1] These are—one cannot repeat this enough—at least a third below market prices. To derive the entire island's trade volume from that of Havana, as Spanish customs *registries* do, one must know all *registered* exports and imports from all ports and add to this the results of illegal trade, I.266 which differs according to place, nature of the merchandise and year-by-year price fluctuations. Only local authorities can undertake such calculations; and what these authorities made public during their skillful battle with the Spanish Courts proves that they do not believe themselves sufficiently prepared for large-scale work that encompasses different elements and variables.

Each year, the *Junta del Gobierno* [governing board] and the *Real Consulado* [royal council] demand that receipts for all registered exports and imports for the port of Havana be compiled. This compilation for Havana alone goes under the name of Trade Balance, *Balanza de Comercio*.[2] These logs distinguish between imports carried on national (Spanish) and foreign ships; I.267 between exports destined for the Peninsula, for Spanish American ports, and for ports outside the realm of the Spanish crown. Merchandise weight and value (*valor por aforos*) and municipal and royal duties are added. Yet, as we have already noted, *official* estimates of prices are far below market rates.[3]

1. In this system, one distinguishes between the real price, the *official value*, and the *declared* or *bona fide value*.

2. These *Balanzas de Comercio* for Havana, some of which are printed with every partial value in minute detail, typically have 25 to 30 folio pages and list more than 1,800 *articles*. I own a large number of them, but in the present *Political Essay on the Island of Cuba*, I publish only those figures that point to general conclusions. I followed the same method in my *Essay politique sur le royaume de la Nouvelle Espagne*.

3. For example, imported blacks are estimated at 150 piasters per head, barrels of flour at 10 piasters. After having given the total value for the purported *trade balance*, I indicated the amount of gold and silver that only *passed through* Cuba. To give an approximate idea of the island's domestic consumption and its demand for manufactured goods from Europe, I listed the same articles in exports and imports.

Year 1816

A.	Imports				13,219,986 p.
	on	339	Spanish ships	5,980,443 p.	
			foodstuffs and merch. 1,032,135 p.		
			African slaves. 2,659,950 p.		
			in gold and silver 2,288,358 p.		
	on	672	foreign ships.	7,239,543 p.	
		1,008	ships	13,219,986 p.	
B.	Exports				8,363,135 p.
	on	497	Spanish ships.	5,167,966 p.	
			for the Peninsula 2,419,224 p.		
			for the Spanish ports in Am. 2,104,890 p.		
			to the African coasts 643,852 p.		
			5,167,966 p.		
	on	492	foreign ships	3,195,169 p.	
		989		8,363,135 p.	

1.268

Of 2,439,991 piasters of imports, registered exports accounted for only 480,840 piasters in gold and silver.

Imports can be broken down into the following figures: 71,807 barrels, or 718,921 piasters, worth of flour; 463,067 piasters worth of wine and liquor from Europe; 1,096,791 piasters of salted meats, provisions, and spices; 127,681 piasters worth of miscellaneous clothing; 282,382 piasters of silks; 3,226,859 piasters worth of canvas; 103,224 piasters worth of broadcloth and other linen items; 267,312 piasters worth of furniture, crystal, and hardware; 61,486 piasters worth of paper; 330,368 piasters worth of iron work; 135,103 piasters worth of leather and hides; and 285,217 piasters worth of planks and other types of lumber (for construction).

1.269

For *exports*, we find: 10,965 barrels, or 145,254 piasters worth of flour; 111,466 piasters worth of wine and liquor; 227,274 piasters worth of salted meats, provisions, and spices; 4,825 piasters worth of sundry clothing; 47,872 piasters worth of silks; 1,529,610 piasters worth of canvas; 29,000 piasters worth of furniture, crystal, and hardware; 20,497 piasters worth of paper; 99,581 piasters worth of iron work; 3,207,792 arrobas, or 3,962,709 piasters, worth of sugar; 370,229 arrobas, or 847,729 piasters, worth of coffee; 22,365 arrobas or 169,683 piasters worth of wax; and 19,978 piasters worth of cured leather.

YEAR 1823

A.	IMPORTS		13,698,735 p.
	on Span. ships	3,562,227 p.	
	on foreign ships	10,136,508	
B.	EXPORTS		12,329,169 p.
	on Span. Ships	3,550,312 p.	
	on foreign ships	8,778,857	

Number of ships docking in Havana: 1,125 with 167,578 tons of portage; number departing from Havana: 1,000 with 151,161 tons of portage.

In this trade registry, local products exported and registered were estimated at:

I.270

95,884	cases of white sugar
204,327	cases of light-brown sugar
672,007	arrobas of first quality coffee
223,917	arrobas of second quality coffee
15,692	arrobas of wax
30,145	bocoyes of molasses
13,879	arrobas of *leaf* tobacco (*en rama*)
71,108	pounds of bundled tobacco (*torcido*)
26,610	pieces of leather from the island of Cuba
3,368	carafes of bee honey

Imported gold and silver coins: 1,179,034 piasters; exported: 1,404,584 piasters.

Imported merchandise and commodities: 213,236 piasters worth of clothing; 2,071,083 piasters worth of canvas and linen; 459,869 piasters worth of silks; 1,021,827 piasters worth of cloth, muslin, etc.; 163,962 piasters worth of broadcloth; 3,269,901 piasters worth of salted meats, rice, other foodstuffs, and spices (including 431,464 arrobas of tasajo valued at 701,129 piasters; 309,601 arrobas of rice valued at 348,301 piasters; and 89,947 barrels of fat valued at 259,941 piasters); 74,119 barrels, or 889,428 piasters, worth of flour; I.271 1,119,437 piasters worth of wine and liquor; 288,697 piasters worth of iron work; 464,328 piasters worth of metal work, furniture, crystal, and porcelain; 35,186 reams, or 158,337 piasters, worth of paper; 53,441 arrobas or 213,764 piasters worth of Castilian soap; 42,512 arrobas or 170,050 piasters worth of tallow (sebo labrado); 353,765 piasters worth of planks and other types of lumber (for construction).

For *exports*, we will single out the following items in addition to the products listed above: 29,526 piasters worth of canvas and linen; 69,049 piasters worth of cotton fabric; 11,316 piasters worth of silks; 9,633 piasters worth of woollens; 8,046 piasters of furniture, crystal, and hardware; 63,149 piasters worth of iron work; 23,453 piasters worth of lumber (for construction); 5,572 reams or 22,288 piasters worth of paper; 49,286 piasters of wine and liquor; 86,882 piasters of salted meats, provisions, and spices; 15,322 reams or 27,772 piasters worth of paper.

Here are the most precise data I could gather on the coming and going of ships in Havana's port. From 1799 to 1803, 905 vessels, including warships, entered in an average year.

1799	883
1800	784
1801	1,015
1802	845
1803	1,020

I.272

Sugar exports weighed in at 40,000 tons. Between 1815 and 1819, the number of vessels entering in an average year was 1,192, of which 226 were Spanish and 966 foreign. In 1820, 1,305 ships docked, of which 288 were Spanish; 1,230 ships departed, 919 of them foreign. In subsequent years, only merchant vessels were counted:

	entered	*departed*	
1821.	1,268	1,168.	There were only 258 Spanish ships among these 1,268. In addition, 95 war ships docked, 53 of which were Spanish.
1821.	1,182	1,118.	Among the 1,182,843 were foreign. In addition, 141 war ships docked, 72 of them Spanish.
1823.	1,168	1,144.	Of these 1,168 (at 167,578 tons), 274 were Spanish, and 708 from the United States. The others included 149 war ships, 61 of which were Spanish, 54 were from the United States, and 34 were British or French.
1824.	1,086	1,088.	Among the 1,086 were 890 foreign ships. 129 war ships, 59 of which were Spanish, also docked in Havana.

I.273 **Exports[1] from the Island of Cuba via Havana, 1815-1819**

Years	Cases of refined sugar (at 184 kg)	Pipas of sugar cane rum	Bocoyes of molasses	Arrobas of coffee (at 11.5 kg)	Arroba of wax (at 11.5 kg)	Skins and hides	Value based on average prices in piasters
1815	214,111	3,000	17,874	918,263	23,398	60,000	11,955,705
1816	200,487	1,860	26,793	370,229	22,365	80,000	10,171,872
1817	217,076		30,759	709,351	20,076	60,000	10,691,219
1818	207,378	3,219	34,994	779,618	24,156	60,000	21,628,248
1819	192,743	2,830	30,845	642,716	19,373	60,000	10,776,997
5-yr Total	1,031,795	10,909	141,265	3,420,177	109,368	320,000	56,224,041
Average year	206,359	2,182	28,253	684,035	22,233	64,000	11,244,808

I.274 Surprisingly, a comparison of the high value of imports with the low value of re-exports in Havana's trade registries shows how considerable domestic consumption already is in a country of only 325,000 whites and 130,000 free people of color.[2] Analyzing different items in light of actual market prices yields the following: 2½ to 3 million piasters in canvas and linen (*bretañas, platillas, lienzos e hilo*); one million piasters worth of cotton cloth (*zarazas musulinas*); 400,000 piasters worth of silks (*rasos y géneros de seda*); and I.275 220,000 piasters worth of broadcloth and woollens. The island's demand for European fabrics—*registered* as exports in the port of Havana—has thus exceeded 4 to 4½ million piasters in the last few years.[3] To Havana's legal imports we should add: more than half a million piasters in hardware and furniture; 380,000 piasters in iron and steel; 400,000 piasters in planks and large pieces of construction lumber; and 300,000 piasters in Castilian soap. Imports of provisions and beverages to Havana alone, it seems to me, are worthy of attention to those who wish to know the true state of the societies known

1. In this table of exports *registered* over 5 years, the price of a case of sugar was successively estimated at 16 and 12 *reales,* 22 and 18 *reales,* 20 and 16 *reales,* and at 20 and 16 *reales;* a *pipa of rum* was estimated at 35 piasters; a bocoy of *molasses* was estimated at 7 *reales;* a quintal of *coffee* was estimated at 15, 15, 12, 16, and 16 piasters; an arroba of *wax* at 16 piasters.

2. It is clearly a numerical error that a recently published work ([Huber and Jameson,] *Aperçu statistique sur l'île de Cuba*, 1826, p. 231) counts 257,000 free people and 395,000 slaves in Cuba. The 130,000 free people of color were thrown into the same class as the 260,000 slaves, and the white population was reduced to 68,000.

3. Fabric imports (*géneros y ropas*) from Veracruz was 9,200,000 at the beginning of this century, before the Mexican Revolution. It must not be forgotten that Mexico has native factories whose products suffice for the needs of the lower classes. See above on consumption compared to Mexico and Venezuela, Vol. IX, pp. 313 and passim.

as *slave or sugar colonies*. Such is the composition of these societies established on the most fertile soil that nature can offer for the nourishment of men; such the management of agricultural enterprises and industry in the Antilles that, in the most fortunate climate of the equinoctial region, the population would fall below subsistence level without free and unimpeded foreign trade. I am not speaking of the importation of wine through Havana's port, whose volume grew (still following customs registries) to 40,000 *barrels* in 1803 and to 15,000 *pipas* and 17,000 *barrels* (a 1,200,000 piasters value). Nor am I speaking of the introduction of 6,000 *barrels* of liquor and 113,000 *barrels* (1,864,000 piasters) of flour from Spain and Holland. This wine, liquor, and flour, worth more than 3.3 million piasters, is consumed by the nation's upper classes. Grains from the United States became a genuine need in an area where corn, manioc, and bananas had been preferred to any other amylaceous [starchy] food for a long time. There would be nothing to complain about the development of European luxury in the midst of Havana's prosperity and growing civilization save for the fact that, in 1816, there were 1½ million piasters and, in 1823, 3½ million piasters of salted meat, rice, and dried legumes imported along with European flour, wine, and liquor. In 1823, rice imports were 323,000 *arrobas* (still following Havana's customs registries and ignoring contraband). Cured and salted meat (*tasajo*) imports, so necessary to the slaves' diet, reached 465,000 *arrobas*.[1]

1.276

1.277

This lack of food characterizes a part of the tropical regions where the Europeans' imprudent activities have upset the natural order. It will diminish proportionately as the inhabitants become more aware of their true interests, are discouraged by the low price of colonial commodities, and diversify their plantings to allow all branches of rural industry to take wing. The principles of narrow-minded and petty politics that dominate the administration on very small islands—veritable workshops dependent upon Europe and inhabited by men who desert the soil as soon as it has sufficiently enriched them—will not do for an island nearly as large as England, covered by teeming cities, and whose inhabitants, going back from son to father for centuries, are far from seeing themselves as strangers on American soil, who in fact cherish it as their true homeland. Through its consumption alone, the population of the island of Cuba, which will probably exceed one million fifty years from now, could open an immense field to native industry. If the slave trade ceases altogether,

1.278

1. In Havana's *balanza de comercio* [trade balance] (1823), even *official rates* reach 755,700 piasters for *tasajo*, 363,600 piasters for rice; 223,000 piasters for pork; 373,000 piasters for lard, butter, and cheese; and 100,000 piasters for the salted cod they give to the blacks along with the *tasajo*.

slaves will gradually pass into the class of free men and the society, having rebuilt itself without any of the violent civil upheavals, will return to the paths nature has traced for all societies that have become populous and enlightened. Sugarcane and coffee cultivation will not be abandoned; but they will no longer serve as the primary basis for national existence, as ▾cochineal does in Mexico, indigo in Guatemala, and cacao in Venezuela. A free and intelligent I.279 yeomanry will gradually replace a slave population lacking in foresight and industry. The capital that Havana's commerce delivered into planters' hands during the past twenty-five years has already begun to change the country. The power of capital, whose influence is ever growing, is necessarily tied to another power equally indispensable to the progress of industry and national wealth: the growth of human intelligence. The destiny of the center of the Antilles depends on the unification of these two powers.

We have seen that, according to Havana's trade tables, registered exports grew by an average of 12,245,000 piasters in island products between 1815 and 1819 and by 13 million piasters in recent years.[1] If registered exports in native I.280 products and foreign merchandise re-exported from Havana and Matanzas in 1823 were taken together—15,139,200 piasters[2]—then it would not be an exaggeration to presume that the entire island must have exported more than 20 to 22 million piasters by legal and illegal means in the same year, 1823, when trade was very active.[3] Estimates naturally vary with merchandise and crop prices. Before Jamaica came to enjoy free trade in 1820, its exports were 5.4 million pounds sterling. It is generally believed that Spain receives forty to I.281 fifty thousand cases of sugar annually from Havana (in 1823, statements show 100,766 *caxas*; in 1825, only 47,547). The United States[4] attracts more than half of Cuba's trade by tonnage and more than a third by price. We estimated the island's total imports to be greater than 22 to 24 million piasters, includ-

1. I record here estimates derived from *market prices* in the port of Havana not from customs prices.

2. In the highly valued work that appeared under the title *Commerce du dix-neuvième siècle* [by Moreau de Jonnès], Vol. I, p. 259, these exports from Havana are estimated at less than 2 million piasters in 1823; but this estimate is due to a numerical error. Registered sugar was 300,211 *caxas*, or 120,084,400 Spanish pounds, and not 6 million pounds. Coffee exports were 22,398,100 Spanish pounds, and not 3 million pounds (Vol. XI, pp. 366, 367, and Vol. XII, pp. 7ff [this edition, pp. 118ff.]).

3. In 1788, exports from the French part of Saint-Domingue were 67 million francs in sugar, 75 million francs in coffee, and 15 million francs in cotton, a total of 51,400,000 piasters.

4. According to official documents, total imports for the United States equaled 62,586,724 dollars in 1820, Great Britain and India having furnished 29 million, the *island of Cuba* 6,584,000, Haiti 2,246,000, and France 5,909,000 dollars.

ing contraband. In 1822, the value of merchandise and products coming from the United States on ships of 106,000 tons[1] was 4,270,600 dollars. According to Mr. Stewart, Jamaica's imports rose to 2 million pounds sterling in British manufactured goods in 1820. I.282

Logged flour imports[2] to the port of Havana:

1797	62,727	*barrels* (at 7.25 *arr.*, or 84 kg)
1798	58,474	
1799	59,953	
1800	54,441	
1801	64,703	
1802	82,045	
1803	69,254	

In 1823, 38,987 barrels of registered imports came from Spanish ships, and 74,119 barrels came from foreign ships to the port of Havana alone, for a total of 113,506 barrels at an average price of 16½ piasters (including duties); 1,864,500 piasters altogether. We have the prudent administration of Governor Don Luis de las Casas[3] to thank for the first direct imports of flour from the United States. Until then, this flour could only be imported *after having passed through European ports*! ▾Mr. Robinson[4] estimates total legal and illegal imports of this commodity to the different parts of the island at 120,000 barrels. He adds something that seems less certain to me: "that the island of Cuba, because of the poor division of black labor, is so lacking in provisions that it would not be able to survive a five-month blockade." In 1822, the United States exported to the island of Cuba 144,980 barrels (more than 12 million kg) whose value in Havana (including duties) reached 2,391,000 piasters. Despite the seven-piaster tax charged for each barrel of flour imported from the United States to the island of Cuba, Peninsular (specifically Santander) flour cannot compete. This competition had begun under the most I.283

1. [Huber and Jameson,] *Aperçu statistique de l'île de Cuba*, 1826 (Table B). ▾Mr. Huber added much important information on trade and Cuba's customs system to the translation of *Letters from the Havana*. Imports of 4,270,600 dollars can be regarded as quite considerable, for in 1824 imports from Great Britain to Mexico, Colombia, Buenos Aires, Chile, or Peru reached only 2,377,110 pounds sterling ([▾Núñez,] *An Account of the United Provinces of Río de la Plata*, 1825, p. 172).

2. In 1820, the United States exported roughly 9,075,000 dollars in wheat and corn flour. Flour exports experience extraordinary fluctuations. In 1803, they were 1,311,853 barrels; in 1817, 1,479,198; in 1823, 756,702 barrels.

3. See Vol. XI, p. 303 [this edition, p. 71].

4. Robinson, *Memoirs of the Mexican Revolution*, Vol. II, p. 330.

1.284 fortunate of auspices: during my stay in Veracruz, 300,000 piasters worth of Mexican flour were already being exported from this port. According to ▾Mr. Pitkin, this amount grew to 27,000 barrels or 2,268,000 kg in 1809. Mexico's political troubles completely disrupted the grain trade between the two countries, both situated in the Torrid Zone but at variable elevations above sea level, which has a significant impact on climate and culture.

Recorded imports of alcoholic beverages in Havana were:

1797.	12,547 *barrels* of wine	2,300	*bar.* of liquor
1798.	12,118	2,412	
1799.	32,073	2,780	
1800.	20,899	5,592	
1801.	25,921	3,210	
1802.	45,676	3,615	
1803.	39,130	3,553	

To complete this exposition on foreign trade, let us consider the author of a report that I have cited many times, who sets forth the island's true situation: 1.285 "In Havana, all the effects of growing wealth are beginning to be felt. Staples doubled in price in just a few years. Manual labor is so expensive that a *bozal* recently imported from Africa's shores earns 4 to 5 reales (2 francs and 13 sols to 3 francs and 5 sols) a day from the work of his hands alone (without having learned a trade). Blacks who practice a mechanical trade, however basic it might be, earn 5 to 6 francs. Patrician families remain tied to the soil: a man who has enriched himself does not return to Europe to spirit away his capital. Some families are so powerful that the recently deceased ▾Don Mateo de Pedroso left more than two million piasters in land. Many commercial houses in Havana buy ten to twelve thousand cases of sugar a year for which they pay a going rate of 350,000 to 420,000 piasters. Business in this place earns more than twenty million piasters annually" ([Valle Hernández,] *De la situación presente de Cuba*, unpublished). Such was the state of public wealth at the end of 1800. Twenty-five years of growing prosperity have elapsed since 1.286–87 then. The island's population has nearly doubled. Before 1800, registered sugar exports had not reached 170,000 cases (31,280,000 kg) in any given year. Recently,[1] they have consistently exceeded 200,000 cases and have even

1. Ever since the Court in Madrid resolved to open several ports in the western part of the island to Spanish and foreign trade, sugar exports registered in Havana customs should not be considered an exact measure of agricultural wealth. The port of Mariel, so useful to the Guanajay district's planters, had already received its *habilitación* [authorization] (a technical term from Spanish commercial law) by the *royal cédula* of October 20, 1817 but it is only in the last five or six years that exports from Mariel have appreciably influenced Ha-

reached 250,000 to 300,000 cases (46 to 55 million kg). Coffee plantations, a new branch of industry furnishing exports worth 3½ million piasters, have sprung up. Guided by greater enlightenment, this industry has been better managed. The tax system that had weighed upon national industry and foreign trade was shaken up after 1791 and has been improved by subsequent changes. Each time that the Spanish metropole, misrecognizing its own interests, wanted to take a step backwards, courageous voices not only among *Habaneros* but as often among Spanish colonial administrators spoke up to defend the cause of free trade in America. Recently the intendant ▾Don Claudio Martínez [de] Pinillos's enlightened efforts and patriotic views have opened a new avenue for investment. Havana has been accorded the status of open port under highly advantageous conditions.[1] I.288

The island's interior traffic, which is difficult and costly, pushes up the price of products in the ports, despite the small distance between the north and south coasts. A canal building plan, bringing together the double advantage of connecting Havana and Batabanó with a navigable pathway and lowering the cost of transport and domestic production, deserves special mention here. The idea of a Güines Canal[2] was conceived over half a century ago with the simple aim of providing construction wood for the carpenters of Havana's arsenal at a more modest price. In 1796, Count Jaruco y Mopox, an amiable and enterprising man whose connections with the prince of Peace gave him great influence, undertook the responsibility to revive this project. In 1798, I.289 two highly skilled engineers, ▾Don Francisco and Don Felix Lemaur, undertook dredging. These officers understood that the entire canal would be 19 leagues (5,000 varas or 4,150 meters) in length, that the departure point would be *Taverna del Rey*, and that there would have to be 19 locks to the north and 21 to the south. The distance between Havana and Batabanó is no more than

vana's. The government has also extended franchises to other ports, for instance Baracoa (December 13, 1816), San Fernando de Nuevitas in the Bagá and Guiros Estuary (April 5, 1819), Guantánamo Bay (August 13, 1819) and San Juan de los Remedios which one can consider the district of Villa Clara's port (September 23, 1819). *Jagua Bay*, where ▾Don Luis de Clouet started a farming and trade settlement by bringing former colonists from Louisiana and other white and free men there, has not yet been *certified* (*Memorias de la Sociedad Económica de la Habana*, no. 34, pp. 287, 293, 297, 300, and 303).

1. *Acuerdos sobre arreglo de derechos y establecimiento de Almacenes de Depósito* (see Suplemento al Diario del Gobierno constitucional de la Habana del 15 de octubre 1822). Without *the port of Havana's fortunate franchise*, Jamaica would have become the center of all mercantile activities with the neighboring continent.

2. Dredging would be (in Burgos feet): 106.2 in Cerro near Zanja bridge; 329.3 in Taverna del Rey; 295.3 in Pueblo del Rincón; 237.3 in Saldívar lagoon, when it is full; 166.1 in Quibicán; and 21.3 in Batabanó village.

8⅓ nautical leagues in a direct line.[1] *Even as a canal for secondary navigation*, the Güines Canal would be of great use for the transportation of agricultural products by steamship,[2] because it would be near the best cultivated lands. Nowhere are roads worse during the rainy season than in this part of the island where the soil consists of a crumbly limestone hardly appropriate

I.290 for ▾railroad construction. Today, sugar transport costs a piaster per quintal from Güines to Havana, a distance of 12 leagues. Besides having the advantage of connecting the island internally, the canal would also give great importance to the Surgidero de Batabanó in which small freighters laden with Venezuelan salted meats (*tasajo*) would enter without needing to round Cape San Antonio. In the bad season and in times of war, when pirates cruise between Cape Catoche, the Tortugas, and Mariel, one is happy to be able to curtail the trip from the mainland by entering Cuba not via Havana but via a port along the southern coast. Construction costs for the Güines Canal were estimated at 1 or 1.2 million piasters in 1796. Today, costs are expected to rise above a million and a half. Products that could pass through the canal were estimated at 75,000 cases of sugar, 25,000 *arrobas* of coffee, 8,000 *bocoyes* of molasses and rum. The first plan, of 1796, intended to connect the canal to the small Güines River that one would take from the Ingenio de la Holanda to Quivi-

I.291 can, 3 leagues to the south of Bejucal and Santa Rosa.[3] Today, this idea has been abandoned, because the Güines River loses its waters to the east to the irrigation of the savannas of the Hato [cattle ranch] de Guanamón. Instead of building the canal to the east from the Barrio del Cerro and to the south from Atarés Fort in Havana Bay, one wanted first to make use of the Chorrera riverbed or the Armendaris River from Calabazar to Husillo, and then of the Zanja Real, not only to allow boats to pass between the *arrabales* [suburbs] and the city of Havana proper but also to provide water for the wells that lack it for three months of the year. I was privileged to visit the plains through which this navigation route should pass several times with the Lemaur brothers. The project's usefulness is incontrovertible, provided a sufficient amount of water can be brought to the summit level during times of extreme drought.

In Havana, people complain about harmful effects of this growth on the *old ways*, as people do wherever trade and the wealth it produces increases

I.292 quickly. This is not the place to compare what Cuba was like when the island was covered with pastureland before the British invaded the capital and what

1. See Vol. XI, p. 219.

2. Steamships already travel along the coast from Havana to Matanzas, and less regularly from Havana to Mariel. The government granted a *steamship* monopoly to ▾Don Juan de O'Farrill (March 24, 1819).

3. Official papers from the *Comisión para el fomento de la Isla de Cuba*, 1799, and ▾Mr. Bauduy's unpublished notes.

it has now become as the metropole of the Antilles. Nor is this the place to contrast the frankness and simplicity of an emerging society's customs with the mores that develop in an advanced civilization. The spirit of trade, carrying with it the cult of wealth, probably leads people to devalue what money cannot buy. Yet, human affairs are fortunately such that man's most desirable, noblest, and freest qualities come from the soul's inspirations and the intellect's development and improvement. The cult of wealth, were it to seize all levels of society absolutely, would inevitably lead to the evil that those lament who look with sadness upon what they call the domination of the industrial system. But trade's expansion itself will furnish remedies for these supposed dangers by multiplying relations between peoples, opening an immense sphere to the workings of the mind and spirit, contributing capital to agriculture, and creating new needs through refinements in luxury. In this complex entanglement of cause and effect, time is needed to reestablish a balance between the classes. Surely we cannot claim that an era's civilization, enlightenment's progress, and the development of public reason can be measured by *tonnage*, the value of exports, or the improvement of industrial techniques. Peoples, like individuals, must not be judged by one stage of their lives alone. They fulfill their destinies only by climbing the entire civilizational ladder appropriate to their national character and their physical circumstances.

I.293

FINANCE—In recent years, the growth of agricultural wealth on the island of Cuba and the rise of fortunes that affect import prices have increased public revenues to four and a half or perhaps even five million piasters. ˇEver since the decrees of free trade went into effect,[1] Havana customs, which had brought in less than 600,000 piasters before 1794 and 1.9 million in an average year between 1797 and 1800, have been contributing more than 3.1 million piasters in net revenue (*importe líquido*) to the general Treasury. Because the colonial government permits the greatest transparency regarding Cuba's finances, one can see from the *budgets* of the *Cajas matrices de la Administración general de Rentas* for the city and the jurisdiction of Havana that public revenue fluctuated between 3.2 million and 3.4 million piasters between 1820 and 1825. If we add to this sum, on the one hand, 800,000 that flow directly into the *Tesorería general* [general treasury] from other revenue streams[2] (*directa entrada*) and, on the other hand, the custom's revenue from Trinidad, Matanzas, Baracoa,

I.294

I.295

1. In Haiti, Port-au-Prince customs yielded the sum of 1,655,764 piasters in 1825; Buenos Aires yielded 1,655,000 piasters in an average year between 1819 and 1821. See *Centinela de la Plata* (September 1822), no. 8. *Argos de Buenos Aires*, no. 85.

2. Lottery, *Renta décimal* [tithes], etc.

and Santiago de Cuba, which had risen to more than 600,000 piasters before 1819, we see that a five-million piaster, or 25 million franc, estimate for the whole island[1] is not exaggerated. Very simple comparisons prove how considerable this sum is relative to the colony's current situation. The island of Cuba still has no more than one forty-secondth of France's population, and half of its inhabitants live in frightful poverty and consume very little. Its revenue nearly equals the republic of Colombia's[2] and exceeds the whole of the United States' customs income[3] before 1795, a period when that federation already had 4.5 million inhabitants while the island of Cuba had no more than 715,000. Customs fees are the primary source of public revenue in this beautiful colony: They alone produce more than three-fifths of that revenue and are largely adequate for internal administration and military defense. If in recent years, spending from Havana's Treasury has risen to more than four million piasters, these excess expenses are due only to the unrelenting fight that the Spanish metropole wished to pursue against the emancipated colonies. Two million piasters were earmarked for the payroll of an army and navy that returned from the American continent to the Peninsula by way of Havana. As long as Spain ignores its true interests by not recognizing the new republics' independence, the island of Cuba, threatened by Colombia and the Mexican Federation, must maintain a military that is ruinous to colonial finances. The Spanish navy, stationed in Havana, generally costs more than 650,000 piasters. The army demands more than 1½ million piasters per year. Such a state of affairs cannot persist for long, if the Peninsula does not lighten the yoke that weighs on the colony.

I.296

I.297

From 1789 to 1797, Havana's customs revenues never rose above 700,000 piasters in an average year, because the royal duties (*rentas reales*) that flowed into to the Treasury's coffers were as follows:

1789	of	479,302	Piasters
1790	—	642,720	
1791	—	520,202	

1. Deputies from the island of Cuba declared to the Spanish Cortes (May 1821) that the total sum of contributions "in the province of Havana alone" rose to five million strong piaster ([Zayas and Benítez,] *Reclamación [hecha por los representantes de la isla de Cuba] contra la ley de aranceles [sobre las restricciones que ésta impone al comercio de dicha isla]*, p. 7, no. 6). In 1818 and 1819, total receipts for the *general Treasury* were already 4,367,000 and 4,105,000 piasters, and spending was between 3,687,000 and 3,848,000 piasters.

2. See Vol. IX, pp. 403 and 404. "In 1530, this Island had revenue of 6,000 gold pesos." Herrera, *Historia de las Indias occidentales*, Vol. IX, p. 367.

3. In 1815, United States customs, which had brought in up to 16 million dollars between 1801 and 1808, took in no more than 7,282,000 dollars. Morse, *New System of Modern Geography*, p. 638.

1792	—	849,904
1793	—	635,098
1794	—	642,320
1795	—	643,583
1796	—	784,689

From 1797 to 1800, royal and municipal duties collected in Havana came to 7,634,126 piasters or 1,908,000 piasters in an average year: 1.298

1797	1,257,017	piasters
1798	1,822,348	
1799	2,305,080	
1800	2,249,680	
1801	2,170,970	
1802	2,400,932	
1803	1,637,465	

Havana customs revenue:

1808	1,178,974	piasters
1809	1,913,605	
1810	1,292,619	
1811	1,469,137	
1814	1,855,117	

The decline in customs revenue in 1808 was attributed to the *embargo* placed on American ships.[1] But in 1809, Spain permitted free entry of neutral foreign ships.[2]

From 1815 to 1819, royal duties in Havana's port were 11,575,460 piasters, and municipal duties were 6,709,347, for a total of 18,284,807 piasters, or 3,657,000 piasters in an average year, of which municipal duties accounted for 56 percent. 1.299

Years	Number of Ships, arrived and departed	Royal Duties	Municipal Duties
1815	2,402	1,851,607 p.	804,693 p.
1816	2,252	2,233,203	971,056
1817	2,438	2,291,243	1,429,052
1818	2,322	2,381,658	1,723,008
1819	2,365	2,817,749	1,781,530

The Havana jurisdiction's *Administración general de Rentas* rose to:

1. *Patriota Americano*, Vol. II, p. 305.

2. [Zayas y Chacón and Benítez,] *Reclamación [hecha por los representantes de la isla de Cuba] contra los leyes de aranceles*, p. 8.

1820	3,631,273	piasters
1821	3,277,639	
1822	3,378,228	

In 1823, royal and municipal import duties reached 2,734,563 piasters. In 1824, the *Administración general de Rentas* of the jurisdiction of Havana re-

1.300 ceived public revenues as follows:

I.	Import taxes		1,818,896	piasters
	▼*Almojarifazgo.*	1,817,950 p.		
	Alcabala	802		
	Armada	144		
II.	Export taxes		326,816	
III.	Coastal trade and various other branches (salt, 27,781 p.; depot fees, 154,924 p.; *media, anata, armadilla,* etc.); total.		188,415	
IV.	*Rentas de tierra* (tax on slaves, 73,109 p.; on selling land or *fincas,* 215,092 p.; other administration, 154,840 p.; shops or *pulperías,* 19,714 p., etc.); total		473,686	
V.	Auxiliary branches of the *Tesorería del Ejercito* (*Almirantazgo, Registros estrangeros,* etc.)		136,923	
VI.	*Consulado, Cuartillo adicional del muelle, Vestuario de milicias,* etc.		80,564	
	Total revenue in 1824		3,025,300	piasters

1.301 In 1825, the revenue for the city and district of Havana amounted to 3,350,300 piasters.

These partial data show that public revenue increased sevenfold from 1789 to 1824. This growth becomes still more remarkable when one focuses on the income of ten administrations in the *Tesorerías subalternas interiores* [interior subtreasuries] (Matanzas, Villa Clara, Remedios, Trinidad, Sancti Spíritus, Puerto Príncipe, Holguín, Bayamo, Santiago de Cuba, and Baracoa). ▼Mr. Barrutia[1] published an interesting table on these provincial administrations, covering a period of 83 years from 1735 to 1818. The total revenue for 10 cases gradually rose from 900 piasters to 600,000 piasters.

1735	898	piasters
1736	860	
1737	902	
1738	1,794	
1739	4,747	
Yearly average	1,840	

1. *Memorias de la Real Sociedad económica de la Habana*, no. 31, p. 220.

1775	123,246	piasters	1.302
1776	114,366		
1777	128,303		
1778	158,624		
1779	146,007		
Yearly average	133,315		

1814	317,699	piasters
1815	398,676	
1816	511,510	
1817	524,442	
1818	618,036	
Yearly average	474,072	

The 83-year total was 13,098,000 piasters, of which Santiago de Cuba contributed 4,390,000, Puerto Príncipe 2,224,000, and Matanzas 1,450,788.

According to the statements of the *Cajas matrices*, public revenue in 1822 for Havana province alone amounted to 4,311,862 piasters of which 3,127,918 came from customs; 601,898 from *ramos de directa entrada* [direct sources], such as lottery, tithes, etc.; and 581,978 were drawn against *Consulado* and *Depósito* funds. In the same year, the outlays of the island of Cuba were: 2,732,738 piasters and 1,362,022 piasters in aid to sustain Spain's fight against the continental colonies that had declared independence. The former category of expenses includes 1,355,798 piasters for the maintenance of the garrison charged with the defense of Havana and its environs and 648,908 piasters for the royal navy in Havana's port. The second category, separate from the local administration, includes 1,115,672 piasters for the maintenance of 4,234 soldiers who, after having withdrawn from Mexico, Colombia, and other formerly Spanish parts of the continent, passed through Havana on their way back to Spain; and 164,000 piasters for the defense of San Juan de Ulúa castle. Cuba's intendant, Don Claudio Martínez de Pinillos, makes the following observation in one of the notes that accompany the 1822 *Estado de las Cajas matrices*: "If one adds to the extraordinary 1,362,022 piasters that were used to finance the Spanish monarchy's general interests the larger part of the 648,908 piasters earmarked on the one hand for the maintenance of the royal navy, whose mission is not limited to Havana's defensive needs, and for expenses from maritime couriers and warships on the other, then it becomes clear that 2,010,930 piasters (nearly half of public revenue) are eaten up by expenses not directly related to the island's domestic administration." How this country's culture and its wealth would progress if during a period of domestic tranquility, more than a million and a half piasters a year were applied toward the public good and especially toward buying the freedom of hardworking slaves,

1.303

1.304

as is already the case in the republic of Colombia, thanks to its sensible and humane legislation!

In documents I gathered in the archives of the Viceroyalty of Mexico, I saw that, at the beginning of the nineteenth century, the Treasury of New Spain provided Havana with annual funds in the following amounts:

Navy	a) for the squadron, the docks, and all of the royal navy's necessities, according to a January 16, 1790 cédula	700,000 p.
	b) for the maritime establishment on the Mosquito coast	40,000
Army	a) for the Havana army on land, according to the cédulas of May 18, 1784, Feb. 4, 1788, and November 1, 1790	290,000
	b) for the Santiago de Cuba army on land	146,000
Fortifications,	according to the royal cédula of February 4, 1788	150,000
Tobacco,	that is, purchase of leaves and manufacture of tobacco destined for Seville, according to the cédulas of August 2, 1744 and December 22, 1767	500,000
	Total	1,826,000 p.

I.305

To this sum of *nine million francs* that are now Havana's responsibility may be added 557,000 piasters that Mexico paid to bolster Louisiana's Treasury; 151,000 piasters to back up Florida's; and 377,000 piasters to bolster Puerto Rico's.

Here, I end the *Political Essay on the Island of Cuba*, in which I have recounted the state of this important Spanish possession as it is today. As a historian of America, I wanted to clarify facts and specify ideas by means of comparisons and statistical data. An investigation of such minute detail seems necessary at a point in time when the enthusiasm that inspires benign gullibility on the one hand and, on the other, spiteful passions that threaten the new republics' safety has given rise the vaguest and most erroneous observations. It was my plan from the start to abstain from all reasonable speculations about future developments or probable changes that a shift in foreign policy might bring about in the Antilles. I have examined only what pertained to the organization of human societies: the unequal distribution of rights and of life's enjoyment and the threats and dangers that legislators' wisdom and voluntary moderation can avert—whatever the form of government. It befits the traveler, who witnessed up close the torment and degradation of humanity, to bring the laments of the wretched to the ears of those who have the power to

I.306

assuage them. I have observed the conditions and circumstances of blacks in countries where laws, religion, and national customs tend to soften their lot. Nevertheless, when leaving America, I still harbored the same hatred for slavery with which I had left Europe. It is in vain that wily writers have invented phrases such as *black peasants of the Antilles*, *black vassalage*, and *patriarchal protection* in order to veil institutionalized barbarity in ingenious linguistic fictions. It is a profanation of the noble arts of the spirit and the imagination to use illusory compromises and misleading sophistries to excuse the excesses that afflict humanity and bring about violent upheavals. Do people really consider themselves exempt from compassion when they compare[1] the status of black people with that of medieval serfs or with the yoke under which certain classes of humans in northern and Eastern Europe still groan? In the times in which we live, the comparisons, the rhetoric, and the disdainful impatience with which some shrug off as chimerical even the hope for slavery's gradual abolition are useless weapons. The great revolutions that the American continent and the Antillean archipelago have seen since the beginning of the nineteenth century have affected ideas and public reason even in countries where slavery exists and is now beginning to change. Many reasonable men with a vested interest in the political stability of the *sugar and slave islands* sense that a free accord among owners, along with measures initiated by those familiar with the localities, can overcome a state of crisis and malaise, whose dangers are exacerbated by indolence and obstinacy. I will conclude this section by forecasting the prospects of such measures, and I will prove, through citations taken directly from official documents, that the local authorities in Havana which are most closely connected to the metropole have, on occasion, been

1.307

1.308

1.309

1. These comparisons calm only those secret partisans of the slave trade who seek to numb themselves to the afflictions of the black race and thus resist, in a manner of speaking, all emotions that might surprise them. Often one confuses a caste's permanent condition, founded upon legal and institutional barbarity, with an excess of power temporarily wielded over a few individuals. This is why ▾Mr. Bolingbroke, who lived in Demerary and visited the Antilles, does not hesitate to repeat "that on board a British warship, the whip is used more often than on the plantations of the British colonies." He adds "that, normally, one seldom whips blacks but has thought up other, more reasonable correctional methods, such as forcing them to eat boiling, heavily spiced soup or to drink a solution of Glauber salt with a small spoon." The slave trade strikes him as a *universal benefit*, and he is convinced that if one were to let the blacks, who enjoyed "all the comforts of a slave's life" in Demerary for twenty years, return to the African coast, they would serve as wonderful recruiters, bringing entire nations under British domination (*Voyage to Demerary*, 1807, pp. 107, 108, 116, 136). Here, we have a clear example of the very stubborn, naïve *colonist's faith*. Nevertheless, as many other passages in his book prove, Mr. Bolingbroke is a moderate man full of kind intentions toward slaves.

favorably disposed toward improving the conditions of black people, long before foreign affairs could have had any influence on their opinions.

Slavery is possibly the greatest evil ever to have afflicted humanity, no matter if one focuses on the individual slave ripped from his family in the country of his birth and thrown into the hold of a slave ship[1] or considers him as part of the herd of black men penned up in the Antilles. Still, there are degrees of suffering and deprivation. What a difference there is between a slave who works in a rich man's house in Havana or Kingston, or who works for himself and gives his master only a daily amount, and a slave who labors in a sugar factory! The threats with which masters attempt to discipline an unruly black show the degrees of human depravity. The *calesero* [coachman] is threatened with the *coffee plantation*, the *cafetal*, while the slave who works on a *cafetal* is threatened with the *sugar plantation*. On the sugar plantation, the black man—affectionate like most Africans are—who has a wife and lives in his own cabin finds comfort after work in the midst of an impoverished family. His lot cannot be compared to that of a slave who is isolated and lost in the crowd. This difference is lost on those who have never set foot on the Antilles. Gradual improvements even among the enslaved caste explain how the masters' luxuries and the possibility of gain through work could have drawn[2] more than 80,000 slaves to Cuba's cities, and how manumission, which sensible laws favor, could have become so effective that, in our day, it has produced more than 130,000 free people of color. The colonial administration will find ways of improving the conditions of the blacks by considering each class' relative position, by rewarding intelligence, love of work, and domestic virtues according to a descending scale of dispossession. Philanthropy does not mean "a little more cod and fewer lashes of the whip." Genuine improvement for the enslaved class must consider the human condition as a whole, both moral and physical.

The European governments that value human dignity and know that injustice carries within it the seeds of destruction can take the lead. Yet, their leadership (regretfully) will have no effect if landowners, colonial assemblies, or *legislatures*, do not adopt the same views and do not act according to well-

1. "If one whips the slaves," said one of the witnesses in a *parliamentary inquiry* in 1789, "to make them dance on the slave ship's deck, or one forces them to sing in choir: *messe, messe, mackerida* (that one should live happily among whites), this only proves the care we show for their health." Such delicate care reminds me of a description I have of an auto-da-fé wherein one boasts of the generosity with which refreshments are distributed to the condemned and of "these steps erected by the inquisition's friends in the midst of the pyre for the comfort of the *relajados* [recidivists]."

2. See Vol. XI, p. 300 [this edition, p. 70].

orchestrated plans, whose final end is slavery's cessation in the Antilles. Until then, one can count the lashes of the whip, decrease the number of lashes that a person can receive at any one time, require the presence of witnesses, and name slave protectors. But all these regulations, created with the best of intentions, are easy to evade. The plantations' isolation renders their enforcement impossible. Such regulations presuppose a system of domestic control that is incompatible with what one refers to, in the colonies, as "acquired rights." The conditions of slavery in their entirety cannot be improved peaceably without the combined efforts of the free men in the Antilles (whites and men of color); of colonial assemblies and *legislatures*; of those who enjoy high moral standing and positions of influence among their fellow-citizens and who, because they understand the locales, know how to calibrate the means of improvement to each island's mores, customs, and circumstances. Preparing this work for a large part of the Antillean Archipelago would benefit from looking back and weighing the circumstances under which many humans were freed in medieval Europe. When one wishes to improve a situation without causing upheavals, it is necessary to let new institutions grow from institutions that have evolved during centuries of barbarism. One day, people will hardly believe that, before 1826, no law existed in any of the Greater Antilles against selling young children and separating them from their parents, or prohibiting the degrading practice of branding blacks with a hot iron simply to be able to identify the human chattel more easily. To remove even the possibility of such barbaric practices, I offer the following as the most urgent subjects for colonial legislators: enact laws that fix the numbers of female blacks, and of blacks in relation to each other, for each sugar plantation; grant freedom to every slave who has served 15 years and every black woman who has raised 4 or 5 children; emancipate both under the condition that they work a certain number of days for the plantation's profit; give slaves some of the net profits to incentivize their interest in the growth of agricultural wealth;[1] and set aside

I.313

I.314

1. Already in 1785, ▾General Lafayette, whose name is connected to everything that promises to contribute to human freedom and the improvement of the human condition through institutions, propagated a plan to buy a settlement in Cayenne to be shared among the blacks who farm it, and whose owner would renounce all gain for himself or his descendents. He had interested preachers from the Holy Spirit Mission, who owned land in French Guiana, in his noble enterprise. A letter to the ▾Marshal of Castries dated June 6, 1785 proves that the unfortunate King Louis XVI, dispensing his compassionate intentions even to blacks and free people of color, had ordered similar experiments to be undertaken at the government's expense. ▾Mr. de Richeprey, charged with dividing the land among the blacks by Mr. de Lafayette, died as a result of Cayenne's climate.

I.315 a certain amount of public funds in the *budget* for buying slaves' freedom and improving their lives.

The Spanish *conquest* of the American continent and the slave trade in the Antilles, in Brazil, and in the southern United States have thrown together in the same place the most heterogeneous of populations. Yet, this strange combination of Indians, whites, blacks, people of different racial admixtures [métis and mulâtres], and *zambos* seems to be part of the dangers that strong and unrestrained passions produce during these hazardous times, when a society, shaken to its foundations, is on the brink of a new era. The *colonial system*'s hateful logic of safety founded upon enmity between castes, which has been propagated for centuries, is now exploding with violence. Fortunately, the number of black people is so insignificant in the new continental states that, except for ▾the cruelties in Venezuela, where the royal faction had armed the slaves, vengeful acts on the part of the enslaved population have not sullied the fight between the independents and the loyalist soldiers. Free men I.316 of color (blacks, mixed-race, and *mestizos*) have embraced the national cause warmly. ▾The copper-colored indigenous race, however, in its timid suspicion and its mysterious indifference, has remained aloof from these movements, even though it will benefit from them in spite of itself. Long before the revolution, Indians were free, poor farmers. Isolated by language and customs, they lived separately from whites. If, in disregard of Spanish laws, the greed of the *chief magistrates*, *corregidores*, and the *missionaries*' meddlesome practices often hampered Indians' freedom, there was still a big difference between this oppressive state, black chattel slavery, and the serfdom that peasants experience in Europe's Slavic regions. The small number of blacks and the freedom of the aboriginal races, of whose representatives America preserves eight and a half million without any admixture of foreign blood, characterizes the former Spanish continental possessions and renders their moral and political predicament entirely different from that of the Antilles, where the *logic of the colonial system* could develop with greater energy because of the dispropor- I.317 tion between free men and slaves. In this archipelago, as in Brazil—two parts of America that have more than three million two hundred thousand slaves—the fear of a reaction among blacks and of the perils that surround whites have been, to this day, the most powerful force behind the metropoles' safety and the ▾survival of the Portuguese dynasty. Can this safety, by its very nature, last very long? Does it justify the inaction of governments that neglect to remedy the evil while there is still time? I doubt it. When fears will have weakened under the influence of extraordinary circumstances, and when countries in which the accumulation of slaves has created an explosive mix of heterogeneous elements are dragged into an external conflict, perhaps despite themselves, civil strife will erupt in all its violence, and European families, who are not responsible for a social order not of their own making, will face the most imminent of dangers.

One cannot praise enough the intelligent legislation of Spanish America's new republics, which, since their inception, have been seriously concerned with slavery's total cessation. In this respect, this vast part of the earth has an immense advantage over the South of the United States, where, during the fight against Britain, whites established their freedom for their own profit and where the slave population, already at a million and six hundred thousand, grows still more rapidly than the white population.[1] If civilization moved ahead rather than just spread; if, following great and appalling upheavals in Europe, the part of America between Cape Hatteras and the Missouri River became the foremost home of Christianity's beacon, what a spectacle this center of civilization would offer when, in freedom's sanctuary, one were able to attend a *slave sale after the death of the master*, hearing parents wail as they are torn away from their children! We hope that the generous principles that have animated *legislatures* in the northern United States for quite some time now[2] will gradually extend toward the south and toward the western regions beyond the Alleghenies and the banks of the Mississippi, where slavery and its iniquities have spread in the wake of ▾the adoption of an imprudent and harmful law.[3] We hope that the force of public opinion, the progress of enlightenment, the improvement of mores, the legislation of the new continental republics, and the momentous and felicitous recognition of Haiti by the French government will have—either through fear and foresight or through more decent and disinterested sentiments—a beneficent influence on the condition of black men and women in the rest of the Antilles, the Carolinas, ▾the Guianas, and Brazil.

1.318

1.319

1.320

To succeed at undoing slavery's ties gradually, one needs the strictest en-

1. See Vol. XI, p. 351 [this edition, p. 90].

2. Already in 1769 (forty-six years before the declaration of the ▾Congress of Vienna, and thirty-eight years before the slave trade's cessation was decreed in Washington and London), the Massachusetts house of representatives had inveighed against "the unnatural and unwarrantable custom of enslaving mankind" (see ▾Walsh, *Appeal to the United States*, 1819, p. 312). The ▾Spanish writer Avendaño is perhaps the first to speak forcefully not only against the commerce in slaves, abhorred even by the Afghanies (▾Elphinstone, *An Account of the Kingdom of Caubul*, p. 245), but also against slavery in general and against "all the iniquitous sources of colonial wealth." [Avendaño,] *Thesaurus indicus*, Vol. I, book 9, chap. 2.

3. ▾Rufus King, *Speeches on the Missouri Bill* (New York, 1819). *North-American Review*, no. 26, pp. 137–68.

forcement of laws against the slave trade, humiliating punishments against those who infringe upon them, the formation of mixed tribunals, and the right of mutual inspections carried out with equitable reciprocity. It is sad to learn that the slave trade, having become crueler for being more hidden, still wrests almost the same number of black people from Africa as it did before 1807, all because of the disdainful and guilty negligence of certain European governments. Yet, one cannot posit, as do the secret partisans of slavery, the practical impossibility of the beneficent measures enacted first in Denmark, the United States, Great Britain, and then in the rest of Europe. What happened between 1807 and the time when France regained possession of some of its former colonies and what is happening today in the nations that sincerely desire the abolition of the slave trade and its abominable practices proves the fallacy of this conclusion. Also, is it reasonable to compare numerically slave imports from 1825 to slave imports in 1806? Considering the activity at the core of all industrial enterprises, what kind of increase would we have seen in the importation of blacks to the British Antilles and the southern United States, had the slave trade, entirely without constraints, continued to deposit new slaves, thus rendering unnecessary any care for the conservation and growth of the former slave population? Is it believable that British trade would have limited itself to the sale of 53,000 slaves, as it did in 1806? That the United States would have restricted itself to the sale of 15,000? We know with enough certainty that the British Antilles alone received more than 2,130,000 slaves ripped from Africa's coasts during the 106 years before 1786. At the time of the French Revolution, the slave trade (according to ▼Mr. Norris) brought in 74,000 slaves per year, of which the British colonies absorbed 38,000 and the French colonies 20,000. It would be easy to prove that the entire Antillean Archipelago, which has barely 2.4 million blacks and persons of mixed race (frees and slaves) today, received five million Africans (*negros bozales*) between 1670 and 1825. These dreadful calculations about the consumption of human beings do not even account for the number of unfortunate slaves, who either died during the ▼Middle Passage or were thrown overboard like damaged goods.[1] By how many thousands would losses have to be increased, if the two peoples who show the most passion and aptitude for commercial and industrial development—the British and the inhabitants of the United States—had continued to take as liberal a part in the slave trade after 1807 as the other peoples of Europe did? Sad experience proved how catastrophic for humanity were the treaties from ▼July 15, 1814 and January 22, 1815, in which

I.321

I.322

1. See Vol. XI, p. 351 [this edition, p. 90]. See also the ▼Duke of Broglie's eloquent discourse (March 28, 1822), pp. 40, 43, 96.

Spain and Portugal still reserved[1] "the privilege of trading black people" for a certain number of years. 1.323

Local authorities—or, to be more precise, the rich landowners who comprise Havana's *Ayuntamiento* [city council], the *Consulado* [merchant guild], and the *Patriotic Society*—have been favorably disposed toward the improvement of the slaves' lot on several occasions.[2] If the metropole's government had known to take advantage of these happy circumstances and of the rise of men of talent among their compatriots instead of fearing even the appearance of innovation, the state of society would have changed progressively and the inhabitants of the island of Cuba would have already enjoyed some of the improvements that had been discussed thirty years ago. The disturbances in Saint-Domingue in 1790 and in ▾Jamaica in 1794 caused so much alarm among Cuba's *hacendados* that means to preserve the country's tranquility were debated passionately at a *Junta económica* [economic summit]. They regulated the pursuit of fugitives,[3] an activity that had, up to then, led to the most shameful of excesses. They proposed to increase the number of female blacks on sugar plantations; to take better care of the raising of children; to curtail slave imports from Africa; to invite white colonists from the Canary Islands and Indian colonists from Mexico; to establish schools in the coun- 1.324–25

1. "Dicen nuestros Indios del Río Caura cuando se confiesan que ya entienden que es pecado comer carne humana; pero piden que se les permita desacostumbrarse poco a poco: quieren comer la carne humana una vez al mes, despues cada tres meses, hasta que sin sentirlo pierdan la costumbre" [During confession, our Indians on the Río Caura admit that they understand that eating human flesh is a sin; but they ask that they be permitted to wean themselves of it gradually: they want to eat human flesh once a month, then once every three months, until they lose the habit without realizing it]. *Cartas de los Reverentes Padres Observantes*, no. 7 (unpublished).

2. [▾Abad y Queipo,] *Representación al Rey de 10 de Julio de* 1799 (manuscript).

3. [▾Zamora y Coronado,] *Reglamento sobre los negros cimarrones de 20 Dec. de 1796*. Before 1788, there were many fugitive slaves (▾*cimarrones*) in the Jaruco Mountains, where they were sometimes *apalancados*, that is, many of these unfortunates built little trenches with tree trunks for their common defense. The maroons, *bozales* born in Africa, are easy to capture, for the majority walk day and night to the east in the vain hope of finding their homeland. They are so exhausted from fatigue and hunger when they are taken that they cannot be saved unless they are fed small amounts of broth over many days. Creole maroons hide in the forest during the day and steal provisions at night. Until 1790, the right to capture fugitive slaves belonged only to the *Alcalde mayor provincial* [provincial mayor], a hereditary position in ▾Count Barreto's family. Today, every inhabitant may seize a maroon, and the slave owner pays 4 piasters per head plus food expenses. If the master's name is unknown, the *Consulado* uses the maroon for public works. These manhunts, which have brought a deplorable renown to Cuba's (as well as Haiti and Jamaica's) dogs, were conducted in the cruelest possible manner prior to the regulation that I cited above.

tryside to improve the mores of the lower classes and to thus mitigate the effects of slavery in an indirect way. These proposals did not have the desired effect. The court opposed all immigration policies, and most owners, given over to old fantasies of safety, did not want to restrict the slave trade at a time when high crop prices fed hopes for extraordinary profits. It would be unfair, however, not to point to the hopes and principles that some of Cuba's inhabitants articulated during this fight between private interests and prudent policies —either in their own name or in the name of rich and powerful corporations. "Our legislation's humanity," Mr. Arango y Parreño[1] gallantly stated in an account from 1796, "bestows four rights (*cuatro consuelos*) upon the slave, which go some way toward alleviting his suffering and which have been constantly denied him in other countries. These rights are: the choice of a less severe master;[2] the right to marry whom he pleases; the possibility of working to purchase his freedom[3] or of receiving it as reward for his good services; the right to own property and to pay for his wife and children's freedom with acquired property.[4] Despite the wisdom and leniency of Spanish legislation, to

1.326

1.327–28

1. *Informe sobre negros fugitivos* (*de* 9 *de Junio* 1796), by Don Francisco de Arango y Parreño, Oidor honorario y síndico del Consulado [honorary magistrate and trustee of the council].

2. It is the right to *buscar amo*, to find a new master. As soon as a slave has found a new master who wishes to buy him, he may leave the first about whom he feels he has something to complain: such is the letter and spirit of a compassionate law that is nevertheless often evaded like all laws that protect slaves. It is in hope of exercising the privilege to *buscar amo* that blacks often ask travelers whom they meet a question that is never posed out loud in civilized Europe, though one sometimes sells one's vote or one's opinion: "quiere Vm. [Vuestra Merced] Comprarme" [Would Your Highness like to buy me]?

3. According to the law, the slave in the Spanish colonies should be put at the lowest price: during my trip, this estimate was 200 to 380 piasters, depending on the locale. We saw above (Vol. XI, pp. 351 and 389 [this edition, pp. 90 and 115]) that the price of an adult slave on the island of Cuba was 450 piasters in 1825. In 1788, the French market offered slaves for 280 to 300 piasters (Page, *Traité d'économie politique des colonies*, Vol. VI, pp. 42 and 43). Among the Greeks, a slave cost 300 to 600 drachmas (54 to 108 piasters), while a worker cost one-tenth of a piaster a day. While Spanish laws and institutions favor every kind of *manumission*, the master in the non-Spanish Antilles pays the treasury five to seven hundred piasters for every freed slave!

4. What a contrast between the humanity of the oldest Spanish laws concerning slavery and the traces of barbarity that one finds on every page of the ▾*Code noir*, and in certain provincial laws in the British Antilles! Barbados's laws, from 1688, and Bermuda's, from 1730, command that the master who kills his slave while punishing him cannot be prosecuted, while the master who kills his slave out of malice will pay 10 pound sterling to the royal treasury. A law in St. Christophe [St. Kitts] from March 11, 1784 begins with these words: "Whereas some persons have *of late* been guilty of cutting off and depriving slaves

how many excesses does the slave remain exposed in the solitude of a farm or plantation where a brutal *capataz* [overseer] armed with a *machete* and a whip exercises absolute authority with impunity! The law limits neither the slave's punishment nor his work's duration. Nor does it dictate the quality or the quantity of his provisions.[1] It is true that the law grants the slave recourse to a magistrate, who can enjoin the master to be more equitable; but this recourse is nearly entirely illusory, because there is another law according to which one must arrest and return to his master every slave discovered more than a league and a half from his plantation without a pass. How can an abused slave, exhausted by hunger and excessive work, appear before a magistrate? If he does, how is he to defend himself against a powerful master, who calls his salaried accomplices as witnesses?" I.329

I will end by quoting another most remarkable extract from the *Representación [. . .] del Ayuntamiento, Consulado y Sociedad Patriótica* from July 20, 1811. "Everything about proposed changes in the conditions of the *enslaved class* has less to do with our fears about declines in agricultural wealth than with the safety of white people, something so easily compromised by imprudent measures. Besides, those who accuse the council and the Havana municipality of stubborn resistance forget that, after 1799, these same authorities unsuccessfully proposed that one should concern oneself with the condition of black people on the island of Cuba (*del arreglo de este delicado asunto* [the mending of this sensitive matter]). What is more: we are far from adopting principles that European nations, which brag about their level of *civilization*,[2] saw as indisputable. For instance, that without slaves, there could not be any I.330

of their ears, we decree that whosoever will have plucked out an eye, torn out a slave's tongue, or cut off his nose, will pay 500 pounds sterling and shall be condemned to six months in prison." I need not add that these British laws, in effect 30 or 40 years ago, have been abolished and replaced by more humane legislation. I wish I could say as much of the legislation in the French Antilles, where six young slaves suspected of wanting to escape had *the tendons in the backs of their knees severed* after being arrested in 1815! (See also Vol. XI, pp. 324ff [this edition, pp. 79ff.]).

1. A *cédula* from May 31, 1789 had attempted to regulate food and clothing, but this *cédula* was never implemented.

2. "Hasta abandono hemos [hecho] de [e]species muy favorable[s] que pasan por inconc[l]usas en esas *naciones cultas*. Tal es la de que sin negros esclavos no pudiera haber colonias. Nosotros contra este dictamen decimos que sin esclavitud, y aún sin negros, pudo haber lo que [por] colonia[s] se entiende, y que la diferencia habría estado en las mayores ganancias o en los mayores progresos" [We have even abandoned favorable ideas that pass for undisputed in those cultured nations. For example, that without black slaves, there could be no colonies. We say against this dictum that, without slavery and even without blacks, there might be recognizable colonies; the difference would be in higher prof-

colonies. On the contrary, we declare that colonies could have existed without slaves, and even without blacks, and that the only difference would lie in greater or lesser profit and in more or less rapid growth in productivity. But if this is our firm conviction, we should recall to Your Majesty that a social organization cannot be changed in unreflecting haste once slavery has been introduced as a factor. We are far from denying that dragging slaves from one continent to another was an evil contrary to moral principles, and that it was a political mistake not to listen to the complaints of Hispaniola's governor, ▾Ovando, against the importation and accumulation of so many slaves alongside such a small number of free men. But once these moral crimes and abuses are already deeply rooted, we must avoid worsening our position and that of our slaves by violent means. What we ask, Sire, is consistent with a wish articulated by one of the most ardent protectors of human rights, the most relentless enemy of slavery. Like he, we wish that civil law deliver us from both abuse and danger at the same time."

I.331

Here is the solution to the problem upon which depends the safety of 875,000 free people (whites and people of color[1]) in the Antilles alone, excluding Haiti, and the adjudication of the fates of 1,150,000 slaves. We have demonstrated that this cannot be achieved by peaceful means without the participation of local authorities, whether *colonial assemblies* or landowner meetings known by names less threatening to the old metropoles. Direct influence from the authorities is indispensable, and it is a grave mistake to believe "one can let time take its course." Time will work simultaneously on slaves, on the relations between the islands' and the mainland's inhabitants, and on events that one cannot control any longer, because one will have waited for them in a state of apathetic inaction. Wherever slavery is long established, civilization's advance influences the treatment of slaves far less than one would care to admit. A nation's civilization rarely extends to a large number of individuals. It does not reach those in workplaces who are in direct contact with black people. Owners—I have known some who are very humane—recoil before the difficulties on large plantations. They hesitate to trouble the established order, to adopt innovations that fall short of their ends—where they are not simultaneously supported by legislation, or, what would be more effective, by the general will—and perhaps worsen the lot of those whom they would like

I.332

I.333

its or greater progress] ([Arango,] *Documentos [. . .] sobre el tráfico y esclavitud de negros*, 1814, pp. 78–80).

1. Specifically: 452,000 whites, of whom 342,000 were in the only two Spanish Antilles (Cuba and Puerto Rico), and 423,000 free people of color, persons of mixed race, and blacks.

to help. Such timid considerations inhibit the good in those men whose intentions are the most benevolent and who suffer under the barbaric institutions that are their sad legacy. Familiar with local conditions, they know that, to produce a basic change in the status of slaves and gradually to lead them to the enjoyment of freedom, the local authorities must have a strong will, support from rich and enlightened citizens, and a comprehensive plan that considers all possibilities for disorder and repressive means. Without this community of action and effort, slavery will sustain itself with its pains and excesses, as it did in ancient Rome,[1] side by side with elegant manners, with enlightenment's much trumpeted progress, with all the glories of a civilization that its presence condemns and that it threatens to devour when the time for revenge will have come. Both civilization, on the one hand, and a slow brutalization of peoples on the other only prepare minds for future events. But to bring about large changes in the social order, the coincidence of certain events, whose moment cannot be determined in advance, is necessary. Such is the complexity of human destiny that the very cruelties that covered the conquest of both Americas in blood are returning before our eyes during a time we believed to be characterized by enlightenment's prodigious progress and by a general softening of mores. The course of one man's life sufficed to see ▾the French Reign of Terror, the Saint-Domingue expedition,[2] and Spain and Naples's political reaction. I could add the ▾Chio, Ipsara, and Missolonghi massacres, the work of barbarians from Eastern Europe that the civilized peoples of the west and north did not believe themselves able to prevent. In slave countries, where long-standing habits tend to legitimize institutions that are the most opposed to justice, one can count on the influence of enlightenment, intellectual culture, or the softening of mores only insofar as they all accelerate gov-

I.334

I.335

1. The argument in favor of slavery derived from Roman and Greek civilizations is very popular in the Antilles, where occasionally one indulges in adorning it with all of the trappings of philological erudition. Because of this, in a 1795 speech before Jamaica's *legislative assembly*, it was argued that the example of elephants used in Pyrrhus's and Hannibal's wars justified importing a hundred dogs and forty hunters to track down maroons. Bryan Edwards, [*The History, Civil and Commercial, of the British Colonies in the West Indies*,] Vol. I, p. 570.

2. [Review of ▾Vastey in] *North American Review*, 1821, no. 30, p. 116. Battles against slaves fighting for their freedom are deplorable not only because of the atrocities they unleashed on both sides, but also because such wars contribute to the confusion of all feelings of justice and injustice once emancipation is achieved. "Certain colonists condemned to death the entire male population above the age of six. They hold that an example before the eyes of those who did not bear arms could become contagious. This lack of moderation is the result of the colonists' long misfortunes." Charault, *Réflexions sur Saint-Domingue*, 1814, p. 16.

ernment initiative and facilitate the implementation of measures once they are adopted. Without the governments' and *legislatures*' guiding action, peaceful change is not to be hoped for. The danger becomes particularly imminent I.336 when a general restlessness seizes the spirit and when the faults and responsibilities of governments are revealed in the midst of political upheavals that agitate neighboring peoples. Then, calm can only be restored by a power that knows how to control events by itself initiating improvements out of a noble sense of its strength and its right.

▾At the end of April, after having finished the observations that we had proposed to make, Mr. Bonpland and I, at the Torrid Zone's northern extreme, were about to depart for Veracruz with Admiral Aristizábal's squadron. But false rumors about ▾Captain Baudin's expedition spread by public broadsheets led us to give up the plan to cross Mexico to arrive at the Philippine Islands. Many newspapers, particularly in the United States, announced that two French warships, the *Géographe* and the *Naturaliste*, had set sail for I.337 Cape Horn, that they would hug the Chilean and Peruvian coasts and proceed from there to New Holland. I was quite agitated by this news. All the projects I had thought up during my trip to Paris, when I pestered the minister of the *Directory* to hasten Captain Baudin's departure, once again flooded my mind. At the time I left Spain, I had made the promise to join this expedition wherever I could link up with it. When one avidly desires something whose outcome might be disastrous, one is easily persuaded that a sense of duty alone motivated the decision that one made. Always enterprising and confident in our good fortune, Mr. Bonpland determined on the spot to divide our herbarium into three parts. So as not to expose to the vicissitudes of a long navigation what we had gathered with great effort on the banks of the Orinoco, the Atabapo, and the Río Negro, we sent one collection to Germany by way of England, another to France via Cádiz. The third collection remained in Havana. We had to congratulate ourselves on these arrangements I.338 that prudence rendered necessary. Each package contained roughly the same specimens, and no precaution had been neglected: if the cases were taken by British warships, they would be delivered to ▾Sir Joseph Banks; if they were taken by French warships, they would be sent to the professors of the Museum of natural history in Paris. Happily, the manuscripts that I had at first wanted to add to the Cádiz shipment had not been entrusted to our friend and travel companion, ▾Fray Juan Gonzáles of the Franciscan Order.[1] This esti-

1. Vol. IV, pp. 58 and 59; IX, pp. [96], 97, 114.

mable young man, whom I have had occasion to mention often, had followed us to Havana in order to return to Spain. He left the island of Cuba a short time after us, but the ship on which he departed perished, bodies and goods, in a storm off Africa's coasts. We lost some of our herbarium's doubles in this shipwreck, and, what constitutes a more significant loss for science, all the insects that Mr. Bonpland had gathered under the most difficult circumstances during our voyage down the Orinoco and the Río Negro. By an extraordinary twist of fate, we remained in the Spanish colonies two years without getting a single letter from Europe. Those we received in the following three years told us nothing about our shipments. One can imagine how anxious I would have been about the fate of a *Journal* that contained astronomical observations and all the measurements of elevation with the help of a barometer, and of which I did not have the patience to make a detailed copy. After having traversed New Granada, Peru, and Mexico, at the very moment of leaving the New Continent, my eyes fell, as if by chance, on a scientific *Journal*'s table of contents in the Philadelphia Public Library. I read the following words there: "Mr. Humboldt's manuscripts arrived at his brother's in Paris." I could barely hide my joy: never had a table of contents delighted me more.

I.339

While Mr. Bonpland worked day and night to divide and organize our collections, I had the misfortune to worry about countless obstacles against such an unplanned departure. There was not a single ship in the port of Havana that wanted to undertake our passage to Portobelo or Cartagena. The persons whom I consulted delighted in exaggerating the difficulties of a passage across the isthmus and the slowness of a trip from north to south, from Panama to Guayaquil and from Guayaquil to Lima or Valparaíso. They reproached me, and perhaps with good reason, for not continuing to explore the vast and rich possessions of Spanish America that, for half a century, had not been open to any foreign travelers. To them, the possibilities of a voyage around the world during which one generally touches only on some islands or on a continent's arid coast did not seem preferable to the opportunity to study the geological relationships of New Spain's interior, regions that by themselves furnish five-eighths of all the silver annually mined in the known world. I countered these considerations with my interest in determining, on a larger scale, the inflection of curves of equal inclination, the decrease in the intensity of magnetic forces from the pole to the equator, and the Ocean's temperature as it varies by latitude according to the direction of currents and the proximity of sandbars. The more I found myself opposed in my designs, the more I rushed to their implementation. Not able to find passage on any neutral ship, I chartered a Catalán schooner that was docked in Batabanó and that would be available to take me either to Portobelo or to Cartagena de Indias, depending on whether the sea or the Saint Martha breezes, which still blow with vio-

I.340

I.341

lence during this season below 12° lat., would permit it. Havana's prosperous trade situation and the multiplying connections that this city has even with the South Sea ports helped me procure funds for many years. ▾General Don Gonzalo O'Farril, equally distinguished by his talent and by the loftiness of his character, lived then in my homeland as Minister of the Spanish Court. I could exchange my revenues in Prussia against part of his in Cuba. And the family of the respectable ▾Don Ignacio O'Farril y Herrera, the General's brother, wanted to support everything that could favor my new plans during my unexpected departure from Havana. We learned on March 6 that the schooner I had chartered was ready to take us on. The path from Batabanó led us one more time through Güines to the Río Blanco plantation, whose owner (Count Jaruco y Mopox) embellished our stay with everything that a taste for pleasures and great wealth can provide. Hospitality, which generally declines with civilization's advance, is still practiced on the island of Cuba with as much assiduousness as in the most remote corners of Spanish America. These simple traveling naturalists want to express to the people of Havana the same appreciation that was granted them by those illustrious strangers[1] who, wherever I could follow their tracks in the New World, left behind the memory of their noble simplicity, their passion for education, and their love of the public good.

I.342

I.343

From Río Blanco to Batabanó, the path goes through uncultivated country half covered by forests. In the clearings, indigo and cotton grow wild. As the *Gossypium*'s capsule opens during the period when storms from the north are most frequent, the down that covers the seeds is blown all over. And the cotton harvest, which is, by the way, of superior quality, suffers much when storms coincide with the fruits' ripening. Many of our friends, among them ▾Mr. [de] Mendoza, Captain of the port of Valparaíso and brother of the famous astronomer who resided in London for a long time, accompanied us to the *Potrero de Mopox* [Mopox cattle ranch]. While gathering herbs farther to the south, we found a new palm tree[2] with fan shaped leaves (*Coripha maritima*) and with free filaments between the interstices of the leaflets. This *Coripha* covers part of the south coast and takes the place of the majestic *Palma Real*[3] and the north coast's crispa Coconut. On the plain, porous limestone (from the Jurassic formation) occasionally appears above ground.

I.344

1. ▾The young princes of the House of Orléans—The Duke of Orléans, the Duke de Montpensier, and the Count de Beaujolois—who came from the United States to Havana by way of the Ohio and Mississippi Rivers, and who stayed on the island of Cuba for a year.

2. See our *Nova genera et species plantarum*, Vol. I, p. 29[8].

3. Oreodoxa regia.

Batabanó was[1] a poor village then, whose church had only been completed a few years earlier. The *Ciénaga* [swamp], marshy terrain that stretches from the Laguna de Cortés to the mouth of the Río Jagua for 60 leagues from west to east, begins half a league away. In Batabanó, it is believed that the sea continues to gain on the land in these regions and that the flooding was particularly extensive during the great landslide[2] at the end of the eighteenth century, when the tobacco farms disappeared and the Río Chorrera changed its course. Nothing is sadder than the appearance of these swamps around Batabanó. Not a single shrub interrupts the monotonous landscape. Amidst large clumps of Juncaceae and Iradaceae, a few stunted palm trunks stand alone like broken masts. As we were spending only one night in Batabanó, I passionately regret not having been able to gather detailed information on the two species of crocodile that infest the *Ciénaga*. The locals call one *cayman* [caiman or alligator], the other *crocodile*, or, as is commonly said in Spanish, *cocodrilo*. They claim that crocodiles are more agile and have longer legs, that they have pointier snouts than *caimans*, and that the two never live together. Crocodiles are fearless, and it is said that they climb onto boats when they can ground their tails. This animal's extreme daring had already been noted during Governor Diego Velázquez's first expeditions.[3] The *crocodile* will go on land for up to one league from the Río Cauto and Jagua's swampy coast to devour pigs in inland areas. One sometimes sees some crocodiles that are 15 feet long, and the most vicious of them will pursue (people say) a man on a horse, as European wolves do, while the animals known exclusively as *caymanes* in Batabanó are so shy that people are not afraid to bathe in places where they live in groups. These behaviors and the name *cocodrilo* given to the most dangerous carnivorous Reptilians on the island of Cuba seem to me to indicate a species different from the large animals of the Orinoco, the Río Magdalena, and Saint-Domingue. Throughout continental Spanish America, colonists, fooled by the exaggerated tales of the Egyptian crocodile's ferocity, insist that only the Nile has *real crocodiles*. Zoologists, on the other hand, have acknowledged that there are *caimans* with blunt snouts and smooth legs and *crocodiles* with pointed snouts and serrated legs. On the old continent, there are both *crocodiles* and *gavials*. Saint-Domingue's *Crocodylus acutus* [American

I.345

I.346

1. On the true astronomical position of Batabanó, see Vol. XI, p. 2[19]. Before then, one situated Batabanó 10′ farther to the south, at lat. 22° 33′ on the most in-depth nautical maps by ▾Bellin, San Martín Suárez, etc. Arrowsmith even put it at 22° 24′ instead of 22° 43′ 24″. Frigate Captain Don Ventura Barcaíztegui and Don Francisco Lemaur made the first good observations of Cuba's southern coast.

2. See Vol. XI, pp. 229 and 230.

3. Herrera, *Historia de las Indias occidentales*, Dec. I, book 9, chap. 4, p. 232.

crocodile], which I still do not know how to distinguish precisely from the crocodile of the large Orinoco and Magdalena Rivers, even has, to borrow an

1.347 expression from ▾Mr. Cuvier,[1] such a startling resemblance to the Nile crocodile [Crocodylus niloticus] (Laurenti, 1768) that it took a meticulous examination of each to prove that ▾Buffon's law about species distribution in the tropical regions of both continents was indeed valid.

As I could not return to Batabanó's *Ciénaga* during my second trip to Havana in 1804, I had the two specimens that the locals call *caymanes* and *cocodriles* sent to me at great expense. Two of the crocodiles arrived alive, the oldest being 4 feet 3 inches in length. It had been difficult to capture them.

1.348 They were transported, muzzled and bound, on a mule. They were strong and quite ferocious. To observe their habits and their movements,[2] we placed them in a large room, where, perched on a very tall piece of furniture, we could watch them attack some fat dogs. Having lived on the Orinoco, Apure, and Magdalena Rivers among crocodiles for six months, we were delighted to observe once more before returning to Europe these singular animals that pass from immobility to the most impetuous movements with surprising speed. The animals we received from Batabanó as *crocodiles* had snouts as pointy as the Orinoco and Magdalena crocodiles (*Crocodilus acutus*, Cuvier). Their color was a little darker, blackish-green on the back and white on the belly; their sides were speckled in yellow. I counted 38 teeth on the upper jaw

1.349 and 30 on the lower, as is the case with all true crocodiles. On the upper jaw, the tenth and the ninth were largest, on the lower, the first and the fourth. The description that Mr. Bonpland and I made in the field expressly stated that the lower fourth tooth *extends freely* along the outside of the upper jaw. The back teeth were flat. These Batabanó *crocodiles* seemed to us absolutely identical to the *Crocodilus acutus*. It is true that everything reported to us about their behavior did not accord well with what we ourselves had observed on the Orinoco. But within the same river carnivorous Reptilians of the same species can

1. Cuvier, *Recherches sur les ossemens fossils*, Vol. V, Pl. II, p. 27. ▾Mr. Geoffrey de Saint-Hilaire could not have recognized this striking analogy before 1803 when General Rochambeau sent a Saint-Domingue crocodile to the Museum of Natural History in Paris (*Annales du Muséum*, Vol. II, pp. 37, 53). Mr. Bonpland and I prepared drawings and detailed descriptions of the same species that inhabit the large rivers of southern America in 1800 and 1801 during our travels of the Apure, Orinoco, and Magdalena Rivers. We made a mistake so common to travelers of not having them sent to Europe right away, in the company of some young specimens.

2. ▾Mr. Descourtilz, who knows crocodile behavior better than any other author who has written about this reptile, saw—as did Dampier and I—the *Crocodylus acutus* frequently bring its snout to its tail. *Voyage d'un naturaliste*, Vol. III, p. 87.

be either gentler and more timid or more ferocious and bolder depending on their habitat.[1] The animal they called *cayman* in Batabanó died en route, and they did not have the foresight to bring it to us so that we could compare both species. Are there real *caimans* on the island of Cuba, with blunt snouts and whose lower fourth teeth fit within the upper jaw? Are these *alligators* that resemble the ones in Florida? What the colonists say about the much longer head of their *Batabanó cocodrilo* makes this possibility a near certainty.[2] If so, people in Cuba would, by lucky instinct, have distinguished between the *crocodile* and the *alligator* with the same accuracy as learned zoologists who established subgroups under the same names. I do not doubt that the crocodile with the pointy snout and the alligator, or cayman, with the blunt snout[3] both inhabit the swampy coastline between Jagua, the *Surgidero* de Batabanó and the Isle of Pines, though in distinct groups. On the Isle of Pines, ▾Dampier, praiseworthy both as a scientific observer and a fearless sailor, was struck by the enormous differences between American *caimans* and *crocodiles*. What he reported on this subject in his Voyage to the Bay of Campeche [*Voyages to Campeachy*] would have aroused the curiosity of scholars more than a century ago, if zoologists had not so often dismissed everything that sailors or other travelers without any scientific background observed about the animals. After having outlined several characteristics that were not equally reliable in distinguishing between *crocodiles* and *caymans*, Dampier comments on these enormous Reptilians' geographical spread. In Campeche Bay, he wrote, "there are abundance of Alligators, where yet I never saw nor heard of any Crocodiles.

1.350

1.351

1.352

1. Vol. VIII, pp. 357 and 358; IX, pp. 99 and passim.

2. I thought I found a slight difference in the position of the large plates (*ridges*) on the neck. The large Batabanó specimen has, near the head, first four small outgrowths or tubercles arranged in a line, and then three rows of two. In the younger specimen, I at first counted a first row of 4 horns and then a single row of 2 followed by a large empty space. After this space, the back ridges began. This arrangement is most common among Orinoco crocodiles. Those from the Río Magdalena have three rows of horns on the neck, the first two having 4 horns, and the last 2. On the *Crocodilus acutus* specimens that the Natural History Museum of Paris received from Saint-Domingue, there are first 2 rows of 4, and then one row of 2 horns. I discuss this characteristic's consistency in the second volume of my work on zoology. On the Batabanó crocodile, the four pockets containing musk (*bolsas de almizcle*) are situated exactly as I drew them on the Río Magdalena crocodile, that is, below the lower jaw and near the anus. But I was singularly surprised not to smell this odor in Havana three days after the animal's death in 30° C weather, while in Mompox on the banks of the Río Magdalena, living crocodiles stank up our apartment. I have since seen that Dampier has also noticed "an absence of smell among Cuba *crocodiles* where *caimans* give out a very strong musky odor."

3. Saint-Domingue's *Crocodylus acutus*. Florida and Mississippi's *Alligator lucius*.

At the Isle *Grand Caymanes*, there are Crocodiles, but no Alligators. At *Pines* by *Cuba*, there are an abundance of Crocodiles, but I cannot say there are no Alligators, tho' I never saw any there."[1] I will add to Dampier's precise observations that the true crocodile (*C. acutus*) is found in the leeward Antilles that are closest to the continent, for instance, in Trinidad, Margarita Island, and plausibly, despite the lack of freshwater, Curaçao.[2] Farther south, they can be found (without my having encountered any of the alligator species that I.353 abound on Guiana's coast[3]) from the Neveri, Magdalena, Apure, and Orinoco Rivers to the confluence of the Casiquiare and the Río Negro (lat. 2° 2′), thus more than 400 leagues from Batabanó. It would be interesting to find out how far the carnivorous Reptilians reach into Mexico's east coast, Guatemala, and the area between the Mississippi and Río Chagre (on the Isthmus of Panama).

We set sail on March 9th before sunrise, a little frightened by our schooner's extremely small size, whose layout scarcely allowed us to sleep anywhere but on deck. The cabin (*the hold* or *cámera de pozo*) did not get any air or light except from above. It was a veritable supply ship, in which we had some difficulty placing our instruments. The thermometer constantly hovered around 32° or 33° C. Luckily, these inconveniences lasted only for 20 days. Traveling by canoe on the Orinoco and on an American vessel filled with several thousand *arrobas* of sun-dried meat was less strenuous than this.

I.354 The Gulf of Batabanó, bordered by a low and swampy coastline, looked like a vast desert. The fish-eating birds, generally at their post before the little land birds and the lazy *zamuros*[4] awakened, appeared only in small numbers. The seawater was greenish-brown, as in certain Swiss lakes, while the sky at sunrise, because of its extreme purity, had the slightly cold pale-blue tint that attracts our landscape painters on the Italian Riviera to the same earlier morning hour and that puts faraway objects in remarkable relief. Our schooner was the only ship in the bay, because Batabanó's harbor was seldom visited except by smugglers or, as people here put it more politely, *tratantes* [slave traders]. When discussing the planned canal from Güines,[5] we noted how important Batabanó could become for communications between the island of Cuba and I.355 the Venezuelan coastline. In its current state, with no *dredging* having been

1. Dampier, *Voyages and Descriptions* (1696), Vol. II, Part I, pp. 30 and 75.

2. ▾Seba, [*Locupletissimi rerum naturalium thesauri accurata descriptio*,] p. CIV, fig. 1–9.

3. Alligator sclerops and Alligator palpebrosus.

4. The equinoctial American Egyptian vulture, *Vultur aura*.

5. See Vol. XII, pp. 49 and passim [this edition, pp. 135 and passim].

attempted, there are barely 9 feet of water.[1] The port is situated at the back of a bay that ends at Gorda Point to the east and Salinas Point on the west. But the bay itself forms only the back (the concave apex) of a large gulf that is more than 14 leagues deep from south to north and which, for a stretch of 50 leagues between the Laguna de Cortés and Cayo Piedras, is enclosed by countless sandbars and keys. A single large island, whose *area* is more than four times Martinique's and whose arid mountains are crowned by majestic Conifers, rises in the middle of this maze. It is the *Isla de Pinos*, which Columbus called *El Evangelista,* The Evangelist; other sixteenth-century seafarers named it *Isla de Santa María*. This island is famous for its excellent mahogany (Swietenia Mahagoni). We made for ESE, taking *San Cristóbal pass* to reach the rocky islet *Piedras Key* and leaving behind this archipelago that Spanish seafarers have called *Jardines y Jardinillos* [gardens and groves] since the *conquest*. The actual archipelago, the *Queen's Gardens*,[2] is closer to Cape Cruz and separated from the *Jardines y Jardinillos* by what I would describe as 35 leagues of open sea. Even Columbus called them this when, during his second voyage in May, 1494, he fought against the currents and the winds between the Isle of Pines and Cuba's Eastern Cape for 58 days. He described these islets as green, full of groves and full of grace—"verdes, llenos de arboledas y graciosos."[3]

I.356

I.357

Indeed, these supposed gardens are quite lovely. The sailor sees a changing scene every second, and the greenery of certain islands seems even more beautiful in contrast with the other keys that offer only arid white sands. The

1. The largest vessels that enter the Batabanó *Surgidero* displace 15 *palmas* (9 Spanish inches). Good passageways include, to the west, the *Puerto Frances Channel* between the western cape of the Isle of Pines and the Laguna de Cortés, and, to the Isle of Pine's east, the four passages of *Rosario*, *Gordas*, the *Savanna de Juan Luis*, and *Don Cristóbal*, between the keys and Cuba's coastline.

2. Even in Havana, there is a great deal of geographical confusion about the former names for the *Jardines del Rey* and the *Jardines de la Reyna*. The description of Cuba in *Mercurio Americano* (Vol. II, p. 388) and in the *Historia natural de la Isla de Cuba* (Chap. 1, paragraph 1), compiled in Havana by Don Antonio López Gómez, places the two groups on the island's south coast. Mr. López says that the *Jardines del Rey* stretch from Cortés Lagoon to Jagua Bay. There is no historical doubt that Governor Diego Velázquez gave this name to the western part of the *Old Bahama-Channel* keys, between Cayo Frances and the Monillo on Cuba's north coast (Herrera, *Historia de las Indias occidentales*, Dec. I, pp. 8, 81, 55, and 232; Dec. II, p. 181). The *Jardines de la Reyna*, situated between Cabo Cruz and the port of Trinidad, are in no way linked to the *Jardines* and *Jardinillos* near the *Isla de Pinos*. Between these two groups of keys are the Paz and Jagua sandbars.

3. ▼Churchill, *A Collection of Voyages and Travels*, p. 560. ▼[Juan Bautista] Muñoz, *Historia del nuevo mundo*, pp. 214, 216.

sands' surface, warmed by the sun's rays, seems as wavy as a liquid's surface. Through contact with layers of air of differing temperatures, the surface produces the most varying levitation and *mirage* effects from 10 in the morning until 4 in the afternoon.[1] In these deserted places, it is still the daystar that enlivens the landscape, animating the objects struck by its rays on the dusty plain, on tree trunks, and on those rocks that jut out into the sea in the form of capes. As soon as the sun rises, these inert masses look as if they were suspended in the air, and, on the neighboring beach, the sand gives the deceptive appearance of a layer of water gently mussed by the winds. A trail of clouds suffices to settle the ground, the tree trunks, and the suspended rocks, to still the plain's waving surfaces and dissipate the marvels that Arab, Persian, and Hindu poets have immortalized as "the gentle deceptions of desert solitude."

I.358

We rounded Matahambre Cape extremely slowly. As the ▾Louis Berthoud chronometer had worked very well in Havana, I took advantage of the occasion that presented itself that day and in the days following to determine the position of *Cayo de Don Cristóbal*, *Cayo de Flamenco*, *Cayo de Diego Pérez*, and *Cayo de Piedras*.[2] I also busied myself examining the influence of a changing sea bottom on surface temperature.[3] At the threshold of so many islets, the surface is calm like a freshwater lake, as the layers from different depths do not mix. The least changes registered by the plummet disturb the thermometer. I was surprised to see that, to the East of the little Cayo de Don Cristóbal, the depths were not characterized by a milky color, as is the case on Vibora bank to Jamaica's south and on so many other banks that I had recognized thanks to a thermometer. At the bottom of Batabanó cove is broken coral. It nour-

I.359

I.360

1. See the extraordinary refraction measurements I made in Cumaná, Vol. IV, pp. 290–306.

2. See my *Recueil d'observations astronomiques*, Vol. II, pp. 109f. Mr. Bauzá connected my observations with Mr. Del Río's in the *Jardines y Jardinillos* sketch, corrected the southern part of my Cuba map, and communicated these corrections to me (see this map's second edition, 1826).

3. I found the following in Réaumur Thermometer degrees:

Sea.	*Air.*	*Depth.*	*Places.*
19.7°	22.3°	10 feet.	8 miles north of Punta Gorda
18.8	23.0	7 ½	Between the keys of Las Gordas and Don Cristóbal
19.7	22.2	10	Around Cayo Flamenco
20.7	22.0	80	Gulf between Cayo Flamenco and Cayo de Piedras
19.6	24.2	9	Eastern side of the gulf; very near Cayo de Piedras
18.2	24.3	8	A little more to the east
21.5	23.0	. . . .	No depth reading, south of Jagua

ishes algae that almost never come to the surface. The water is greenish, as I already remarked, and the absence of milkiness is probably due to the perfect calm that reigns in these regions. Wherever disturbances propagate themselves at certain depths—very fine sand or limestone particles suspended in the water—make it murky and milky. There are nevertheless sandbars that distinguish themselves neither by color nor by low water temperature, and I think that these phenomena depend on the nature of a *hard* and rocky *bottom* lacking sand and coral, the sloping form of the ledges, the speed of the currents, and the lack of movement in lower water layers. The cold that the thermometer most frequently picked up at the surface of deep water is due both to the water molecules that warming and nightly cooling drive down from the surface to the depths—the lower depths break their descent—and to the intermingling of very deep water layers that climb the bank's shelves like an inclined plane to mix with the surface layers. I.361

Despite the small size of our craft and our captain's vaunted wisdom, we often scraped the sea bottom. Because the bottom was soft, there was no danger of a shipwreck. However, at sunset, near *Don Cristóbal pass*, we preferred to drop anchor. The first part of the evening was admirably serene. We saw countless shooting stars over the land, all following the same direction: against the easterly wind that rules the lower atmospheric regions. Nothing today approaches the solitude of these places, which, during Columbus's times, were inhabited and frequented by a large number of fishers. ▾Cuba's natives used a small fish to reel in huge sea turtles. They attached a long line to the tail of the *revés* (the name that the Spanish gave to this species in the Echeneis family).[1] The *fisher fish* [suckerfish], which has a flat disk with suction cups on its head, attaches itself to the shells of the sea turtles that abound in the narrow and tortuous channels of the *Jardinillos*. "The *revés*" said Christopher Columbus, "would rather allow itself to be pulled to pieces than to release involuntarily the body to which it is attached." With the I.362

1. Known to the native Cuban peoples as *guaicán*. In characteristic fashion, the Spanish named it *revés*, as if to say, "fish placed on its back, oriented backwards." At first glance, one confuses the position of the back and the belly. Anghiera says: "called Born Backwards, because the back comes first." I examined a *remora* from the South Seas during the trip from Lima to Acapulco. As it lived for a long time out of water, I attempted some experiments on how much weight it could carry before the sucker that attached the animal to the plank would give way; but I lost this part of my journal. It is probably fear of danger that keeps the *remora* from releasing its hold when it feels itself being tugged by a cord or a man's hand. The sucet that Columbus and Martin d'Anghiera talked about was probably the Echeneis Naucrates and not the Echeneis Remora (see my *Recueil d'observations de zoologie*, Vol. II, pp. 192[f]).

same line, Indians would pull up the *suckerfish* and the turtle. When Gómara and ▾Petrus Martyr d'Anghiera, the learned secretary of Emperor Charles V, spread this fact, which came from the mouths of Columbus's companions, in Europe, the public likely took it for a *seaman's yarn*. There is, in fact, some quality of the marvelous in Anghiera's story, which begins with these words: "Non aliter ac nos canibus gallicis per aequora campi lepores insectamur, incolae (Cubae insulae) venatorio pisce pisces alios capiebant" [Just as we pursue rabbits through the open field with Gallic dogs, the inhabitants (of the Cuban island) catch fish by means of hunter-fish.][1] Thanks to the testimony compiled by ▾Captain Rogers, Dampier, and Commerson,[2] we know today that the same technique for hunting turtles observed in the *Jardinillos* is used by the inhabitants of Africa's east coast, near Cape Natal, Mozambique, and Madagascar. Men, whose heads were covered by large gourds pierced with holes, captured ducks in Egypt, on Saint-Domingue, and on the lakes and in the valleys of Mexico by hiding underwater and seizing the birds by the legs. Since high antiquity, the Chinese have used Cormorants, a bird in the Pelican family, which they sent fishing on the coastline with rings around their necks so that they could not swallow their prey and hunt on their own account. In the least developed civilizations, all of man's ingenuity is deployed in hunting and fishing schemes. People who have probably never had contact with each other show the most striking similarities in how they exert their power over animals for their own gain.

I.363

I.364

—— END OF THE FIRST VOLUME ——

II.5

We could not leave the *Jardines y Jardinillos* maze for three days. Every night we weighed anchor. During the day we visited the islets and keys that were easiest to approach. The farther east we went, the more troubled the sea became, and the depths took on a milky white color. On the edge of a kind of trough between Cayo Flamenco and Cayo de Piedras, we found that the surface temperature of the ocean suddenly rose from 23.5° to 25.8° C. The geognostic composition of the rocky islets that encircle the *Isle of Pines* captured

II.6

1. Fernando Colombo, in Churchill, *A Collection [of voyages and travels]*, Vol. II, Chap. LVI, p. 560; Petrus Martyr, [*Oceanica*,] 1532, Dec. I, p. 9; Gómara, *Historia de las Indias*, 1553, fol. XIV; Herrera, *Historia de las Indias occidentales*, Dec. I, p. 55.

2. Dampier, *Voyages*, Vol. II, Pl. III, p. 110. ▾Lacépède, *Histoire naturelle des poisons*, Vol. III, p. 164.

my attention because I had always found it hard to believe in the existence of these structures of lithophytic coral in Polynesia which are said to rise up from the Ocean's very abysses to the water's surface. It seemed more likely to me that these enormous masses of coral had a base of some primitive or volcanic rock to which they adhered in shallow waters. The formation, partly compact and lithographic and partly bubbly *Güines Limestone*,[1] had accompanied us to Batabanó. It is quite similar to Jurassic Limestone. And, to judge from their simple external appearance, the *Cayman* islets are made of the same rock. If the *Isle of Pines*' mountains, covered with (as the first historians of the conquest say) both *pineta* and *palmeta*, that is, areas of pine and palm trees[2] visible at 20 nautical leagues,[3] then they must be more than 500 toises high. I have been told that they are also made of a limestone entirely like that of Güines. These facts made me believed that I found this same rock (Jurassic) in the *Jardinillos*. Yet, while investigating the keys, which typically rise 5 to 6 inches above the water's surface, I saw only *fragmentary rock*, in which angular madrepore pieces are glued together by quartz sand. Sometimes the fragments were one or two cubic feet in volume, and the quartz grains disappeared such that one would be tempted to suspect lithophytic polyparies [stems] throughout several layers. The total rock mass of these keys appeared to be genuine *limestone agglomerate* quite similar to the tertiary limestone of the Araya Peninsula[4] near Cumaná, but far more recent. The uneven areas of this coral rock are covered by a shell and madrepore *detritus*. Everything above the water's surface is composed of broken and cemented lime carbonate shot through with quartz sand grains. Would one find polypary structures still alive at great depth below this fragmentary coral rock? Are these polyparies affixed to the Jurassic formation? I do not know. Mariners believe that the water table lowers in these passages, perhaps because they see the keys grow larger and higher through alluvial deposits either washed ashore by the waves or created through successive agglutination. It is not impossible, moreover, that over the course of centuries, the enlargement of the Bahama Channel (through which the *Gulf-stream*'s waters pass) caused a slight lowering of the water tables to Cuba's south, and especially in the Gulf of Mexico, which is the center of the massive wheel that is part of the pelagic stream that skirts the United States and deposits tropical plants on the Norwegian coast.[5] The

II.7

II.8

1. Vol. XI, pp. 235 and 236.

2. Petrus Martyr, *Oceanica*, Dec. III, book 10, p. 68.

3. Dampier, *Discourse of Winds, Breezes and Currents*, 1699, Chap. VII, p. 85.

4. Cerro Barrigón.

5. "The Gulf-stream between the Bahamas and [East] Florida is not much wider [and perhaps not much deeper] than the Behring's Strait; and yet the water rushing through

II.9 coastline's configuration, certain currents' direction, force, and duration, and changes in barometric pressure due to variations in the predominant winds, are the elements whose convergence can change the sea's equilibrium over a long period of time and within rather circumscribed limits of width and height.[1] Wherever the coast is so low that the ground level does not rise up II.10 more than a few inches for a league inland, these increases and decreases of the sea level vex the locals.

Cayo Bonito, which we visited first, deserves its name[2] because of its abundant vegetation. Everything indicates that it has been above the Ocean's surface for a long time, including the fact that the *Cayo*'s interior is almost no lower than its shoreline. On a 5 to 6 inch bed of sand and crushed shells which covers the fragmentary madreporic rock rises an entire Mangrove (Rhizophora) forest. From a distance, height and foliage suggest that these are laurels. The intertwining roots of the Avicennia nitida commonly known as black mangrove, Batis or Turtleweed, little Euphorbia, and some grasses an- II.11 chor the shifting sands. But it is ▼Jacquin's magnificent Tournefortia gnaphalodes with its silver leaves which distinguishes the *coral island* flora; we saw it here for the first time.[3] It is a sociable plant [it prefers to grow in bushes], a veritable bush 4 and a half to 5 feet in height, whose flowers spread a very pleasant fragrance. It also adorns Cayo Flamenco, Cayo Piedras, and probably most of the *Jardinillos* lowlands. While we were busy gathering herbs,

---

this passage is of sufficient force and quantity to put the whole northern Atlantic in motion, *and to make its influence felt in the distant Strait of Gibraltar and on the more distant coast of Africa*" ([Barrow,] *Quarterly Review*, February 1818, pp. 216–17 [emphasis is Humboldt's]). For the same influence on the Canary Islands, see *Relation historique*, Vol. IX, pp. 177 and 178.

1. I do not presume to explain, by the same causes, the grand phenomena on Sweden's coast where the sea appears to have inconsistently lowered between 3 and 5 feet over 100 years (▼Bruncrona and Hällström, in *Poggendorff's Annalen der Physik* [series 2, volume 2,] 1824, part 11, item VII, pp. 308–28; ▼Hoff, *Geschichte der Erdoberfläche*, Vol. I, pp. 405–6). The great geologist Mr. Leopold von Buch generated new interest in these phenomena by examining whether, instead, certain parts of the Scandinavian continent had imperceptibly risen (*Reise durch Norwegen*, Vol. II, p. 291). An analogous hypothesis was advanced about Dutch Guiana (Bolingbroke, *Voyage to Demerary*, p. 148).

2. *Bonito*, lovely.

3. We gathered: Cenchrus myosuroides, Euphorbia buxifolia, Batis maritima, Iresine obtusifolia, Tournefortia gnaphalodes, Diomedea glabrata, Cakile cubensis, Dolichos miniatus, Parthenium hysterophorus, etc. This last plant, which we found in Caracas valley and on the temperate plateaus of Mexico between 470 and 900 toises of elevation, covers all of Cuba's fields. The inhabitants use it for aromatic baths and to chase away fleas so common in tropical climes. In Cumaná, the leaves of many Cassia species, because of their smell, are used against pesky insects.

our sailors looked for spiny lobsters. Irritated at not finding any, they took out their disappointment by climbing the Mangroves and inflicting terrible carnage on the young *Alcatraz* grouped by twos in their nests. In Spanish America, the brown Pelican, the size of Buffon's swan, is known by this name. Exhibiting the foolish confidence and recklessness that is typical of large sea birds, the Alcatraz builds its nests with only a few tree branches. We counted II.12 four or five of these nests on the same Mangrove trunk. The young birds valiantly defended themselves with their enormous 6 to 7 inch beaks. ▾The adult birds hovered above our heads, letting out hoarse and plaintive cries. Blood dripped from the tree tops, for the sailors were armed with large clubs and *machetes*. We tried to chastise them for their lack of empathy and for their pointless cruelties. Condemned to long obedience on the solitary seas, sailors are more than pleased to impose a cruel dominion on animals as soon as the opportunity arises. The ground was covered with wounded birds in the throes of death. When we arrived, a deep calm had reigned in this little corner of the earth. Now, everything seemed to cry out: Man was here.

The sky had been covered with reddish vapors that thinned out toward the southwest. We waited in vain to see the *Isle of Pines* bluffs. These places have a charm lacking in most of the New World. They memorialize the greatest II.13 names associated with the Spanish monarchy, those of Christopher Columbus and Fernando Cortés. During his second voyage, on the southern coast of the island of Cuba between Jagua Bay and the *Isle of Pines*, the Admiral was astonished to see "this mysterious king who spoke to his subjects only through signs and this group of men wearing long white tunics resembling the monks of *la Merced*, while the rest of the people were naked." During his fourth voyage, Columbus encountered in the *Jardinillos* the large canoes of Mexican Indians, loaded with the Yucatán's rich products and merchandise. Seduced by his lively imagination, he thought that he heard from the mouth of these seafarers "that they had come from a country where the men were mounted on horses[1] and wore gold crowns on their heads. Already, "Cathay II.14

1. Compare the ▾*Lettera rarissima di Cristoforo Colombo di 7 Julio 1503,* p. 11, with Herrera, *Historia de las Indias occidentales,* Dec. 1, pp. 125, 131. There is nothing more moving and more pathetic than the expression of melancholy that pervades this letter from Columbus, written in Jamaica and addressed by the Admiral to King Ferdinand and Queen Isabella. I recommend to anyone who wishes to study the personality of this extraordinary man the account of the nocturnal vision in which, in the midst of a storm, a heavenly voice reassures the old man with these words: "Iddio maravigliosamente fece sonar tuo nome nella terra. Le Indie que sono parte del mondo cosi ricca, te le ha date per tue; tu le hai repartite dove ti è piaciuto, e ti dette potenzia per farlo. Delli ligamenti del mare Oceano che erano serrati con catene cosi forte, ti donò le chiave, etc." [God has marvelously made

II.15 (China), the Great-Khan's empire, and the Ganges' mouth" seemed so near to him that he hoped soon to make use of the two Arab interpreters whom he had picked up in Cádiz on his way to America. Other memories associated with the *Isle of Pines* and the *Jardines* concern Mexico's conquest. When Fernando Cortés was preparing for his grand expedition, he shipwrecked on one of the *Jardinillos* sandbars while sailing in his ship *Capitana* from the port of Trinidad to Cape San Antonio. Cortés was deemed lost for five days, when (in November 1518) the valorous Pedro de Alvarado sent three vessels to search II.16 for him from Carenas[1] port (Havana). Later, in February of 1519, Cortés convened his entire flotilla west of Batabanó across from the Isle of Pines near Cape San Antonio, probably in the place that still bears the name *Ensenada de Cortés*. From there, believing he could better escape Governor Velázquez's traps, he clandestinely left for the Mexican coast. How strange are human affairs! ▾Moctezuma's empire was conquered by a handful of men who sailed to the Yucatán coast from Cuba's far west. Today, three centuries later, the same Yucatán, as part of the new Confederation of the free States of Mexico, is in a position to threaten the west coast of the island of Cuba.

On the morning of March 11, we visited Cayo Flamenco. I calculated the latitude at 21° 59′ 39″. This islet's center is low, only 14 inches above sea level.

---

your name resonate throughout the earth. India, which is part of such a rich world, he gave it to you; you divided it as you pleased, and he granted you the power to do so. I give you the keys to the Ocean sea's ligaments which have been locked with such strong chains, etc.]. This nugget, full of elevation and poetry, comes to us only by way of an old Italian tradition, because the original Spanish cited in ▾Don Antonio León's *Biblioteca nautica* is lost. I might add other simple phrases from the mouth of him who discovered a new world: "Your Highness may believe me," said Columbus, "that the earth's globe is far from being as large as common people assert. For seven years, I was at your royal court, and for seven years I was told that my enterprise was madness. Now that I have opened the path, even tailors and cobblers ask the privilege to discover new lands. Persecuted and forgotten as I am, I cannot remember Hispaniola or Paria without my eyes brimming with tears. I have been in the service of Your Highness for twenty years; not one of my hairs has not turned white; my body is enfeebled; I can cry no more, *pianga adesso il cielo e pianga per me la terra; pianga per me chi hà carità, verità, giustizia*" [now the sky cries and the earth cries for me; he who has charity, truth, justice cries for me]. *Lettera rarissima,* pp. 13, 19, 34, 37 (*see* Vol. VIII, p. 307 and passim).

1. During this period there were still two settlements, one at Puerto de Carenas, in the old Indian province of Havana (Herrera, [*Historia de las Indias occidentales*,] Dec. I, pp. 276, 277); and the other, larger, in the town of San Cristóbal de Cuba. Only in 1519 were the two settlements united, and then Puerto de Carenas took the name of San Cristóbal de la Habana. See Vol. III [Q] p. 400 [this edition, pp. 45, 84]: "Cortés," Herrera said, "went to the town of San Cristóbal, which was at that time *on the south coast*, and then went to Havana" ([*Historia de las Indias occidentales*,] Dec. II, pp. 80 and 95).

Its water is barely salty. Other keys even have freshwater. Cuba's sailors, like the inhabitants of Venice's lagoons and some modern naturalists, attribute the freshness of the water to the sand's filtering effect on seawater. But what is this effect whose existence has not been verified by any chemical analogy? Besides, the keys are made of rocks not sand, and their smallness makes it hard to believe that rainwater collects there in permanent pools. Perhaps the keys' freshwater comes from the neighboring coastline, or even from Cuba's mountains, through hydrostatic pressure. This would prove an extension of the Limestone layers below sea level and the superimposition of coral rock on the limestone.[1] It is a common misconception that every freshwater or salt-water source is a distinct local phenomenon. Much like the rivers that furrow the globe's surface, water currents circulate subterraneously across great distances, between layers of rock of a certain density or texture. The learned engineer Don Francisco Lemaur, the same man who would later show such an energetic steadfastness in the defense of San Juan de Ulúa castle, reported to me that, in Jagua Bay, a half-degree to the *Jardinillo's* east, one can see freshwater sources bubbling up in the midst of the sea, two and a half leagues from the coast. The force with which these waters gush forth is so great that it causes a shockwave that is often dangerous for little dinghies. Ships that do not want to enter Jagua sometimes take on water at this salty source. This water is fresher and colder the farther down one gets. Their instinct has led manatees (*manatís*) to this region of nonsalty water. Fishermen, who are fond of the flesh of these herbivore cetaceans,[2] find them there in abundance and kill them out in the ocean.

II.17 II.18 II.19

Cayo Flamenco is a half mile to the east. We scraped two rocks just beneath

1. See Vol. XI, p. 236 and passim. The ancients knew of eruptions of freshwater in the midst of the sea, near Baiae, Syracuse, and Aradus (in Phoenicia). ▾Strabo, Book XVI, p. 754. The coral islands that surround Radak, especially the low island of Otdia, also have freshwater (Chamisso in Kotzebue, *Entdeckungsreise*, Vol. III, p. 108). One cannot encourage voyagers enough to examine these phenomena of the sea carefully.

2. Do they feed on seaweed, as we saw them on the shores of the Orinoco and the Apure feed on many Panicum and Oplismenus (*camalote*) species [grasses] (Vol. VI, pp. 234 and passim)? It seems, moreover, that it is a somewhat common phenomenon around river mouths on the Tabasco and Honduras coastline to find manatees swimming in the ocean, as crocodiles sometimes do. Dampier even differentiates between *Freshwater Manatí* and *Sea Manatí* (*Voyages and Descriptions*, Vol. II, pl. II, p. 109). Among the *Cayos de las doce leguas* to Jagua's east, there are some islands known as *Meganos del Manatí*. I have already noted elsewhere that the observations that we have just reported on crocodile and manatee behavior have great interest for the geognost who finds himself often baffled when he sees land animal fossils and pelagic productions in the same terrain.

the surface on which the waves crashed with a roar. They are[1] the *Piedras de* II.20 *Diego Pérez* (lat. 21° 58′ 10″). Here, the surface sea temperature goes down to 22.6° C, the water depth being no more than 6 and one-half feet. At night, we approached *Cayo de Piedras*, two reefs connected by shoals and oriented from NNW to SSE. As these reefs are very isolated—they form the *Jardinillos*'s eastern limit—many ships are lost there. *Cayo de Piedras* is almost entirely lacking in bushes, because the shipwrecked, in their distress, cut them down to make signal fires. The islet's edges are very steep on the seacoast. Toward the middle, there is a little freshwater basin. We found a block of madrepores over three cubic feet in size embedded in the rock. We had no more doubts that this limestone formation, which sufficiently resembled Jurassic limestone from afar, was only a fragmentary rock. Hopefully, one day field geognosts will examine this chain of keys surrounding the island of Cuba to determine what II.21 is caused by animals whose activities still continue in the sea depths and what belongs to real tertiary formations whose age goes back to that of the common Limestone abounding in remains of lithophytic coral. What is above water is generally only the peak or aggregate of madreporic fragments held together by lime carbonate, broken shells, and sand. It is important to examine on what this peak rests in each key. Does it cover structures of still living mollusks or, rather, secondary or tertiary rocks that seem to belong to our day because of the appearance of well preserved coral remains? The gypsum found on the keys across from San Juan de los Remedios on Cuba's northern coast deserves much attention. Its age probably goes back to prehistoric times, and no geognost would believe it to be the work of mollusks from our seas.

From *Cayo de Piedras*, we began to see the high mountains that rise beyond Jagua Bay toward ENE. We remained at anchor one more night. The following II.22 day (March 12), emerging by way of the passage between the northern Cape of *Cayo de Piedras* and Cuba's coastline, we entered a sea free of reefs. Its deep indigo color and its higher temperature proved how much deeper the sea was here. The thermometer, which we had often seen at 22.6° C when the water was 6 and a half to 8 feet deep, presently hovered at 26.2° C. During these experiments, the air temperature was 25° to 27° C during the day, as it had been in the *Jardinillos*. By following the variable land and sea winds, we tried to sail back east toward the port of Trinidad, thereby avoiding problems with the northeasterly winds that blew at that time, and make the crossing to Cartagena, whose meridian falls between Santiago de Cuba and Guan-

1. The Cayos de Flamenco, Diego Pérez, Don Cristóbal, and Piedras are situated 2′ farther north in the table of positions published by Mr. Espinosa (*Memorias de los navegantes españoles*, Vol. II, p. 65).

tánamo Bay. After passing the swampy *Camareos* coastline where in 1514, ▾Bartolomé de las Casas, famous for his humanity and his noble courage, had been offered[1] a good-sized *repartimiento de Indios* [allotment of Indians] by his friend, Governor Velázquez, we arrived (at 21° 50′ of latitude) at the meridian of the entrance to the *Bahía de Jagua*. The chronometer gave me a longitude of 82° 54′ 22″, nearly identical with what has meanwhile been published (in 1821) on the map of the *Depósito hidrográfico de Madrid*. II.23

Jagua is one of the island's most beautiful, but least frequented, ports. The *chief Chronicler* ▾Antonio de Herrera[2] had already said that "there is probably no other like it in the world." The landfill and defensive structures that Mr. Lemaur built during Count Jaruco's commission proved that Jagua's anchorage deserves the fame that it has enjoyed since the early days of the *conquest*. There is only a small group of houses and a small fort (*castillito*) to keep the British navy from repairing the vessels in the bay, as they had undisturbedly done during their wars with Spain. To the east of Jagua, the mountains (*San Juan Hills*) draw close to the shoreline and take on a more and more majestic appearance, not from their height, which does not seem to surpass 300 toises,[3] but from their steep, cleft slopes and their general shape. The coastline, I have been told, is so *steep* that a frigate can get close to it all the way to the mouth of the Río Guaurabo. When the temperature fell to 23° C at night and breezes from the land began to blow, we smelled the exquisite honey and flower aroma that characterizes Cuba's anchorages.[4] We hugged the coast for two or three miles. On March 13, a little before sunset, we found ourselves across from the mouth of the San Juan River, a place sailors fear because of its countless *mosquitos* and *zancudos* [crane flies]. Ships with heavy loads would be able to enter the ravine, if a sandbar (*placer*) did not block the passage's II.24 II.25

1. He renounced it the same year because of moral qualms he experienced during a brief stay in Jamaica.

2. [*Historia de las Indias occidentals*,] Dec. I, Book IX, p. 233.

3. Estimated distance: 3 nautical leagues. Angle of height not corrected for the earth's curvature and refraction: 1° 47′ 10″. Height: 274 toises.

4. See Vol. III [Q], p. 330 [Vol. XI, p. 141]. I already noted (this edition, vol. I, p.124) that Cuban wax, which is an important object of trade, comes from European bees (genus Apis, Latr.). Christopher Columbus expressly stated that Cuba's natives did not harvest wax at that time. The large loaf of this substance that he found on the island during his first voyage and that he presented to King Ferdinand at the celebrated Barcelona audience was later recognized to have been carried from the Yucatán on Mexican canoes (Herrera, [*Historia de las Indias occidentales*,] Dec. I, pp. 25, 131, 270). It is curious to observe that *Melipones' wax* was the first Mexican product to fall into Spanish hands since November 1492. See my *Recueil d'observations de zoologie*, Vol. I, p. 251; and *Essai politique sur le royaume de la Nouvelle-Espagne*, Vol. II, p. 455.

opening. A few hour angles gave me 82° 40′ 50″ for this port frequented by traffickers from Jamaica and even pirates from Providence. The mountains above the port barely rise to 230 toises.[1] I spent a good portion of the night on deck. What a deserted coastline! Not a single light announced a fisherman's cabin. Not a single village for 50 leagues from Batabanó to Trinidad. There II.26 are barely two or three pig *sties* or cow pens. However, this land has been inhabited since Columbus's time, even along the coast. When one digs in the earth to make a well or when torrents scar the earth during flash floods, stone axes and some copper tools often come to light,[2] artifacts of America's ancient inhabitants.

At sunrise, I had the captain take a sounding. The sea floor was at 60 fathoms; also, the Ocean's surface was warmer here than anywhere else. It was 26.8° C, 4.2° higher than what we found near the Diego Pérez *shoals*. Half II.27 a mile from the shore, the water temperature was no more than 25.5° C. We did not have a chance to take soundings, but the seafloor had lifted, no doubt about it. On March 14, we entered the Guaurabo River, one of *Trinidad de Cuba's* two ports, to drop off the *práctica* [pilot] from Batabanó, who had guided us across the *Jardinillo* sand-bars, running us aground several times. We were hoping to find a pack boat (*correo marítimo*) that could take us to Cartagena. I disembarked around nightfall and set up the Borda incline compass and the artificial horizon to observe the passage of some stars across the meridian. But we had barely begun our preparations when some Catalán storekeepers (*pulperos*), who had dined onboard a recently arrived foreign ship, invited us with much gaiety to accompany them into the city. These fine folks had us mount two by two onto the same horse, and, as the heat was excessive, we did not hesitate to accept such a generous offer. The mouth of the Río Guaurabo is nearly four miles northwest of Trinidad. The path passes II.28 through a plain that, one would suspect, has been leveled by water over the course of many years. The plain is covered with beautiful vegetation, which the *Miraguama*, a palm tree with silver fronds that we saw here for the first time, gave its particular character.[3] This fertile land, although it is *tierra colo-*

1. Distance: 3½ miles. Angle of height to the culminating point of Serranía: 3° 56′.

2. Probably Cuban copper. This metal's abundance in its natural state must have given Cuba's and Haiti's Indians the idea to smelt it. Columbus said that pieces of native copper weighing 6 *arrobas* could be found in Haiti, and that the Yucatán canoes he encountered on Cuba's south coast carried, among other Mexican merchandise, "cauldrons for melting copper" (Herrera, *Historia de las Indias occidentales*, Dec. I, pp. 86 and 131).

3. Corypha Miraguama. See the *Nova Genera*, Vol. I, p. 298. It is probably the same species whose bearing struck ▾Mr. John and Mr. William Fraser (father and son) in the Matanzas area. A few weeks before my departure for Cartagena, these botanists, who have in-

rada, only waits for the hand of man to be cleared and to bring forth abundant harvests. The west offers a picturesque view of the *San Juan Hills*, a limestone mountain range that reaches elevations of 1,800 to 2,000 feet, with steep inclines to the south. These bare and arid summits are sometimes rounded, sometimes craggy[1] and slightly inclined. Despite the dramatic temperature drops during the *Nortes* [Northwind] season, there is no snow on these mountains, nor on Santiago's, only, at times, ice and hoarfrost (*escarcha*). I have already mentioned elsewhere that the lack of snowfall is hard to explain.[2] When we left the forest, a curtain of mountains appeared whose southern slope was covered with houses. This is the city of Trinidad, founded, in 1514, by Governor Diego Velázquez, who was motivated by the "rich gold mines" that were said to have been discovered in the small Arimao River valley.[3] Trinidad's roads are very steep. People here complain, as they do in most parts of Spanish America, of the poor choice of terrain on the part of the *Conquistadores*, the founders of new cities.[4] At the city's northern end, there is the church of *Nuestra Señora de la Popa*, a famous pilgrimage site. This point appeared to me to be elevated 700 feet above sea level. Here, as from most of the streets, one has a magnificent view of the Ocean, of the two ports—*Puerto Casilda* and *Boca Guaurabo*—, of a palm tree forest, and of the high mountain range of San Juan. The next day, since I had forgotten to have my barometer and the rest of my instruments brought to the city, I tried, as an alternative, to determine the *Popa's* elevation by measuring the sun's height above the horizon of the sea and an artificial horizon. I had already tested this method[5] at Murviedro castle, in the Sagunto ruins [in eastern Spain], and on Cabo Blanco, near Guaíra. But the sea horizon was shrouded in fog and, in some places,

II.29

II.30

II.31

troduced a large number of precious plants to Europe's gardens, were shipwrecked when they were heading for Havana from the United States and saved themselves with difficulty on the keys around the entrance to the Old Bahama Channel.

1. Wherever rock was exposed above ground, I saw a compact, grayish-white limestone, in part smoothly porous, in part fissured, as is the case in the Jurassic formation. Vol. XI, pp. 229 and passim.

2. Vol. XI, p. [256].

3. This river enters into the Bahía de Jagua toward the east.

4. Should the town Velázquez founded have been situated on the plains closer to Casilda and Guaurabo ports? Some inhabitants think that fear of French, Portuguese, and British pirates compelled a choice of an inland site, on the slope of a mountainside, from which one could see, as from a crow's nest, the enemy's approach. But these fears, it seems to me, could not have been felt before Hernando de Soto's government. Havana was sacked by French pirates for the first time in 1539.

5. Vol. IV, p. 121 and passim. It is a method of finding the horizon's dip by means of a reflecting instrument.

broken up by blackish streaks that portend either gentle air currents[1] or an extraordinary play of refractions. In the *Villa* (today *Ciudad*) of Trinidad, Mr. [José Antonio] Muñoz, the administrator of the Real Hacienda, received us in his home with the utmost hospitality. I made observations for most of the night and, under conditions that were also not so favorable, I found the latitude near the cathedral through the ▾Spica Virginis, α of the Centaurus and β of the Southern Cross: 21° 48′ 20″. My chronometric longitude was 82° 21′ 7″. When I passed through Havana a second time upon my return from II.32 Mexico, I learned that this longitude was almost identical to what Frigate Captain Don José del Río, who had lived in this spot for a long time, had found, but that he had put Trinidad's latitude at 21° 42′ 40″. I have discussed this discrepancy elsewhere.[2] Suffice it to mention that Mr. de Puységur found 21° 47′ 15″ and that four stars of the Great Bear that Gamboa had observed in 1714 gave Mr. Oltmanns 21° 46′ 35″ (when he determined the ▾declination on the basis of Piazzi's catalogue).

Trinidad's *Teniente Gobernador*, whose jurisdiction extends to Villa Clara, Príncipe, and Sancti Spíritus, was the nephew of the famous astronomer ▾Don Antonio Ulloa. He threw a big party in our honor which was attended by some of the French émigrés from Saint-Domingue, who had brought their intelligence and their industry to these regions. Trinidad's sugar exports, according only to customs registries, still did not exceed 4,000 cases. People II.33 complained about "the obstacles that the central government put up against farming and trade in the island's central and eastern regions out of its unfair preference for Havana." They complained of "a large concentration of wealth, population, and power in the capital, while the rest of the country was nearly deserted. Several small centers, distributed at equal distances across the island, would be preferable to the current system, which had gathered luxury, moral corruption, and yellow fever in a single spot." These exaggerated indictments, complaints of provincial towns against the capital, are the same everywhere. There is no doubt that general well-being, in both political and physical organizations, depends on life being spread out evenly. There are dif-

1. Vol. IV, pp. 295 and 296. According to the great naturalist ▾Mr. Wollaston, whom I had the pleasure of consulting on this curious phenomenon, these black streaks are perhaps closer to the Ocean's surface when the wind begins to ruffle it. In this case, differences in color would render invisible to the eye the true horizon, which is farther away.

2. *Recueil d'observations astronomiques*, Vol. II, p. 72. In my Map of Cuba, I adopted the position that my March 14, 1801 observations gave me. In the map of *Depósito de Madrid*, published in Paris in 1824, Mr. Del Río's result was preferred (Espinosa, *Memorias,* Vol. II, p. 65).

ferences, however, between that which grows naturally and that which results from government intervention.

The advantage of having two ports is a topic of frequent debate in Trinidad. Perhaps it would be better if the municipality, which has few available funds, focused on improving only one of them. The distance from the town to Puerto de Casilda and Puerto Guaurabo is nearly the same. But transport costs are higher in Puerto de Casilda. The Mouth of the Guaurabo River, which is guarded by a new battery, offers safe anchorage but is less sheltered than that of Puerto Casilda. Only ships with shallow drafts or vessels that can be lightened enough to clear the sandbar can sail upriver to within a mile of the town. *Correos*, pack boats that come through Trinidad de Cuba on their way from the ▾Tierra Firme, generally prefer the Río Guaurabo, where they can anchor in complete safety without needing a pilot. Puerto de Casilda is a more enclosed place, more embedded in the land, but it cannot be entered without a pilot because of the Mulas and Mulatas shoals (*arrecifes*). The large harbor pier, built of wood and very useful to commerce, was damaged when pieces of artillery were unloaded. It was entirely destroyed, and people were uncertain whether it was better to rebuild it out of masonry, according to ▾Don Luis de Bassecourt's plans, or to dredge the Guaurabo sandbar. The greatest inconvenience of Puerto de Casilda is its lack of freshwater. Ships are forced to look for water a league away by rounding the western point and exposing themselves to pirates during wartime. We were told that Trinidad's population, including the farms that surround the city within a 2,000-toise radius, reached 19,000. Sugar and coffee cultivation have expanded prodigiously. European grains are cultivated only farther north, near Villa Clara.

II.34

II.35

We passed a very pleasant evening in the house of one of the wealthiest citizens, ▾Don Antonio Padrón. All of Trinidad's high society attended this informal gathering, a *tertulia*. We were once more struck by the playfulness and vivacity of spirit that distinguish Cuban women both in the provinces and the capital. These are fortunate natural gifts to which the refinement of European civilization might add even greater charm but which are nevertheless pleasing in their original simplicity. We left Trinidad on the night of March 15; our exit from the city hardly resembled the entrance we had made on horseback with the Catalán shopkeepers. The municipality had us taken to the mouth of the Río Guaurabo in a beautiful carriage lined in old crimson damask. And, to add to our embarrassment, a clergyman, the local poet, entirely dressed in velvet despite the heat, celebrated our Orinoco voyage in a sonnet.

II.36

On the way to the port, we were singularly struck by a spectacle that our two years' stay in the hottest part of the tropics should have made familiar. Nowhere else have I seen so many phosphorescent insects.[1] The ground cover, the branches, and the foliage were aglow with these creatures' moving reddish lights whose intensity varied according to the will of the animal that II.37 generates it. It was as if the firmament's starry vault had lowered itself onto the savanna! The countryside's poorest inhabitants would place about fifteen *cocuyos* in a gourd with holes to look for things in their cabins at night. Shaking the gourd forcefully stimulated the animals to brighten the phosphorescent disks located on either side of their thoraxes. People say, with sincerity, that the gourds full of *cocuyos* are lanterns that always burn. In fact, they do not go out unless the insects are either sick or dead. They are easy to feed with a little sugarcane. A young woman in Trinidad de Cuba told us that she had relied on the *cocuyos*' phosphorescence every time she breastfed her child at night during a long and difficult crossing to the Tierra Firme. The ship's captain did not want to have any other light on board, out of fear of pirates.

As the northeast breeze continued to freshen, we wanted to avoid the Cayman Islands, but the current dragged us toward them. By setting sail II.38 for the S¼SE, we lost sight of the palm-dotted shoreline, the hills in the town of Trinidad, and Cuba's high mountains. There is something solemn in the appearance of a land sinking, little by little, beneath the sea's horizon as one pulls away. This image had a special poignancy during a period when Saint-Domingue, a center of political unrest, threatened to engulf the other islands in one of these bloody fights that reveals the cruelty of the human species. Luckily, these threats and fears proved groundless. The storm has died down, even in the places where it was born, and a free black population, far from troubling the peace of the neighboring Antilles, has made some progress toward refining their manners and establishing good civil institutions. Puerto Rico, Cuba, and Jamaica, with 370,000 whites and 885,000 people of color, surround Haiti with its 900,000 blacks and mixed-race people, emancipated by the force of their will and their arms. Given more to subsistence agriculture than to cultivating colonial commodities, these blacks multiply II.39 with a speed surpassed only by the United States' population. Will the tranquility that the Spanish and British islands have enjoyed during the twenty-six years since the first Haitian revolution continue to inspire a deceptive sense of safety among whites who disdainfully oppose any improvement in the status of the enslaved class? In the Antillean Mediterranean, to the west and to the south, in Mexico, Guatemala, and Colombia, new laws work passionately to extinguish slavery. One can only hope that the confluence of

1. *Cocuyo* (Elater noctilucus).

these imposing circumstances will favor the good intentions of some European governments that would like to improve the slave's lot gradually. Fear of danger will wrest some concessions demanded by the eternal principles of justice and humanity.

On the Consumption of Sugar in Europe

II.40

One of the most interesting problems of political economy is to determine the consumption of commodities that, in the current state of Europe's civilization, are the principal objects of colonial production. One can approximate exact results, *upper and lower numerical limits*, in two different ways: (1) by discussing exports from regions that provide the largest quantity of these commodities, which are, for sugar, the Antilles, Brazil, the Guianas, Île de France [Mauritius], Bourbon [Réunion], and the East Indies; (2) by examining Europe's colonial imports and then comparing annual consumption with each region's population, wealth, and national habits. When there is only a single source for a product, as there is for tea, this kind of research is easy and quite certain. But difficulties increase for tropical regions, all of which produce more or less considerable amounts of sugar, coffee, and indigo. In this case, to establish a *lower numerical limit* for consumption, one must begin by paying attention to the larger quantities. If one knows from customs statements that the British, Spanish, and French Antilles annually export 269 million kg sugar, then it is of little concern whether the Dutch and Danish Antilles produce 18 or 22 million. If Brazil, Demerary, Berbice, and Essequibo export 155 million kg sugar, doubts about Surinam or Cayenne's production—less than a total of 12 million kg—have very little impact on how one meets Europe's overall consumption. The same applies to sugar imports from the East Indies to England, about which all sorts of exaggerated notions have been spread. Completely neglecting these imports would not lead one to err about current European consumption by more than one-forty-third, and a single one of the Lesser Antilles—for example, Grenada, Barbados, or St. Vincent—sends more sugar to Europe than all British possessions in the East Indies. I have already considered elsewhere (*Relation historique*, Vol. V, p. 296) the problem whose solution I discuss in this note. I had thought back then, on the basis of less extensive and less precise materials, that European sugar consumption in 1818 amounted only to 450 million pounds. Even for that time, this number would seem to fall short by a fifth or a fourth, but one must also not forget that the price of American sugar fell 38 percent between 1818 and 1823, and that consumption stands in an inverse relation to price (*Table of Prices* in Tooke, *Appendix to Part* IV, *idem,* 1824, p. 53; and [Powell,] *Statistical Illustrations*

II.41

II.42

of the British Empire, 1825, p. 56). In France, for instance, consumption increased by more than 40 percent between 1788 and 1825. It was 21 million kg in 1788, 34 million in 1818, and more than 50 million in 1825. Because of the rapid growth of colonial trade and European prosperity, it is important to pin down in numbers the state of things at a given time. Work of this type provides comparison points whose importance will be vividly felt by those who would want to follow ▾Mr. Tooke's footsteps into another century and trace both worlds' economic progress.

I. PRODUCTION. We will examine the state of agriculture only insofar as it contributes its products to European and United States trade. From this point of view, the Antillean Archipelago, Brazil, British and Dutch Guiana, II.43 Louisiana, Île de France [Mauritius], Bourbon [Réunion], and the East Indies are the only regions worthy of our attention. Between 1802 and 1804, Mexico exported 5 to 5.5 million kg sugar per year from Veracruz. Specifically:

in	1802.	439,132	arrobas at	1,476,435	p.
	1803.	490,292		1,514,882	
	1804.	381,509		1,097,505	
	1810.	121,050		272,362	
	1811.	101,016		251,040	
	1812.	12,230		30,575	

But the decline in price (from 3 piasters per *arroba* in 1823 to 1.6 piasters in 1825), the costliness of transportation from Cuernavaca, Puente de Istla, and Valladolid de Michoacan to the port of Veracruz, and political troubles have entirely devastated Mexican sugar exports. Exports from Venezuela, Cayenne, Guayaquil, and Peru are only for the coastal trade, part of the exchange of goods between the many parts of Spanish America.

We pointed out elsewhere (Vol. XI, p. 378) that, according to customs reg- II.44 istries between 1823 and 1825, the entire Antillean Archipelago exported at least 287 million kg sugar per year, three-fourths unrefined, one-fourth refined (in that discussion, we subtracted goods for the illegal trade). The island of Cuba alone contributes 56 million kg *azúcar blanco y quebrado* [white and brown sugar] to the legal trade. By dividing between the Greater and Lesser Antilles the 287 million kg sugar that the entire Archipelago produces, we find that the proportions are about the same as during a period when the production from sugarcane cultivation barely exceeded domestic consumption on the island of Haiti. Together, Cuba and Jamaica, whose combined surface area is 4,400 square nautical leagues and whose slaves number 623,500, export 136 million kg (150 million with contraband). The Lesser Antilles, with 940 sq. lg. and 524,000 slaves, export 144 million kg.

When comparing the countries that currently contribute the most to European and United States trade, we find that, on a scale of agricultural production, they can be placed in the following order:

BRAZIL	125	millions kg
(Saint-Domingue produced more than 80 million kg in 1788.)		
JAMAICA (area, 460 square nautical leagues).	80	
CUBA (area, 3615 s. l.), including contraband	70	
According to customs registries, 56 million kg		
BRITISH GUYANA	31	
GUADELOUPE (area, 55 square leagues).	22	
MARTINIQUE (area, 30 square leagues)	20	
ÎLE-DE-FRANCE [Mauritius] (area 108 square leagues)	14	
LOUISIANA (questionable result)	13	
BARBADOS or SAINT-VINCENT, each island	12.5	
Area of the first, 13 s. l.; of the second, 11 s. l		
GRENADA and ANTIGUA, each island	11	
Area of the first, 15 s. l.; of the second, 7 ½ s. l		
SURINAME	10	
EAST INDIES	10	
TRINIDAD (area, 139 square leagues)	9	
ÎLE-DE-BOURBON [Réunion] (area, 190 square leagues)	8	
SAINT-CHRISTOPHE [St. Kitts] and TOBAGO, each island	6	
Area, from 5 to 12 square leagues		
DOMINICA, NEVIS, and MONTSERRAT, each island less than	2	

II.45

II.46–47

Years	Imports from the British Antilles to Great Britain's ports	Exports from Great Britain		
		To Ireland	To other countries	Total re-exports
1761.	1,517,727 cwt	130,811 cwt	444,228 cwt	575,039 cwt
1762.	1,428,086	100,483	366,327	466,810
1763.	1,765,838	159,230	398,407	557,637
1764.	1,488,079	125,841	371,453	497,294
1765.	1,227,159	152,616	191,756	344,372
Average annual quantity	1,485,377	133,796	354,434	488,230
1771.	1,492,096	207,153	82,563	289,716
1772.	1,829,721	189,555	48,678	238,233
1773.	1,804,080	200,886	37,323	238,209
1774.	2,029,725	224,733	55,481	280,214
1775.	2,021,059	272,638	190,568	463,206
Average annual quantity	1,835,336	218,993	82,922	301,915
1781.	1,080,848	162,951	114,631	277,582
1782.	1,374,269	96,640	49,816	146,456
1783.	1,584,275	173,417	177,839	351,256
1784.	1,782,386	142,139	222,076	364,215
1785.	2,075,909	210,939	223,204	434,143
Average annual quantity	1,579,537	157,217	157,513	314,730
1791.	1,808,950	141,291	267,397	408,688
1792.	1,980,973	115,309	508,821	624,130
1793.	2,115,308	145,223	360,005	505,228
1794.	2,330,026	153,798	792,364	946,162
1795.	1,871,368	147,609	551,788	699,397
Average annual quantity	2,021,325	140,646	496,075	636,721
1801.	3,729,264	113,915	862,892	976,807
1802.	4,119,860	179,978	1,747,271	1,927,249
1803.	2,925,400	144,646	1,377,867	1,522,513
1804.	2,968,590	153,711	762,485	916,196
1805.	2,922,255	153,303	808,073	961,376
1806.	3,673,037	127,328	791,429	918,757
Average annual quantity	3,389,734	145,480	1,058,336	1,203,816
1809.	3,974,185	272,943	1,223,748	1,496,691
1810.	4,759,423	102,039	1,217,310	1,319,349
1811.	3,897,221	335,468	355,602	690,870
Average annual quantity	4,210,276	236,816	932,220	1,168,970

I repeat: the British quintal, or *cwt*, is equal to 50.8 kg. The preceding table was compiled under ▾Mr. William Irving's direction in the *Office of the Inspector-general of the Custom-house* in London. Between 1812 and 1815, exports from the British Antilles, Demerary, Berbice, and Essequibo were II.48

in	1812	3,551,449	cwt
	1813	3,500,000	
	1814	3,408,793	
	1815	3,493,116	

Only British Guyana contributed no more than 340,000 cwt per year during this period ([Powell,] *Statistical Illustrations*, p. 56). The following table, taken from the *Parliamentary Returns*, details sugar exports from the Antilles and from Guyana to different ports in Great Britain between 1816 and 1824.

II.49

British Antilles	Slaves in 1823	1816 (cwt)	1817 (cwt)	1818 (cwt)	1819 (cwt)	1820 (cwt)	1821 (cwt)	1822 (cwt)	1823 (cwt)	1824 (cwt)	Average export from 1816 to 1824 (cwt)
Jamaica.	342,382	1,389,411	1,717,259	1,653,303	1,614,346	1,769,124	1,679,720	1,413,717	1,417,746	1,451,332	1,567,328
Antigua	30,985	197,300	179,370	228,308	209,395	162,573	207,548	102,938	135,466	222,207	182,789
Barbados	73,345	288,623	239,732	249,076	282,456	179,951	211,371	156,682	314,630	245,828	240,928
Dominique	16,554	47,035	31,678	33,820	42,896	45,932	38,119	41,650	39,013	42,329	40,275
Grenada	25,580	266,055	196,959	220,958	204,565	184,551	216,367	199,178	247,369	227,613	218,180
Montserrat	6,593	28,981	31,214	36,919	37,168	32,815	33,282	27,071	24,466	30,648	31,396
Nevis	9,261	71,655	45,852	82,368	63,154	36,395	66,023	31,696	44,283	40,734	53,573
St. Christophe [St. Kitts]	19,817	124,757	125,977	130,218	141,501	89,501	128,436	89,682	76,181	132,585	115,426
S. Lucia.	13,794	69,830	56,401	42,006	78,719	50,220	77,971	92,060	62,148	73,100	66,939
S. Vincent	24,252	263,433	242,413	254,446	262,033	216,679	233,448	261,159	232,575	246,821	245,890
Tobago	14,314	139,157	132,387	112,930	132,544	109,194	108,243	100,725	113,015	123,868	119,118
Tortola	6,460	51,092	42,934	43,573	36,421	15,225	23,459	22,170	21,583	20,559	30,780
Trinidad	23,537	132,893	128,433	138,153	166,591	156,041	162,257	178,491	186,891	180,093	158,872
Total of Brit. Antilles	606,876	3,070,222	3,170,609	3,226,078	3,271,789	3,048,201	3,186,244	2,717,219	2,915,366	3,037,717	3,071,494

Guyana.	Slaves in 1823	1816 (cwt)	1817 (cwt)	1818 (cwt)	1819 (cwt)	1820 (cwt)	1821 (cwt)	1822 (cwt)	1823 (cwt)	1824 (cwt)	Average exports. (cwt)
Demerary	77,370	323,443	377,796	420,186	480,933	536,561	492,146	530,948	607,858	613,990	487,095
Berbice	23,356	15,308	14,158	17,764	29,967	37,696	53,257	55,357	55,995	64,608	38,235
Total for Brit. Guyana	100,726	338,751	391,954	437,950	510,900	574,257	545,403	586,305	663,853	678,598	525,330

Exports to Ireland's ports are not included here. According to information kindly communicated to me by Mr. Charles Ellis (today Lord Seaford), they were: II.50

1821 from Jamaica 21,785 cwt; from the other Brit. Antilles 123,037 cwt; from Brit. Guyana 24,843 cwt.
1822 from Jamaica 15,715 cwt; from the other Brit. Antilles 93,406 cwt; from Brit. Guyana 22,327 cwt.
1823 from Jamaica 28,490 cwt; from the other Brit. Antilles 149,994 cwt; from Brit. Guyana 21,605 cwt.
1824 from Jamaica 30,472 cwt; from the other Brit. Antilles 155,197 cwt; from Brit. Guyana 31,508 cwt.

From all this, one sees that production nearly doubled in Demerary and Berbice between 1816 and 1820; that it shrank over these last few years by nearly one eighth in Jamaica; but that increases in production on many of the Lesser Antilles, especially Trinidad, Antigua, and Saint-Lucia, rendered this decline less consequential for Great Britain's trade.

According to ▾Baron Delessert's researches, *Brazil*'s exports, which reached only 90 million kg during years of great drought, rose to 130 million. II.51

Louisiana (with more than 75,000 slaves) now probably exports nearly 13 million kg sugar. In 1810, Mr. Pitkin estimated production at 5 million kg, but it is said that, in 1815, the total harvest rose to 40,000 boucauts, or hogshead, (at 1,000 pounds per boucaut).

Exports from British and Dutch Guiana may be estimated at 40 million kg. The colony of Surinam alone yields

1820	18,086,000	pounds
1821	18,549,000	
1822	17,964,000	
1825	20,266,000	

In Île de France and Bourbon, sugarcane cultivation is making extraordinary headway. Although one should concede that Bourbon has become important only since 1814, its sugar exports were already at

in	1820 from	4,541,000	kg
	1821	4,926,000	
	1822	6,995,000	
	1823	5,608,800	

I owe these official figures to ▾Count des Bassayns de Richemont, a former intendant of the colony. A windstorm on February 24, 1823 reduced harvest for that year. Given the financial commissioner's reports, it was thought II.52

that production might increase to 8 million kg for 1825. But it must not be forgotten that the administration tends to exaggerate the island's wealth so as to justify tax increases, while the consultative Committee tends to make revenues appear less significant to prove that they are out of proportion with the tax burdens. In his compelling study of the *Commerce extérieur de la France et la question d'un Entrepôt à Paris* (1825, p. 150), Mr. Rodet puts Bourbon's sugar exports to the metropole only at 13,503,000 kg for the four years 1820–1823. ▾Sir Robert Farquhar, a former [British] governor of Île de France, saw this colony's exports rise from 8 million pounds in 1820 to 15 million in 1821, and to 25 million in 1822. It is believed that today exports exceed 30 million pounds. Because British customs tables conflate sugar from Île de France and the East Indies, and because sugar exports from the East Indies to all of Great Britain's ports before 1822 were no higher than 14 million kg (an amount that corresponds to 1820), it is likely that exports from ▾India's three Presidencies did not exceed nine to ten million kg. Moreover, all of the three Presidencies' sugar does not surpass the amount that flows from Île de France alone to Great Britain's ports. For instance, according to reports made on the state of external trade from Calcutta and Bombay from 1814 to 1821, these ports exported 24,411,000 rupees worth of sugar from the British East Indies during these seven years: 10.5 million to England, 2 million to the rest of Europe, and 5½ million to the United States. Exports from the three Presidencies to Great Britain's ports, which were valued at 1,139,400 rupees in 1815, had risen to 2,097,800 rupees in 1821 (▾[Larpent,] *On Protection to West-India Sugar*, 1823, p. 154).

II.53

II. CONSUMPTION. We can precisely enough determine sugar destined for Europe and the United States, or, rather, the quantities of sugar exported from and registered in America, from the islands Bourbon, Île de France, and in the East Indies. But it is much more difficult to ascertain how this sum breaks down from country to country. We will soon see that consumption is known with certitude only in Great Britain, France, and the United States, three countries that altogether consume 230 million kg. Statistical data gathered in the German states, in Holland, and in Italy offer unsatisfactory evidence: re-exports are in part mixed up with domestic consumption, and the complication of borders invite illegal trade. Comparing the population, well-being, and habits of the British and the French with the same elements in the rest of Europe, one can barely imagine how they use this enormous amount of sugar (495 million kg, or 9,744,000 cwt) exported each year from the ports of the Antilles, Brazil, the Guianas, America's islands, and the Indian Peninsula.

II.54

Great Britain's current domestic consumption is 142 million kg. Twice, in 1810 and 1811, it even reached 182,321,000 and 163,932,000 kg. It grew in the following progression after the end of the seventeenth century:

	average year				
from	1690 to 1699.	200,000	cwt or	10,160,000	kg
	1701 to 1705.	260,000		13,208,000	
	1771 to 1775.	1,520,000		77,216,000	
	1786 to 1790.	1,640,000		83,312,000	
	1818 to 1822.	2,577,000		130,912,000	

Sugar consumption has thus increased nearly thirteenfold in 124 years ([Liverpool East India Association,] *Report of a Committee of the Liverpool East India Association*, 1822, p. 41; [Powell,] *Statistical Illustrations*, p. 57), while the population has more than doubled (see Vol. XI, 62 and 63). England's population was 5,475,000 in 1700; Ireland's was 2,099,000 twelve years later; and Scotland's was probably 1.5 million in 1700. The United Kingdom had a total population of nearly 9 million in 1700 and more than 21,200,000 souls in 1822. Combining sugar consumption for all of the British isles (in Great Britain and Ireland), we get the following figures for an average year: II.55

from	1761 to 1765.	1,130,943	cwt or	57,452,000	kg
	1771 to 1775.	1,752,414		89,023,000	
	1781 to 1785.	1,422,024		72,239,000	
	1791 to 1795.	1,525,250		77,483,000	
	1801 to 1806.	2,331,398		118,435,000	
	1809 to 1811.	3,288,122		167,036,000	

The following table shows the relation between total imports in all of Great Britain's ports (without Ireland) and the small amounts of sugar from India.[1] II.56

1. The differences between duties for Antillean sugar and Indian sugar in Great Britain's ports is the primary cause keeping trade in the latter crop from gaining in importance. This inequality dates to a 1787 act of parliament, and it was exacerbated by 1813 and 1821 edicts. It is 10 shillings per cwt, or 50.79 kg. "If encouragement were given, through a reduction of duties in India and in this country," Mr. Cropper wrote, "in ten years, or perhaps less, India would produce more sugar than all Europe could consume" (▾*Letter to William Wilberforce*, p. 48).

QUANTITY OF SUGAR, IMPORTED TO RE-EXPORTED AND CONSUMED IN GREAT BRITAIN

Years	Total imports cwt	Sugar imports from the East Indies cwt	Re-exports: Raw Sugar cwt	Re-exports: Refined sugar cwt	Re-exports: Total cwt	Re-exported sugar from the East Indies cwt	Domestic consumption cwt
1810	4,808,663	49,240	616,896	413,209	1,319,350	7,095	3,489,314
1811	3,917,627	20,320	519,177	100,997	1,690,870	4,032	3,226,758
1812	3,762,182	72,886	674,314	284,617	1,158,162	6,964	2,604,020
1813	4,000,000	50,000	850,5[00]	450,000	1,615,500	10,000	2,384,500
1814	4,035,323	49,849	1,058,040	555,335	2,002,109	41,311	2,033,215
1815	3,984,782	125,629	870,992	609,247	1,906,712	68,422	2,078,070
1816	3,760,548	127,203	670,508	584,182	1,663,620	102,056	2,096,930
1817	3,795,550	125,894	486,693	697,087	1,671,740	95,494	2,123,809
1818	3,965,948	162,395	486,614	711,185	1,695,62[7]	110,325	2,270,322
1819	4,077,009	205,527	409,308	525,220	1,302,179	88,214	2,774,830
1820	4,063,540	277,228	504,303	679,565	1,659,156	186,603	2,404,385
1821	4,200,857	269,162	482,812	645,357	1,589,915	147,283	2,610,942
1822	3,643,127	226,476	411,159	374,784	1,048,297	102,467	2,594,830
AVERAGE	4,001,165	135,000	618,000	510,000	1,486,402	74,000	2,514,763

II.57 In this table, total unrefined sugar exports were estimated, using the formula that 34 cwt of unrefined sugar yields 20 cwt of refined sugar. In 1813, a fire destroyed London customs registries. The figures for that year were taken from [Powell's] *Statistical Illustrations* from 1825 (p. 56, 57). Compare [Tooke's] *Thoughts and Details on the High and Low prices*, 1824, *Appendix*, IV, p. 72.

In 1823, Great Britain's imports were 4,012,144 cwt, or 203,817,000 kg, and domestic consumption was 2,807,756 cwt, or 142,634,000 kg. When ▾Mr. Huskisson estimated this consumption at 3,000,130 cwt, or 152,406,000 kg, in an excellent parliamentary speech (from March of 1824), he probably meant the United Kingdom's total consumption. It must not be lost from view that the amount of sugar marked as *home consumption* in official registries is only the difference between imported and exported quantities, without regard for sugar that builds up in storage from year to year. The average value of imports, varying with market prices and commercial activity, had risen (from 1813 to 1815) to 10 and 12 million pounds sterling. In recent years, from 1820 to 1823, II.58 they were only 6 million pounds sterling. It follows that partial consumption of Indian sugar in Great Britain rose as follows:

1808 to	23,526	cwt
1809	9,313	
1810	42,145	
1820	90,625	
1821	121,859	
1822	124,009	

It has, therefore, grown nearly sixfold in twelve years (see also [Larpent,] *On Protection to West-India Sugar*, 1823, pp. 9, 148). Today, production from the British Antilles alone would amply satisfy the needs of the British people. Yet, this population forms only 7 percent of the total European population, while Great Britain's sugar consumption accounts for nearly 30 percent of all European sugar imports.

France consumed no more than one-fifth (at most one-fourth) of its colonies' sugar in 1788. ▾Mr. Peuchet (*Statistique élémentaire de la France*, p. 406) estimated the kingdom's consumption during this period at 21,266,000 kg refined sugar. According to ▾Mr. Chaptal, it was still no more than 25,220,000 kg in 1801. But, according to customs statements between 1816 and 1821, France received, in kg: II.59

Years	Sugar from the French Colonies	Sugar from Foreign Countries	Total
1816	17,530,000	7,049,000	24,579,000
1817	31,102,000	5,443,000	36,545,000
1818	29,809,000	6,277,000	36,086,000
1819	34,360,000	5,400,000	39,760,000
1820	40,752,000	8,467,000	49,219,000
1821	41,702,000	2,649,000	44,351,000

This amounts to 32,542,000 kg sugar from France's colonies in an average year and 5,881,000 kg foreign sugar, for a total of 38,423,000 kg. If we limit ourselves to results from the last four years, 1820 to 1823, France's average imports are 48,019,636 kg sugar, 40,367,452 of which come from the French Antilles and Cayenne, 3,375,888 from Bourbon [Réunion], and 4,276,296 from India, Brazil, and Havana. Of these 48,019,636, 1,123,158 kg refined sugar are re-exported in an average year, as are 3,707,507 kg molasses. France's consumption was therefore nearly 44 million kg a year between 1820 and 1822 (Rodet, *Du commerce extérieur*, p. 154). According to notes that the ▾Count Saint-Cricq, president of the bureau of commerce, was kind enough to give me, the amount of sugar imported to France in the last four years was II.60

in	1822	55,481,004	kg
	1823	41,542,856	
	1824	60,031,122	
	1825	56,081,506	

In 1825, 3,264,734 kg refined sugar were re-exported, along with 4,856,775 kg molasses. Taking into account the sugar contained in molasses, it follows that France consumed more than 51 million kg unrefined sugar. Between 1788 and 1825, sugar consumption grew in France and England in the proportions of 10 to 24.4 and 10 to 17.3. But, from 1819 to 1825, growth was much speedier in France, where sugar consumption rose from 39.8 million to 51 million kg.

II.61 According to information for which I am indebted to ▾Mr. Gallatin, the *United States*, over the three years 1800, 1801, and 1802, averaged 116,644,000 pounds of sugar imports, including light-brown sugar or cassonade, and 71,676,000 pounds of re-exports, resulting in a total consumption of 44,668,000 pounds (*Essai politique sur la royaume de la Nouvelle-Espagne*, pp. 846–47). Mr. Pitkin (*Statistical View,* 1816, p. 249) estimates this consumption at 70 million British pounds, or 31,500,000 kg, for 1815. However, for the ten-year average (1803–1812), ▾Mr. Seybert (*Annales statistiques*, 1820, p. 129), using customs registries, derives for the beginning of the nineteenth century no more than 120,613,130 pounds of sugar imports and 66,243,660 of re-exported sugar, for an average consumption of 54,369,470 pounds. This estimate does not include molasses, consumed at a rate of 7,355,000 pints during the same period. From 1821 to 1825, sugar exports to the United States were 75 million pounds in an average year, of which 4.3 million pounds came from the East Indies, Île de France, and Bourbon. Re-exports reached 18 million pounds a year during the same period. The United States, therefore, consumed 57 million pounds of sugar from the Antilles and the East Indies, 15 million pounds from Louisiana, and 8 million pounds of maple sugar, for a total of 36 million kg.

When we correlate the population of the island of Cuba with that of Great II.62 Britain, the United States, and France with the amount of raw sugar each consumed annually, we find that sugar consumption decreases depending on degrees of well-being and, above all, national habits.

Country	Annual Consumption in kg of Raw Sugar	Free Population	Annual Sugar Consumption per Head
Island of Cuba	11 million	450,000	24.4 kg
Great Britain	142 million	14,500,000	9.8 kg
United States of America	36 million	9,400,000	3.8 kg
France	52 million	30,600,000	1.8 kg

I mentioned (Vol. XI, pp. 368 and 369) the prodigious amount of sugar consumed in the tropical regions of America by people of the Spanish race. I focused on the free population only. But black slaves also consume raw sugar in factories during production. Figures for Ireland being insufficiently precise, I gave only Great Britain's consumption in the above table, estimated today at approximately 2.8 million cwt. Ireland's direct imports (Vol. XII, p. 170 [this edition, p. 180]), leads one to believe that this country, with an impoverished population of 6.8 million inhabitants, consumes no more than 12 million kg annually, which comes to 1.8 kg per head. In 1825, United States consumption, for both free and slave populations (probably 11,138,000), would still show 3.2 kg per individual, or a third more than in France. Mr. Pitkin's estimate (31.5 million kg for 1825) was probably too high. This estimate would amount to 4.6 kg for a free population of 6,983,000 at that time. II.63

Today, relative consumption for Cuba, Great Britain, France, and the United States shows roughly the following proportions: 13.6 to 5.4 to 2.1 to 1. If one hypothesizes that the United Kingdom's consumption (Great Britain and Ireland) is 152.5 million kg, which is not entirely accurate, then one arrives at 7.2 kg per person for a population of 21.3 million inhabitants of vastly different prosperity. Before adding some speculations about consumption in other parts of our continent to the more or less accurate data for the United States, Great Britain, and France, we will state the total amount of sugar infused into the trade per year: II.64

Antillean Archipelago	287	million kg
British Antilles	165	million kg

Elsewhere (Vol. [XI], pp. 3[72] and 3[73]), we estimated Jamaica's average exports to Great Britain and Ireland's ports—exports should not be confused with production—between 1816 and 1824 as 1,597,000 cwt, or 81,127,000 kg. For the rest of the British Antilles, they were 1,634,000 cwt, or 83,007,000 kg, for a total of 3,231,000 cwt, or more than 164 million kg. Focusing on the last five years (1820–1824) would show 1,573,000 cwt, or 79,908,000 kg for Jamaica in an average year, according to the same data. For the other British Antilles, 1,564,000 cwt, or 79,451,000 kg, for a total of 159,359,000 kg. The difference, depending on whether one takes the averages after 1816 or after 1820, is, therefore, no more than 4.5 million kg, or 88,500 cwt, an amount much less than the variations in Jamaica's sugar exports to Europe in any two consecutive years. If we rank the British Antilles according to the amount of sugar they contribute to trade, they appear in the following order: Jamaica, St. Vincent and Barbados (nearly equal in production), Grenada, Antigua, Trinidad, Tobago, St. Croix, St. Lucia, Dominica, Nevis, Montserrat, and Tortola.

II.65

SPANISH ANTILLES	62	million kg

This number includes only registered amounts: if we add in contraband, Cuba alone exports more than 70 million kg.

FRENCH ANTILLES	42	million kg

The slave population of the French and the Spanish Antilles stands in the same exact relation to their sugar exports, which proves that Cuba's soil is very fertile; for nearly one-third of this island's slaves live in the large towns (Vol. XI, p. 300; and Vol. XII, pp. 4, 5 and 6 [this edition, pp. 109, 117]).

II.66

DUTCH, DANISH AND SWEDISH ANTILLES	18	million kg
	287	million kg
BRAZIL	125	million kg

In 1816, exports were 5.2 million kg higher, but I have already noted above that exports declined to 91 million kg during years of extreme drought.

BRITISH, DUTCH, AND FRENCH GUYANA	40	million kg

Over the last five years (1820–1825), Demerary, Essequibo, and Berbice or British Guyana exported 30,937,000 kg. Crops from this part of Guiana increase as crops from the British Antilles decrease. Between 1816 and 1824, British Guyana averaged 525,000 cwt, or 26.5 million kg, which indicates an annual export growth of 4.5 million kg, or one-eighth, while exports from the British Antilles declined by 4.5 million kg, or one thirty-fifth, according to a comparison of averages between 1816 and 1824, and between 1814 and 1824.

II.67

LOUISIANA	13	million kg
EAST INDIES, ÎLE DE FRANCE, AND BOURBON	30	million kg

Île de France, 12 million kg; East Indies, more than 10 million kg; Bourbon, 8 million kg. Here exports to the United States are included with exports to

Europe, as they are throughout. For the East Indies to displace the British Antilles, their sugar exports would have to be 16 times larger.

Total	495	million kg

I have meticulously noted the sources from which I derive my overall results; without attribution, research of this sort has little value. The reader must be brought to a level where he can examine the partial data. Today, doubts remain only about small amounts (for example, Puerto Rico, Curaçao, and St. Thomas exports), and about Brazil's fluctuating sugar production. Estimating these variations, or the margin of error, at 35 million kg, total exports would still change by only one fourteenth. Subtracting 38 million kg for the consumption of the United States and British Canada, 457 million kg sugar remain for annual export to Europe (seven-eighths unrefined, one-eighth refined). This is a *lower limit*, because all the variables in this calculation are from customs registries and do not include fraudulent trade. Dividing the total amount of unrefined sugar consumed in Europe by the number of inhabitants (208.5 million), we get 2.2 kg per capita. But this figure is merely a bloodless mathematical abstraction that is as pointless as attempts to divide the population contained in the cultivated regions of the United States or Russia by a total area of 174,000 and 616,000 nautical leagues. Europe has 55 percent, or 106 million, inhabitants who, distributed across the British Empire, the Netherlands, France, Germany proper, Switzerland and Italy, consume an enormous amount of sugar, and 33 percent, or 73 million, is dispersed throughout Russia, Poland, Bohemia, Moravia, and Hungary, countries where the poverty of the majority of the population renders consumption singularly low. These are the extreme points on the scale of society's luxuries or artificial needs. To illustrate the German population's prosperity, I note here that in the port of Hamburg alone, nearly 45 million kg sugar were imported in 1821, while in 1824, 44,800 cases or 29,120,000 kg were imported from Brazil, 23,800 cases (or 4,379,000 kg) came from Havana, and 10,600 barrels (or 8,480,000 kg) came from London, for a total of 41,979,000 kg. In 1825, they imported 31,920 cases (or 20,748,000 kg) from Brazil; 42,255 cases (or 7,774,900 kg) from Havana; and 20,506 barrels (or 16,404,800 kg) from England; for a total of 44,927,000 kg. In 1825, Hamburg's imports were therefore only one sixth less than all of France's. The port of Bremen imported nearly 5 million kg in 1825; Antwerp imported 10,758,000 kg in the same year. In Germany's south, where sugar consumption is also quite considerable, the complexities of transportation and the illegal trade make statistical research very difficult. For example, how can we believe ▾Mr. Memminger that 1,446,000 in-

II.68

II.69

habitants consume no more than 980,000 kg sugar a year in the kingdom of Württemberg, which is enjoying great prosperity?

Subtracting from the 457 million kg unrefined sugar imported into Europe the 204,500,000 kg unrefined sugar consumed in France and the three parts of the United Kingdom, and assuming consumption of 2 kg per capita (a very high estimate) for the 76 million people in the Netherlands, Germany proper, Switzerland, Italy, the Iberian Peninsula, Denmark, and Sweden, nearly 100.5 II.70 million kg remain for Asia Minor, the ˺Barbary coast, Siberia's western districts, and the parts of Europe inhabited by the Slavic, Hungarian, and Turkish races. Yet, Morocco, Algeria, Tunisia, and Tripoli's populations are rather considerable, adding up to a total of 24 million. Asia Minor has more than 4 million people. If one counts only the population on the coast, which is dotted with large commercial towns, it would be no exaggeration to assume that the coasts of Africa, Asia Minor, and Syria receive 10 million kg in exports. From these data, one should conclude that Slavic, Magyar, and Turkish Europe's 80 million inhabitants (Russia, Poland, Bohemia, Moravia, Hungary, and Turkey) still consume 1.13 kg per capita. This result is somewhat surprising, if we compare these regions' current level of civilization with that of France. One would expect much lower consumption. Rather than being exaggerated, however, estimates of sugar exported from America and the East Indies to Europe and the United States are probably well below actual numbers. If customs fraud makes the consumption in France and Great Britain (two countries that have served as benchmarks in the preceding discussion) higher than one would expect, and if one wishes to assume that the French II.71 and the British still consume more than 1⅘ and 9⅘ kg per capita, one must not forget that the same sources of error exist for export estimates from America and the East Indies. In 1810, when Great Britain consumed nearly 177.5 million kg, the rate was 12.2 kg per capita. It would be desirable for someone who has the habit of precision in quantitative research and who could draw on reliable sources to write a book about the important problems in European sugar, coffee, tea, and cacao consumption over a given period of time. This study would take many years, for many documents are unpublished and could be obtained only through active correspondence with the largest commercial houses of Europe. I could not fully devote myself to this research. The time is approaching when colonial commodities will, in large measure, be produced in independent countries rather than in colonies, not by islands but by the large American and Asian continents. The history of trade between peoples lacks quantifiable data for the whole of society, and this gap cannot be filled until such a time when large revolutions threaten the industrial world and someone has the courage to gather scattered materials and submit them to rigorous analysis. II.72

I conclude my research by comparing sugarcane, beet sugar, and wheat production in the tropics and in the middle region of Europe. On the island of Cuba, one hectare produces 1,330 kg refined sugar, valued locally at 870 francs, assuming that a case of sugar (or 184 kg) costs 24 piasters (Vol. XI, pp. 396, 397, 398, 414, 415, and 416). In Havana and Matanzas, land prices are considered extremely high when a *caballería* costs between 2,500 and 3,000 piasters; yet, this is no more than roughly 1,000 francs per hectare, since a *caballería* has 13 hectares. It is known that land prices are as high as that—between 2,500 and 3,000 francs—in the Paris region. Land of average fertility yields 500 kg unrefined beet sugar per hectare, a value of 450 francs. People claim, however, that, on very fertile land in Beauce and Brie, a hectare yields more than 1,200 kg. Assuming that harvests are increasing eightfold in France, a hectare of land yields 1,600 kg wheat valued at 288 francs, pegging 100 kg wheat at 16 to 20 francs. ▼Lavoisier estimated a kilogram of wheat at 4 sous, which equals 20 francs for 100 kg. In the Antilles, one hectare thus yields within about one fifth the amount of sugarcane that wheat would produce in the temperate zone. Amylaceous grain seeds weigh only 270 kg per hectare II.73 more than the crystallized sugar extracted from sugarcane in the tropics. In all of France, an adult consumes between 1.5 and 1.75 pounds of bread per day, or 200 kg wheat per year. Lavoisier figured that a population of 24,676,000 (Peuchet, *Statistique de la France*, p. 286) consumed 11,667 million pounds of wheat, rye, and barley, which comes to about 230 kg a year per person. In Paris, bread consumption is only 168 kg a year (Chabrol de Volvic, *Recherches Statistiques*, 1823, p. 73). France consumes 125 times more wheat than sugar per capita, England barely 23 times. In Paris, spending on bread is estimated at more than 38 million francs, while annual spending on sugar—a large portion of sugar is re-exported to the departments—amounts to 27 million francs (*Budget et Comptes de la ville de Paris pour 1825*, p. xvi).

Above, I listed beet production estimates for the Paris region according to processes that were commonly used 4 to 5 years ago. As this crop continues to provoke lively curiosity in the Antilles, I will report here the most recent data that ▼Mr. de Beaujeu outlined in a fascinating report to the Academy of Sciences in August of 1826. This great agriculturalist was kind enough to give II.74 me an excerpt from his report. And since his results are far superior to those obtained through older methods, I will insert them here:

"In considering the cultivation of the sugar beet at large, especially that of the *yellow variety*, in the parts of France particularly suitable for it, such as

Beauce, Brie, parts of Normandy, and the kingdom's northern plains," said Mr. de Beaujeu, "I estimate a hectare's typical yield at 30,000 kg,[1] which is based on my own experience. In less fertile countries, 20,000 kg is rather a high estimate. This same *yellow variety* should yield at most 5, at least 4 percent unrefined sugar, *including that which is furnished as a byproduct of molasses refining*. Assuming, then, 30,000 kg tubers per hectare for the *fertile parts* of France, 1,200 to 1,500 kg unrefined sugar can be extracted from roots that are well stripped and well worked during the right season; refining will yield 750 kg loaf sugar, 450 kg brown sugar, and 300 kg molasses for rum production, which amounts to 50 percent loaf sugar, 30 percent brown sugar, and II.75 20 percent molasses. Given the current improvements in native sugar production, one can accept an average of 1,000 to 1,200 kg unrefined sugar per hectare.

"Beets from fertile terrain yielding 30,000 kg per hectare should yield 75 percent[2] squeezed beet juice, which, in turn, yields 5⅓ to 6⅔ percent of unrefined sugar including the juice derived from molasses reboiling which has become quite profitable since the improvements in syrup production. To the best of my knowledge, there are currently (1826) no more than 50 beet sugar factories in France that can produce more than 500,000 kg unrefined sugar of various qualities. But, the largest of these factories is far from yielding 50 percent loaf sugar. One always assumed, in 1812, that there were 200 factories that could produce a million kg unrefined sugar. But many of these factories succeeded only in producing syrups or moscovade of the worst quality which is very difficult to use. In fertile soil, it is easy to get a good beet harvest every II.76 three years. For a long time, I had one every two years where the soil was best for this crop. If France currently consumed 56 million kg unrefined sugar, no more than 168,000 hectares of good land, with one-third, or 56,000 hectares, devoted to beet cultivation each year, would be required to provide all the sugar that the whole kingdom needs."

1. Compare above [p. 194 in this edition], Vol. XI, pp. 396, 397, and 398.
2. Vol. XI, p. 416, note.

METEOROLOGICAL OBSERVATIONS MADE IN HAVANA'S BOTANICAL GARDEN IN 1825 BY ▼DON RAMON DE LA SAGRA, PROFESSOR OF NATURAL HISTORY

	Barometer			Thermometer Centigrade			Hygrometer			
Month	Max.	Min.	Averages	Max.	Min.	Averages	Max.	Min.	Averages	Wind Directions
	po li	po li	po li							
January	28 5.5	27 11.8	28 1.8	26.5°	15.0°	21.42°	97.0°	69.0°	73.29°	E and ENE, 8. SSE and SW, 19. NNE and NW, 12.
February	28 0.5	27 11.5	28 4.5	26.5	15.0	22.85	95.0	70.0	80.45	SW, S and SE, 38 NE, N and NW, 21. E and ENE, 15.
March	28 1.9	27 9.3	27 11.92	29.5	19.0	23.72	98.0	73.2	88.47	S and SE, 65. N and NE, 12. E and ENE, 10.
April	28 2.5	27 10.0	28 1.32	30.2	19.0	24.15	98.0	66.0	84.94	S and SE, 34. N and NW, 15. E And ESE, 23.
May	28 1.5	28 0.1	28 1.09	30.2	21.9	25.06	97.0	75.2	83.54	S and SE, 17. NE, 12. E and ESE, 18.
June	28 2.1	27 10.3	28 0.45	31.0	23.0	28.12	96.0	77.3	87.41	S and SE, 33. NE and NNE, 16. E, ESE and ENE, 21.
July	28 2.8	28 0.2	28 1.79	31.7	20.0	28.22	96.0	71.8	85.19	SW, SSE, 37. NE, 11. E, ESE, 22.
August	28 1.7	28 0.0	28 1.42	31.6	21.0	.	96.2	78.0	86.98	S and SE, 40. NE, 18. E and ESE, 23.
September	28 0.7	27 10.5	27 11.31	31.4	23.9	28.52	96.0	82.1	88.65	S and SE, 48. NE and NW, 22. E and ENE, 5.
October	28 1.8	27 7.5	28 0.24	30.4	24.1	27.35	99.0	81.0	90.42	S and SE, 22. NE and NW, 45. E and ENE, 16.
November	28 2.9	27 11.8	28 1.24	27.8	19.0	23.54	99.0	75.0	87.26	S, SE, 32. NE, 19. NNE, ESE, 22.
December	28 4.9	28 0.3	28 2.45	28.0	15.4	21.62	99.0	71.0	84.24	S, SE and SW, 26. N, NE and NW, 44. E and NNE, 14.
Annual	po li	po li	po li							
Average.	28 5.5	27 7.5	28 1.05	31.7°	15.0°	24.9°	99.0°	66.0°	85.45°	SW, S, SSE and SE, 407. NE, N and NW, 259. ENE, ENE, E and ESE, 197.

II.78 *January*, 7 days of rain. *February*, 9 days of rain. The barometer reached its highest level during this month and during the two that preceded it. *March*, heavy showers for 7 days; hail. *April* and *May*, little rain. *June*, 8 days of rain. *July*, beginning of the southern storm season; storms; 8 days of rain. *August*, calm with winds from the S and SE; 7 days of rain. *September*, calm preceding the southerly gales (*chubascos*); heat wave; 13 days of rain. *October*, heavy showers under a menacing sky. Trinidad de Cuba suffered greatly from a hurricane on October 1. On the same day, there was a precipitous drop in barometric pressure. *November*, little rain; thick clouds to the south and southwest. *December*, N and NW winds dominate; some gales; overcast and hazy sky. 75 days of rain over the entire year. Comparing this year of temperature observations in Havana with averages from three years of Ferrer's observations (Vol. XI, p. 264), we find the following:

Average annual temperature in 1825, 24.9°; from 1810 to 1812, 25.7°.
Average temperature of the hottest month, 28.5°; 28.8°.
Average temperature of the coldest month, 21.4°; 21.1°.

The instruments were compared with those at the royal Observatory of II.79 Paris. The *barometer* is divided into inches and lines (old French divisions). The *thermometer* is centesimal. The *hygrometer*, following ▾Saussure's design, uses hairs. Figures attached to wind directions do not indicate length but show how many times the wind blew from such and such a direction. Averages are taken from all of the observations made three times a day. Hour variations in measurements were 0.7 to 1.7 lines.

▾On the temperature in different parts of the Torrid Zone at sea level.

Exact information on the climate of Havana and Rio de Janeiro, both situated below the Tropics of Cancer and Capricorn, round out the notions that we have developed on average temperatures in different parts of the equinoctial region. This region most likely offers the *maximum* mean yearly temperature below the equator. The temperature decreases almost imperceptibly from the equator to 10° lat. It decreases more quickly from the 15th to the 23rd parallel. What strikes the traveler moving from the equator to the tropics is not so much the decrease in mean annual temperature but the unequal distribution of heat over the course of the year. It is incontestable that the numerical constants of tropical Climatology have still not been determined with uniform precision. One must constantly work at perfecting them. But already in the current state of the science, one can assign certain margins of error to these figures that subsequent observations will probably not change. We mentioned elsewhere (Vol. XI, p. 253) that the mean temperatures of Havana, Macao, and Rio de Janeiro, three places situated at sea level on the edge of the equatorial zone in both hemispheres, are 25.7°, 23.3°, and 23.5° C, and that these differences stem from the unequal distribution of neighboring lands and seas. What is the equator's mean temperature? This question was recently raised in a report that ▾Mr. Atkinson published in the second volume of *Memoirs of the Astronomical Society of London* (pp. 137–83), which contains many very thoughtful considerations of various important Meteorological points. Using my own observations and the most rigorous mathematical tools, the learned author attempts to deduce that the mean temperature for the equator is at least 29.2° C (84.5° F) and not 27.5° C (81.5° F) as I had assumed in my essay *On Isothermal Lines*. ▾Kirwan had suggested 28.8° C; Mr. Brewster, in his climatic Formulas, 28.2° (*Edinburgh Journal of Science*, 1829, no. 7, p. 180).

II.80

II.81

If the issue here were to determine an average temperature for an equatorial belt encircling the entire globe and delimited by the parallel of 3° N and 3° S, one would have to examine above all the equatorial Ocean temperature, for only one sixth of the globe's circumference in this belt belongs to the Tierra Firme. Yet, the Ocean's mean temperature within the limits that we have just defined generally fluctuates between 26.8° and 28° C. I say generally because one sometimes finds within these limits *highs* restricted to zones barely a degree wide, and in which the temperature rises from 28.7° to 29.3° C in different longitudes. I measured the latter temperature, considered extremely high for the Pacific Ocean, to the east of the Galapagos Islands. Furthermore, ▾Baron Dirckinck de Holmfeldt, a very knowledgeable officer in the Danish navy who made a large number of thermometric observations at my request, recently detected a surface temperature of 30.6° almost on the Punta Guascama parallel (lat. 2° 5′ N; long. 81° 54′ W). Not even the equator has such *highs*. One observes them sometimes to the north, sometime to the south of the equator, often between 2.5° and 6° latitude. The great circle that transects the points where the sea's waters are warmest cuts the equator at an angle that seems to vary with the sun's angle. In the Atlantic Ocean, we ourselves passed at various times from the northern temperate zone to the southern temperate zone without seeing the temperature climb above 28° C *in the belt of warmest water*. *Highs* were 28.2° C for ▾Perrins, 28.7° C for Churruca, 28.6° C for Quevedo, 28.8° C for Rodman, and 28.1° C for John Davy. The air above these equatorial waters is 1° to 1.5° C colder than the Ocean. It follows from these facts that the pelagic equatorial belt, rather than having a mean temperature of 29.2° C (84.5° F), probably does not reach 28.5° C on five-sixths of the globe's circumference. Mr. Atkinson himself agrees (p. 171) that the mix of oceanic and continental regions tends to diminish the equator's mean temperature. But, limiting himself to southern America's continental plains, this learned man, using different theoretical assumptions, settles on 29.2° or 31° C for the equatorial zone (from 1° N to 1° S). He bases this conclusion on the fact that the mean temperature is already 27.6° C at 10° 27′ lat. in Cumaná and that the mean temperature for the equator should be at least above 29.2° C, according to the law of increasing heat from the pole to the equator (an increase that depends on the square of the cosine of latitudes). Mr. Atkinson confirmed this result by reducing to sea level many of the temperatures that I measured on the slopes of the Cordilleras, sometimes at elevations as high as 500 toises. Even while making corrections that he believes are due to latitude and to the gradual decline in temperature on a vertical plane, he admits how much the location of places on vast plateaus or in narrow valleys renders some of his corrections questionable ([in] *Memoirs of the Astronomical Society*, Vol. II, pp. 149, 158, 171, 172, 182, 183).

When one studies the problem of heat distribution across the globe's surface in all its aspects and one discounts peripheral local considerations (such as effects of composition, color, and geognostic nature of the soil, the predominance of certain winds, the sea's proximity, the frequency of clouds and fog, nighttime radiation cooling under more or less clear skies, etc.), one finds that a station's mean temperature depends on the different ways in which the influence of the sun at its meridian height [at noon] manifests itself: The sun's height simultaneously determines the duration of semidiurnal arcs; the length and the diaphaneity of the part of the atmosphere that the rays cross before reaching the horizon; the amount of radiation absorbed or transferred into heat (an amount that rises rapidly when the surface-level angle of incidence increases); and, finally, the quantity of solar radiation that a given horizon contains. ▾Mayer's law, with all the modifications introduced over the last thirty years, is an empirical law that approximates the generality of certain phenomena, and often in a satisfactory manner. But one cannot really use it to invalidate direct observations. If the globe's surface, from the equator to the Cumaná parallel, were a desert like the Sahara, or a savanna uniformly covered by grass, like the Plains of Calabozo and of Apure, there would probably be a rise in mean temperature from 10.5° lat. to the equator. It is quite likely, however, that this increase would not reach three fourths of a degree on a centigrade thermometer. Mr. Arago, whose important and inspired research extends across all branches of Meteorology, realized, through direct experimentation, that the amount of reflected light is about the same from the perpendicular impact to 20° zenith distance. He also found that the photometric effect of sunlight barely varies in Paris from noon to three o'clock at night in August, despite changes in the distance that sunrays must travel when crossing the atmosphere.

II.84

II.85

I settled on a round number—27.5° C—for the equator's mean temperature to attribute Cumaná's mean (27.7° C) to the equatorial zone proper (between 3° N and 3° S). This town, surrounded by arid sand and situated under a perpetually calm sky where the water vapor almost never condenses into rain, has a more scorching climate than any of its surroundings, which are also at sea level. When one travels across South America toward the equator on the Orinoco and Río Negro, the heat drops, not because of elevation which, from the fort of San Carlos, is negligible, but because of the jungle, the frequency of rain, and the lack of diaphaneity in the atmosphere. It is regrettable that travelers, even the most meticulous of them, are poorly equipped to advance Meteorology by increasing our knowledge of mean temperatures. They do not stay long enough in the countries whose climate one would like to know better. For yearly averages, they can only draw upon observations made by others, quite often at times, and with the assistance of instruments, that are far from yielding exact results. Because of the constancy of atmospheric phenom-

II.86

ena in the zone closest to the equator, a short period of time probably suffices to approximate mean temperatures at different elevations above sea level. I have always jumped at this kind of research. But the only precise result that I can report, from observations I made twice a day, is from Cumaná (on the degree of the reliability of that mean temperatures compare, *Relation historique*, Vol. III, pp. 145, 146; IV, pp. 101, 102, 190, 191, 306–27; V, pp. 175, 176; VII, pp. 307, 308, 309, 421, 422; XI, pp. 7–2[7], 247–63). Educated persons who have lived in different places around the world for a great many years can determine true numerical constants of climatology. In this respect, the intellectual resurgence stirring in free equatorial America, all the way from the coast, at elevations of two thousand toises, to the ridges and slopes of the Cordilleras, and between the parallels of Chiloe Island and San Francisco in New California, will have the most felicitous impact on the physical sciences.

If we compare what we knew about the equatorial region's mean temperature forty years ago with what we know of it today, it is surprising how slow the progress of empirical Climatology has been. To this day, I know of only II.87 one mean temperature observed with any appearance of accuracy between 3° N and 3° S: San Luis de Maranhão (lat. 2° 29′ S) in Brazil. From observations he made in 1821 three times a day (at 8 p.m., 4 a.m., and 11 a.m.), ▾Colonel [Antonio Bernardino] Pereira [do] Lago derived 27.4° C (*Annaes das Ciências, das Artes e das Letras*, 1822, Vol. XVI, Pl. 2, pp. 55–80). This is still 0.3° C less than Cumaná's mean temperature. Below the 10.5° lat., we know the mean temperatures only for

Batavia (lat. 6° 12′ S)	26.9°	C
Cumaná (lat. 10° 27′ N)	27.7°	

Between 10.5° lat. and the edge of the Torrid Zone:

Pondicherry (lat. 11° 55′)	29.6°
Madras (lat. 13° 4′ N.)	26.9°
Manilla (lat. 14° 36′ N.)	25.6°
Senegal (lat. 15° 53′ N.)	26.5°
Bombay (lat. 18° 56′)	26.7°
Macao (lat. 22° 12′ N.)	23.3°
Rio de Janeiro (lat. 22° 54′ S.)	23.5°
Havana (lat. 23° 9′ N.)	25.7°

Bearing in mind Colonel Pereira's observations:

Maranham (lat. 2° 29′ S.)	27.4°

It seems to follow from these data that the only place in the equinoctial region where the mean temperature exceeds 27.7° C is situated at 12° lat. The climate of Pondichéry [Southeast India] can no more serve to characterize the entire equatorial region than the Mourzouk Oasis can characterize north Africa's climate in the temperate zone. In the Mourzouk Oasis, the unfortunate ▾Ritchie and Captain Lyon claim to have seen the Réaumur thermometer oscillate between 38° and 43° for months at a time (perhaps because of the sand suspended in the air). The largest expanse of tropical land is found between 18° and 28° in the northern latitudes, and thanks to the existence of so many wealthy commercial towns, we have the most extensive meteorological data about this zone. The three or four degrees closest to the equator are a *terra incognita* for Climatology. We still do not know mean temperatures for Grand Para, Guayaquil, and even Cayenne. II.88

When one considers only the high temperatures in the northern hemisphere during a certain part of the year, one finds the hottest climates just below the tropics, and a little above. In Abusheer, for example (lat. 28.5°), July's mean temperature is 34° C. On the Red Sea, the centigrade thermometer reaches 44° C at noon and 34.5° C at night. In Benares (lat. 25° 20′), highs reach 44° during the summer, while dropping to 7.2° C in the winter. These observations from India were made with ▾Six's excellent *maximum* thermometer. Benares's average temperature is 25.2° C. II.89

The extreme heat that one observes in the temperate zone's southern regions between Egypt, Arabia, and the Persian Gulf is the aggregate effect of the neighboring land's configuration, its surface condition, the constant unlimited diaphaneity of air deprived of water vapor, and the length of days which increases in higher latitudes during certain times of the year. In the tropics themselves, extreme heat is rare and generally does not exceed 32.8° C in Cumaná and Bombay and 35.1° C in Veracruz. It is practically needless to point out that we have included only observations taken in the shade far from the earth's glare. On the equator, where the two solsticial heights reach 66° 32′, the sun's passages through the zenith are separated from each other by 186 days. In Cumaná, the summer's solsticial height is 76° 59′, and the winter's 56° 5′, with the passages through the zenith (April 17 and August 26) 131 days apart. Farther to the north, in Havana, one finds a summer solsticial height of 89° 41′, a winter height of 43° 23′, and 19 days between passages (June 12 and July 1). If these passages do not always register with equal evidence over the course of months, it is because their influence is masked, in some places, by the rainy season's beginning and other electrical phenomena. In Cumaná, the sun is lower than at the equator during 109 days, or more precisely, during 1,275 hours (from October 28 to the following February 14). II.90

But during this period, its *maximum* zenithal distance still does not exceed 33° 55′. The slowing down of the sun's movement closer to the tropics increases the heat in places situated farthest from the equator, especially near the border between the torrid and the temperate zones. Near the tropics, for example in Havana (lat. 23° 9′), the sun takes 24 days to move a degree each side of the zenith. Below the equator, it takes no more than five days. In Paris (lat. 48° 50′), where the sun sets at 17° 42′ on the winter solstice, the summer solsticial height is 64° 38′. In Paris, between May 1 and August 12, an interval of 103 days or 1,422 hours, the warm star is therefore as high as it is in Cumaná at a different time of year. Comparing Paris between March 26 and September 17 (175 days or 2,407 hours) to Havana shows that the Parisian sun is as high as it is during another season below the Tropic of Cancer. During II.91 this 175-day interval between 1806 and 1820, the hottest month (July) had an average temperature of 18.6° C, according to records from the royal Observatory of Paris. In Cumaná and Havana, notwithstanding the longer nights, when the sun sets—at 56° 5′ and at 43° 23′, respectively—the coldest month still has an average temperature of 26.2° C in Cumaná and 21.2° in Havana. In all zones, the preceding seasons influence the temperature of the following. Temperature drops are slight in the tropics; mean temperatures over the previous months are steady at 27° C in Cumaná and 25.5° in Havana.

Given all of the considerations that I have just outlined, it does not at all seem likely to me that equatorial temperatures can reach 29.2° C, which is what the knowledgeable and esteemed author [Atkinson] of a report on *Astronomical and other Refractions* claims. ˺Father de Bèze, the first voyager who advised making observations during the coldest and warmest hours of the day, already believed that he had discovered "that heat is no greater on the equator than at 14° lat.," according to comparisons between Siam, Malacca, II.92 and Batavia in 1686 and 1699. I think that there is a difference but that it is very slight and obscured by the effect of so many other causes that simultaneously impact the mean temperatures of a place. Observations gathered so far do not suggest a progressive temperature increase from the equator to Cumaná's latitude.

ˇSUPPLEMENT

II.93

When I published the *Political Essay on the Kingdom of New Spain* after my return to Germany, I also disseminated some of the material I have on South America's territorial riches. I compiled this picture comparing population, agriculture, and commerce throughout the Spanish colonies at a time when civilization's march had foundered on imperfect social institutions, the tariff system, and other dreadful detours away from sound political principles. Since I expounded on the immense resources that the peoples of both Americas might enjoy domestically and in their trade relations with Europe and Asia through the thoughtful use of their freedom, one of those great revolutions that occasionally shakes up humankind has changed the state of society in the vast countries through which I traveled. Today, the continental part of the New World finds itself in effect divided among three peoples of European heritage: one, the most powerful, is of the Germanic race; the other two are part of Latin Europe by virtue of their language, literature, and customs. The westernmost parts of the old world—the Iberian Peninsula and the British Isles—also have the largest colonies. Four thousand leagues of coastline inhabited by the descendents of the Spanish and Portuguese alone attest to the preeminence that the peninsular peoples attained over the other seafaring peoples through their maritime expeditions in the fifteenth and sixteenth centuries. One might say that their languages, resounding from California to the Río de la Plata, on the ridge of the Cordilleras and in the Amazonian jungles, are monuments to national glory that will survive all political revolutions.

II.94

At this time, the number of the inhabitants of Spanish and Portuguese America taken together form a total population two times greater than that of Anglo-America. The New Continent's French, Dutch, and Danish possessions are small. To complete the general picture of the peoples who could influence the other hemisphere's destiny, we should forget neither the Slavic colonists attempting to gain a foothold from Alaska to California, nor Haiti's free Africans who fulfilled the prophecy that the ˇMilanese traveler Benzoni

II.95

had made in 1545. That these Africans reside in the middle of the Antillean Mediterranean, on an island that is two and a half times larger than Sicily, testifies to their political importance. All friends of humanity wish for the success of a civilization that, after so much blood and fury, is making surprising advances. Russian America resembles less a farming colony than one of those trading posts that Europeans established on Africa's coastlines to the great detriment of the natives, nothing but military posts and depots for fishermen

II.96 and Siberian hunters. It is no doubt vexing to find the Greek Church in parts of America and to watch two nations from Europe's eastern and western extremes—the Russians and the Spanish—adjoin one another on a continent where they arrived by different routes. But Okhotsk and Kamchatka's wild and sparsely populated coastlines, the lack of help from Asia's ports, and political arrangements in the Slavic colonies are obstacles that will keep them in their infant state for a long time. It follows that if, in political-economic research, one is used to considering only large groupings, then one will not realize that the American continent is actually shared among more than the three great nations of English, Spanish, and Portuguese provenance. The first of these three nations, Anglo-Americans, is also—after the British in Europe—the one that covers the largest area of sea with its ensign. Without remote colonies, their trade has surpassed commercial activity in all of the Old World,

II.97 except for the nation that brought to America's north its language, the glory of its literature, its love of work, its inclination toward freedom, and some of its political institutions.

British and Portuguese colonists settled only the coastlines facing Europe. In contrast, the Castilians crossed the Andes and established themselves in the westernmost regions since the beginning of the conquest. It is only there, in Mexico, in Cundinamarca, Quito, and Peru that they found traces of an ancient civilization, agricultural peoples, and thriving empires. This circumstance, together with the growth of a native, mountain-dwelling population, the nearly exclusive possession of great mineral riches, and the establishment of trade relations with the Indian Archipelago since the beginning of the sixteenth century all gave the Spanish possessions in equinoctial America their own character. In the eastern regions, shared among British and Portuguese colonists, the natives were nomads and hunters. Far from becoming part of the farming and laboring population as on the Anahuac plateau,

II.98 in Guatemala, and in Upper Peru, they generally retreated before the whites' advance. The need for work, the preference for sugarcane, indigo, and cotton cultivation, and the greed that accompanies and often degrades industry all gave birth there to the infamous slave trade, whose consequences have been equally grave for the two worlds. Fortunately, in the continental part of Spanish America, the number of African slaves is so trivial that it stands in a 1 to 5 ratio with Brazil's and the southern United States' enslaved population. All of Spanish America, including the islands of Cuba and Puerto Rico, does not have as many slaves in an area at least one fifth the size of Europe as the single state of Virginia. The union between New Spain and Guatemala is the only example in the Torrid Zone of a nation of 8 million people governed by European laws and institutions, cultivating sugar, cacao, wheat, and grapes with practically no slaves torn from African soil.

The New Continent's population barely surpasses that of France or Germany. In the United States, it doubles every twenty-three or twenty-five years; in Mexico, it doubled in forty or forty-five years, even under the metropole's governance. Without giving in to hopes that are too exalted, it is likely that America's population will equal Europe's in less than a century and a half. Far from impoverishing the old continent to the benefit of the new, as has been fashionable to predict, this noble cultural, industrial, and commercial rivalry will increase consumption, productivity, and trade. It is likely that after the great revolutions that change the state of human societies, public wealth—civilization's common heritage—will be distributed differently among the peoples of both worlds. But, little by little, balance returns, and it is a grave, I would almost dare say impious, prejudice to consider the growing prosperity of a completely different part of our planet as a calamity for old Europe. The colonies' independence will not contribute to isolating them; rather, it will bring them closer to the civilized peoples of old. Trade tends to unify what jealous politics have long separated. There is still more: it is civilization's nature to carry on without obliterating the place where it first sparked. Its gradual march from east to west, from Asia to Europe, does not disprove this axiom. A vivid light keeps its brightness even when it illuminates a larger space. Intellectual culture, a fertile source of national wealth, travels from neighbor to neighbor. It spreads without moving. Its movement is not a migration. If this appears to be the case in the East, it is because barbarian hordes [the Ottomans] have seized Egypt, Asia Minor, and the formerly free Greece, the abandoned cradle of our ancestors' civilization.

II.99

II.100

The coarsening of peoples is the consequence of oppression inflicted either by a domestic despot or a foreign conqueror. It is always accompanied by gradual impoverishment and a decline in public wealth. Free and strong institutions, operating in everyone's interests, limit these perils. And the world's growing civilization and competition in work and trade do not ruin states whose well-being flows from a natural source. Productive and commercial Europe will benefit from the new order of things emerging in Spanish America, as it would profit, through an increase in consumption, from events that

II.101

would bring barbarism to an end in Greece, on Africa's north coast, and in other countries subjected to Ottoman tyranny. There is nothing that poses a threat to the old continent's prosperity except the prolongation of internecine struggles that halt production and simultaneously reduce the ranks of consumers and their demands. In Spanish America, this struggle, which began six years after my departure, is gradually nearing its end. We will soon see independent peoples inhabiting both sides of the Atlantic Ocean—governed according to a wide array of forms but unified by the memory of a common origin, linguistic continuity and the needs that always give birth to civilization. II.102 One might say that the immense strides in navigation have shrunk the size of the oceans. Already, the Atlantic Ocean appears to our eyes as a narrow channel that does not divide Europe's trading states from the New World any more than the Mediterranean basin separated Peloponnesian Greeks from Greeks in Ionia, Sicily, and Cyrene, when navigation was in its infancy.

I thought it was necessary to mention these general considerations on future relations between the two continents before sketching my political tableau of Venezuela's provinces. I show in it different human races, cultivated and wild crops, variations in the soil, and inland communications. These provinces, governed by a Captain-general who resided in Caracas until 1810, are currently linked with the former viceroyalty of New Granada or Santa Fé under the name of the republic of Colombia. I will not prematurely provide a description of New Granada. However, to render my statistical observations on Venezuela more useful to those who would like to judge this II.103 country's political importance and the advantages it can provide to European trade, even in its underdeveloped cultural condition, I will illustrate intimate relations that the *United Provinces of Venezuela* have with Cundinamarca and New Granada and show how these provinces are part of the new state of Colombia. This exposition will necessarily have five parts: territory, population, production, trade, and public revenue. Because some of the data that will help fill out this picture already appear elsewhere [in my *Personal Narrative*], I will be able to be quite succinct about broad results. Mr. Bonpland and I spent nearly three years in the countries that currently form the republic of Colombia's territory, specifically, sixteen months in Venezuela and eighteen in New Granada. We traversed this territory in all its vastness: first, from the Paria Mountains to Esmeralda on the Upper Orinoco and to San Carlos del Río Negro situated near Brazil's borders; second, from the Sinú River and II.104 Cartagena de Indias to the snowy peaks of Quito, and then on to the port of Guayaquil on the Pacific Ocean and the Amazon banks in Jaén de Bracamoros province. Such a long stay and a 1,300 nautical league inland journey, for more than 650 of which were in a boat, provided me with a sufficiently detailed knowledge of the local conditions. I would not, however, pretend to have gathered statistical materials on Venezuela and New Granada as numerous and as reliable as those I collected during a much shorter stay in New Spain. One is less inclined to discuss questions of political economy in purely agricultural countries that have numerous power centers than in places where civilization is concentrated in a large capital and where the immense output from mines accustoms men to estimate natural wealth in numbers. In Mexico and in Peru, I found some of my data in official documents. It was not like this in Quito, Santa Fé, and Caracas where interest in statistical research will only develop at the pleasure of an independent government. Those who are used to examining figures before accepting their truth know that newly founded states like to exaggerate increases in public wealth, while the old colonies focus on the growing list of evils attributed to the tariff system. It is almost an act of revenge against the metropole to exaggerate commercial stagnation and the slowness of population growth. II.105

I am aware that travelers who have recently visited America regard this growth as much more rapid than the numbers on which I have settled in my statistical research would seem to suggest. They indicate that Mexico, whose population supposedly doubles every twenty-two years, will have 112 million inhabitants by 1913. In the same year, the United States will have 140 million.[1] I admit that these numbers do not concern me in the same way that they would alarm ▾Mr. Malthus's zealous followers. It is possible that two or three hundred million men will one day find their subsistence in the New Continent's wide-open spaces between Lake Nicaragua and Lake Ontario. ▾I assume that the United States will have more than 80 million inhabitants a hundred years from now, allowing a gradual change in the period it takes for it to double (from 25 to 35 to 40 years). But despite the essentials for prosperity present in equinoctial America, and despite the wisdom I would very much like to attribute to the new republican governments formed to the equator's south and north, I doubt that population increases in Venezuela, Spanish Guyana, New Granada, and Mexico can generally be as rapid as they are in the United States. ▾Situated entirely in the temperate zone and lacking in high mountain ranges, the United States offers a wide expanse ideal for the development of agriculture. Hordes of Indian hunters flee before both the colonists whom they hate and the ▾Methodist missionaries who oppose their inclination to lassitude and the nomadic life. In all likelihood, Spanish America's more fertile soil produces more food on a similarly sized area. On the equinoctial region's plateaus, each grain of wheat probably yields another 20 to 24 grains. But the Cordilleras—riven by nearly impassible crevices, arid and bare steppes, and forests impervious to axe and fire and filled with poi- II.106 II.107

1. Robinson, *Memoirs of the Mexican Revolution*, Vol. II, p. 315.

sonous insects—will present significant obstacles to agriculture and industry for a long time to come. The most enterprising and hardy colonists will not be able to advance into the mountainous districts of Mérida, Antioquia, and Los Pastos, into Venezuela and the Guaviare Plains, into the forests of the Río Magdalena, Orinoco, and the Esmeraldas provinces to the west of Quito, as they did when they expanded their agricultural conquests into the wooded plains to the West of the Alleghenies, and from the springs of the Ohio, Tennessee, and Alabama Rivers to the banks of the Missouri and the Arkansas. Recalling the narrative of my journey to the Orinoco, readers will appreciate the obstacles that a powerful nature mounts against man's efforts in a hot and humid climate. In Mexico, large areas of land lack sources of water. Rain is very rare there, and the lack of navigable rivers slows communication. As the former native population is agricultural, and was so for a long time before the Spaniards' arrival, the accessible and easily cultivated land already has owners. Fertile and extensive land either available to settlers on a first-come basis or susceptible to being divided up and sold off for the state's benefit is less common than is assumed in Europe. It follows that colonization cannot move as swiftly and freely in Spanish America as it has to this day in the western provinces of the Anglo-American Union. This Union's population is composed only of whites and blacks who, either torn from their homeland or born in the New World, have become instruments of the whites' industry. By contrast, in Mexico, Guatemala, Quito, and Peru, there are currently more than five and a half million natives of the coppery race. In spite of the attempts at *de-indianizing* them, their partially forced, partially chosen isolation, their attachment to former habits, and their defiant stubbornness of character prevent them from participating in the progress of public prosperity.

II.108

II.109

I am emphasizing these differences between the free states of temperate America and those of equinoctial America to show that the latter have to struggle against obstacles to their physical and moral condition and to recall that countries lavished by nature with the most varied and precious products do not always accommodate easy, rapid, and even settlement. If population limits were exclusively dependent on the amount of food that the earth can produce, then the simplest calculation would prove the primacy of societies established in the Torrid Zone's beautiful regions. But political economy, or empirical political science, disdains figures and vain abstractions. An empty continent could acquire more than 8 hundred million inhabitants through the multiplication of a single family. And yet, these estimates, founded on the hypothesis of the *constancy of doubling* over twenty-five or thirty years, are contradicted by the history of every people that is already advanced on civilization's course. The destinies that await Spanish America's free states are

II.110

too weighty to beautify them with gratifying illusions and chimerical calculations.

AREA AND POPULATION—To draw the reader's attention to the political importance of the former *Capitanía general* of Venezuela, I will begin by comparing the large bodies into which the diverse peoples of the New Continent are grouped today. From a more general perspective, the statistical data that show the variables behind prosperity and national power should be of greater interest. Of the 34 million inhabitants who consist of the *three predominant races* and are spread across the vast surface of *continental America* (an estimate that includes the original free natives), 16 million live in *Spanish-American* possessions, 10 million in *Anglo-American* possessions, and nearly 4 million in *Portuguese-American* territories. The populations in these three large categories currently stand in a 4 to 2.5 to 1 ratio, while the territory that these populations inhabit stands in a ratio of 1.5 to 0.7 to 1. The *area* of the United States is nearly one fourth larger than that of Russia to the west of the Ural Mountains, and Spanish America is larger than all of Europe by the same ratio. The United States[1] has five eighths of the population of the Spanish possessions; however, its *area* is less than half. Brazil encompasses such empty countryside to the west that its population is a quarter of that of Spanish-America in a territory only one third smaller. The following tableau shows results of research on the estimated size of various American states that I undertook in collaboration with ▾Mr. Mathieu, a member of the Academy of sciences and the Bureau des Longitudes [in Paris]. We used maps whose borders had been revised according to data that I published in my *Recueil d'observations astronomiques*. We typically worked on a large enough scale to accommodate spaces of 4 to 5 sq. lg. We believed that we could push precision this far in order not to compound the uncertainty of the geographical data with inaccuracies in measuring triangles, trapezoids, and sinuous coastlines.

II.111

II.112

1. To avoid fastidious circumlocutions, I continue in this study to designate the countries inhabited by *Spanish-Americans* by the name *Spanish-America*, despite the political changes that the colonies have undergone. I call the *United States*—without adding *of north America*—the country of *Anglo-Americans*, although other *United States* have formed in South America. It is awkward to speak of peoples who play such an important role on the world scene but who lack collective names. The word *American* may no longer be applied exclusively to citizens of the United States of North America, and it would be desirable if this nomenclature for the independent nations of the New Continent could be fixed in a way that would be at once convenient, consistent, and precise.

II.113

	Principal political divisions		Surface in square leagues (20 to a degree equinoctial)	Population (1823)
I.	SPANISH-AMERICAN POSSESSIONS		371,380	16,785,000
	Mexico or New Spain		75,830	6,800,000
	Guatemala		16,740	1,600,000
	Cuba and Puerto Rico		4,430	800,000
	Colombia	Venezuela	33,700	785,000
		New Granada and Quito	58,250	2,000,000
	Peru		41,420	1,400,000
	Chile		14,240	1,100,000
	Buenos Aires		126,770	2,300,000
II.	PORTUGUESE-AMERICAN POSSESSIONS (BRAZIL)		256,990	4,000,000
III.	ANGLO-AMERICAN POSSESSIONS (UNITED STATES)		174,300	10,220,000

Explanations

Taking the eastern edge of Panama province as its border, I found that South II.114 America's area was equal to 571,290 sq. lg. The Spanish part—Colombia (without the Isthmus of Panama and Veragua province), Peru, Chile, Buenos Aires (without the Magellan territories)—encompasses 271,774 sq. lg.; the Portuguese possessions 256,990 sq. lg.; British, Dutch, and French Guiana 11,320 sq. lg.; and the Patagonian territories to the south of Río Negro 31,206 sq. lg. The following numbers, concerning large surface areas, can supply terms for comparison[1]: Europe, 304,700 sq. lg.; the Russian Empire in Europe and Asia, 603,160 sq. lg.; the European part of the Russian Empire, 138,116 sq. lg.; the United States of America, 174,310 sq. lg. All these estimates are in square leagues, 20 to a degree from the equator, or 2,855. I adopted this unit of measurement in the *Personal Narrative* of my voyage, because nautical leagues, three nautical miles apiece, would be easier to introduce as a uniform geographical measure among the commercial peoples of Spanish America than Spain's *le-* II.115 *guas legales* or *leguas comunes*, which are 26.5 and 19 to a degree, respectively. In the *Political Essay on the Kingdom of New Spain*, area is indicated in square leagues at 25 to a degree, as in most statistical works published in France. I am recalling these standards because many contemporary authors confused leagues that are 25 to a degree with nautical and geographical leagues when

1. See note B at the end of the 9th Book.

copying surface area estimates from my *Political Essay on the Kingdom of New Spain*, a confusion no less distressing than mixing up centigrade with the octogesimal degrees on the Réaumur scale. Next to nonvariables such as *area* (which depends on the precision of the maps that I drew), I placed a very variable element: population. The following data will clarify a subject that has been known for a long time (and with good reason) as *plenum opus aleæ* [a game of dice]. There are figures in studies of political economy that are much like meteorological data and astronomical tables. They acquire precision only gradually, and more often than not, it is necessary to make do with *approximations*.

A. Population

II.116

MEXICO. I believe that I already proved elsewhere, through empirical data, that the viceroyalty of New Spain, including the *internal Provinces* and the Yucatán but not the *Captaincy-general* of Guatemala, had at least 5,840,000 inhabitants in 1804—2.5 million copper-skinned natives, 1 million Spanish Mexicans, and 75,000 Europeans. I even predicted (*Essai politique sur le royaume de la Nouvelle-Espagne*, I, pp. 65, 76–77) that by 1808, the population would reach 6.5 million, of whom two to three fifths (or 3,250,000) would be Indians. The civil conflicts that have long agitated the districts of Mexico City, Veracruz, Valladolid, and Guanajuato have likely slowed Mexico's annual population growth, which was probably more than 150,000 during my stay there (*Essai politique sur le royaume de la Nouvelle-Espagne*, I, pp. 62–64). The birthrate appears to be 1 in 17, the death rate 1 in 30. Positing population growth of no more than a million inhabitants over 18 years, I believe that I have sufficiently accounted for the effects of the internal strife that disrupted mining, commerce, and agriculture. Field research has recently proven that the estimates on which I settled 12 years ago are not far from the truth. ▾Don Fernando Navarro y Noriega published results from his wide-ranging study of the number of *curatos y missiones* [parishes and missions] in Mexico. He estimated the country's population at 6,128,000 in 1810 (*Catálogo de los curatos y misiones que tiene la Nueva España en cada una de sus diócesis*, 1813, p. 38; and *Rispuesta de un Mexicano al no. 200, del Universal*, p. 7). The same author, whose post as a financial minister of the crown (*Contador de los ramos de arbitrios*) put him in a position to examine statistical data in the field, believes (*Memoria sobre la población del Reino de Nueva España, México* 1814; and *Semanario político y literario de la Nueva España*, no. 20, p. 94) that New Spain's 1810 population, not including Guatemala's provinces, breaks down into the following components:

II.117

II.118

1,097,928	Europeans and Spanish-Americans
3,676,281	Indians
1,338,706	Castes or mixed race [race mixte]
4,229	Lay preachers
3,112	Ecclesiastics of regular clergy
2,098	Nuns
6,122,354	

I am inclined to believe that New Spain currently has almost 7 million inhabitants. This is also the opinion of a respectable prelate, the Archbishop of Mexico, ▾Don José de Fonte, who traveled through a large part of his diocese and whom I recently had the honor of receiving in Paris.

GUATEMALA. This country, which was considered a kingdom until now, comprises the four parishes of Guatemala, León de Nicaragua, Chiapas or Ciudad Real, and Comayagua or Honduras. The census that the secular government undertook in 1778, and which ▾Mr. Del Barrio (deputy to the Spanish Courts before Mexico's declaration of Independence) obligingly conveyed to me, puts the population at no more than 797,214. But ▾Don Domingo Juarros, the knowledgeable author of the *Compendio de la historia de la ciudad de Guatemala* published serially from 1809 to 1818 (Vol. I, pp. 9 and 91), demonstrates that this figure is very imprecise. Censuses that the bishops ordered during the same period come up with a count more than a third higher. During my stay in Mexico, the population of Guatemala, where Indians are extremely numerous, was estimated to be 1.2 million, according to official documents. People with extensive local knowledge estimate it to be 2 million today. Always concerned to settle on figures *that err on the low side*, I put the population at no more than 1.6 million.

II.119

CUBA AND PUERTO RICO. Little is known about the population of the large island of Puerto Rico. It has grown quite a bit since 1807. At that time, it had no more than 136,000 inhabitants, of whom 17,500 were slaves. As we already reported (Vol. I, p. 335), Cuba's 1811 census counted 600,000 inhabitants, 212,000 of whom were slaves ([Arango,] *Documentos de que hasta ahora se compone el expediente sobre los negros de la isla de Cuba*, Madrid, 1817, p. 139). A much more recent official document ([Zayas y Chacón and Benítez,] *Reclamación hecha por los Representantes de [la isla de] Cuba contra la ley de aranceles*, Madrid, 1821, p. 6) estimates the total population at 630,980.

II.120

COLOMBIA. According to the materials that I gathered, the seven provinces once known collectively as the *Capitanía general* of Caracas had 800,000 inhabitants at the beginning of the nineteenth century when the revolution broke out. These materials are not a comprehensive census undertaken by the secular authorities. They are only partial estimates based on information from missionaries and priests and on a consideration of consumption and the degree of (agri)cultural development. Employees of the intendancy of Caracas, and in particular ▾Don Manuel [de] Navarrete, an officer in the royal treasury at Cumaná who is extremely well versed in financial matters, were of great help in this work. The period in question is of great interest, and one day, people will measure the population growth after independence and freedom from this point forward. Quite conceivably, this population growth will make itself felt only when domestic peace returns to these beautiful regions. It is possible that the population was a little smaller in 1800 than when this study was published. The armies were not very large, but they ravaged the best cultivated land on the coast and in the neighboring valleys. The March 26, 1812 earthquake (see Vol. V, pp. 14–24), the 1818 fever epidemics (Vol. VIII, p. 418), the arming of black people so unwisely undertaken by the royalists, the immigration of many rich families to the Antilles, and long commercial stagnation all have increased public misery.

II.121

Provinces of Cumaná and Barcelona	110,000 souls
I have results from a census conducted in 1792, which is at least one sixth off and which counts 86,083 souls, 42,615 of whom are Indians; that is: 27,787 *de doctrina* or habitants of villages with a lay preacher; and 14,828 in *missions* or governed by missionary monks. In 1800, I counted 60,000 in the province of Cumaná or New Andalucía; 50,000 in the province of Barcelona.	
Province of Caracas	370,000
Surveyed in 1801: Caucagua valley and savannahs of Ocumare, 30,000; city of Caracas and valleys of Chacao, Petare, Mariches and Los Teques, 60,000; Portocabello, Guayre, and all of the coast from the cape Codera to Aroa, 25,000; valleys of Aragua, 52,000; le Tuy, 20,000; districts of Canora, Barquisimeto, Tocuyo, and Guanare, 54,000; San Felipe, Nirgua, Aroa and the neighboring plains, 34,000; Llanos of Calabozo, San Carlos, Araure, and San Juan Bautista del Pao, 40,000. These partial estimates embracing almost all of the habited parts add up to a total of no more than 315,000.	
Province of Coro	32,000
Province of Maracaibo (with *Mérida* and *Trujillo*)	140,000

II.122

II.123

Province of Barinas	75,000
Province of Guyana	40,000
A census in 1780, the results of which I found in the archives of Angostura (Santo Tomé de la Nueva Guyana), counted 19,616 inhabitants; that is, 1,479 whites, 16,499 Indians, 620 blacks, 1,016 *pardos* and *zambos* (people of mixed color).	
Margarita Island	18,000
Total	785,000

It is possible that population figures for Caracas and Maracaibo and for Margarita island (▾[Charles] Brown, *Narrative*, 1819, p. 118) are a little exaggerated, even for the time period on which I focused. ▾Mr. De Pons, however, who also had access to the surveys that priests presented to the bishops, came up with 500,000 for the province of Caracas, including Barinas province ([De Pons,] *Voyage à la partie orientale de la terre-ferme*, Vol. I, p. 177). The II.124 villages in the provinces of Maracaibo are very populous, both on the banks of the lake and in the mountains of Mérida and Trujillo. Out of 780,000 to 800,000 inhabitants estimated for the *Capitanía general* of Caracas in 1800, 120,000 were probably pure Indians. Official documents[1] count 25,000 for Cumaná province (15,000 in the missions of Caripe); 30,000 for Barcelona province (24,700 in the missions of Píritu); 34,000 for Guyana province (17,000 in the missions of Caroni, 7,000 in the missions of Orinoco, and 10,000 living free on the Orinoco Delta and in the jungle). These data suffice to prove that the number of bronze Indians in the *Capitanía general* is neither 72,800 nor 280,000, as has been erroneously reported recently (De Pons, Vol. I, p. 178; ▾Malte-Brun, *Géographie*, Vol. V, p. 549). The former, who estimates the population at only 728,000 instead of 800,000, has singu- II.125 larly overestimated the number of slaves. He puts it at 218,400 (Vol. I, p. 241). This number is almost four times too large (see Vol. IV, p. 15[9]). According to partial estimates by three people familiar with the place, ▾Don Andrés Bello, Don Luis López, and Don Manuel Palacio Fajardo, there were at most 62,000 slaves in 1812:

10,000	in Caracas, Chacao, Petare, Baruta, Mariches, Guarenas, Guatire, Antímano, La Vega, Los Teques, San Pedro, and Budare.
18,000	in Ocumare (las Sabanas), Yare, Santa Lucía, Santa Teresa, Marín, Caucagua, Capaya, Tapipa, Tacarigua, Mamporal, Panaquire, Río Chico, Guapo, Cupira, and Curiepe.
5,600	in Guayos, San Mateo, Victoria, Cagua, Escobal, Turmero, Maracay, Guácara, Guïgüe, Valencia, Puerto Cabello, and San Diego.

1. See note C at the end of the IXth Book.

3,000 in Guaíra, Choroní, Ocumare, Chuao, and Borburata.
4,000 in San Carlos, Nirgua, San Felipe, Llanos de Barquisimeto, Carora, Tocuyo, Araure, Ospino, Guanare, Villa de Cura, San Sebastián, and Calabozo. II.126
22,000 in Cumaná, Nueva Barcelona, Barinas, Maracaibo, and in Spanish Guyana.

The number of Spanish Americans is probably no more than 200,000. The number of whites born in Europe is likely no more than 12,000, which means that the entire population of the former *Capitanía general* of Caracas is 51 percent mixed-race castes (mulâtres, zambos, and mestizos), 25 percent Spanish-Americans (creole whites [blancs creoles]), 15 percent Indians, 8 percent blacks, and 1 percent Europeans.

As for the kingdom of New Granada, I refer to the 1778 surveys that counted 747,641 for the Santa Fé Audiencia and 531,799 for the Quito Audiencia. Yet, assuming that no more than a seventh has been omitted and positing an annual growth rate of no more than 1.8 percent, one arrives at over 2 million for 1800, even with moderate estimates. ˅Mr. Caldas, who is very well informed about the political state of his homeland in other respects, puts it as high as 3 million for 1808 (*Semanario de Santa-Fé*, no. 1, pp. 2–4). But I fear that this scholar has exaggerated the number of free Indians significantly. II.127 I find, after a thorough examination of all the materials I have at hand, that the population of the republic of Colombia is 2,785,000. This number is lower than the 3.5 million that the President of the Congress proposed in his January 10, 1820 proclamation. It is slightly higher than what officially appeared in the *Gazeta de Colombia* on February 10, 1822, of which I learned only from Buenos Aires newspapers.

Departamientos	Provinces	Population
Orinoco	Cumaná	70,000
	Barcelona	44,000
	Guyana	45,000
	Margarita	15,000
		174,000
Venezuela	Caracas	350,000
	Barinas	80,000
Zulia	Coro	30,000
	Trujillo	33,400
	Mérida	50,000
	Maracaibo	48,700
		162,100

II.128 These three districts form the old *Capitanía general* of Caracas, with a population of 766,100.

Boyacá	Tunja	200,000
	Socorro	150,000
	Pamplona	75,000
	Casanare	19,000
		444,000
Cundinamarca	Bogotá	172,000
	Antioquia	104,000
	Mariquita	45,000
	Neiva	50,000
		371,000
Cauca	Popayán	171,000
	Chocó	22,000
		193,000
Magdalena	Cartagena	170,000
	Santa Marta	62,000
	Río Hacha	7,000
		239,000

In the same year (1822), the two Colombian provinces whose deputies had II.129 not yet arrived in Congress had the following:

Panama	50,000
Veragua	30,000
	80,000

The four provinces of Boyacá, Cundinamarca, Cauca, and Magdalena constitute, together with Panama and Veragua, the former *Audiencia de Santa Fé*, that is, New Granada, not including the *Presidencia de Quito*. Total population: 1,327,200.

Former *Presidencia* of Quito.	Quito	230,000
	Quixos and Macas	35,000
	Cuenca	78,000
	Jaén de Bracamoros	13,000
	Maynas	56,000 (!)
	Loja	48,000
	Guayaquil	90,000
		550,000

These data from Colombia's official Gazette imply the following for the former parts of the Santa Fé viceroyalty:

Venezuela	766,000
New-Granada	1,327,000
Quito	550,000
	2,643,000

II.130 This total estimate accords within one forty-sixth with what I had published twelve years ago in my *Essai politique sur le royaume de la Nouvelle-Espagne* (Vol. II, p. 851). It does not come from a real census but from "reports that the deputies of each province submitted to Colombia's congress to help revise the election law" (*El Argos de Buenos Aires,* no. 9, November 1822, p. 3, and [Walker,] *Colombia, Being a Statistical Account of that Country*, 1822, Vol. I, p. 375.) Because the Congress could not consult with Quito's deputies, this *Presidencia*'s population was probably underestimated. The official Gazette practically used the 1778 numbers, while the Santa Fé *Audiencia*'s estimate indicated growth of more than 70 percent over 43 years. One must hope that an accurate census will soon dispel the doubts that I am raising here about Colombia's statistics. It seems likely to me that, despite the ravages of war, the total population is greater than 2.9 million.

PERU. The population numbers in the overview are not too high. Works printed in Lima as much as thirty years ago ([ᵀHipólito Unanue,] *Guía* II.131 *política del Virreynato del Perú para el año* 1793, *publicada por la Sociedad académica de los Amantes del país*) had the population at a million inhabitants, 600,020 of whom were Indians, 240,000 mixed, and 40,000 slaves. The inhabited part of the country is only 26,220 sq. lg., and a large, fertile part of Upper Peru belonged to the viceroyalty of Buenos Aires after 1778.

CHILE. The census from 1813 counted 980,000 souls. ᵀMr. d'Irisarri, who holds an important post in Chile's government, thinks that the population may already have reached 1.2 millions.

BUENOS AIRES. According to official documents given to ᵀMr. Rodney, one of the commissioners whom the United States president had sent to Río de la Plata in 1817, the population was 2 million. At that time, it was found to be 965,000, not including Indians. The number of natives is quite large in Upper Peru, that is, in the *Provincias de la Sierra* which belong to the state II.132 of Buenos Aires. Official censuses estimate that there are 130,000 Indians in Buenos Aires province, 25,000 in that of Córdoba, 371,000 in the intendancy of Cochabamba, 230,000 in that of Potosí, and 154,000 in that of Charcas. There are 400,000 inhabitants of all castes—Indians, mixed-race [métis], and whites—in La Paz province alone.

These data show that the census included all castes in some districts, while in others it counted only whites and mixed-race people [mulâtres et métis], excluding the natives of the coppery race. Yet, in the eight provinces in the first category alone (Buenos Aires, Córdoba, Cochabamba, Potosí, Charcas, Santa Cruz, La Paz, and Paraguay), there are already 1,805,000 souls. The provinces and districts of Tucumán, Santiago de Estero, Valle de Catamarca, Rioja, San Juan, Mendoza, San Luis, Jujuy, and Salta are not included in this number. As other censuses indicate nearly 330,000 souls, not includ-

II.133

ing Indians, the total population of the former viceroyalty of Buenos Aires, or La Plata, has probably reached 2.5 million inhabitants of all castes ([Monroe], *Message from the President of the United States at the Commencement of the Session of the Fifteenth Congress*, Washington, 1818, pp. 20, 41, and 44). The detailed estimates[1] I obtained from ▾Mr. Brackenridge, secretary of the United States mission to Buenos Aires, were published in a study replete with philosophical insights which suggests 1,716,000 for Upper Peru alone, that is, for the four intendancies of Charcas, Potosí, La Paz, and Cochabamba.

UNITED STATES. Given current growth rates, the United States' population at the beginning of 1823 must be 10,220,000, 1,623,000 of whom are slaves. It was determined that:

1700	262,000	(uncertain)
1753	1,046,000	(*idem,* Mr. Pitkin)
1774	2,141,307	(*idem,* ▾Gov. Pownall)
1790	3,929,328	(first certain census)
1800	5,306,032	
1810	7,239,903	
1820	9,637,999	

II.134

The 1820 census counted 7,862,282 whites, 1,537,568 slaves, and 238,149 free people of color. According to a very interesting study by ▾Mr. Harvey (*Edinburgh Philosophical Journal,* January, 1823, p. 41), the population of the United States grew 35, 36.1, and 32.9 percent each decade from 1790 to 1820. The slowdown in growth is thus still no more than 2 to 3 percent, or one-eleventh of the total, over 10 years.[2]

1. See note D at the end of the IXth Book.
2. See note E at the end of the IXth Book.

BRAZIL. Until today, people have settled on 3 million,[1] but the estimate I give in the Table below is based on unpublished official documents for which I have ▾Mr. Adrien Balbi of Venice to thank. His long stay in Lisbon put him in a position to spend much time with statistics about Portugal and the Portuguese colonies. According to an 1819 report to the king of Portugal on the population of his overseas possessions and various surveys written up by the Captains-general, the provincial governors, in keeping with the decrees of Rio de Janeiro from August 22 and September 30, 1816, Brazil had a population of 3,617,900 inhabitants around 1818. II.135

1,728,000	black slaves (*pretos captivos*).
843,000	whites (*brancos*).
426,000	frees, of mixed blood (*mestissos, mulatos, mamalucos libertos*).
259,400	Indians of different tribes (*Indios de todas as castas*).
202,000	slaves of mixed blood (*mulatos captivos*).
159,500	free blacks (*pretos foros de todas as naçoes africanas*).
3,617,900	

All the censuses were not conducted at the same time, and provincial surveys should be seen as pertaining to the years 1816 and 1818. Brazil's population growth must nevertheless have been considerable over the last 4 to 5 years, both because of natural increases, or an accelerated birth rate, and the fateful import of African blacks. Documents presented in the House of Commons in London in 1821 show that the port of Bahia received 6,070 slaves and the port of Rio de Janeiro 18,032 between January 1, 1817 and January 7, 1818. During 1818, Rio received 19,802 blacks ([African Institution,] *Report made by a Committee [...] to the Directors of the African Institution, on the 8th of May*, 1821, p. 37). I do not doubt that Brazil's population exceeds 4 million today. They therefore overestimated it in 1798 (*Essai politique sur le royaume de la Nouvelle-Espagne*, II, p. 855). Using old censuses that he was able to study carefully, ▾Mr. Correia de Serra believed that Brazil's population was 1.9 millions in 1776, and this statesman's authority carries quite a lot of weight. A population table reported by ▾Mr. [Auguste] de Saint-Hilaire, a correspondent of the Institute, estimates Brazil's 1820 population at 4,396,132. But in this table, as the knowing traveler can see, the number of uncivilized and *baptized* Indians (800,000) and free men (2,488,743) is much too high, while the number of slaves (1,107,389) is far too low (see Veloso de Oliveira's statistics on Brazil in *Anais Fluminenses de Ciências*, 1822, Vol. I, §4). II.136 II.137

Having continued to pursue, over the last few years, painstaking research

1. Brackenridge, *Voyage to South-America*, Vol. I, p. 141.

on population in the new states of Spanish America, the Antilles, and on the Indian tribes that roam the two Americas, I believe myself capable of attempt-

II.138 ing another table showing total population for the New World for 1823.

I.	CONTINENTAL AMERICA, NORTH OF THE PANAMA ISTHMUS		19,955,000
	British Canada	550,000	
	United States	10,525,000	
	Mexico and Guatemala	8,400,000	
	Veragua and Panama	80,000	
	Autonomous Indians, perhaps	400,000	
II.	INSULAR AMERICA		2,826,000
	Haiti (Saint-Domingue)	820,000	
	British Antilles	777,000	
	Spanish Antilles (not including Margarita)	925,000	
	French Antilles	219,000	
	Dutch, Danish, etc. Antilles	85,000	
III.	CONTINENTAL AMERICA, SOUTH OF THE PANAMA ISTHMUS		12,161,000
	Colombia (not including Veragua and Panama)	2,705,000	
	Peru	1,400,000	
	Chile	1,100,000	
	Buenos Aires	2,300,000	
	British, Dutch, and French Guyanas	236,000	
	Brazil	4,000,000	
	Autonomous Indians, perhaps	420,000	
	TOTAL (in 1823)		34,942,000

II.139 The total population for the Antillean archipelago is probably not below 2,850,000, although the partial distribution of this population across different island groups might change with new research. This kind of follow-up is especially necessary for the free inhabitants of the British Antilles, for the Spanish part of the republic of Haiti, and for Puerto Rico.

B. Area

It is practically needless to note the precautions that Mr. Mathieu and I took when calculating surface area: We broke up the irregular shapes of new states into trapezoids and ▾well-*conditioned* triangles; we measured sinuous external borders by means of tiny squares traced on transparent paper; and we corrected large-scale maps. Despite these precautions, undertakings of this sort can produce very different results, (1) if the maps that one uses are based on less precise astronomical data; (2) if one takes too seriously the varying claims of adjoining states when tracing borders; and (3) even if one recognizes legally established borders and makes use of sufficiently precise astronomically determined coordinates, if one excludes from the *area* regions that are *entirely uninhabited* or are occupied by simple peoples. The first problem tends to affect area measurements where the border stretches north to south, as with Peru along the Cordilleras. It is well known that longitude mistakes are more frequent and more pronounced than latitude mistakes. The latter would cause a 4,600-square-league variation in the republic of Colombia's *area*, if one assumed[1]—as one once did—that the fort of San Carlos del Río Negro on Spanish Guyana's southern border with Brazil is situated below the equator. When making observations at the rock of Culimacari, I found that the fort was at 1° 53′ 41″ southern latitude. The second cause of inaccuracy, having to do with political border disputes, is of great importance wherever Portuguese territory abuts Spanish-American territory. Unpublished maps from Rio de Janeiro or Lisbon barely resemble those from Buenos Aires and Madrid. In the 23rd Chapter [of the *Personal Narrative*],[2] I discussed the interminable operations attempted by the *border commissions* that were established over forty years in Paraguay on the banks of the Caqueta and in the *Capitanía general* of Río Negro. According to my study of this great diplomatic controversy, the most disputed areas are the following: that between the sea[3] and the Río Uruguay; the banks of the Guaray and the Ibicuy, the Iguaçu, and the San Antonio rivers; the area between the Paraná and the Paraguay rivers; the banks of the Chichuy to the southeast of the Portuguese fortress of Nova Coimbra;[4]

II.140

II.141

II.142

1. Vol. VIII, pp. 45–47, and note F at the end of the IXth Book.

2. Vol. VII, pp. 365 and passim.

3. Since the annexation of the territory of Montevideo by the Portuguese, the borders between the state of Buenos Aires and Brazil have undergone large changes in the *banda oriental* or the *cisplatina* province, that is, on the north bank of the River Plate, between this river's mouth and the left bank of the Uruguay. The Brazilian coastline, from 30° to 34° southern latitude, resembles Mexico's coast between Tamiagua, Tampico, and the Río del Norte. It is formed by narrow peninsulas behind which there are large lakes and salt water marshes (Laguna de los Patos [Lagoa dos Patos], Laguna Merín [Lagoa Mirim]). The two Portuguese and Spanish *marcos* are situated near Lagoa Mirim's southern edge where the tiny Río Tahym (lat. 32° 10′) runs. The plain between the Tahym and the Chuy was regarded as neutral territory. Santa Teresa fort (lat. 33° 58′ 32″ after ˺Don Joseph Varela's unpublished map) was the southernmost post that the Spanish had on the Atlantic Ocean south of the equator.

4. Nova Coimbra (lat. 19° 55′) is a *presidio* founded in 1775. It is probably the southernmost Portuguese settlement on the Río Paraguay. On different Spanish and Portuguese maps, the Yaguary (Menici, Monici), a large tributary of the Paraná, is consistently enough marked as the eastern border between Paraná and Paraguay. To the west, the border some-

II.143 the eastern borders of the Spanish provinces of Chiquitos and los Moxos, the banks of the Aguapehy, the Jaurú and the Guaporé, and a little to the east of the isthmus that separates the Paraguay and Madeira River tributaries, near Villa Bella (lat. 15° 0′); the area to the Amazon's south and north, the entirely unknown land between the Río de la Madeira and the Río Javari (lat. 10.5°–11° south); the plains between the Putumayo and the Japurá, between the Apoporis, which is a tributary of the Japurá, and the Uaupés which flows into the Río Negro;[1] the jungles to the southwest of the mission of Esmeralda, between the Mavaca, the Pacimoni, and the Cababuri;[2] and finally, the northern portion of the Río Branco and the Uraricuera, between the northern part of the Portuguese fort of São Joaquim and the sources of the Río Caroni[3] (lat. 3° 0′–3° 45′). A few *piedras de marco* [flagstones] indicate the boundary between Spanish America and Portuguese America. They were decorated[4] with the absurd inscription: *Pax et Justitia osculatæ sunt. Ex pactis finium regundorum Madridi Idibus Jan.* 1750 [Peace and Justice embraced one another. After the accords on the settings of boundaries of Madrid on January 13, 1750,]. But these II.144 scattered points were never connected, so that there was never any delineation and formal acceptance of borders. Everything done until today has been seen as provisional, and in the meantime, the two neighboring nations carry on peaceably, without, however, renouncing their territorial claims.

We noted above that *inland navigation* would open up between the mouths II.145 of the Orinoco and the River Plate, and between Angostura and Montevideo, if the Villa Bella portage (15.5°) between the Río de la Madeira and Río Paraguay were replaced with a canal 5,300 toises long.[5] The southerly course of

---

times runs along the Chichuy (Xexuy) and the Ipane near the old Belêm mission (lat. 23° 32′), sometimes along the Mboymboy (lat. 20° 27′), across from the ruins of the Itatiny mission, and sometimes along the Río Mondego in Mbotetey [Miranda] (lat. Yaurú 19° 35′), near the ruined city of Jerez. All of these rivers are tributaries on the east bank of the Río Paraguay. The border closest to Nova Coimbra, the Río Mboymboy, was generally considered provisional by Brazil and the former viceroyalty of Buenos Aires.

1. Vol. VII, p. 411.

2. Vol. VIII, pp. 5 and 200.

3. Vol. VIII, pp. 116 and 448.

4. For example, at the point where the Río Jaurú flows into the Paraguay. See the *Patriota do Rio Janeiro*, 1813, no. 2, p. 54.

5. Properly speaking, the portage (*varadoiro*) is between the tiny Aguapehy and Alegre rivers. The first empties into the Jaurú, which is a tributary of the Paraguay. The Río Alegre flows into the Guaporé, a tributary of the Madeira. The sources of the Río Topayos are also very near to Villa Bella and to the sources of the Paraguay. This region, which forms a *land peninsula* between the Amazon and River Plate basins, will one day be of the greatest importance for inland trade in South America.

the large rivers will perhaps provide a *natural border* between Portuguese and Spanish possessions. This border would follow the Orinoco, the Casiquiare, the Río Negro, the banks of the Amazon for a length of 20 leagues, the Madeira, the Guaporé, the Aguapehy, the Jaurú, the Paraguay and the Paraná or the River Plate, and would form a more-than-860-league line of demarcation. To the east of this border, Spanish Americans possess Paraguay and part of Spanish Guyana; to the west of this line, Portuguese Americans occupy the land between the Javari and Madeira Rivers, and between the Putumayo and the Río Negro. It is not only from Brazil's coast and from Peru that civilization has spread to the central regions. It has penetrated these regions by three other pathways: the Amazon, the Orinoco, and the River Plate, following the tributaries of these three rivers and their secondary branches. The intersection of these rivers and their various courses has created territorial configurations and snaking borders so difficult to pin down astronomically that they hinder interior trade. II.146

In addition to these two sources of imprecision in surface area estimates that we have just discussed—errors in astronomical geography and border disputes—there is a third source, which is the most significant one. When people talk about the *area* of Peru or the *area* of the former *Capitanía general* of Caracas, it is up for debate whether these names designate only regions in which Spanish-Americans have settled, and which are therefore subject to their political and religious hierarchy, or whether they include the jungles and savannas in part empty, in part inhabited by savages—peoples both native and free—along with the regions governed by whites—magistrates, or corregidores, leaders of military posts, and missionaries. We saw above that in inland areas simple mistakes of 1° in latitude or 2° in longitude[1] can increase or decrease the surface area of new states by 12,000 sq. lg. along borders of II.147

1. I note only mistakes in *relative longitude*, for example, the differences in longitude between the coast and the Río Mamoré valley or Upper-Javari. I am not concerned with mistakes in *absolute longitude* which sometimes exceed 3° to 4° without impacting surface measurements. The new longitude I derived for the city of Quito (81° 5′ 30″ to the west of Paris) caused a significant change in the western part of America on the most recent maps. This figure differs by 0° 50′ 30″ from the longitude accepted before my return to Europe (*Connaissance des temps pour l'année* 1808, p. 236). According to ▾D'Anville, South America's length, from Cayenne to Quito, is 30 leagues too short. Errors in *relative longitude* that alter *area* calculations arise from *unequal partial displacements*. La Cruz [y] Olmedilla, whose large map has been copied and gradually distorted, places Santa Fé de Bogotá 0.5°, San Carlos del Río Negro 2.5°, and the mouth of the Apure 0.25° too far to the east. The distance from Cumaná to the Esmeralda mission on the Upper Orinoco is 2.25° too small, in La Cruz's estimate. Before my voyage, the entire Orinoco and Río Negro systems were typically placed 1° to 1.5° too far to the south and 2° too far to the east.

300 leagues. But the most significant changes result from demarcation lines II.148 that are arbitrarily drawn through regularly inhabited terrain or through land that is empty or occupied by uncivilized tribes. The *borders of civilization* are more difficult to draw than *political boundaries*. Small missions run by monks follow the length of a river; they are, in a manner of speaking, advanced outposts of European culture. Run by close-knit and adaptable groups, these outposts have at times penetrated more than a hundred leagues into jungles and deserts. Should one count as Peruvian or Colombian territory everything between these isolated villages, or between the crosses planted by Franciscan monks and surrounded by a few Indian huts? The hordes that roam the edges of missions on the Upper Orinoco, the Caroni, the Temi, the Japurá, the Mamoré (a tributary of the Madeira River) and the Apurimac (a tributary of the Ucayali) barely even know of whites. They are unaware that the land where they have lived for centuries lies within the borders of Venezuela, New Granada, or Peru, in keeping with the political dogma of *closed territory*.

II.149 In the current state of affairs, *contiguity of cultivated terrain*, or, better put, *contiguous Christian settlement*, exists in only very few places. Brazil abuts Venezuela only around the group of missions on the Río Negro, the Casiquiare, and the Orinoco; it abuts Peru only around the missions on the Upper Maragnon and those in Maynas province, between Loreto and Tabatinga. Little strips of cleared land conjoin the various states of the New World. Between the Río Branco and the Río Caroni, between the Javari and the Huallaga, the Mamoré and the Cuzco mountains, land inhabited by savages, on which whites never trod, separates the civilized parts of Venezuela, Brazil, and Peru like the branches of an inland sea (compare Vol. IV, pp. 146–53). European civilization has fanned out, like sunrays, from the coastline and from the mountains bordering the coastline to South America's center, and government influence diminishes as one moves away from the coast. Missions en- II.150 tirely dependent on monastic power and exclusively inhabited by the bronze race form a vast belt around regions that were cleared a long time ago. And these missions are located on the edge of savannas and jungles, between the agricultural and pastoral life of the colonists and the nomadic life of the hunters. Often, in maps from Lima, the easternmost Peruvian provinces—Tarma and Cuzco—do not extend to the borders of Grand Pará and Mato Grosso. Only the regions under white rule (*tierras conquistadas*) are considered Peru. The rest have vague names, such as *países desconocidos, comarca desierta, tierras de Indios bravos e infieles* [unknown lands, deserted country, country of fierce and faithless Indians]. All of Peru, up to the Portuguese borders, is 41,420 square nautical leagues. If we subtract unknown and Indian lands between the Brazilian border and the eastern banks of the Bení and Ucayali

Rivers, we are left with no more than 26,220 sq. lg. We will soon see that the differences are even greater in the former viceroyalty of Buenos Aires, known II.151 today as the *United States of the Río de la Plata*. Likewise, one can consider Brazil either 257,000 or 118,000 sq. lg., depending on whether one includes the entire surface of the country from the coast to the banks of the Mamoré and the Javari, or if one stops at the Paraná and Araguay Rivers, excluding most of Mato Grosso, Río Negro, and Portuguese Guiana—three sparsely populated provinces more than a third the size of Europe.

It follows from these considerations that one should not be surprised if different geographers, who calculate surface area with the same degree of precision in consultation with sufficiently accurate maps, arrive at results that differ by a quarter, a third, and sometimes even by more than half. Regions that are either empty or inhabited by free natives do not have borders that are easy to fix. Missions follow rivers into the midst of these wild regions. Calculations of surface area vary depending on whether one counts only the territory already conquered by missionaries or if one adds the jungles in between the conquered areas. Thus, the lack of agreement between the preceding table II.152 and that which Mr. Oltmanns derived in 1806 results entirely from "the exclusion of regions not subject to white rule." The old calculations are necessarily smaller than the new ones, which give total *area*. By converting common leagues to nautical leagues in the *Essai politique sur le royaume de la Nouvelle-Espagne* (Vol. II, p. 511), I arrived at only 299,810 sq. lg. (20 to a degree) for all of Spanish America; 30,628 for Venezuela, the former *Capitanía general* of Caracas; 41,291 sq. lg. for New Granada; 19,449 sq. lg. for Peru's settled areas (according to borders on the *Map of Intendancies* that Don Andrés Baleato published in Lima in 1792); 14,447 sq. lg. for Chile; and 91,528 sq. lg. for the United Provinces of the River Plate, the former viceroyalty of Buenos Aires. What I have just explained about the difficulties of calculating the surface area of Spanish America also applies to the United States, whose western borders at various times were the Mississippi, the Rocky Mountains, II.153 and the Pacific coast. For a long time, the *Missouri* and *Arkansas territories* had no western frontiers; in this, they resembled Chiquitos province in South America. In the tableau I present here, I have adopted a method of calculation that is distinct from the one I had previously used. I estimated the territory that each state's growing population is expected to settle over the coming centuries. I have traced division lines (*líneas divisorias*) on my unpublished Spanish and Portuguese maps in accordance with received traditions and rights derived from long, peaceful ownership. Where the two nations' maps differed substantially, I took these differences into account by using an average. The numbers on which I have settled in the preceding table thus show the *maxi-*

II.154 *mum* surface area available for development in Colombia,[1] Peru, and Brazil. But since, at a given time, a state's political power derives not so much from the relation between its surface area and its population as from the density of its population centers, I treated inhabited and uninhabited regions separately. My task was made less daunting by respectable persons from the new governments in Spanish America who also wanted to know total and partial surface area for the purposes of domestic administration. It is likely that province designations will still undergo frequent changes, as is the case in all emerging societies. II.155 One experiments with different combinations before achieving a state of balance and stability. And if this kind of innovation has been less frequent in the United States (at least to the east of the Alleghenies), the reason for this is not exclusively national character but, rather, the fortunate position of the Anglo-American colonies, which, with their excellent political institutions, had liberty before they had independence.

NEW SPAIN. Mr. Oltmanns carefully calculated this vast country's surface in keeping with the borders marked on my large map of Mexico. There will likely be some changes to the north of San Francisco and beyond the Río Norte [Rio Grande], and between the mouth of the Sabine River and Texas's Colorado River. ▾Major Pike's *Voyage*, published in Philadelphia in 1810, amply verified what I added to my Mexican map—drawn up in 1804 and published in 1809—about the identification of the Napestle and Pecos Rivers with Louisiana's Arkansas and Red Rivers of Natchitoches. II.156

GUATEMALA. This little-known country comprises the provinces of Chiapa, Guatemala, Verapaz (or Tezulutlan), Honduras (towns: Comayagua, Omoa, and Trujillo), Nicaragua, and Costa Rica.[2] Guatemala's coast stretches

1. In the declaration of ▾the Congress of Venezuela from December 17, 1819, which is regarded as the republic of Colombia's *foundational law*, the republic's territory is estimated (in article 2) at 115,000 sq. lg., without specifying what kinds of leagues. If they are nautical leagues (which seems quite likely), then the estimate is too high by 25,000 leagues (one and a half times France's *area*). They must have consulted maps that had not been corrected in accordance with the astronomical observations on the southern and eastern borders. To this day, all *area* estimates published in the new states of America are very imprecise, except for the partial data from the *Abeja argentina* (1822, no. 1, p. 8), an interesting journal published in Buenos Aires.

2. Juarros, *Compendio de la historia de la ciudad de Guatemala*, published in Guatemala City, 1809, vol. I, pp. 5, 9, 31, 56; Vol. II, p. 39. ▾José Cecilio [del] Valle, *Periódico de la Sociedad económica de Guatemala*, Vol. I, p. 38.

along the South Sea from Barra de Tonalá and (lat. 16° 7′, long. 96° 39′) to the east of Tehuantepec to Punta de Burica, or Boruca (lat. 8° 5′, long. 85° 13′) to the east of Costa Rica's Golfo Dulce. From this point, the border runs as follows: to the north, following the Colombian province of Veragua, toward Cape Careta (lat. 9° 35′, long. 84° 43′), which juts out into the Antillean sea, a little to the west of the beautiful port of Boca del Toro; to the NNW along the coast to the Bluefields or Nueva Segovia River [Escondido River] (lat. 11° 54′, long. 85° 25′) on Mosquito Indian territory; to the NW along the Nueva Segovia River for 40 leagues; and finally, to the N, to Cape Camarón (lat. 16° 3′, long. 83° 31′) between Cape Gracias a Dios and Trujillo port. From Cape Camarón, the coastline of Honduras runs to the W and the N, delineating the border until the Sibun River's mouth (lat. 17° 12′, long. 90° 40′). From there, the border follows the course of the Sibun to the E, crosses the Río Sumasinta [Usumacinta], which flows into the Laguna de Términos, and extends to the Río Tabasco or Grijalva right up the mountains that dominate the Indian city of Chiapa [de Corzo] and then turns to the SW to link up with the South sea coast at Barra de Tonalá. II.157

CUBA AND PUERTO RICO. For Puerto Rico, the surface *area* calculation is based on the maps of hydrographic Depot of Madrid. For the island of Cuba, they are based on the map that I drew in 1820 according to my own astronomical observations and on the data published by Mr. Ferrer, Mr. Robredo, Mr. Lemaur, Mr. Galiano, and Mr. Bauzá up to then.

COLOMBIA. Here are the current borders of the republic of Colombia, according to information that I gathered in this place, particularly on the southern and western reaches, that is, on the Río Negro, in Quito, and in Jaén de Bracamoros province: on the north coast of the Antillean sea, from Punta Careta (lat. 9° 36′, long. 84° 43′) via the eastern border of Costa Rica province (belonging to the state of Guatemala) up to the Morocco and Pomeroon rivers[1] to the east of Cape Nassau. From this point along the coast (lat. 7° 35′, II.158 II.159

1. Vol. VIII, pp. 408, 409, and 410. A number of uncertainties remain about the astronomical position of the easternmost point of Colombian territory. Longitudes between the Orinoco's mouth and British Guyana are even more poorly delineated, as they are not linked chronometrically. The mouth of the Pomeroon, or Poumaron, River depends both on the position of Punta Barima and on that of the Essequibo (Esquivo) River. Put differently, Cape Barima is situated half a degree too far to the east on Mr. Arrowsmith's large map of southern America. This geographer fixes Puerto España on Trinidad with adequate precision (63° 50′), but he holds that the difference in longitude between Puerto Es-

II.160 long. 61° 5′?), Colombia's border crosses savannas pocked with little granite outcroppings to the SW and then runs to the SE toward the confluence of the Cuyuni and the Mazaruni, where there was once a Dutch outpost across from the Caño Tupuro.[1] After crossing the Mazaruni, the border follows the west banks of the Essequibo and the Rupununi up to the point where the Pacaraimo cordillera (at 4° northern latitude) allows passage to the Rupununi, a tributary of the Essequibo. Then, following the southern slope of Mount Pacaraimo, which separates the Caroni from the Río Branco, the border runs to II.161 the W, through Santa Rosa (about lat. 3° 45′, long. 65° 20′) to the Orinoco's sources (lat. 3° 40′, long. 66° 10′!); toward the SW to the sources of the Mavaca and Idapa Rivers (lat. 2°, long 68°), and, after crossing the Río Negro, to San José island (lat. 1° 38′, long. 69° 58′), near San Carlos del Río Negro, to the WSW through completely unknown plains to the *Gran Salto del Japurá*, or *Caqueta,* situated near the mouth of the Río de los Engaños (south. lat. 0° 35′). Finally, it abruptly detours to the SE at the confluence of the Río Yaguas and the Putumayo or Ica (lat. 3° 5′ south.) where the Spanish and Portuguese

paña and Punta Barima is 1° 52′—in truth, it is no more than 1° 31′, as Churruca determined with great precision (Vol. VIII, p. 373, and Espinosa, *Memorias [...] de los Navegantes Españoles*, Vol. 1, no. 4, pp. 80–82). The southeast bank of the Orinoco's delta is at 8° 40′ 35″ lat. and 62° 23′ long. If one were to determine the Essequibo River's delta by taking its widely accepted difference in longitude from Cape Barima (1° 22′-1° 30′), the Essequibo would be around 60° 53′. This is practically the position on which ▾Mr. Buache settled in his map of Guyana (1797), a map that also expertly fixes the longitude of Cape Barima (62° 28′). Many geographers, like ▾Captain Tuckey (*Maritime Geography*, Vol. IV, p. 733), believe that the middle of the Essequibo's mouth is situated at 60° 32′ - 60° 41′ and that this mouth was likely connected to the position of Suriname or of Stabroek, the bustling capital of Demerary. When one sails along the coastline from Cayenne to Cape Barima and then to Trinidad, where the current pulls violently to the NW, measurements of differences in longitude tend to shrink. The longitude of the tiny Morocco River's mouth, situated near the Pomeroon's delta and serving as a border between the British colony of Guyana and Colombian territory, depends on the longitude of the Essequibo River, from which it is 45′ to the west, according to Bolingbroke, and 30′ to 35′, according to other recently published maps. An unpublished map of the Orinoco's deltas in my possession has no more than 25′. It follows from these meticulous discussions that the longitude of the Pomeroon's mouth fluctuates between 60° 55′ and 61° 20′. I shall repeat here a wish that I have already expressed elsewhere: that Colombia's government link—chronometrically and by means of uninterrupted navigation—the Essequibo's mouth, Cape Nassau, Punta Barima (old Guyana and Angostura), the Orinoco's *small mouths*, *bocas chicas*, Puerto España, and Punta Galera, the northeast cape on the island of Trinidad.

1. One must not confuse this post with the former Spanish *destacamento* [military outpost] on the right bank of the Cuyuni at its confluence with the Curumu.

missions of Lower Putumayo touch. From this point, Colombia's border runs to the south, crossing the Amazon near the mouth of the Río Javari between Loreto and Tabatinga, following the east bank of the Javari until it is 2° away from its confluence with the Amazon. It then runs to the west, crossing the Ucayali and the Río Guallaga between the villages of Yurimaguas and Lamas (in Maynas province 1° 25′ south of the confluence of the Guallaga and the Amazon) and then to the WNW, crossing the Río Utucubamba, near little Bagua, across from Tompenda. From Bagua, the border stretches SSW to a point on the Amazon (lat. 6° 3′) between the villages of Choros and Cumba, between Colluc and Cujillo, a little below the mouth of the Río Yauca. Then it turns W, crossing the Río de Chota, toward the Cordillera of the Andes near Querocotillo, and to the NNW, following and then crossing the Cordillera between Landaguate and Pucara, Guancabamba and Tabaconas, Ayavaca [Ayabaca] and Gonzanama [Ecuador] (lat. 4° 13′, long. 81° 53′) to reach the mouth of the Río Tumbez (lat. 3° 23′, long. 82° 47′). The Pacific coast delimits Colombia's territory at 11° lat. to the western edge of Veragua province or Cape Burica (lat. 8° 5′ north, long. 13° 18′). From this cape, the border runs to the north (across the continental isthmus between Costa Rica and Veragua) and links up with Punta Careta on the Antillean coast to the west of Lake Chiriquí, our departure point for this tour of the vast territory comprising the republic of Columbia.

II.162

These coordinates can help revise maps, even the most modern of which—for instance, the one published under ▾Mr. Zea's auspices that *claims* to be based on materials I gathered[1]—only vaguely reflects the story of long and peaceful settlements of neighboring states. Customarily, the entire southern bank of the Japurá, from the Salto Grande to the interior delta of the Abatiparaná, is considered Spanish. A *border marker* is located there, on the Amazon's northern bank, at lat. 2° 20′ and long. 69° 32′, according to Portuguese astronomers ([▾Requena,] *Carte manuscrite de l'Amazone, par don Francisco Requena*, border commissioner for S.M.C., 1783). The Japurá and Caqueta Spanish missions, commonly referred to as *missiones Andaquiés*, only extend to the Caguán River, a tributary of the Japurá below the ruined San Francisco Solano mission. The rest of the Japurá south of the equator, from the Río de los Engaños and the Great Cataract, is held by natives and by the Portuguese. The Portuguese even have some tentative footholds in Tabocas, San Joaquin de Cuerana, to the south of the Japurá,

II.163

II.164

1. [Neele and Howard,] *Colombia from Humboldt and Other Recent Authorities*, London, 1823.

and in Curaots on the northern tributary of the Japurá, the Apoporis.[1] It is at the mouth of the Apoporis—at 1° 14′ south. lat. and 71° 58′ long. (still to the west of the Paris meridian)—that the Spanish commissioners wanted to place the border stone in 1780, suggesting that they not wish to keep the *marco* at Abatiparaná. The Portuguese commissioners contested what was taken to be the Apoporis's border, claiming that the new *marco* should be placed at *Salto Grande del Japurá* (lat. south. 0° 33′, long. 75° 0′) to include the Brazilian possessions on the Río Negro. In Putumayo or Ica, the southernmost Spanish missions (*missiones bajas*) of the Popayán and Pasto clergy do not extend to the confluence of the Amazon but only to 2° 20′ on the southern latitude. The little villages of Marive, San Ramón, and Assumpción are situated there. The Portuguese control the mouth of the Pu- II.165 tumayo. To arrive at the *Lower* (or *Bajo*) *Putumayo*, the Pasto's monks are forced to travel down the Amazon to the mouth of the Napo at Pevas, go north on land from Pevas to the Yaguas's *Quebrada* or *Caño*, and to enter the Río Putumayo from this *Caño*. It is also not clear how the left bank of the Amazon, from Abatiparaná (long. 69° 32′) to the Pongo [Gorge] de Manseriche at the western edge of Maynas province, can be considered New Granada's border. The Portuguese have always controlled both banks up to the east of Loreto (long. 71° 54′), and even Tabatinga's position to the Amazon's north where the Portuguese have their last outpost sufficiently proves that they have never regarded the left bank of the Amazon, between Abatiparaná's mouth and the border near Loreto, as Spanish territory. Likewise, to prove that the Amazon's south bank westward from the Javari's mouth has never marked the border with Peru, I need only point out the existence of numerous villages belonging to Maynas province, which is situated from II.166 the Guallaga to the area just beyond Yurimaguas, 28 leagues to the south of the Amazon. The border's remarkable meandering between the Upper Río Negro and the Amazon derives from the fact that the Portuguese settled the Río Japurá by pushing up to the NW, while the Spanish went down the Putumayo. From the Javari, the Peruvian border passes beyond the Amazon, because the missionaries of Jaén and Maynas, having come from New Granada, moved into these nearly untamed regions by traveling on the Chinchipe and the Río Guallaga.

Calculating the republic of Colombia's surface area in accordance with the borders we have just traced, we arrive at 91,952 sq. lg. (still 20 to a degree). To wit:

1. Vol. VII, pp. 412–16.

Political Divisions	Square Leagues	Square Leagues
I. *Venezuela*		33,701
New Andalucía or Cumaná	1,299	
New Barcelona	1,564	
Orinoco Delta	18,793	
Spanish Guyana	652	
Caracas	5,140	
Barinas	2,678	
Maracaibo	3,548	
Margarita Island (not including *Laguna*)	27	
II. *New Granada* (with Quito)		58,251
Republic of Colombia		91,952

II.167

Whatever alterations Venezuela's territorial divisions may undergo, either because of changing administrative priorities or because of the desire for innovation always so active during a period of political awakening, exact knowledge of the *area* of old provinces will help with approximating the new ones. A close consideration of territorial divisions over the last ten years shows that the various attempts at *rebuilding societies* are simply reconfigurations of the same elements in search of a stable equilibrium.

II.168

Partial borders:

A. Former Capitancy General of Caracas:

a) THE CUMANÁ ADMINISTRATION, containing the two provinces of New Andalucía and Barcelona, is a little smaller than the state of Pennsylvania, which is 46,000 squares (at 69.2 to a degree). The border to the south and southwest is formed by the course of the Lower Orinoco to its main delta[1] (*boca de Navíos*) and to the north by the Atlantic Ocean and the Antillean Sea, from long. 62° 23′ to the mouth of the Unare River (long. 67° 39′). From this mouth to the south, the border between Caracas and Barcelona provinces first follows the Unare to its spring in the slightly mountainous region to the west of Pariaguan village and then follows the Orinoco between the mouths of the Río Sauta [Cauta] and the Río Caura, 24′ to the east of Alta Gracia, to which old maps refer as Ciudad Real. I established the longitude of this point

II.169

1. Vol. VIII, pp. 373 and 381. I nevertheless calculated separately the nearly uninhabited Orinoco delta, between the main river and the Mánamo Grande, the westernmost of the *bocas chicas*. This swampy delta is three times the average size of a département in France.

on the Orinoco in my calculation (Atlas, Pl. xv) by reducing it to the longitude of the Caura. It is about 68° 3′ to the west of the Paris meridian. Other geographers, for example, López [de Vargas] in his map of Caracas province, draw the border at Raudal de Camiseta, 8 leagues to the east of the Río Caura. In an unpublished map that I copied in Cumaná's archives, the border is near Muitaco at the mouth of the Río Cabrutica, 3 leagues to the east of the Río Pao. Cumaná's governors have long claimed that their authority extends well beyond the mouth of the Río Unare to the Río Tuy, and even to Cabo Codera.[1] On this assumption, they would draw a line to the south, 15 leagues to Calabozo's east between the sources of the Río Orituco and those of the Río Manapire, following the latter until it joins the Orinoco 4 leagues to the east of Cabruta.[2] This westernmost border would add 400 sq. lg. to Barcelona province, including the *Valle de la Pascua*. ˇLa Cruz and Caulín mark it on their maps with the words: "*terreno que disputan las dos provincias de Barcelona y de Caracas*" [terrain that the two provinces of Barcelona and Caracas dispute]. When I calculated the *area*, I followed the border along the Unare River, because it delimits the *current ownership* for the neighboring provinces. ˇ*Cumaná's gobierno* has four *ciudades*—Cumaná, Cariaco, Cumanácoa, Nueva Barcelona—and four *villas*—Aragua, La Concepción del Pao, La Merced, Carúpano.[3] New towns will probably rise on the edge of Paria Gulf, the *Golfo triste*, as on the banks of the Areo and the Guarapiche: these are the points that offer significant advantages for commercial enterprises in New Andalucía.

II.170 II.171

b) Prior to ˇthe July 5, 1811 revolution, SPANISH GUYANA was administered by a governor who resided in Angostura (Santo Tomé de la Nueva Guyana). It has more than 225,000 English square miles, exceeding the *area* of all the *Atlantic Slave-States*, Maryland, Virginia, both Carolinas, and Georgia. More than nine-tenths of this province is still uncultivated and practically uninhabited. I delineated the borders to the east and south from the Orinoco's main delta to San José Island, on the Río Negro, when I described the republic of Colombia's general shape. To the north and west, Spanish Guyana's border follows the Orinoco, from Barima Cape to San Fernando de Atabapo, and

1. Vol. VIII, p. 137.

2. Vol. VIII, p. 33.

3. Vol. VI, p. 393; VII, pp. 1–39, 156, 208–29, 345–405; VIII, p. 128. See Vol. IX, p. 53. I do not know the Villa de Merced's real position; it is marked on the unpublished map in Cumaná's archives. Píritu and Manapire also claim the title of *villas* (Caulín, p. 190).

then a line that runs from the north to the south of San Fernando to a point situated 15 leagues to the west of the fort of San Carlos. This line crosses the Río Negro a little above Maroa.[1] Its northeast border with British Guyana merits the greatest attention because of the political importance of the Orinoco delta, which I discussed in chapter 24 of this work [the *Personal Narrative*]. The sugar and cotton plantations had already crossed the Pomeroon River when the Dutch ruled; they spread beyond the mouth of the tiny Morocco River where there is a military post (see the very interesting map of the *Essequibo and Demerary colonies*, published in 1798 by ▾Major F. de Buchenröder). Far from recognizing the Pomeroon or Morocco rivers as their border, the Dutch placed this border at the Río Barima, near the Orinoco delta, and drew a line of demarcation, NNW to SSE, from there to Cuyuni. They had even militarily occupied the east bank of the tiny Río Barima before the British (1666) destroyed the fortresses of New Zealand and New Middlebourg on the right bank of the Pomeroon. These forts, along with the fort ▾Kyk-over-al at the confluence of the Cuyuni, Mazaruni, and Essequibo, were never rebuilt. During my stay in Angostura, people familiar with this area—which will one day create a conflict between England and the republic of Colombia—maintained that this area to the Pomeroon's west is swampy but highly fertile. Guyana's towns, or, rather, those places that enjoy the status[2] of *villas* and *ciudades*, are Angostura, Barceloneta, Upata [Venezuela], Guirior (a simple military outpost at the confluence of the Paraguamusi [Paravamusi] and the Paragua, a tributary of the Caroni), Borbón, Real Corona or Muitaco, La Piedra, Alta Gracia, Caycara, San Fernando del Atabapo, and Esmeralda (a few Indian huts around a church).

II.172

II.173

c) CARACAS PROVINCE is 61,000 English square miles, about one seventh smaller than the state of Virginia. Northern border: the Antillean Sea, from the mouth of the Río Unare, long. 67° 39′ to the Río Maticores (long. 73° 10′) toward the Gulf, or *Saco*, of Maracaibo, to the east of San Carlos Castle. Western border: a line to the S, between the mouth of the Río Motatán and the town of Carora, through the sources of the Río Tocuyo and the Páramo de las Rosas,[3] between Boconó and Guanare; to the ESE, between the Portuguesa and the Río Guanare where the Caño de Ygues, a tributary of the Portuguesa, divides Barinas and Caracas provinces; to the SE, between San Jaime and

II.174

1. Vol. VII, pp. 243–77, 434, 445; VIII, pp. 46 and 48.
2. Vol. VIII, p. 331.
3. See my *Atlas géographique*, Pl. XVII.

Oritucu, to a point on the left bank of the Río Apure across from San Fernando. Southern border: first, the Río Apure, from lat. 7° 54′, long. 70° 20′ to its confluence with the Orinoco near Capuchino (lat. 7° 37′, long. 69° 6′); and then the Lower Orinoco, toward the east, to the western border of the Gobierno de Cumaná near the Río Suata to the east of Alta Gracia. Cities: Caracas, La Guaira, Portocabello, Coro, Nueva Valencia, Nirgua, San Felipe, Barquisimeto, Tocuyo, Araure, Ospino, Guanare, San Carlos, San Sebastián, Villa de Cura, Calabozo, and San Juan Bautista de Pao. II.175

d) BARINAS PROVINCE has an *area* of 32,000 English square miles, a little smaller than the state of Kentucky. Eastern border: from the southern edge of the Páramo de las Rosas and the sources of the Río Guanare to the SE, at Caño de Ygues. From there, it runs ESE between the Portuguesa and Río Guárico to the mouth of the Apure, and then S along the Orinoco's left bank from lat. 7° 36′ to the mouth of the Río Meta. Southern border: the Meta's north bank until just beyond Las Rochellas de Chriricoas, between the mouths of the Caño Lindero and the Macachare (perhaps long. 70° 45′). Western border: from the Meta's left bank first to the NW, across the Casanare plains between Guasdualito and the Villa de Arauca, and then to the NNW above the Quintero and the mouth of the Río Nula, which flows into the Apure after Río Uribante, toward the headwaters of the Río Canagua, and to the foot of Páramo de Porquera. Northern border: the southeast slope of the Mérida Cordillera from Páramo de Porquera, between the Grita and the Pedraza up to La Vellaca ravine on Los Callejones road between Barinas and Mérida. From there, it runs to the sources of the Río Guanare NNW of Boconó. Towns: Barinas, Obispos, Boconó, Guanarito, San Jaime, San Fernando de Apure, Mijagual, Guasdualito, and Pedraza. Comparing my map of Barinas province with those of Cruz, 'López, and Arrowsmith, one will see the confusion that still reigns about the maze of the rivers that form the tributaries of the Apure and the Orinoco. II.176

e) MARACAIBO PROVINCE (with Trujillo and Mérida) is 42,500 square English miles, a little smaller than the state of New York. Northern border: the Antillean sea from the Caño de Oribono (to the west of Río Maticores) to the mouth of the Río Calancala, a little to the east of the Grand Río del Hacha. Western border: a line calculated from the coast, first to the S, between Villa de Reyes, also known as Valle de Upar [Valledupar], and the small mountain range (Sierra de Perijá) that rises toward the west of Lake Maracaibo, to the Catatumbo River. Then it runs east from Salazar to the Río Zulia, a little above San Faustino, and then finally E, at Páramo de Porquera, situated to the NE of the Grita. The southern and eastern boundaries extend to the south of Mérida's snowy mountains, across the La Vellaca ravine, to the eastern base of Páramo de las Rosas toward the sources of the Río Tocuyo, and from there, between the mouth of Río Motatán and the city of Carora, toward the Caño Oribono, as I explained when I described the borders of Barinas and Caracas provinces. The westernmost part of Maracaibo *Gobierno*, comprising Cape La Vela, is called the *Provincia de los Guajiros* (Guahiros) because of the wild Indians who go by this name and live between the Río Socuyo and the Río Calancala. A free tribe, the Cocinas, lives in the south. Towns: Maracaibo, Gibraltar, Trujillo, Mérida, and San Faustino. II.177

B. ▾FORMER VICEROYALTY OF NEW GRANADA contains New Granada proper (Cundinamarca) and Quito. The western borders of Maracaibo, Barinas, and Guyana circumscribe the viceroyalty's territory to the east. To the south and the west the borders are shared with Peru and Guatemala. Just to correct map errors, we note here that the following belong to New Granada: Upar Valley [Valledupar] or Villa de Reyes, Salazar de las Palmas, El Rosario de Cucuta, famous for being the seat of Colombia's constituent assembly in August of 1821, San Antonio de Cucuta, la Grita, San Cristóbal, and Villa de Arauca, as well as the confluence of the Casanare with the Río Meta and the confluence of the Inírida with the Río Guaviare. Casanare province, under Santa Fé de Bogotá, stretches north just beyond the Uribante. To the northeast, New Granada's easternmost province, called *Río Hacha*, is separated from Santa Marta province by the Río Enea. In 1814, the Río Guaytara divided Popayán province from the Presidencia of Quito, to which Pastos province belonged. The Isthmus of Panama and Veragua province have always been under the jurisdiction of the Audiencia of Santa Fé. II.178

PERU. To arrive at 41,500 sq. lg. (20 to a degree) for Peru's current *area*, the eastern border was (1) the course of the Río Javari from 6° to 9.5° southern latitude; (2) the 9.5° parallel from the Río Javari to the left bank of the Río Madeira, cutting other Amazon tributaries, that is, the Jatahi (Hyutahy), the Jurua, and the Tefé, that seem to be the Tapy of Acuña, the Coary, and the Purus; (3) a line that follows first the Río Madeira, then the Mamoré, from the Salto de Theotino to the Río Maniquí[1] between the confluence of the Gua- II.179

1. See the rather rare map *Misiones de Mojos de la Compañía de Jesús*, 1713. The Río Maniquí flows into the Yacuma, which is how ▾Mr. Haenke traveled from *Pueblo de Reyes* to

poré (Itonamas for the Jesuits) and the Santa Ana mission (about 12.5° lat.); (4) the course of the Maniquí, extending to the Río Bení which geographers had thought a tributary either of the Río Madeira or of the Río Purus; (5) the II.180 right bank of the Río Tequieri which flows into the Bení below Pueblo de Reyes, and from the Tequieri's sources a line that crosses the Río Inambari, runs to the SE toward the high Cordilleras[1] of Vilcanota and Lampa and separates the Peruvian districts of Paucartambo and Tinta from the Apolobamba district and from the Lake Titicaca's basin (Chucuito); (6) from 16° south. lat., the Andes western range, bordering the Lake Titicaca's basin to the east and dividing, below the 20th parallel, the tributaries of the Desagüadero and the tiny Paria Laguna from those of the Pilcomayo and the rapids that flow into the South Sea. According to these borders, Peru is 200 leagues wide in the north (toward the Río Javari); toward the Madeira and the Mamoré, it is 260 leagues wide along the parallels; in the south, the average width of the country is no more than 15 to 18 leagues. The *partido* of Tarapacá (part of the Arequipa intendancy) touches the Atacama desert where the mouth of the II.181 Río de Loa—Malaspina's expedition placed it at 21° 26′ south. lat.—forms the boundary line between Peru and the viceroyalty of Buenos Aires. By annexing from Peru the four intendancies of La Paz, Charcas or La Plata, Potosí, and Cochabamba, one has subjected to a government on the banks of the Río de la Plata not only provinces, whose waters drain to the southeast, and the vast region where the tributaries of the Ucayali and the Madeira—two tributaries of the Amazon—originate. One has also created an inland river system that, on the ridge of the Andes and in a longitudinal valley, is blocked on both ends by the *anchoring mountains* of Porco and Cuzco. This river system feeds the alpine Lake Titicaca. Despite these arbitrary divisions, the memories of the Indians who inhabit the lakeshore and the cold regions of Oruro, La Paz, and Charcas, tend to be drawn toward Cuzco, the center of the Incas' ancient imperial grandeur, rather than to the savannas of Buenos Aires. The Tiahuanaco plateau—where the ▾Inca Maita Capac erected buildings and gigantic statues, whose origin date back to before Cuzco's founding—was taken from Peru. II.182 Such an attempt at erasing a people's historical memory is tantamount to saying that the shores of Lake Copais are no longer called Greece. Hopefully, the many political confederations that are emerging in our day will not determine

the Río Mamoré. Modern geographers give the Río Maniquí a significant role in the story about the Lake Rogaguado and the forks of the Bení.

1. The *partidos* of Paucartambo and Tinta are under the administration of Cuzco. Apolobamba district and the basin of Lake Titicaca belong to the former viceroyalty of Buenos Aires.

lines of demarcation according to bodies of water alone but will also consider peoples' cultural roots. Upper Peru's fragmentation should provoke regrets among all those who know how to appreciate the importance of the Andean plateau's native population. If one were to draw a line from the southern edge of Maynas province, or from the banks of the Guallaga, to the confluence of the Apurimac and the Bení (a confluence that gives birth to the Río Ucayali) and from there to the west of the Río Vilcabamba and the Paucartambo plateau, toward that point where the southeastern border cuts the Río Inambari, one would cut Peru into two unequal parts: one (26,220 sq. lg.) the center of the civilized population, the other (15,200 sq. lg.) wild and almost entirely uninhabited.

BUENOS AIRES. The editors of the excellent journal *El Semanario* (Vol. I, p. 111) are right to say that no one on the banks of the River Plate knows the true borders of the former viceroyalty of Buenos Aires. The Portuguese contest the borders between the Paraná and the Río Paraguay, between the sources of the Paraguay and the Guaporé, a tributary of the Madeira. To the south, it is not clear if one should extend the borders beyond the Río Colorado to the Río Negro whose tributary is the Río Diamante (*Abeja Argentina* 1822, no. 1, p. 8, and no. 2, p. 55). Keeping in mind these doubts, which the fragmentation of Paraguay and the *Cisplatina* Province only magnify, I have calculated the *area* of the viceroyalty's immense territory on the basis of Spanish maps from before the 1810 revolution. From the east coast, the first *flagstone* is placed to the N[orth] of Santa Teresa fort, at the mouth of the Río Tahym; from there, the borders run NNW along the sources of the Ibicuy and along the Juy (cutting the Uruguay at 27° 20′) to the confluence with the Paraná and the Iguaçu; to the N, along the Paraná's left bank to the south. lat. 22° 40′; to the NW, following the Ivinhema toward the Nova Coimbra Presidio (lat. 19° 55′) founded[1] in 1775; to the NNW near Villa Bella and the isthmus that divides the waters of the Aguapehy (a confluence of the Paraguay River) from those of the Guaporé, toward the merging[2] of the Guaporé with the Mamoré, below the fort of Príncipe (south. lat. 11° 54′ 46″); to the SW, following the Mamoré and the Maniquí, as I had indicated above when tracing the border between Peru and the viceroyalty of Buenos Aires. Between 21° 26′ and 25° 54′ south. lat. (between the Río de Loa and the Punta de Guacho), the viceroyalty's territory extends across the Andes Cordilleras and follows the shores of the South Sea for 90 leagues. That is where Atacama Desert is,

II.183

II.184

1. *Patriota do Rio Janeiro*, 1813.
2. [*Patriota do Rio Janeiro*,] p. 40.

and the little port of Cobija, which will, one day, be very important for the trade of products coming from the Sierra or from Upper Peru. To the west, the Andes range extends to 37° lat.; to the south, either the Río Colorado—sometimes designated Desagüadero de Mendoza (lat 39° 56′)—or, according to the most recent authorities, the Río Negro, separates Buenos Aires from Chile and from the Patagonian coast.

II.185

As it is possible that ˈParaguay, *Entre-Ríos* Province, and *Banda Oriental*—or the *Cisplatina Province*[1]—will remain separate from the state of Buenos Aires, I felt justified in calculating separately the *area* of each of these disputed territories. Within the borders of the former viceroyalty, *between the Ocean and the Río Uruguay*, I calculated 8,960 sq. lg.; *between the Uruguay and the Paraná (Provincia Entre-Ríos)* 6,848 sq. lg.; between the *Paraná and the Río Paraguay* (the actual province of Paraguay) 7,424 sq. lg. These three regions to the east of the Río Paraguay from New Coimbra to Corrientes, and to the east of the Paraná, from Corrientes to Buenos Aires, form an area of 23,232 sq. lg.,[2] almost 1.5 times larger than France. Adding 18,300 sq. lg. for the *Pampas*, or savannas, to my previous calculations for the three parts that make up the former viceroyalty of Buenos Aires, I arrive at the following figures:

II.186

Northern Region of Upper Peru, from the Tequieri and Mamoré to Pilcomayo, between 13° and 21°southern latitude	37,020	naut. sq. leagues
Western Region or the lands between Pilcomayo, Paraguay, The River Plate, Río Negro, and the Cordilleras of the Andes (Tarija, Jujuy, Tucumán, Cordoba, Santa Fé, Buenos Aires, San Luis de la Punta, and Mendoza)	66,518	
Eastern Region, that is, everything east of the Río Paraguay and the Paraná	23,232	
	126,770	

II.187

Buenos Aires's government can compensate itself for what it may lose in the northeast with the 5,054 sq. lg. between the Río Colorado and the Río Negro. The Patagonian plains down to the Straits of Magellan cover another 31,206 sq. lg., and nearly two-thirds of that area has a much more temperate

1. The space of territory contained between the sea, the River Plate, Uruguay, the Missions and the Brazilian Captaincy of Río Grande (Auguste de Saint-Hilaire, *Aperçu d'un voyage dans l'intérieur du Brésil*, 1823, p. 1).

2. About 36,300 sq. lg. at 25 to a degree, not 50,263 of these leagues, as Buenos Aires newspapers say.

climate than is generally supposed. Saint Joseph Bay could very well tempt some European maritime power.

In the part of the viceroyalty of Buenos Aires to the east of Uruguay, which Brazil occupies, one must distinguish[1] between the borders recognized before the 1801 occupation of *Missions Province* to the north of the Río Ibicuy, and the borders established in the 1821 treaty between the *Cabildo* of Montevideo and the captaincy of Río Grande. *Missions Province* extends from the left bank of the Uruguay, the Ibicuy, the Toropi (a tributary of the Ibicuy), the San Javier Sierra, and the Río Juy (a tributary of the Uruguay). Its territory even extends a little beyond the Juy, toward the plains where the northernmost mis- II.188 sion of San Angel is situated. Beyond are jungles inhabited by free Indians. When the Franco-Spanish alliance spurred England to force the Portuguese into declaring war against Spain in February 1801, the Spanish Missions province was easily invaded. Hostilities did not last long. Although the Spanish court contested the occupation's legitimacy, the Missions province remained in Portuguese hands. The treaty of 1777 should serve as a basis for the borders between the viceroyalty of Buenos Aires and the Captaincy of Río Grande. These borders were formed by a line stretching from the Río Guaray (Arrowsmith's Guaney) and from the sources of the tiny rivers of Ibirapuitã, Nanday, and Ibycuimerim, which flow into the Ibicuy, (lat. 29° 40′) first to the confluence of the Río de Ponche Verde with the Ibicuy. Then, still to the southeast to the sources of the Río Negro (a tributary of the Uruguay) crossing Lake Mirim, the line stretches to the mouth of the Itahy, popularly referred to as the Tahym. The southernmost Portuguese *flagstone* is found at this river delta on the seacoast. The country between the Tahym and the Río Chuy, a little to the II.189 north of Santa Teresa, was neutral and was known as *Campos neutraes*. Despite diplomatic agreements, however, it was already largely settled by Portuguese farmers in 1804. ▾The French invasion of Spain and the Buenos Aires revolutions allowed the Brazilians to push their conquests to the mouth of the Uruguay such that the new internal borders between old Brazil and recently occupied territory were established by deputies from the Montevideo *cabildo* and the Río Grande Captaincy in 1821, without interference from the congress of Buenos Aires. It was agreed that Brazil's *Cisplatina Province—Banda oriental* according to Spanish geographical nomenclature—would be delimited to the north along the confluence of the Uruguay and the Arapay (Arrowsmith's Ygarupay), to the east along a line that, beginning in Angostura, 6 leagues to the south of Santa Teresa, passes through the San Michel

1. These clarifications are based on the unpublished notes that Mr. Auguste de Saint-Hilaire gathered in the field. I owe them to his friendship.

marshes, follows the Río San Luis to its mouth in Lake Merín, extends along II.190 this lake's eastern shore for 800 leagues, passes through the mouth of the Río Sabuaty, climbs to the mouth of the Rio Jaguarão, follows this river's course to the Acegua Hills, crosses the Río Negro, and then, still curving to the northwest, rejoins the Río Arapay.

The territory between the Arapay and the Ibicuy, the southern border of Missions Province, belongs to the Río Grande Captaincy. Portuguese-Brazilians have not yet attempted to establish settlements in *Entre-Ríos* province, between the Paraná and the Paraguay, ▾a region devastated by Artigas and Ramírez.

The natural borders of Potosí and Salta, that is, Upper Peru and Buenos Aires, tend to get into the savannas (*pampas*) which, like an arm of the sea, spread from Santa Fé in the north between Brazil's mountains and those of II.191 Córdoba and Jujuy.[1] Chichas and Tarija are considered Upper Peru's southernmost provinces. The Manso plains, between the Pilcomayo and the Río Grande, or Bermejo,[2] like Jujuy, Salta, and Tucuman, belong to the state of Buenos Aires proper. Upper Peru's eastern border is no more than an imaginary line drawn across uninhabited savannas. It cuts the Andes at the tropic of Capricorn, and, from there, it crosses first the Río Grande, at 26 leagues below Santiago de Cotagaita, and then the Pilcomayo, at 22 leagues below its confluence with the Cachimayo, which comes from the Plata or Chuquisaca, and finally the Río Paraguay, at 20° 50′ southern latitude. Even if Lake Titicaca's basin and the mountainous part of Upper Peru, where the language of the Incas predominates, were once again linked to Cuzco, the Chiquitos and II.192 Chaco plains could still remain part of Buenos Aires Pampas territory.

CHILE. Its borders are formed by the Atacama Desert to the north and the Andes to the east, where, according to Mr. Espinosa and Mr. Bauzá's 1794 barometric readings, the mail route between Mendoza and Valparaiso runs 1,987 toises above sea level.[3] To the south, I put the border[4] at the opening of

1. According to ▾Mr. Redhead (*Memoria sobre la dilatación [. . .] del aire atmosférico*; Buenos Aires, 1819, pp. 8 and 10), this city is 700 toises above sea level. Already the absolute height of San Miguel del Tucumán is 260 toises, according to the barometric readings of the same author (who lives in Salta).

2. The real name of this river, whose banks were once inhabited by the Abipons, is Iñate (see ▾Dobrizhoffer, *Historia de Abiponibus*, 1784, Vol. II, p. 14).

3. This is still 440 toises less than the Azuay's route's highest point between Quito and Cuenca, which I measured in 1802. See my *Observations astronomiques*, Vol. II, p. 385, no. 209.

4. *Essai politique sur le royaume de la Nouvelle-Espagne*, Vol. I, p. 4; Vol. II, p. 831.

the Gulf of Chiloe, where the fort of Maullín (lat 41° 43′) is Spanish America's southernmost continental possession. The bays of Ancud and Reloncaví have no permanent European settlements. Free, not to say savage, Indians known as Juncos live in these parts. It follows from these data that European settlement extends much farther down the continent's west coast than its east coast. West coast settlements have already crossed the parallel of the Río Negro and the Puerto de San Antonio by one degree of latitude. The capital of Santiago de Chile is situated on a plateau almost as high as Caracas.[1] II.193

BRAZIL. Columbia's southern borders, Peru's eastern borders, and Buenos Aires's northern borders determine Brazil's boundaries to the north, west, and south. To calculate the *area*, I consulted unpublished maps given to me by Rio de Janeiro's government during the diplomatic squabbles between French and Portuguese Guiana over the extremely vague wording of article 8 in ▾the treaty of Utrecht and article 107 of the congress of Vienna.[2] By draw- II.194

1. According to Mr. Bauzá, 409 toises: that is, three hundred toises lower than the town of Mendoza on the opposite side of the Andes Mountains (unpublished notes of ▾Don Louis Née, botanist of the Malaspina expedition).

2. Vol. VIII, p. 503. Brazil's borders were demarcated by: in the district of Río Negro, the astronomers ▾José Joaquim Victorio da Costa, José Simoes de Carvalho, Francisco José de Lacerda, and Antonio Luiz Pontes; in Grand Pará, especially between the Araguari and the Calsoene (perhaps the Río Carsewene in the *Carte des côtes de la Guyane*, published by the Naval Office in 1817), by the astronomer José Simoes de Carvalho and the ingenious colonel Pedro Alexandrino de [Pinto de] Souza. For a long time, French claims extended beyond the Calsoene, near the North Cape. Today, the border has been rolled back to the mouth of the Oyapok. This river's principal tributary, the Canopi and the Tamouri, which is a tributary of the latter, approach within a league (at 20° 30′ lat.?) of the Maroni's source, or, rather, one of its branches, the Río Araguari, near the Aramichaun Indian village. As the Portuguese wanted to draw the border between the banks of the Oyapok and the Araguari (Araouari), they had colonel [Pinto] de Souza carefully measure the latitude of the latter's springs. They found that it was farther north than its mouth, which would have put the border on the same parallel as that of the Calsoene. The Río de Vicente Pinçon, made famous by a serious diplomatic row, has disappeared from the new maps. According to an old unpublished Portuguese map I have of the coastline between San José de Macapa and the Oyapok, the Río Pinçon would be identical with the Calsoene. I suspect that the unintelligible terms of article 8 in the Treaty of Utrecht ("the line of the *Japoc River or Vicente Pinçon* that should cover the Cape's and the North's possessions") derives from the fact that the name North Cape is occasionally used for Orange cape (see Laet, *Novus orbis*, Nov. 1633, p. 636). ▾Mr. La Condamine, whose sharp mind misses nothing, already noted in the *Relation abrégée d'un voyage fait dans l'interieur de l'Amérique Méridionale* (p. 199) that: "The Portuguese have their reasons for confusing the bay (?) of Vincent Pinçon, near the western mouth of the Río Arawari (Araguari), lat. 2° 2′, with the Oyapok River, lat. 4° 15′.

II.195 ing a north-south line from the mouth of the Tocantins River, following the course of the Araguari 40 leagues to the west of Villaboa to the point where the Río Paraná crosses the tropic of Capricorn, one divides Brazil into two II.196 parts. The westernmost part includes the Captaincies of Grand Pará, Río Negro, and Mato Grosso. The area is nearly uninhabited, Europeans having settled only along the rivers: the Río Negro, the Río Branco, the Amazon, and the Guaporé, which is a tributary of the Río Madeira. Its size is 138,156 sq. lg. (20 to a degree). The eastern part, comprising the coastal Captaincies of Minas Gerais and Goiás, is 118,830 sq. lg. My measurements are consistent with those of a highly distinguished geographer, Mr. Adrien Balbi, who puts the entire Brazilian empire at 2,250,000 Italian square miles (250,000 square nautical leagues), not including (as I have not) the Cisplatina and Missions Provinces to the east of the Uruguay ([Balbi,] *Essai statistique sur le Portugal*, Vol. II., p. 229).

UNITED STATES. I had already noted elsewhere (*Essai politique sur le royaume de la Nouvelle-Espagne*, Vol. I, p. 153) that the United States' surface area has been somewhat difficult to measure ever since the acquisition of Louisiana, whose northern and western borders have remained uncertain for so long. ▾These borders were fixed by the convention that ended in London on II.197 October 20, 1818, and by the treaty of the Floridas, signed in Washington on February 22, 1819. I therefore thought to reopen this question with fresh evidence. I undertook this endeavor with special care, given that contemporary writers have estimated the surface area of the United States, from the Atlantic Ocean to the South Sea, at 125,400, 137,800, 157,500, 173,400, 205,500, and 238,400 square nautical leagues, at 20 to a degree. Given these divergent figures, which vary by more than 100,000 sq. lg. (six times the *area* of France), it seemed to me impossible to choose one to which I could compare the surfaces areas of the new free states of Spanish America. Sometimes the same author gives the most varying estimates of the same territory at different times, assuming it is delimited by the two seas, Cape Hatteras and the Columbia River, the Mississippi deltas and the Lac des Bois [Minnesota]. In an 1816

The peace of Utrecht made of them a single river." This 2° 2′ lat. brings the imaginary Vincent Pinçon River closer to the Majacarí and the Calsoene but pulls it more than a degree away from the Araguari, which is at north. lat. 1° 15′. Mr. Arrowsmith, whose map provides excellent material on the Amazon delta, situates the Río de Vicente Pinçon to the south of Majacarí, where the Matario disappears into a bay across from the tiny Tururi Island, lat. 1° 50′. As the Araguari connects to the Matario and forms a delta around the flooded terrain of Carapaporis to the northwest, it is possible that Mr. La Condamine took the tiny river across from the island of Tururi for the Araguari's western branch.

map, ▾Mr. Melish estimated the United States at 2,459,350 square miles (69.2 to a degree), 1,580,000 miles of which belongs to the Missouri territory alone. In his *Travels through the United States of America* (1818, p. 561), he settles on 1,883,806 square miles, 985,250 of them for the Missouri territory. Later II.198 on in the *Geographical Description of the United States* (p. 17), he increases this number once more, to 2,076,410 square miles. These fluctuating opinions on the United States' surface area cannot be attributed to the differing ways in which borders are drawn. Most mistakes in the measuring of the size of the *area* between the Mississippi and the Rocky Mountains, and between the mountains and the South Sea, are due to simple mathematical errors. Taking the average of several estimates from maps by Arrowsmith, Melish, ▾Tardieu, and Brué, I find that:

I. East of the Mississippi or 930,000 *square miles.*		77,684 s. n. l.
a.) Atlantic region east of the Alleghenies or 324,000 *square miles.* The chain of the Alleghenies extends north toward Plattsburg and Montreal; to the south along the Apalachicola; in this way, most of Florida belongs to this part of the Atlantic seaboard.	27,064	
b.) Between the Alleghenies and the Mississippi or 606,000 *square miles.*	50,620	
II. West of the Mississippi or 1,156,800 *square miles.*	96,622	
a.) Between the Mississippi and the Rocky Mountains, lakes included. or 868,400 *square miles.*	72,531	
b.) Between the Rocky Mountains and the coast of the South Sea, taking as southern and northern limits the 42° and 49° parallels (Western Territories) or 288,400 *square miles.*	24,091	
Territory of the United States, between the two Oceans, 2,086,800 *square miles,* or		174,306 s. n. l. 20 to deg.

II.199

All of the United States' territory, from the Atlantic Ocean to the South Sea, is II.200 therefore a little larger than Europe west of Russia. The Atlantic portion alone is comparable to Spain and France together; the part between the Alleghenies and the Mississippi to Spain, Portugal, France, and Germany; the part to the Mississippi's west to Spain, France, Germany, Italy, and the Scandinavian kingdoms. The Mississippi thus separates the United States into two large regions. The first or eastern region (the east), advancing rapidly in terms of culture and civilization, is about the size of Mexico. The western part, almost entirely wild and uninhabited, is about the size of the republic of Colombia.

Statistical research on several European countries has led to important conclusions about comparisons between the *relative population* of maritime and inland provinces. In Spain,[1] this population ratio is 9 to 5; in the *United Provinces of Venezuela*, particularly in the former *Capitanía general* of Caracas, it is 35 to 1. However powerful the influence of trade is on states' prosperity and on peoples' intellectual development, one would be wrong to attribute to this single cause the differences we have just noted for America and Europe. In Spain and Italy, not including the fertile plains of Lombardy, the inland regions are arid and either mountainous or on high plateaus. The meteorological conditions determining the soil's fertility are not the same on the coast and in the interior. In America, colonization typically began on the coast and only slowly made its way inland. Such is its gradual progression in Brazil and Venezuela. It is only where the coasts are disease-ridden, as in Mexico or New Granada, or sandy and without rain, as in Peru, that the population has concentrated in the mountains and inland plateaus. Such local conditions were too often neglected in discussions on the future of the Spanish colonies. They add a particular quality to some of the countries where analogies between the physical and the moral state are less striking than one would generally think. Considered from the perspective of *population distribution*, the two territories that now form a single body politic—New Granada and Venezuela—offer the most dramatic contrast. Their capitals—and the position of capitals always signals the area of the highest population density—are situated at such unequal distances from the commercial coastline of the Antillean Sea that the city of Caracas would need to be transplanted southward, to the confluence of the Orinoco and the Guaviare where the San Fernando de Atabapo mission is located, to be on the same parallel as Santa Fé de Bogotá.

II.201

II.202

The republic of Colombia is the only state in Spanish America[2]—along with Mexico and Guatemala—with coastlines that face both Europe and Asia. From Cape Paria to the western edge of Veragua province, it is 400 nautical leagues. From Cape Burica to the mouth of the Río Tumbez, it is 260 nautical leagues. Colombia's coastlines on the Antillean Sea and on the Pacific Ocean are therefore the same length as the coastline from Cádiz to Danzig, or

II.203

1. ▾Antillón, [*Elementos de la*] *Geografía astronómica, natural y política*, 1815, p. 145.

2. The former viceroyalty of Buenos Aires also extended into a small part of the coast on the South Sea, but we have seen above (Vol. XI, pp. 229 and 230 [this edition, pp. 238–39]) how deserted this portion is.

from Ceuta to Jaffa. To this inestimable resource for national industry must be added another, whose importance has as yet not been sufficiently appreciated. The Isthmus of Panama is part of Colombia's territory. If this strip of land were crossed by fine roads and populated with camels, it would serve as a *portage* for world commerce, even though neither the Cúpica plains, nor Mandinga Bay, nor the Río Chagre are suitable for a canal fit for ships going from Europe, or the United States, to China, or from the United States to America's northwest coast.

Examining over the course of this study the influence that a country's configuration (that is, its shape and the form of its coastline) exerts everywhere on civilization's progress and the destiny of peoples, I have often outlined the disadvantages presented by these large, triangular continental masses that lack gulfs and interior seas, like Africa and most of South America. Beyond a shadow of a doubt, the existence of the Mediterranean was intimately linked with the first glimmer of human culture among the people of the west, and the land's *articulated form*, the frequency of bottlenecks and the succession of peninsulas, favored the culture of Greece, Italy, and perhaps all of Europe west of the latitude of the Propontis. In the New World, the noninterrupted coastline and the monotony of its rectilinear progression is particularly striking in Chile and Peru. Colombia's coast offers slightly more varied forms, such as spacious gulfs—Paria, Cariaco, Maracaibo, and Darien—that were already more densely settled than the rest at the time of the first discovery and that stimulated the exchange of goods. This coastline (and this is an incalculable benefit) is washed by the Antillean Sea, a kind of inland sea with a number of outlets, the only one on the New Continent. This basin, whose opposing shores belong to the United States, the republic of Colombia, Mexico, and some European naval powers, has created a peculiar commercial system that is wholly American. Southeast Asia, with its neighboring archipelago, the Arabian Gulf, and the Mediterranean at the time of the Phoenician and Greek colonies, has proven what a positive influence the close proximity of opposite coasts, with different goods and nations of diverse races, has for trade and for intellectual culture. The importance of the inland sea of the Antilles, which Venezuela borders to the south, will increase even more as population grows steadily on the banks of the Mississippi. For this river, along with the Río del Norte and the Magdalena, is the only large navigable river that flows into the Antilles basin. The depth of America's rivers, their marvelous branches, and the use of steamboats facilitated by the proximity of forests will partially compensate for the obstacles to civilization's development caused by the coastline's uniform direction and the continent's general configuration.

II.204 II.205 II.206

By comparing the territory's size to the actual population in the tables we presented above, we will obtain the relation of these two elements of public wealth, a relation that constitutes the *relative population* of each state in the New World. For each square nautical league we would find: 90 inhabitants in Mexico, 58 in the United States, 30 in the republic of Colombia, and 15 in Brazil. Meanwhile, Asian Russia has 11 inhabitants per square nautical league, the entire Russian empire 87, Sweden and Norway 90, European Russia[1] 320, II.207 Spain 763, and France 1,778. But these estimates of relative population applied to countries of immense size, vast portions of which are entirely uninhabited, are mathematical abstractions that tell us little. In evenly settled II.208 countries, like France,[2] for instance, the number of inhabitants per square league for each département is typically no more than a third greater or smaller than the relative population of the total of all départements. Even in Spain, fluctuations are rarely half or double the average.[3] In America, on the

1. In 1805, the *area* of European Russia, without Finland and the Grand Duchy of Warsaw, was 138,000 square leagues (at 20 to a degree) with 36.4 million souls, according to Mr. Hassel's statistical tables (*Statistischer Umriss der sämtlichen Europäischen Staaten*, Vol. I, p. 10). According to the same tables, the *area* of the entire Russian monarchy, in 1805, was 603,160 square leagues with a population of 40 million. These numbers would give us no more than 264 and 66 inhabitants per square league, respectively. Assuming, with Mr. Balbi (see his interesting research on Russia's population in the *Compendio di Geografia universale*, pp. 143 and 163, and his *Essai statistique sur le royaume de Portugal*, Vol. II, p. 253), that the *area* of European Russia, including Finland and the kingdom of Poland, is 169,400 square leagues; that the *area* of the entire Russian monarchy in Europe and Asia is 686,000 square leagues; and that in 1822, the absolute populations were 48 and 54 million, respectively, there would be 283 and 78 inhabitants per square league. In my recent research on Russia's *area*, I settle on 616,000 square leagues for the entire empire including Finland and Poland; for the European portion, including the former Kazan and Astrakhan kingdoms but excluding the government of Perm, 150,400 square leagues, which gives us the *relative populations* of 320 and 87 cited in the text. See also Gaspari, *Vollständiges Handbuch der Erdbeschreibung*, Vol. XII, p. 210.

2. In 1817, the Land Registry Office estimated France's *area* (not including Corsica) at 51,910,062 hectares, or 5,190 square myriameters, or 26,278 common square leagues at 25 to a degree. For Corsica, ▾Mr. Coquebert de Montbret estimates 442 common sq. lg., so that France, including Corsica, is 26,720 common square leagues, or 17,101 square nautical leagues (20 to a degree). The population having been 30,407,907 in 1820, there were 1,778 inhabitants per square nautical league. The average size of a département in France is 198 square nautical leagues, for an average population of 353,600. The average number of inhabitants per square league is, for most departments, 1,000, 1,200, 2,400, and 2,600. Averaging the 5 least and most populated départements and districts of France and Russia, one arrives at 1 to 3.7 and 1 to 11.2 for the *lower and upper* relative population *limits* in the former and the latter countries.

3. Antillón, *Geografía*, p. 141.

contrary, only the Atlantic states (from South Carolina to New Hampshire) have populations that are now distributed somewhat evenly. In this, the most civilized part of the New World, there are 130 to 900 inhabitants per square league, while the relative population for all Atlantic States together is 240. The proportion between extremes (North Carolina and Massachusetts) is 1 to 7, almost as in France,[1] where it is 1 to 6.7 (the departments of Hautes-Alpes and of Nord). In the civilized countries of Europe,[2] population numbers fluctuate only within narrow margins. By contrast, they fluctuate much more no-

II.209

II.210

1. In continental France, excepting Corsica. The former département of Liamone is still less populated than Hautes-Alpes. In 1804, the département of Nord had a population of 774,500 on 178 square leagues (20 to a degree); in 1820, 904,500. In 1804, the département of Hautes-Alpes had a population of 118,322 on 160 square leagues; in 1820, 121,400. There are, therefore, 5,082 and 758 inhabitants per square nautical league in these two respective départements.

2. *Europe*, bounded by the Yaik River, the Ural Mountains, and the Kara River, has 304,700 square nautical leagues. Assuming 195 million inhabitants, there is a relative population of 639 per square league, a little less than the département of Hautes-Alpes and a little more than Spain's inland provinces. By comparing this *total average* of 639 to the *partial averages* of European countries that have more than 600 square leagues, excluding Lapland and four Russian jurisdictions (Archangel, Olonez, Vologda, and Astrakhan), one arrives at 160 souls per square league for the most sparsely settled regions of Europe and at 2,400 per square league for the most densely settled ones. These figures produce a ratio of 1 to 15. According to my latest calculations, *America* has 1,184,800 square nautical leagues from Cape Horn to 68° lat. (including the Antilles). Estimating its population at 34,284,000, as we have done above, we get barely 29 inhabitants per square league. To find the most populous and continuously inhabited area of 600 square leagues in America, one needs to turn either to the Mexican plateau or to the New England tristate area of Massachusetts, Rhode Island, and Connecticut, which in 1820 had an absolute population of 881,594 on 12,504 square English miles, that is, about 840 souls per square nautical league. In the Antilles, where the population is quite concentrated, one would need to turn to the Greater Antilles, because the Lesser Antilles (or the eastern Caribbean islands), from Culebra and Saint Thomas to Trinidad, amount together to no more than 387 square leagues. Jamaica has practically the same relative population as the three states of New England just mentioned, but its *area* is less than 500 square leagues. Saint-Domingue (Haiti), which is five times larger than Jamaica, has only 266 inhabitants per square league. Its relative population barely reaches New Hampshire's. I will not speculate on the partial number that might express the *lower limits* of the relative population in the New World, for instance, in the savannas between the Meta and the Guaviare, or in Spanish Guyana between the Esmeralda, the Erevato, and the Caura, or, finally, in North America between the Missouri's sources [in Montana] and ▼Slave Lake. The proportion of the extremes that we put at 1 to 15 for Europe is probably 1 to 8,000 in the New World, even without the Llanos and the Pampas.

II.211 tably in Brazil, the Spanish colonies, and even in the federation of the United States, if one considers the latter in its full extent. In Mexico, there are some intendancies (Sonora and Durango) that have 9 to 15 inhabitants per square league, while others, on the central plateau, have more than 500. The relative population of the region between the east bank of the Mississippi and the Atlantic States is barely 47, while in Connecticut, Rhode Island, and Massachu- II.212 setts it is more than 800. To the west of the Mississippi—as in the interior of Spanish Guyana—there are fewer than two inhabitants per square league in areas larger than Switzerland or Belgium. These regions resemble the Russian Empire, where the relative population in some Asian jurisdictions (Irkutsk and Tobolsk) stands in a relation of 1 to 300 to Europe's most densely settled regions.

The enormous discrepancy in newly settled areas between territorial size and population makes partial estimates obligatory. When one learns that New Spain and the United States have 90 and 58 inhabitants per square league, respectively (assuming their total territory to be 75,000 and 174,000 square nautical leagues, respectively), one has no more of a precise idea of the population distribution upon which a peoples' political power depends than one II.213–14 does of a country's climate—that is, temperature variations over the course of the year—from knowing the annual mean temperature.[1] If one stripped the United States of all its possessions to the west of the Mississippi, its population would be 121 instead of 58 per square league, much higher than New Spain's. Strip this last country of its *Provincias internas* (to Nueva Galicia's north and northeast), and you arrive at 190 instead of 90 souls per square league.

1. I would get too far afield if I pushed this comparison further and discussed how far *total averages* can enlighten us as to a country's temperature or population distribution. I endeavored to prove elsewhere (*On Isothermal Lines*, pp. 62 and 71) that mean winter temperatures in the *European climatic system* sink below the freezing point only where the mean temperature during an entire year decreases by at least 10° C. The lower the average annual temperature, the greater the difference is between winter and summer temperatures. Similarly, a fairly large country's very small relative population generally indicates an emerging agricultural sector that causes vast inequalities in population distribution. The climates that Buffon—with the felicity of expression that characterizes his style—dubs *excessive climates* (inland continental climates where harsh winters are followed by hot summers) correspond (so to speak) to unequally dispersed populations. Two phenomena of vastly different natures thus offer remarkable analogies when considered as simple qualitative values.

Here is partial data for Venezuela and New Granada, based on figures that, we have reason to believe, are the most accurate:

Republic of Columbia Six times as big as Spain, almost the same size as the United States west of the Mississippi. *Area:* 91,950 s. l. Actual population: 2,785,000.	30	inhabitants per s. n. l.
A. *New Granada* (with the province of Quito) Not quite four times as big as Spain. *Area:* 58,250 s.l. Actual population: 2 million.	34	
B. *Venezuela* or ancient *Captaincy-general* of Caracas More than twice as big as Spain, a size almost that of the *Atlantic states* of North America. *Area:* 33,700 s. l. Actual population: 785,000.	23	
a. Cumaná and Barcelona *Area:* 3,515 s. l. Actual population: 128,000.	37	
b. Caracas (with Coro) *Area:* 5,140 s. l. Actual population: 420,000.	81	
c. Maracaibo (with Mérida and Trujillo) *Area:* 3,548 s. l. Actual population: 140,000.	40	
d. Barinas *Area:* 2,678 s. l. Actual Population: 75,000.	28	
e. Guyana (Spanish Guyana) *Area:* 18,793. Actual population: 40,000.	2	

II.215

It follows from these data that the northern maritime provinces of Caracas, Maracaibo, Cumaná, and Barcelona are the former *Capitanía general*'s most populous region. But, if we compare their relative population to that of New Spain, where the two intendancies of Mexico and Puebla alone, on a territory barely that of the province of Caracas, have an absolute population that surpasses that of the entire republic of Colombia, we see that Mexican intendancies that occupy the seventh or eighth tier when it comes to a concentration of culture (Zacatecas and Guadalajara) have more inhabitants per square league than the province of Caracas. The average relative population of Cumaná, Barcelona, Caracas, and Maracaibo is 56. Half of these four provinces' combined territory, or 6,200 sq. lg., are virtually uninhabited

II.216

II.217–18 steppes[1] (*Llanos*). If one discounts the steppes' *area* and their small populations, one arrives at 102 inhabitants per square league. Analogous logic produces a relative population of 208 for the province of Caracas alone, that is, one-seventh less than the population of North America's *Atlantic States*.

As numerical data become instructive only when compared to analogous facts (as is the case throughout the domain of political economy), I have carefully examined relative populations that, given the current state of both continents, would be considered small or medium sized in Europe and very large in America. I have chosen examples only from those provinces that have more than 600 sq. lg. of contiguous territory so as to rule out *accidental accumulations* around large cities, for example, along Brazil's coastline, in the valley of II.219 Mexico, on the plateaus Santa Fé de Bogotá and Cuzco, or, finally, in the archipelago of the Lesser Antilles (Barbados, Martinique, and Saint Thomas). The latter's relative populations are 3,000 to 4,700 inhabitants per square league and thus equal to that of the most fertile regions in Holland, France, and Lombardy.

1. The steppes' *area* in these four provinces is 6,219 square leagues at 20 to a degree. Here are the data necessary to evaluate the agricultural state of these regions where the steppes present large obstacles to rapid population growth (Chap. XXV, pp. 72–80).

Province of *Cumaná:*		
Mountainous part of Caripe and coastal Cordilleras	393	s. l.
Llanos or savannahs	1,558	
Including the swampy Orinoco delta, 652 s. l.		
	1,951	
Province of *Barcelona:*		
Slightly mountainous part with forests toward the north	223	
Llanos	1,341	
	1,564	
Province of *Caracas:*		
Mountainous part	1,820	
Llanos, including Carora and Monai	3,320	
	5,140	

These calculations give me 6,219 square leagues of steppe or savanna, 130 of which are to the west of the Río Portuguesa. Barinas *Llanos*, between this river, the Apure, and the mountains of Pamplona, Mérida, and Páramo de las Rosas, are 1,664 square leagues. It follows that the immense basin of the *Llanos*—between the Sierra Nevada de Mérida, the delta of the *bocas chicas* inhabited by the ▾Guaraon [Warao] Indians, and the northern banks of the Apure and the Orinoco—comprises an *area* of 7,753 square leagues, half the size of Spain. The current population of the savannas of Caracas, Barcelona, and Cumaná appears to rise to more than 70,000 because of some scattered populous towns.

II.220

Europe's most sparsely populated areas			America's most densely populated areas		
The 4 least populated districts of *Europ. Russia:*			The central part of the intendancies of Mexico and Puebla,[1] above	1,300	per s. l.
Archangel	10	per s. l.	In the United States, Massachusetts, although it has a surface area of only 522 s. l	900	
Olonez	42		Massachusetts, Rhode Island, and Connecticut together	840	
Vologda and Astrakhan	52		The entire intendancy of Puebla	540	
Finland	106		The entire intendancy of Mexico	460	
The least populated province of *Spain* (Cuenca)	311		When combined the two Mexican intendancies are almost one-third the size of France, and have enough population (in 1823 almost 2,800,000 souls). The towns of Mexico and Puebla do not noticeably impact the relative population figures. The northern part of the province of Caracas (without the Llanos)	208	
The dukedom of Luxemburg (because of wetlands)	550				
The least populated department of continental France (Haut.-Alpes)	758				
The departments of France with medium-size populations (Creuse, Var, and Aude)	1,300				

II.221

This table shows that the places in America that we consider the most densely settled exceed the relative populations of the kingdom of Navarre, Galicia, and Asturias,[2] which have the most inhabitants per square league

1. Is there any part of the United States the size of 600 to 1,000 s. l. whose relative population exceeds the maximum population of New Spain, which has 1,300 inhabitants per square nautical leagues, or 109 per square mile, 69.2 to a degree? Massachusetts's relative population, which is 75.5 per square mile and is considered to be quite large, has made me doubt it up to now. To answer this question, one would have to be able to compare the area of a certain number of bordering districts whose censuses are published by the Congress in Washington. The relative population of the States of New York, Pennsylvania, and Virginia appears so small (240, 204, and 168 per square nautical leagues) only because, in distributing the population across the entire expanse of territory, one has to take into account that each State has partly deserted regions to the West of the Alleghenies. These regions influence the total average almost in the same way that the Llanos of Caracas and Cumaná do. According to ▼Mr. Jomard, only 1,408 of Egypt's 11,000 square leagues are inhabited.

2. Per square nautical league: 1,860 for the kingdom of Valencia and 2,009 for Guipúzcoa. The latter, however, having no more than 52 square leagues, ought to be excluded on the basis of the principles that I have adopted for this type of research. Galicia has an absolute population of 1.4 million; the kingdom of Valencia, which has only half of Galicia's *area*, 1.2 million.

in all of Spain, except for Guipúzcoa and the kingdom of Valencia. This II.222 American *maximum* is below all of France's relative population (1,778 per square league) and would be regarded there as no more than a very modest population. If of all of America's territories, we focus on the *Capitanía general* of Venezuela, which is our special concern in this chapter, we find that the most populous of its jurisdictions—the province of Caracas in its totality including the *Llanos*—has so far only the relative population of Tennessee. Situated on the more than 1,800 sq. lg. in its northern part that includes the *Llanos*, this province has the relative population of South Carolina. A center for agriculture, these 1,800 sq. lg. are twice as populous as Finland; but they are still a third less populated than the province of Cuenca, the most sparsely settled province in Spain. One cannot linger over this figure without giving in to troubling sentiments. 300 years of colonial policy and an unreasonable II.223 public administration have left a country whose natural endowments are as marvelous as anything on the face of the earth in such shambles that to find a comparably depopulated region, one must look either to the frozen northern regions or west of the Allegheny Mountains to the forests of Tennessee, where the first clearings were made only half a century ago!

The most cultivated area in the province of Caracas, the Lake Valencia basin, popularly known as *Los Valles de Aragua*,[1] had more than 2,000 inhabitants per square league in 1810. If one assumes the relative population to be four times smaller and subtracts from the surface area of *Capitanía general* the nearly 24,000 sq. lg. of *Llanos* and Guyanese jungle that present large obstacles to farming, one would still get a population of 6 million for the remaining 9,700 sq. lg. Those who, like me, have lived beneath the beautiful tropical skies for a long time will find nothing exaggerated in these calculations. For those areas most suitable for cultivation, I posit only a relative II.224 population equal to that of the intendancies of Puebla and Mexico,[2] which are filled with arid mountains that stretch toward the South Sea over regions that are almost entirely empty. If, one day, the territories of Cumaná, Barcelona, Caracas, Maracaibo, Barinas, and Guyana were to become so lucky as federated states as to enjoy good provincial and municipal institutions, then it would not take a century and a half for them to reach a population of 6 million. Even with 9 million, Venezuela, the eastern part of the *Republic of Colombia*, would still not have a larger population than Old Spain. Doubtlessly, the most fertile and easily cultivated part of this country, that is, the 10,000 sq. lg. that remain when the savannas (*Llanos*) and the nearly impenetrable jungles

1. These valleys have no more than 30 square leagues. See Vol. V, pp. 142, 143.

2. These two intendancies together nevertheless cover 5,520 square leagues and have a relative population of 508 inhabitants per square nautical mile.

between the Orinoco and the Casiquiare are discounted, will be able to feed as many people under the beautiful tropical sky as the 10,000 sq. lg. of Extremadura, Castile, and other provinces of the Spanish plateau! These predictions are not at all wild guesses, because they are based on physical analogies and the productive capacity of the soil. But to realize these hopes, one must consider another factor less susceptible to calculation: the peoples' wisdom, which calms hateful passions, stifles the seeds of civil discord, and solidifies free and strong institutions. II.225

PRODUCTS—When one considers the soil of Venezuela and New Granada, one realizes that no other country in Spanish America produces such variety and abundance of crops. With Caracas's harvests added to Guayaquil's, the Republic of Colombia by itself produces practically all the cacao that Europe needs per year. The union between Venezuela and New Granada has placed in the hand of one people most of the ˅cinchona that the New World exports. The temperate mountains of Mérida, Santa Fé, Popayán, Quito, and Loja produce the finest quality of this antipyretic bark that has been known to date. I could lengthen the list of these precious products to include the coffee and indigo of Caracas, which have been prized for some time now; sugar; cotton; Bogotá's varieties of flour; ˅ipecacuanha from the banks of the Río Magdalena; Barinas tobacco; Caroni's *Cortex Angosturae*; balsam from the plains of Tolú; leather and cured meats from the *Llanos*, pearls from Panama, the Río Hacha, and the Margarita; and, finally, gold from Popayán and the platinum that is found in abundance only in Chocó and Barbacoas. But following my plan, I should limit myself to the former *Capitanía general* of Caracas. In previous chapters, I considered each crop in its turn. It remains for me only to review succinctly the statistics about the peaceful period that immediately preceded this country's political turmoil. II.226

Cacao. Total production, 193,000 *fanegas*, at 110 Spanish pounds per fanega. Venezuela exports: 145,000 *fanegas* (including contraband). Total value: more than 5 million piastras fuertes. Number of trees in 1814: nearly 16 million. It is cacao that once brought the most fame to this stretch of Terra Firme. Cacao cultivation declined in proportion to the growth of coffee, cotton, and sugar; it spread from west to east. Cacao is important not only for export but also as food for the people. Domestic consumption will therefore grow with the population, and it is to be hoped that national prosperity will soon once again encourage cacao producers (see Vol. III, pp. 240–44; Vol. V, pp. 281–302). Caracas, Barcelona, and Cumaná cacao—the most prized varieties being from Oritucu (near San Sebastián), Capiriqual, and San Bonifacio—is far better than Guayaquil's. Only Soconusco's is superior to it, as is the cacao from II.227

Gualán, near Omoa, but the latter barely enters into European trade (Juarros, II.228 *Compendio de la historia de la ciudad de Guatemala*, 1818, Vol. II, p. 77).

Coffee. The pervasive low plateaus, between 250 to 400 toises high, in the Caracas and Cumaná provinces (in the coastal range and in Caripe), offer temperate sites extremely favorable for this crop. In 1812, when cultivation went back only 28 years, production already rose to nearly 60,000 quintals (on European coffee consumption, see Vol. V, pp. 79 and passim).

Cotton. Although cotton from the Aragua and Maracaibo valleys and the Cariaco Gulf is of very high quality, average exports were no more than 2.5 million pounds in 1809 (Vol. III, pp. 86, 127, 128, 240; Vol. V, pp. 149–52, and ▾Urquinaona, *Relación documentada del origen y progresos del trastorno de las provincias de Venezuela*, 1820, p. 31).

Sugar. The beginning of this century saw beautiful plantations in the Aragua II.229 and Tuy valleys near Guatire and Caurimare [Venezuela]. Exports, however, came to almost nothing (Vol. V, pp. 100–104 and 215–221). During the course of this study, I have repeatedly drawn the reader's attention to the ascendancy that colonial products from continental Spanish America will gradually gain over crops from the comparatively small Antilles.

Indigo. Between 1787 and 1798, this extremely important crop decreased far more than cacao did. It is cultivated profitably only in Barinas province (for example, between Mijagual and Vega de Flores) and along the banks of the Táchira. The value of Caracas indigo rose to 1.2 million piasters during the most prosperous times. In 1794, exports totaled 900,000 pounds in Guaíra; in 1809, 7,000 *zurrones* (Vol. III, pp. 78–82; V, pp. 144, 145, 228).

Tobacco. Venezuela's tobacco is not only far superior to Virginia's; it is second in quality only to tobacco from Cuba and Río Negro. The establishment of II.230 the *royal monopoly* in 1777 hindered the development of this crop that could have been very important for commerce in Barinas and in the Aragua and Cumanácoa valleys. At the beginning of the nineteenth century, the total revenue from the sale of tobacco was 600,000 piasters (Vol. III, pp. 71–77; V, p. 201; VII, p. 450). When, during the government of ▾Don Diego Gardoqui, the king of Spain declared in his September 31, 1792 *cédula* that he would agree to lift the government monopoly (*estanco*), it was suggested that it be replaced either with a general tax on the monopoly of sugarcane rum (*aguardiente de caña*) or with other less onerous taxes. These projects failed, and the tobacco monopoly continued.

Grains. Basing their opinions on the vaguest and most imperfect sense of local conditions, people often indulge in contrasting the eastern and western parts of Colombia. It is said that New Granada is *mining and wheat country*, while Venezuela is given over to *colonial crops*. In these slightly arbitrary distinctions, people consider only the *tierra fría y templada*, that is, the regions where mean annual temperatures[1] are between 13° and 18.5° C—the large mountainous plateaus of Quito, Los Pastos, Bogotá, Tunja, Vélez, and Leyva. They forget that the entire northern and western part of New Granada is low, wet terrain with mean temperatures between 26° and 28° C and thus appropriate for the cultivation of crops known in Europe exclusively as colonial. Venezuela (by which I always mean[2] the former *Capitanía general* of Caracas) also has a cold, temperate climate; it is a *country of bananas and wheat*. European grains were already cultivated on the Mérida and Trujillo mountains (in Puerta near Santa Ana south of Carchi), in the Aragua valleys near Victoria and San Mateo, and in the slightly mountainous countryside between Tocuyo, Quíbor and Barquisimeto, which form a *territorial divide* between

II.231

II.232

II.233

1. Between 800 and 1,600 toises above sea level. Surprisingly, areas in tropical America where mean annual temperatures are even higher than they are in Milan or Montpellier, are known as *cold regions*. But one should not forget that the mean summer temperature in Milan and Montpellier is 22.8° and 24.3° C, whereas in Quito, for instance, daytime temperatures throughout the year typically range from 15.6° to 19.3° C and nighttime temperatures from 9° to 11° C. Temperatures never exceed 22° C, lows never dip below 6°. The *tierras frías* at Santa Fé's elevation (1,365 toises) and at Quito's (1,492 toises) are like Paris in May for the whole year. As temperature variation is so dramatic over the course of the year and too different in the tropical and the temperate zones, it makes more sense to derive a more precise idea of the climate in a place near the equator by comparing this climate to a single month's average temperature in Europe's temperate zone.

2. The name *Venezuela* was also used in this sense during the installation of the congress at Angostura on February 15, 1819, which included deputies from Caracas, Barcelona, Cumaná, Barinas, and Guyana. La Cruz and López's maps use the provinces of Caracas and Venezuela synonymously. The Captain-general, who resides in Caracas and governs the area from the mouth of the Orinoco to the Río Táchira, was called *Capitán general de la provincia de Venezuela y Ciudad de Caracas.* In his Statistic, Mr. De Pons distinguishes the *Captaincy-General of Caracas* from the *district of Venezuela*, which, according to him, only includes the province of Caracas. The *Republic of Venezuela*, founded on July 5, 1811 and restored on August 16, 1813, was joined with the Republic of Cundinamarca (December 17, 1819) under the name *Colombia*. Ever since this union, the name Venezuela has been once more officially limited (February 1822) to a *district* that encompasses the provinces of Caracas and Barinas. Amidst these fluctuations, one risks confusing a country twice the size of Spain with another not even as big as the state of Virginia, if one does not clarify the precise sense in which one uses the name *Venezuela*. Equating this word with the *Capitanía general de Caracas* gives one a collective name for the entire eastern part of Colombia, and *Venezuela* will then assume the same meaning that Mexico, Chile, or Peru have.

the tributaries of the Apure, the Orinoco, and those of the Antillean Sea. In many of these places—and this is worthy of note—wheat is cultivated amid coffee and sugarcane at elevations no greater than 270 to 300 toises above sea level, where the mean annual temperature is at least 25° C. In Mexico's and II.234 New Granada's tropical regions, our grains are produced at elevations where they could not be grown in Europe at 42° and 46° lat.[1] By contrast, in Venezuela and on the island of Cuba, the *wheat's lower growth line* quite unexpectedly drops to the scorching coastal plains. To this day, Venezuela's grain production is insignificant. Barquisimeto and Victoria produce no more than 12,000 quintals per year. And, as these same places of low elevation are also suited to sugarcane, coffee, and cotton production, wheat cultivation could never take off.

Caracas is not the only province in Venezuela with *temperate regions*, that is, regions where the centigrade thermometer drops below 16°, 14°, or even 12.5° C at night. Cumaná province also has a mountainous area rarely visited II.235 to this day, which could become rather important for some new kinds of tropical agriculture. As I have explored a large part of Venezuela with a thermometer in hand, I believe that I can pinpoint the regions that deserve the designation *tierras templadas*,[2] some of which are too cold even for growing coffee, though they are well suited for grain. As my focus here is strictly agricultural, I list only high valleys or plateaus of sufficiently large size. The Páramo de Mucuchíes, which belongs to the ▾*Sierra Nevada* de Mérida, the Caracas Hill in the coastal range, and the Duida Peak in the missions of the Upper Orinoco, is 2,100, 1,340, and 1,280 toises high. But in practice, these mountain slopes can- II.236 not be cultivated. It is the same with all the high mountains of secondary limestone, mica schist, and granite gneiss that stretch along the length of Venezuela's coast from Cape Paria to Lake Maracaibo. This coastal range is not massive enough to have on its ridges extensive plateaus of the sort that support every kind of European crop in Quito and Mexico. The *Capitanía general* of Caracas has the following *temperate regions* (that is, above 300 toises): (1) the mountainous area around the Chaymas missions[3] in New Andalucía, that is the Cerro del Imposible (297 toises), the Cocollar and Turimiquire savannas (400–700), the Caripe valleys (412 toises), and the Guardia de San Augustín

1. At an elevation of 900 and 1,100 toises, the fields of wheat and rye disappear in the Sea Alps and in the Provence. See my research on the requisite temperatures for ▾plant cultivation in *Distributione geographica plantarum*, 1817, p. 161.

2. By the slightly vague terms *tierras calientes, templadas*, and *frías*, I mean terrain from the coast to 300 toises, from 300 to 1,100 toises, and from 1,100 to 2,460 toises, respectively. The third term, the tropical region's snowline, defines the limit of plant life.

3. Vol. III, pp. 108–22, 85, 118–34, 139–52, 199 and 200.

(533 toises); (2) the slopes (*faldas*) of Bergantín,[1] between Cumaná and Barcelona whose elevation, not precisely known, seems to exceed 800 toises; (3) the small Venta Grande plateau, between Guaíra and Caracas (755 toises); (4) the Caracas valley[2] (460 toises); (5) the mountainous and wild countryside between Antímano and the Tuy Hacienda, or Higuerote, and Las Cocuizas[3] rising to nearly 850 toises; (6) the granite plateaus[4] of Yusma (320 toises), Guácimo, Guiripa, Ocumare, and Panaquire, between the *Llanos* and Venezuela's southern coastal mountain range; (7) the divide between the tributaries of the Antillean Sea and the Apure, or the group of plateaus and hills between 350 and 550 toises that connects[5] the coastal range to the Sierra de Mérida and the Sierra de Trujillo; that is, Santa María Mountain to the west of El Torito, the Picacho de Nirgua, the Altar, and the Quíbor, Barquisimeto, and Tocuyo environs; (8) the Trujillo plateau (above 420 toises) and the *tierras frías* of Páramos de las Rosas, of Boconó, and of Niquitao, between the springs of the Motatán, the Portuguesa, and the Guanare Rivers; (9) all the mountainous terrain that surrounds the *Sierra Nevada* de Mérida between Pedraza, La Vellaca, Santo Domingo, Mucuchíes, the Páramo de los Conejos, Bailadores, and La Grita (700–1,600 toises); (10) perhaps some sites in the Parime Cordillera that separates the basin of the Lower Orinoco from the Amazon basin, for example, the group of granite mountains in the Cerro Sipapo and the Cerro Marahuaca.[6] Mr. Bonpland and I did not visit Barinas's cold region, the slopes of the *Sierra Nevada* of Mérida or the *Páramos* to the north of the Trujillo—which probably have elevations between 1,700 and 2,100 toises, according to similar observations that I made in the Pasto and the Quito Andes. I cannot determine how much of the valleys and plateaus in Venezuela's western region will one day be suitable for the cultivation of European grains. As we have already pointed out, knowledge of absolute elevations will not help us resolve agricultural questions. When sites that enjoy cold or temperate climates have slopes that are too steep to be easily cultivated, the price of native flours becomes too high to compete with flour from the United States, Mexico, or Cundinamarca. Much as Italy and Greece brought in their wheat from Egypt and Mauritania on the opposite side of our Mediterranean, Venezuela and the coast of New Granada currently import their flour from the United States on the opposite side of the Antillean Mediterranean. In an official letter

II.237 II.238 II.239

1. Vol. II, pp. 258–381; III, pp. 1–18, 120 and 121.
2. Vol. IV, pp. 135, 192, 193.
3. Vol. V, pp. 94–98.
4. Vol. VI, p. 8.
5. Vol. V, pp. 304 and 305.
6. Vol. VIII, pp. 197, 198, and 25[4].

addressed to the United States Secretary of State in Washington, ▾Don Manuel Torres estimates flour exports from north America to Colombia at 20,000 barrels per year ([Monroe], *Message from the President of the United States*, 1822, p. 48. See also Vol. V, pp. 127–29, 134 and 135). Dramatic progress in the art of navigation, coupled with free trade, is exposing domestic agriculture to dangerous competition with the most distant countries. The fields of the Crimea provision the markets of Livorno and Marseille with flour. The United States supplies it to Europe. The Mexican plateau sends it to Spain, Portugal, and England during times of shortage. Regions, some of which barely produce ▾grain of the sixth or seventh quality, and other grain of the twentieth or twenty-fifth qualities, compete with each other, and the problem of the usefulness of this crop is complicated by the variable effects of soil fertility and labor costs. Because of the size of its mountains and plateaus, Colombia's western region (New Granada) will always enjoy great advantages in grain production over Colombia's east (Venezuela). Regions situated to the Orinoco's north will have to be wary of competition from Socorro and Bogotá's grains being shipped down the Río Meta. Wherever temperate regions abut hot ones, it is as possible to cultivate sugar, coffee, and grain between elevations of 300 and 500 toises, as it is in the temperate locations of the Cumaná and Caracas provinces. Experience typically proves that sugar and coffee are preferred crops because they are considered more lucrative.

II.240

Cinchona. The Cuspare, or *Cortex Angosturæ* of Caroni, incorrectly called Orinoco cinchona, was made famous by the work of the Capuchin-Catalán monks. It is not a Rubiaceae like the Cinchona but a plant from the Diosmeae or Rutaceae family. Until today, this precious plant has been exported only from Spanish Guyana, although it is also found in Cayenne. We still do not know to which genera the Cuspa, or *Cumaná cinchona*, belongs, but its eminently antifebrile properties could make it an important object of trade (Vol. III, p. 33). Beautiful specimens of true cinchona (*Cinchonae, corollas hirsutis*), common in New Granada, were discovered in the western part of Venezuela. Cinchona's antifebrile bark (*buenas quinas* or *cascarillas*) is gathered on both slopes of the *Sierra Nevada* of Mérida, on the old Barinas trail to Páramo de Mucuchíes, called Los Callejones trail, a little above the La Vellaca ravine, and also between Biscucuy and the city of Mérida.[1] To this day, these are the variety of true cinchona (Cinchonae) found farthest to the east in South America. No species of Cinchona, not even from the related genre Exostema, has been found in the mountains of the Silla de Caracas, where Befaria, Aralia, Thibau-

II.241

II.242

1. *Travel journals* by Mr. Palacio-Fajardo.

dia, and other alpine bushes from the Cordillera of New Granada thrive; nor does it seem to grow in the mountains of Turimiquire, Caripe, or French Guiana.[1] The complete absence of the Cinchona and Exostema genera from the Mexican plateau and from South America's eastern regions north of the equator (if indeed it proves to be as complete as it currently appears) is all the more surprising given that the Antilles do not lack cinchona with smooth petals and protruding stamens. To this day, traveling botanists have found only very few specimens of true Cinchona—a genus where the fruit separates from the Macrocnemum in a distinct fashion—in the southern hemisphere in the temperate parts of Brazil. According to Mr. Auguste de Saint-Hilaire's fine discovery, the Cinchona ferruginea grows in the temperate regions of the Captaincy of Minas Gerais where it is used under the name *quina da serra*.

II.243

By way of concluding this note on plant products from Venezuela that may one day become objects of trade, I will briefly mention the Quassia Simaruba from the Río Caura valley; the Unona febrifuga from Maypures, known as *Fruta de Burro*; the *Zarza*, or sarsaparilla, from the Río Negro; oil from the coconut—a plant that can be considered Cumaná province's answer to the olive tree; the oily almonds from Juvia (Bertholletia); the resins and precious gums of the Upper Orinoco (*Mani* and *Caraña*); rubber that resembles Cayenne's,[2] or underground rubber (*dapiche*); the aromatics of Guyana, like the *Tonga bean*, or Coumarouna fruit; *Pucheri* (Laurus Pichurim); the *Varinacu*, or false cinnamon (*L. cinnamamoides*); vanilla from Turiamo and the Orinoco's great Cataracts; the beautiful coloring agents that Casiquiare Indians pound into a paste (*Chica* or *Puruma*); Brazilwood; Dragon's blood; *Maria oil*; the prickly pear that produces Carora cochineal; woods that are valuable for cabinet making, such as mahogany (*cahoba*), cedrela odorata (*cedar* [*cedro*]), the Sickingia Erxthroxylon (*Aguatire roxo*), etc.; superb lumber from the Laurineae and Amyris families; and ropes from the *Chiquichiqui* palm, so remarkable for being lightweight (see Vol. III, pp. 93, 251, 344–46; V, pp. 9[4], 302, 312; VI, pp. 317, 370, 371; VII, pp. 201, 316, 348–51; VIII, pp. 178–87).

II.244

We showed elsewhere[3] how, because of the peculiar distribution of land in Venezuela, the three zones of agriculture, pasturage, and hunting succeed each other from north to south, from the coast to the equator. Advancing in this direction, one traverses in space the different stages through which hu-

1. See Vol. III, pp. 35–42; V, pp. 301–304; VIII, pp. 425–27. ▾Lambert, *A Description of the Genus Cinchona*, 1821, p. 57. The so-called Cinchona brasiliensis from Willdenow's herbarium, which has calices the length of corollas and grows in the warm regions of Grand Pará, is perhaps nothing more than a Machaonia.

2. See note G at the end of the 9th Book.

3. Vol. IV, pp. 147–50.

manity has passed over the course of centuries while advancing toward cultivation and laying the foundations of civil society. The coastal region is the agricultural heartland; the *Llanos* serve only as pastureland for the animals that Europe has bequeathed America, which live there in a half-wild state. Each of these regions encompasses seven to eight thousand square leagues. Farther south, between the Orinoco delta, the Casiquiare and the Río Negro, there is a vast expanse as big as France, which consists of *horrida sylvis, paludibus fœda* [terrifying jungles and foul swamps] inhabited only by hunters. The crops that we have enumerated above come from each end of the country. The savannas in the middle, to which cattle, horses, and mules were introduced as early as 1548, feed millions of these animals. During the time of my voyage, Venezuela's annual exports to the Antilles rose to 30,000 mules, 174,000 cow hides, and 140,000 arrobas (at 25 pounds per arroba) of *tasajo*,[1] dried and lightly salted meat. Cattle *herds* have diminished so considerably over the last twenty years neither because of agriculture's advancement nor the gradual invasion of pastureland but, rather, because of the general disorder and the lack of security on the estates. The impunity with which cattle were stolen and the growing number of vagabonds in the savannas preceded the increase in the destruction of cattle at the hands of armies whose needs, along with their ravages, inevitably reached dreadful proportions during civil wars. The number of goats skins exported from Margarita Island, Araya, and Coro is considerable. Sheep are plentiful only between Carora and Tocuyo (Vol. I, pp. 375, 376; IV, pp. 71–77; V, pp. 255–59; VI, pp. 96, 97, 160, and 161; VII, p. 94; VIII, pp. 326–28, 417–20). Because meat consumption is tremendous in Venezuela, the reduction in animals has a more significant effect on the inhabitants' welfare than it does elsewhere. Caracas, whose population was one thirteenth that of Paris at the time, consumed more than half of the beef eaten annually in the capital of France.[2]

II.245 II.246 II.247

1. Meat from the back is cut into narrow strips. An ox or an adult cow of 25 arrobas yields only 4 to 5 arrobas of *tasajo*, or *tasso*. In 1792, the port of Barcelona alone exported 98,017 arrobas to the island of Cuba. The average price is 14 *reales de plata* and varies from 10 to 18. (The strong piaster equals 8 of these reales.) ▾Mr. Urquinaona estimates that Venezuela's total exports came to 200,000 arrobas in 1809.

2. The following table shows how large meat consumption is in the cities of South America that border the *Llanos*:

Towns	Years	Population	Cattle
Caracas	1799	45,000	40,000
Nueva Barcelona	1800	16,000	11,000
Portocabello	1800	9,000	7,500
(Paris	1819	714,000	70,800)

I might add to the list of Venezuela's animal and plant products mineral deposits whose exploitation is worthy of the government's attention. But, having since my youth been devoted to the practical operation of mines entrusted to my care, I know how vague and uncertain judgments about mineral wealth in a particular region can be, if they are based on the simple appearance of rocks and mineral *outcroppings*. Only after conducting experiments in pits and galleries can one pronounce on the usefulness of a given operation. Everything carried out in this line of research under the direction of the metropole has left the question entirely open, and it is with a highly blameworthy carelessness that the most exaggerated notions about the wealth of Caracas's mines have recently been spread in Europe. Conflating Venezuela with New Granada under the name Colombia has no doubt facilitated these illusions. Without a doubt, *panning* for gold in New Granada did yield more than 18,000 gold marks during these past few years of domestic tranquility. Chocó and Barbacoas abound in platinum. Santa Rosa valley in Antioquia province, the Quindiu and Guazum Andes near Cuenca, abound in mercury sulfite. The Bogotá plateau (near Zipaquirá and Canoas) is full of gem or rock salt and pit coal. But in New Granada itself, actual belowground mining of silver and gold veins has been very rare up to now.[1] I am a long way from wanting to discourage this country's miners. I simply think that to prove Venezuela's political importance to the old world, it is unnecessary, in a country whose prodigious territorial wealth is founded on agriculture and on raising livestock, to present as realities and industrial breakthroughs developments founded only on hope and more or less uncertain probabilities. The republic of Colombia also has famous pearl fisheries along its coastline, on Margarita Island, Río Hacha, and in the Gulf of Panama. Currently, however, these pearls are as insignificant for Venezuela's export as its metals.

II.248

II.249

II.250

Beyond the shadow of a doubt, there are metal deposits at many points along the coastal range. At the beginning of the conquest, gold and silver mines were in operation in Buria near the city of Barquisimeto, in the province of Los Mariches, in Baruta to the south of Caracas, and in Real de Santa

---

In Mexico City, whose population is four or five times smaller than that of Paris, consumption does not exceed 16,300 cattle: it therefore does not seem much larger than Paris. But one should not forget that (1) Mexico City sits on a plateau where grain is grown and which is far away from pastureland; (2) one-fourth of this town's inhabitants are Indians who eat very little meat; and (3) Mexico City's consumption of lamb and pork is 273,000 and 30,000 [pounds], while in Paris, it was no more than 329,000 and 65,000 in 1819, despite the enormous difference in population. See Vol. IV, pp. 196–98; IX, pp. 93 and 94; and my *Essai politique sur le royaume de la Nouvelle-Espagne*, Vol. I, p. 199. Chabrol, *Recherches statistiques sur la ville de Paris*, 1823, table 72.

1. *Essai politique sur le royaume de la Nouvelle-Espagne*, Vol. II, pp. 586, 587, and 625.

Bárbara near Villa de Cura. Gold dust is found throughout the mountainous terrain between the Río Yaracuy, the town of San Felipe, and Nirgua, as well as between Güigüe and San Juan de los Moros. During the long voyage that Mr. Bonpland and I made through the granite gneiss territory along the Orinoco [in Spanish Guyana], we saw nothing that would confirm the ancient belief in this region's mineral wealth. Many historical clues, however, almost certainly suggest that there are two areas with alluvial gold deposits, one between the sources of the Río Negro, the Uaupés, and the Iquiare, the other between the sources of the Essequibo, the Caroni and the Rupununi. I dare flatter myself that, should Venezuela's government wish to undertake an in-depth examination of *metal deposits* within its territory, the responsible parties will find helpful the geognostic observations in Chapters XIII, XVI, XVII, XXIV, and XXVII of this work [*Relation historique*], because they are based on a detailed knowledge of these places.[1] To this day, Aroa has been the sole area of extraction in Venezuela; in 1800, it furnished nearly 1,500 quintals of first-rate copper. The *grünstein* rocks from the Tucutunemo formation (between Villa de Cura and Parapara) contain malachite and copper pyrite deposits. The ochreous and magnetic iron traces in the coastal range, Chuparipari's native alum, Araya's salt, Silia's kaolin, jade from the Upper Orinoco, petroleum from Buen Pastor, and sulfur from the eastern part of New Andalucía equally deserve the administration's attention.[2]

II.251

There are clearly some promising mineral deposits, but further study is needed to determine whether they are easy enough to reach and large enough to defray the costs of extraction.[3] Even in the eastern portion of South America, gold and silver are so widely dispersed that the European geognost is overcome with surprise. But this dispersal—the veins that divide and tangle, the metals that appear only in nodules—makes extraction very expensive. The example of Mexico proves, besides, that the interest in mining takes nothing

II.253

1. Vol. IV, pp. 269, 270, 281, 282; V, pp. 305–8; VI, pp. 8, 9, 15–17; VII, pp. 264, 265, 383, 418–21; VIII, pp. 32, 33, 144, 145, 201, 469, 487, 514 and 524.

2. Vol. II, pp. 323–29, 337–46; III, pp. 129–31, 230, 256 and 257; V, p. 62; IX, pp. 126–32.

3. In 1800, labor costs for a simple day laborer (*peón*) working the earth was 15 soles in the province of Caracas, food included (Vol. V, p. 154). A man who cut timber in the coastal jungles of Paria was paid between 45 and 50 soles a day in Cumaná, without food. A carpenter in New Andalucía made 5 to 6 francs a day. In Caracas, three cassava pies (that country's equivalent for bread) about 21 inches in diameter, 1 and a half lines thick, and weighing 2.25 pounds cost half a *real de plata*, or 6.5 soles. An adult male eats no more than 2 sols worth of cassava a day, this food being constantly mixed with bananas, cured meat (*tasajo*), and *papelón* [hardened sugarcane pulp] or raw sugar. For commodity prices, compare Vol. V, p. 296; VI, p. 161; VII, p. 187.

away from agriculture and that these two types of industry can stimulate each other. The futility of efforts under the intendancy of Don José Avalo should be attributed only to the ignorance of the people employed by the Spanish government who tragically mistook mica and amphibole for metallic substances. If the government commits itself to examining the former *Capitanía general* of Caracas over a long stretch of time, if it is lucky enough to engage men as distinguished as ▾Mr. Boussingault and Mr. Rivero, who are currently setting up a mining school in Bogotá and who join a deep understanding of geognosy and chemistry to practical experience in extraction, one should expect the most satisfying results.

TRADE AND PUBLIC REVENUE—The description that we just gave[1] of Venezuela's products and the development of its coast suffices to underline the importance of trade to this rich region. Even when fettered by the colonial system, the revenue of exports from farming and gold panning in the region currently united under the name Republic of Colombia reached 11 to 12 million piasters. At the beginning of the nineteenth century, exports from the *Capitanía general* of Caracas, not including the precious metals that are regularly extracted, equaled 5 to 6 million piasters (including proceeds from contraband). Cumaná, Barcelona, Guaíra, Portocabello, and Maracaibo are the most important coastal ports. Those situated farther east enjoy the advantage of easier communication with the Virgin Islands, Guadeloupe, Martinique, and St. Vincent. Angostura, whose real name is Santo Tomé de la Nueva Guaíana, may be considered the richest port in the province of Barinas. The majestic river, along which the town is built, is highly advantageous for European trade because of its connections to the Río Apure, the Meta, and the Negro.[2]

II.254

II.255

If one wishes to form a precise idea of Venezuela's importance from the perspective of exports and the consumption of goods from the old world, it is necessary to go back to a time of external peace that preceded the Spanish American revolution by twelve to fifteen years. It was then that Guaíra's trade knew its greatest splendor. Here are official results from customs registries that shed some light on these regions' commercial status and which Mr. De Pons and ▾Mr. Dauxion-Lavaysse did not publish in their *Voyage à la partie orientale de la terre-ferme* and *Voyage aux îles de Trinidad*.

1. Vol. IX, pp. 244–47, 268 and 269.
2. Vol. VI, pp. 384–89; VIII, pp. 151–53, 252, 254, 336–38, 370–72.

I. La Guayrá trade in 1789

Imports, value 1,525,905 p.,			160,504 p. in duties
Exports 2,232,013			167,458
A. Imports:			
Goods Spanish	777,555	piasters	
foreign	748,350		
B. Exports:			
Gold and silver coins	103,177		
Products	2,128,836		
Among those:			
Cotton	170,427	pounds	
Indigo	718,393		
Tobacco	202,152		
Cacao	103,855	fanegas	
Coffee	23,371	pounds	
Hides	12,347	pelts	
Buckskin	2,905		
Maroquins	1,388		

II.256

II. La Guayrá trade in 1792

Import			3,582,311 piasters
Export, value			2,315,692
A. Import:			
from American ports	60,348	piasters	
from Spain	1,855,278		
from other parts of Europe	1,666,685		

B. Export:

	Indigo, pounds	Cotton, pounds	Cacao, fanegas	Coffee, pounds	Hides, pelts
For Spain	669,827	225,503	100,592	138,968	15,332
For the foreign colonies	10,402	33,000	....	9,932	70,896
	680,229	258,503	100,592	148,900	86,228

II.257

III. La Guayrá trade in 1794

A. Exports:

	Indigo, pounds	Cotton, pounds	Cacao, fanegas	Coffee, pounds	Hides, pelts
For Spain	875,907	431,658	111,133	307,032	5,305
For the foreign colonies	22,446	....	....	57,606	49,308
	898,353	431,658	111,133	364,638	54,613

B. Imports:

a. Manufactured goods and foodstuffs:		
Spain	1,111,709	piasters
Foreign from Europe	868,812	
from the United States	75,993	
from the Antilles	13,415	
	2,069,929	
b. Silver coin	60,000	
Total imports	2,129,929	

IV. La Guayrá trade in 1796

II.258

A. Exports, value 2,403,254 piasters.

THAT IS:

	Indigo, pounds	Cotton, pounds	Cacao, fanegas	Coffee, pounds	Tobacco, pounds	Hides, pelts	Copper, pounds
For Spain	709,135	483,250	70,280	482,000	454,723	1,531	31,142
For the United States	132		5,258	162			
For the foreign colonies of the Antilles	28,699	53,928		2,500		79,777	
	737,966	537,178	75,538	484,662	454,723	81,308	31,142

B. Imports:

II.259

a. from Spain,		
nationally produced goods	1,871,571	piasters
foreign	1,429,487	
b. from the foreign colonies in America	179,002	
Total imports	3,480,060	

Entrance and departure fees paid to customs 587,317 piasters

V. La Guayra trade in 1797

II.260

A. Exports, value 1,113,695 piasters

THAT IS:

	Indigo, pounds	Cotton, pounds	Cacao, fanegas	Coffee, pounds	Tobacco, pounds	Sugar cases	Furs, pelts	Leather, pounds
For Spain	61,785	50,285	46,075	153,699	....	....	671	2,000
For the United States	2,256	....	4,024		....	738	....	....
For the foreign colonies of the Antilles	56,894	57,711	20,733	155,813	175,719	638	286	400
	120,935	107,996	70,832	309,512	175,719	1,376	957	2,400

II.261

A. Imports, value		
a. from Spain	98,388	piasters
b. from the foreign countries,		
the United States	76,560	
the Antilles	389,844	
Total import	564,800	piasters
Entrance and departure fees paid to customs	242,160	piasters

Comparing these data from Guaíra customs to information I have on Spanish ports (Vol. V, 294) shows that, according to ships' bills of lading, less Caracas cacao entered Spain than was loaded in Guaíra and bound for Spain. The decline in imports and exports in 1797 does not indicate a commercial decline up to the revolution.[1] It is the result of renewed naval war, Spain having enjoyed felicitous neutrality up to that point. The customs statements that I just mentioned for the years 1789, 1792, 1794, and 1796 show 2,678,000 strong piasters for average imports to Guaíra, which is Venezuela's principal port; average exports came to 2,317,000 piasters. If one considers the years 1793–1796 alone, exports add up to 3,060,000 piasters, while the war years (1796 to 1800) show an average of only 1,610,000 piasters (De Pons, Vol. II, p. 439). This means that, in 1809—shortly before the Caracas revolution—the trade balance at Guaíra once more seemed much like it was in 1796. In a journal from Santa Fé de Bogotá ([Caldas, *Semanario* [*del Nuevo Reino de Granada*], Vol. II, p. 324), I found an official excerpt from customs registries for the first six months of 1809. During this period, imports from Spain equaled 274,205 piasters, and foreign imports reached 768,705 piasters, for a total of 1,042,910 piasters. Exports to Spain equaled 778,802 piasters and 623,805 piasters for

II.262

II.263

1. Here are the principal stages of this revolution: the *supreme Junta* of Venezuela gathered on April 19, 1810, declaring itself in favor of king Ferdinand VII's rights and deporting the Captain-general and the members of the *Audiencia*. The *Congress*, which succeeded the *supreme Junta* on March 2, 1811, declared Venezuela's independence on July 5, 1811. The congress held its sessions in Valencia, in the Aragua valleys, in March, 1812. The earthquake that destroyed most of Caracas on March 26, 1812 (Vol. V, p. 13) gave Spain once again control of the country in August, 1812. ▾General Simón Bolívar retook Caracas and entered the town in victory on August 16, 1813. Royalists became the masters of Venezuela in July, 1814, and of Bogotá in June, 1816. In the same year, general Bolívar marched on Margarita Island, Carúpano, and Ocumare. The second Venezuelan congress was installed in Angostura on February 15, 1819. *The fundamental law* that joined Venezuela to New Granada under the name republic of Colombia was proclaimed on December 17, 1819. The armistice between the generals Bolívar and Morillo dates from November 25, 1820. The constitution of the republic of Colombia dates from August 30, 1821. The United States recognized this republic on March 8, 1822.

the world at large, for a total export value of 1,402,607 piasters. One can thus take 2.7 million piasters as an average figure for the port of Guaíra's exports at the beginning of the nineteenth century, during a year when the country enjoyed both domestic and foreign peace.[1] II.264

At the time of the revolution, the two ports of Cumaná and Nueva Barcelona annually exported 22,000 quintals of cacao, one million pounds of cotton, and 24,000 quintals of salt meat for 1,200,000 piasters (including illegal trade). If we add to the export from Guaíra, Cumaná, and Nueva Barcelona one million piasters for Angostura and Maracaibo's commercial revenue, and 800,000 piasters for the mules and cattle loaded in Portocabello, Carúpano, and other small ports along the Antillean sea, then we arrive at almost six million piasters for the total value of exported goods from the former *Capitanía general* of Caracas. It is quite likely that consumption of goods from Europe and other parts of America reached the same total during the peaceful period II.265 immediately before the revolution. As nothing is vaguer than estimated trade balances based on customs registries, coupled with uncertainty over whether contraband with the Antilles would increase official amounts by one fourth, one third, or one half, it might be useful to check the results we have just obtained against a partial estimate of the population's needs. For 1800, meticulous calculations in the field show that consumption of foreign products[2] came to only 102 piasters a year for each adult from the richest class of urban dwellers in the *Gobierno* of Cumaná. For each adult slave, it was 8 piasters; for non-Indian children below the age of twelve, 0.75 piasters; for each adult Indian in the most civilized communities (*de doctrina* [believers]), 10 piasters; for an II.266 Indian family composed of 4 stark naked individuals, as they are found in the Chaymas missions, 7 piasters. Using these data, assuming that the provinces of Cumaná and Barcelona have no more than 86,000 inhabitants, 42,000 of whom are Indians, and adding the yearly outlay necessary for church decoration and services, for the maintenance of religious communities, and for schooner supplies, Mr. Navarrete estimates the value of foreign merchandise

1. In 1795, I sent these precise and detailed figures about registered merchandise in Spanish customs for the ports of Tierra Firme to Mr. Dauxion-Lavaysse, who included them in his *Voyage aux îles de Trinidad*, Vol. II, p. 464. I had taken these data from the Count de Casa Valencia's very instructive account of ways to revitalize Caracas's commerce. Mr. Urquinaona (*Relación documentada*, p. 31) estimates Venezuela's exports at 8 million piasters in 1809.

2. [Navarrete,] *Informe de Don Manuel Navarrete, Tesorero de la Real Hacienda en Cumaná, sobre el estanco de tabaco y los medios de su abolición total* (manuscript). In this essay on consumption, the phrase "*foreign effects*" designates all merchandise that did not originate in Venezuela.

to be 853,000 piasters, about 10 piasters per individual irrespective of age and class. Undoubtedly, luxury dramatically increased in certain populous Venezuelan towns during the period of domestic upheavals, because of more frequent contact with European nations; however, Spanish America's urban population comprises only a trivial portion of the general populace. Given the abstemious habits maintained by the multitudes that inhabit the rural areas far from the coastline, I think that the 785,000 inhabitants whom we currently posit for Venezuela will need more than seven million piasters worth of foreign merchandise when the country once more enjoys perfect tranquility.

II.267

To bring us to more general considerations, let us linger over these numbers for a moment. Industrious Europe, bursting with factories, seeks outlets for its manufactured goods. The lack of factories is so pronounced in South America's emerging societies that Venezuela's population, which is at the most that of two average-sized French départements,[1] needs 35 million francs in foreign-made goods and commodities each year for domestic consumption. More than four-fifths of these goods come from European markets by various routes. Venezuela's population, however, is poor, frugal, and little advanced in civilization. If import registries show Venezuela to be quite focused on consumption and if its demand feeds the economy of mercantile nations, this is because it is entirely lacking in manufactures and because the simplest mechanical arts are barely even beginning to be practiced. Morocco leather and hides from Carora, hammocks from Margarita Island, and woolen blankets from Tocuyo are not very important products, not even for the domestic market. All fine fabrics and dyed linens that Venezuela needs come from abroad. Before 1789, when trade between France and the American colonies was at its most robust, France exported to its colonies 80 million francs in French agricultural and manufactured goods. This sum is barely larger than the value of Colombia's total consumption of foreign goods. I insist on the importance of these considerations to prove how interested the peoples of the old world are in the prosperity of the free states that are emerging in tropical America. If these states, harassed from without, remain agitated within, a civilization with shallow roots will gradually give way. For a long time, Europe, without having the advantage of a metropole able either to pacify or reconquer its colonies by force, will be deprived of a market fit to enliven its trade and manufacture.

II.268

I will add to my reflections some little-known statistical data from a recent report from the *Consulado de la Veracruz*. This document shows that Venezuela, because it has no factories at all and only a small number of Indians,

1. See [Vol. IX] p. 251, note 1.

consumes, proportionate to its population size, a larger number of foreign goods than New Spain. According to customs registries, imports[1] from Veracruz rose to 259,105,940 piasters during the twenty-five year period from 1796 to 1820, 186,125,113 of which came from the metropole. During the same period, New Spain's consumption of European goods equaled 224,447,132 piasters, or 8,977,885 piasters per year. The insignificance of this sum is striking when measured against the needs of 6 million people. This is how ▾Mr. Quirós, the secretary of the *Consulado de la Veracruz*, could conclude that contraband exports rose to more than 12 to 15 million piasters in an average year. According to these calculations from people familiar with this locale, Mexico would consume between 21 and 24 million piasters in foreign goods, that is, with eight times the population, it would barely consume four times as much as the former *Capitanía general* of Caracas. Such a difference between two markets on the Mexican and Venezuelan coasts, both open to European trade, will seem less remarkable, I think, if one recalls that there are more than 3.7 million pure-blooded Indians[2] among New Spain's 6.8 million inhabitants and that this beautiful country's manufacturing sector was already so advanced in 1821 that the value of native wool and cotton fabrics had risen to 10 million piasters a year.[3] Subtracting from Venezuela and Mexico's total population the Indian population—whose needs are almost entirely limited to the products of the land on which they live—one finds that consumption of foreign manufactured goods equals 10 piasters per individual of all ages and sexes in the former, and 8 piasters in the latter. The closeness of the results for the two countries shows that, if one considers only large entities, the state of society seems to be almost the same, despite the varied influences of physical and moral factors in Spanish America's most remote regions.

II.270 II.271 II.272

Thanks to the beauty of its ports,[4] the tranquility of the sea, and its superb

1. In the trade registries published in Veracruz, imports and exports made *on the government's account* are not included. For example, in 1802, commercial activity (exports plus imports) is listed as 26,445,955 strong piasters. If one had added the 19.5 million piasters on the King's account and the value of the mercury and cigar paper received on the *Real Hacienda*'s account, then total commercial activity should have been 82,047,000 piasters for 1802, and 43,897,000 piasters, instead of 34,349,634, for 1803 (*see* my *Essai politique sur le royaume de la Nouvelle-Espagne*, Vol. II, pp. 702 and 708). During the 25 years before 1820, Mexico minted 429,110,008 piasters in gold and silver.

2. See [Vol. IX], p. 162.

3. *Balanza de Comercio recíproco hecho por el puerto de Vera-Cruz con los de España y de América en los últimos* 25 *años* (*De orden del Consulado de Vera-Cruz, el 18 de abril 1821*).

4. Here is the list of anchorages, docks, and ports that I know from Cape Paria to the Río Hacha: Ensenada de Mejillones; the mouth of the Río Caribe; *Carúpano, Cumaná* (see Vol. II, pp. 267 and 268); Laguna Chica to the south of Chuparipari (Vol. IX, pp. 119–

II.273 timber forests, Venezuela's coastline has enormous advantages over that of the United States. Nowhere else are anchorages so close, nor ports so conducive to the establishment of military settlements. The sea along this coastline is always as calm as it is between Lima and Guayaquil. Storms and hurricanes from the Antilles never reach the *Costa firme* [mainland coast]. And after the noonday sun begins its decline, fat clouds full of lightning gather in the coastal range. But to sailors experienced in these waters, this often threatening appearance of the sky portends nothing more than a wind squall, which barely makes them trim or raise their sails. The virgin forests near the sea in New Andalucía's east are valuable resources for the establishment of dock yards. Forests on Paria Mountain rival those on the island of Cuba, Coatzacoalcos, II.274 Guayaquil, and San Blas. At the turn of the last century, the Spanish government had focused its attention on this important resource. Naval engineers were directed to select and mark the most beautiful specimens of Brazilwood, Acajou [a Mahogany], Cedrela, and Laurel between Angostura and the Orinoco Delta, and along the Gulf of Paria, commonly known as *Golfo triste*. The idea was not to build docks and calles [roads] on site but, rather, to cut and shape the wood into the individual pieces needed for ship building—a bit like a kit—and send them to Caraque near Cádiz on the King's vessels. Although there were no trees appropriate for mast construction in the region, the Spanish government nevertheless expected that this project would considerably reduce timber imports from Sweden and Norway. The project was attempted in an extremely disease-ridden locale[1] in the Quebranta Valley near Guirie. I have written elsewhere about the reasons for its failure. The unhealthiness of II.275 the place would probably have diminished the farther away the settlements were from the virgin forest (*el monte virgen*). To cut the wood, it would have been necessary to employ not whites but people of color; it should have been remembered that costs would have declined as soon as the logging trails (*arrastraderos*) were cleared, and that labor costs would have gradually diminished once the population grew. Only ship builders who know the place can

24); *Laguna Grande del Obispo* (Vol. III, p. 26; Vol. IX, pp. 132 and 133); Cariaco (Vol. III, p. 248); Ensenada de Santa Fé; Puerto Escondido; *Mochima Port* (Vol. IV, pp. 66 and 67; Vol. IX, p. 133); *Nueva Barcelona* (Vol. IV, pp. 71 and 72; Vol. IX, p. 96); the Río Unare delta; Higuerote (Vol. IV, pp. 81 and 82); Chuspa; Guatire; *La Guaíra* (Vol. IV, p. 95); Catia; Los Arrecifes; Puerto-la-Cruz; Choroní; Ciénaga de Ocumare; Turiamo; *Borburata*; Patanemo (Vol. IV, p. 121); *Puerto-Cabello* (Vol. V, p. 245); Chichiriviche (Vol. V, pp. 249–51); Puerto del Manzanillo; *Coro*; *Maracaibo*; Bahía Honda; El Portete; and Puerto Viejo. Margarita Island has three good ports: Pampatar, Pueblo de la Mar, and Bahía de Juan Griego. (Italics indicate the busiest ports.)

1. Vol. III, p. 108.

judge whether, under present circumstances, transporting large amounts of semifinished lumber to Europe on merchant ships is not far too expensive. Still, Venezuela clearly has immense ship-building resources on its coastline and along the Orinoco. The superb ships launched from the docks of Havana, Guayaquil, and San Blas are probably more expensive than ships from European docks. But thanks to the properties of tropical wood, they have the advantage of lasting for a long time.

We have just analyzed Venezuela's commercial goods and their values. It remains for us to consider the *means of transport*, which are limited to inland waterways and external navigation in a country that has neither large roads nor carriages. The even temperature throughout most of the provinces results in the same basic crops, so that the need for exchange is less pronounced here than it is in Peru, Quito, and New Granada, where great differences in climates occur within a small area. Flour from grain is almost a luxury for most of the population. Each province has a share of the *Llanos*, that is, pastureland, and subsists on food from its own soil. The variability of corn harvests (which depend on the frequency of rain), the transportation of salt, and the prodigious consumption of meat in the most populated districts probably spur trade between the *Llanos* and the coast. But the most important object of commercial activity in Venezuela's interior is the transport of goods such as cacao, cotton, coffee, indigo, dried meat, and hides destined for export to the Antilles and Europe. It is surprising to see that, despite the numerous horses and mules that roam the *Llanos*, no one uses the large wagons that have crossed the Pampas between Córdoba and Buenos Aires for centuries. I did not see a single wagon in the Tierra Firme. All transport takes place on the backs of mules or by water. It would nevertheless be quite easy to clear a road for wheel carriages from Caracas to Valencia through the Aragua valleys and from there to the Calabozo *Llanos* via Villa de Cura, as well as from Valencia to Portocabello and from Caracas to Guaíra. The *Consulados* of Mexico and Veracruz overcame much greater difficulties when they built beautiful roads from Perote to the coast and from the capital to Toluca.

II.277

As for Venezuela's inland waterways, it would be needless to repeat here what we have said above about the forks and junctions of the large rivers. We want to draw the reader's attention to the two large *navigable waterways* that run from west to east (by way of the Apure, the Meta and the Lower Orinoco) and from south to north (by way of the Río Negro, the Casiquiare, and the Upper and Lower Orinoco). The first of these carries products from Barinas province[1] toward Angostura on the Portuguesa, the Masparro, the Río Santo

II.278

1. Vol. VI, pp. 165, 243–45.

Domingo, and the Uribante, and it carries goods from *Los Llanos* province and the Bogotá plateau on the Río Casanare, the Río Guaurabo, and the Río Pachaquiaro.[1] The second line of navigation, starting at the Orinoco forks, flows to the southernmost edge of Colombia to San Carlos on the Río Negro and to the Amazon. In Guyana's current state, navigation to the south of the Orinoco's great Cataracts[2] is practically nonexistent, and its usefulness for inland communication both with Pará, or the Amazon delta, and with the Spanish provinces of Jaén and Maynas is based at best on vague hopes. For Venezuela, these communication lines are like those from Boston and New York to the Pacific coast, across the Rocky Mountains, for the inhabitants of the United States. Replacing the portage of Guaporé[3] with a 6,000-toise canal would open a line of inland navigation between Buenos Aires and Angostura. Two other canals would be even easier to build: one, connecting the Atabapo to the Río Negro[4] via the Pimichin, would free boats of the need to detour through the Casiquiare; the other would remove the dangers of the Maypures rapids.[5] But, and I repeat, any consideration of trade south of the large Cataracts would require a state of civilization that seems as yet quite distant; the four great tributaries of the Orinoco—the Caroni, the Caura, the Padamo, and the Ventuari[6]—would have to become as famous as the Ohio and the Missouri are west of the Alleghenies. The longer line of navigation between the west and the east is the only one that commands the attention of the locals, and even the Río Meta does not yet have the same importance as the Apure or the Santo Domingo. On this 300-league line,[7] steamboats will be most use-

II.279 II.280 II.281

1. Vol. VI, pp. 383–89.
2. Atures and Maypures.
3. Vol. VI, p. 55.
4. Vol. VII, pp. 207–9.
5. Vol. VII, pp. 318–20.
6. Vol. VIII, pp. 151–53, 252–54. On the importance of the Guaviare, see also Vol. VII, pp. 264–66; on the Rupununi Isthmus and the portages between the Río Branco, the Essequibo, and the Caroni, see Vol. VIII, pp. 114–18; on the road that goes from the Upper to the Lower Orinoco, from the Esmeralda to the Erevato, see Vol. VIII, pp. 215–17.
7. The title of a recently published book (*Journal of an Expedition 1400 miles up the Orinoco, and 300 up the Arauca*, by H. Robinson, 1822) dramatically exaggerates the length of the Lower Orinoco and its tributaries to the west. A voyage of seventeen hundred thousand English miles would have taken the author well beyond the South Sea. An even more extraordinary geographical error is found in a book that is almost entirely put together from fragments of my *Relation historique*, and accompanied by a map that bears my name, although I look in vain for the town of Popayán. ▾It is said in Walker's *Geographical, Statistical, Agricultural, Commercial and Political Account of Columbia* (1822), Vol. II, p. 28 "that the Casiquiare, believed for a long time to be a branch of the Orinoco, was recently

ful for returning from Angostura to [Los] Torunos, the port of Barinas province. It is difficult to imagine the sheer amount of muscular force that boatmen would have to use, both when lugging their cargo and when pushing up against the riverbank[1] with their oars (*palanca*), on the Apure, the Portuguesa, and the Río Santo Domingo during the periods of heavy flooding. The *Llanos* between the Río Pao and Lake Valencia, and between the Río Mamo and the Guarapiche, have a grade so low that one could establish communication lines by building canals and improve inland commerce by connecting the basin of the Lower Orinoco to the shore of the Antillean Sea and the Gulf of Paria.[2] II.282

To the navigation of Venezuela's interior, which is of purely local interest, might be added another matter that is intimately connected to the prosperity of the mercantile peoples in both hemispheres. Among the five points that seem to offer the possibility of opening direct navigation between the Atlantic Ocean and the South Sea, three are on Colombia's territory. I will not repeat here what I have already said about this important subject in the first volume of my *Essai politique sur le royaume de la Nouvelle-Espagne*,[3] where I demonstrated that one will have to examine all these points before beginning work on any one of them. Only by envisaging the challenges of hydraulic construction in the most general terms will one be able to meet them in a satisfying manner. Ever since I left the New Continent, not a single barometric reading or geodesic leveling has been undertaken to determine the *highest point* over which the projected canals would need to pass. Different works published during the Spanish colonies' wars of independence were based on the same data[4] that I had published in 1808. Only through the connections I mentioned II.283 II.284

found by Mr. Humboldt to be a branch of the Río Negro." The same assertion is repeated in the *Vollständiges Handbuch der neueren Erdbeschreibung*, Vol. XVI, p. 58, compiled by Mr. Hassel [and Gaspari], a man of great merit. It is, however, already almost 25 years since I traced the Casiquiare from south to north.

1. There are twists (*vueltas*) and countercurrents (*barancas y laderas*) in the Portuguesa and the Apure that sometimes detain ships for an entire day. The Tuy and the Yaracuy are partially navigable.

2. Vol. V, pp. 180–84; IX, pp. 62–64.

3. Vol. I [Q], pp. LX and 11; Vol. VII, pp. 462–64. See also my *Atlas géografique et physique sur la Nouvelle-Espagne*, Plate IV.

4. I exempt from this Mr. Davis Robinson's useful information on anchorages in Coatzacoalcos, the Río San Juan, and Panama. *Memoirs of the Mexican Revolution*, 1821, p. 263. See also [Walton in] *Edinburgh Review*, 1810, January; Walton in *Colonial Journal*, 1817 (March and June); *Bibliothèque Universelle de Genève*, 1823, January, p. 47; *Bibliotheca Americana*, Vol. I, pp. 115–29. "Boats can have a draft of 23 feet in the Río Coatzacoalcos delta. There is a good place to lay anchor, and the port can admit the largest ships. There

with inhabitants of rarely visited regions have I been able to acquire new information. I will restrict myself here to the most important political and commercial concerns.

The five points that offer the possibility of communication between the two seas are found between 5 and 18 degrees northern latitude. All of them are therefore part of states bounded by the Antillean Sea, that is, territories belonging to either the Mexican or Colombian federations, or, to use former geographical designations, the Oaxaca and Veracruz intendancies, and to the provinces of Nicaragua, Panama, or Chocó. They are:

THE TEHUANTEPEC ISTHMUS (lat. 16°–18°), between the sources of the Río Chimalapa and the Río del Pasco, which flows into the Río Huasacualco, or Coatzacoalcos;

THE NICARAGUA ISTHMUS (lat. 10°–12°), between San Juan in port Nica- II.285 ragua and the mouth of the Río San Juan, Lake Nicaragua and the coast of the Gulf of Papagayo, near the volcanoes of Granada and Mombacho;

THE PANAMA ISTHMUS (lat. 8° 15′–9° 36′);

THE ISTHMUS OF DARIEN OR CÚPICA (lat. 6° 40′–7° 12′);

THE RASPADURA CHANNEL, between the Atrato and Chocó's Río San Juan (lat. 4° 58′–5° 20′).

The fortunate position of these five points, the last of which will probably always be a *system of secondary navigation*—inland transportation by means of small capacity ships— is that they are all at the center of the New Continent, equidistant from Cape Horn and the northwest coast famous for its fur trade. All of them are opposite the Chinese and Indian seas (within the same parallels), an important circumstance in waters where trade winds prevail. II.286 Ships coming from Europe and the United States can easily approach this area since one knows the precise positions of Bajo Nuevo [Petrel Islands], Roncador Bank, and Serrana Bank.

The northernmost isthmus, that of Tehuantepec, to which Fernando Cortés had already referred as the *secret of the strait* in a letter to Emperor Charles V (October 30, 1520), has of late attracted much attention from sailors. During New Spain's political troubles, the Veracruz trade was shared

is a 12-foot draft on the Río San Juan on Nicaragua's east coast; at one point, there is a tight pass 25 feet deep. The Río San Juan is 4 to 6 fathoms, and Lake Nicaragua 3 to 8 fathoms (British measurement). The San Juan is navigable for brigantines and schooners." Mr. Davis Robinson adds that Nicaragua's west coast is not as stormy as it was represented to me during my trip on the South Sea, and that a canal ending in Panama would have the great disadvantage of having to be extended for two leagues *into the sea*, because there are no more than a few feet of water until one reaches the Flamenco and Perico islets.

among the small ports of Tampico, Tuxpan, and Coatzacoalcos.[1] It has been calculated that the route from Philadelphia to Nootka and to the mouth of the Columbia River, which is nearly 5,000 nautical leagues when sailing around Cape Horn as people usually do, would be shortened to at least 3,000 leagues, if one passed from Coatzacoalcos to Tehuantepec by way of a canal. As I had the archives of the viceroyalty of Mexico at my disposal, reports by the two engineers[2] responsible for mapping the isthmus allowed me to form a rather II.287 precise idea of local conditions. It seems likely to me that the ridge that forms a *dividing line* between the two oceans is transected by a valley that would be perfect for a canal. Recently, it has been claimed that during the rainy season when the water levels rise to the point of flooding, this valley fills with enough water to become a natural passageway for the natives' boats. I do not, however, find any reference to this intriguing notion in any of the official reports to the viceroy, ▾Don Antonio Bucarely. Similar passageways exist during periods of heavy flooding between the basins of the Saint Lawrence Stream and the Mississippi, that is, between Lake Erie and Lake Wabash, and between Lake Michigan and the Illinois River.[3] The Huasacualco Canal, planned under ▾Count Revillagigedo's able administration, would connect the Río Chimalapa to the Río del Pasco, a tributary of the Coatzacoalcos. This canal would be no more than about 16,000 toises long. According to engineer ▾Cramer, who has a great II.288 reputation, it is conceivable that the canal would require neither locks nor underground galleries or inclined planes. One should not forget, however, that no one has done barometric or geodesic leveling in the area between the ports of Tehuantepec and San Francisco de Chimalapa, between the sources of the Río del Pasco and Cerros de los Mixes [in Oaxaca]. A quick glance at the regional map that I sketched shows that the difficulty—which will endlessly occupy the Mexican Government—is not so much the dredging of a canal but, rather, how to render the Río Chimalapa navigable for large vessels up to the seven rapids of the Río del Pasco, between the older *pier* north of the Tarifa jungles and the Río Saravia's mouth near the new *pier* of de la Cruz. Because of the isthmus's breadth (more than 38 leagues), the river's twists and turns and the riverbed would present problems for opening a canal navigable for II.289 oceangoing vessels used in the trade with China and America's northwest coast. Establishing a trade route for small ships and improving the land route from Chihuitán to Petapa are thus of the highest importance. This land route

1. *Balanza del comercio maritimo de la Veracruz correspondiente al año de 1811*, p. 19, no. 10.

2. Don Agustín Cramer and ▾Don Miguel del Corral.

3. Vol. V, pp. 182 and 183; VIII, pp. 108–10.

was opened in 1798 and 1801, and, for a long time, Guatemalan indigo, cochineal, and cured meat followed this path to Veracruz and onwards to the island of Cuba.

The Isthmus of Nicaragua and that of Cúpica always struck me as being the most favorable for *large-dimensioned canals* similar to the Caledonian Canal, which is 103 feet wide (French measurement) at the waterline (not including the barriers that prevent landslides), 47 feet wide at the bottom, and 18.5 feet deep. Establishing an oceanic connection capable of unleashing a commercial revolution can be a matter of building a system of interior navigation through locks that are 16 to 20 feet wide between walls, as is the case with

II.290 the canals of Languedoc, Briare, Grand Junction, or Forth and Clyde. Some of these canals have been gigantic undertakings for quite some time, certainly in comparison with smaller canals whose average depth[1] is no greater than 6 or 7.5 French feet. Unlike the Caledonian Canal, those canals cannot accommodate merchant vessels at greater tonnage or frigates with 32 cannons. But the possible passage of a large vessel is at issue in discussions about cutting through an isthmus in America. That the Languedoc Canal supposedly *connected two seas* did not eliminate a 600-league detour around the Spanish Peninsula. However impressive this hydraulic work may be for accommodating 1,900 flatboats of between 100 and 120 tons per year, it should be seen as

II.291 nothing more than a means of *inland transport*, because it barely reduces the number of ships that pass through the Straits of Gibraltar. Connecting two neighboring ports somewhere in equinoctial America, whether through the Isthmuses of Cúpica, Panama, Nicaragua, or Coatzacoalcos (Tehuantepec), by means of a *small canal* (between 4 and 7 feet deep) will no doubt stimulate trade. This canal would act like a *railroad*: small as it would be, it would enliven and shorten communications between the western seaboards of America and those of the United States and Europe. If people, even at times of war, generally preferred to make the long and dangerous trip around Cape Horn to transport Chilean copper, Peruvian cinchona bark and vicuña wool, and Guayaquil's cacao to warehouses in Panama and Portobelo, it is only because of the lack of infrastructure and the extreme poverty that surrounds two towns that flourished at the beginning of the conquest. The difficulties I mentioned

II.292 before only increase when it comes to the transport of goods from Cartagena de Indias or the Antilles to Quito and to Lima. In the north-south direction,

1. ▾Andréossy, *Histoire du Canal du Midi, ou Canal de Languedoc*, p. 364. Huerne de Pommeuse, *Des canaux navigables*, 1822, pp. 64, 264, 309. ▾Dupin, *Mémoires sur la Marine et les Ponts et Chaussées de France et d'Angleterre*, pp. 65 and 72. ▾Dutens, *Mémoires sur les travaux publics de l'Angleterre*, p. 295.

one has to battle against the current of the Río Chagre and against the winds and current of the Pacific Ocean.

Channeling the Chagre, employing steamboats, building *rail-ways*, importing camels from the Canary Islands—they had already begun to breed them in Venezuela during my voyage[1]—and digging small canals on the Isthmus of Cúpica or on the strip of land that separates Lake Nicaragua from the South Sea will contribute to the prosperity of the Americas' economy; but it will only indirectly further the general interests of civilized peoples. The direction of trade in Europe and the United States with the *fur coast* (between the mouth of the Columbia and the Cook Rivers), with the Sandwich Islands—rich in sandalwood—and with India and China will not change. Distant communications will require ships with heavy tonnage that can carry a lot of merchandise, natural or artificial passageways of an average depth between 15 and 17 feet, and uninterrupted navigation without having to unload any ships. All of these conditions are indispensable, and it would be confusing the matter if one mixed up canals that, because of their size, are suitable only for internal and coastal navigation—canals such as the Canal of Languedoc between the Mediterranean and the Atlantic Ocean and the Forth and Clyde Canal between the Irish Sea and the North Sea—with lock chambers that can accommodate ships for the Canton trade. In this matter which touches upon the interests of all peoples who have made some steps along the path of civilization, we must be more explicit than we have been about the fact that the successful solution to this problem depends on which places one chooses. I repeat: it would be imprudent to begin somewhere without having examined and leveled other places. It would be especially unfortunate if such a project were undertaken on too small a scale, for, in this kind of work, expenses do not increase proportionally with the length of the canal or with the width of the locks.

II.293

II.294

The erroneous notion that geographers, or, rather, cartographers, have spread for centuries about the equal height of the American Cordilleras, their extension in unbroken ridges, and, finally, about the absence[2] of any valleys cutting through the central range has created the general belief that it was much more difficult to connect the oceans than there is actually reason to believe. Apparently, there is no mountain range, not even a ridge or discernible[3]

1. See Vol. I, pp. 165, 221–23; V, pp. 221–25, and *Essai politique sur le royaume de la Nouvelle-Espagne*, Vol. II, p. 689.

2. I discuss the source of these errors in Vol. VI, pp. 49–52; VII, pp. 47–51; VIII, pp. 92–95, 108–10, 166–98.

3. These claims depend on how easy it is to cut the canal. I know that a shallow hill of 40 to 50 toises can be imperceptible due to its gradual incline. I discovered that Lima's

II.295–96 foothills, between Cúpica Bay, on the South Sea shore, and the Río Naipi, which flows into the Atrato about fifteen leagues above its delta. A pilot from Biscay, Mr. Gogueneche, has first called the government's attention to this point in 1799. Highly reliable people, who accompanied him on the journey from the coastline of the Pacific Ocean to Naipi dock, have assured me that they did not see a single hill in this isthmus's alluvial lowlands, which it took them 10 hours to cross. ▾Don Ignacio Pombo,[1] a merchant from Cartagena de Indias with a lively interest in all Statistics having to do with New Granada, wrote to me in February of 1803: "Ever since you went up the Magdalena River to go to Santa Fé and Quito, I have continually gathered information about the Isthmus of Cúpica. It is no more than 5 or 6 leagues from this port II.297 to the Río Naipi pier: this entire area is flat (*terreño enteramente llano*)." Given these facts, it is clear that this part of northern Chocó is of the greatest importance for solving the problem at hand. But to get a clear sense of this absence of mountains on the southern edge of the Isthmus of Panama, we have to bear in mind the general shape of the Cordilleras. At 2° and 5° latitude,[2] the

---

large square was 88 toises above the South Sea; however, passing from Callao to Lima, a distance half that between Cúpica and the Río Naipi pier, this difference in elevation is practically imperceptible. The geographical position of Cúpica is as uncertain as the confluence of the Naipi with the Atrato. And this uncertainty will seem less strange if one remembers that the Bay of Cúpica stretches along the Panama Isthmus's entire south coast, and that the shoreline between Cape Charambira and Cape San Francisco Solano has never been surveyed by sailors with accurate instruments in sight of shore. Cúpica is a port in the little-known province of Biruquete, which maps in Madrid's *Depósito hidrográfico* situate between Darien and the Chocó del Norte. It took its name from a ▾Cacique called Birú or Biruquete, who controlled the lands that neighbored San Miguel and fought as ally of the Spanish in 1515 (Herrera, [*Historia de las Indias occidentales*,] Dec. II, p. 8). I did not find the port of Cúpica on any Spanish map but, rather, *Puerto Quemado ó Túpica*, at 7° 15′ lat. ([Bauzá?] *Carta del Mar de las Antillas*, 1815. [Malaspina?] *Carta de la costa occidental de la América*, 1810). An unpublished sketch of Chocó province that I have in my possession confuses Cúpica with Río Sabaleta, lat. 6° 30′; however, according to maps from the *Depósito*, the Río Sabaleta is situated to the south, not to the north, of Cape San Francisco Solano and there lies 45′ south of Puerto Quemado. According to the map of the province of Cartagena by ▾Don Vicente Talledo (London 1816), the confluence of the Río Napipi (Naipi?) is at 6° 40′ lat. Hopefully, observations made in the field will soon remove these uncertainties about positions.

1. A friend of the famous ▾Mutis and author of a little work on the cinchona trade ([Pombo,] *Noticias varias sobre las quinas oficinales, Carth[agena] de Indias*, 1814) that I have had occasion to cite from time to time.

2. The Suma Paz, Chingasa, and Guachaneque eastern chain runs between Neiva and the Guaviare basin, and between Santa Fé de Bogotá and the Meta basin. The Guanacas, Quindío, and Herveo basin middle range runs between the Río Magdalena and the Río

Andes mountain range is divided into three secondary chains. Two longitudinal valleys separate these secondary chains form the Magdalena and Cauca basins. The eastern branch of the Cordilleras inclines to the northeast and is connected to the *Sierra Nevada de Mérida* and to Venezuela's coastal range by the Pamplona and Grita mountains. The middle and western branches—Quindío and Chocó—merge in Antioquia province between 5° and 7° latitude, forming a mountain group of quite considerable width stretching toward Caceres and the high savannas of Tolú through the *Valle de Osos* and the *Alto del Viento*. Farther west, in *Chocó del Norte* on the Atrato's left bank, the mountains lower to such a degree that they entirely disappear between the Gulf of Cúpica and the Río Naipi. It is the isthmus's astronomical position, and the distance between the mouth of the Atrato at its junction with the Río Naipi,[1] that we must determine precisely. We do not know if schooners can get this far.

II.298

II.299

After Lake Nicaragua, after Cúpica and Coatzacoalcos, it is the Isthmus of Panama that deserves the most serious attention. The possibility of building a canal for oceangoing vessels on this isthmus depends both on the elevation of the highest ridge and on the configuration of the coast, that is, on their *maximum* proximity to each other. Depending on its orientation, such a narrow strip of land could have escaped the destructive influence of the maelstrom, and the assumption that the highest mountain elevation should correspond to the *minimum* distance between the coasts would not be justifiable today, even if one followed the principles of a purely systematic geology. Since I published my first study on connecting the seas, we have unfortunately re-

II.300

Cauca, between the River Plate and the Popayán, and between the Ibague and Cartago. The western chain runs between the Río Cauca and the Río San Juan, between Cali and Novita, and between Cartago and the Tadó (see my *Atlas géographique*, Pl. XXIV). The latter chain, which separates the provinces of Popayán and Chocó, is generally quite low. It is said, however, that it rises precipitously on Torá mountain to the west of Calima. Pombo, *Noticias varias sobre las quinas oficinales*, p. 67.

1. This part of America's geography between the mouths of the Atrato, Cabo Corrientes, Cerro del Torá, and Vega de Supía is in the most deplorable state. It is only farther east, in Antioquia province, that ▾Don José Manuel Restrepo's work offers some points whose positions are astronomically fixed. The distance by land between Cúpica and Cape Corrientes is between 12 and 14 (?) nautical leagues. It takes 7 days by sea to reach the mouths of the Atrato from Quibdó (Zitara), where the *Teniente Gobernador* [deputy governor] resides (because the magistrate lives in Novita). Placing Zitara either 1° too far north, sometimes even at the mouth of the Atrato, sometimes at its confluence with the Naipi, is a widespread error on modern maps (excepting Mr. Talledo's). It is no more than one day's journey from San Pablo, situated a few leagues below Tadó on the right bank of the Río San Juan, to Quibdó or Zitara.

mained ignorant about the height of the ridge that the canal would need to cross. With a precision superior to anything I could attempt in this line of research, two knowledgeable travelers, Mr. Boussingault and Mr. Rivero, have measured with a level the Caracas Cordilleras in Pamplona and from there to Santa Fé de Bogotá. But to the northwest of Bogotá, from the Quindío and Antioquia Andes, leveled by Mr. Restrepo and myself, to the Mexican plateau at 12° lat. of *Central America*, not a single measurement of elevation has been taken since my return to Europe. It is most unfortunate that, toward the middle of the last century when French academicians were crossing the Isthmus of Panama, it did not occur to them to take out their barometers at the summit point. Some barometric measurements that Ulloa reported almost randomly led me to conclude that there is a 210 to 240 feet difference between the mouth of the Río Chagre and the Cruces pier. From the Venta de Cruces II.301 to Panama, one first climbs and then descends into ravines toward the South Sea. The canal would therefore need to cross the divide between this port and Cruces—that is, if we were to persist in directing it this way. I remind the reader that to enjoy the view of both Oceans simultaneously, the mountains in the isthmus's ridgeline would need only 580 feet of elevation, that is, only a third more than the elevation of the Naurouze elevation in the Corbières mountain range, which is the highest point of the Languedoc Canal. The simultaneous view of two seas is considered something remarkable in some parts of the isthmus. From this, one may conclude, I think, that the mountains here are generally not 100 toises high. On the basis of some sketchy notations on temperatures in this area and on the geography of native plants, I would be disposed to believe that the ridge to Cruces over which the road II.302 from Panama passes, does not reach 500 feet.[1] Mr. Robinson believes it to be more than 400 feet.[2] According to the claims of another traveler,[3] who describes what he saw in simple terms, the hills that make up the isthmus's central range are separated from each other by valleys that "let water pass through freely." Engineers should, above all, focus their efforts on finding and exploring these transversal valleys. Every country has examples of natural gaps in ridges. The mountains between the Saône and Loire basins, which the Central Canal would have had to cross, were eight to nine hundred feet high. But

1. For example, near Chepo and Penomene village (*Mss. du curé Don Juan Pablo Robles*). The mountains seem to rise toward Veragua province, where even wheat is cultivated in the Chiriquí del Guami district near Palma village, a Franciscan mission connected to ▾Panama's Propaganda college.

2. *Memoirs on the Mexican Revolution*, p. 269.

3. ▾Lionel Wafer, *A New Voyage and Description of the Isthmus of America*, 1729, p. 297.

a gorge, or break in the range, near Lake Long-Pendu, provided an opening at 350 feet lower elevation.

While our knowledge of the elevations of the Isthmus of Panama has not advanced at all, recent accounts by ▾Mr. Fidalgo and a few other Spanish sail- II.303 ors have at least given us more precise data on its configuration and *minimal* width. This *minimum* is not 15 miles, as the earliest maps from the *Depósito* II.304 *hidrográfico*[1] had indicated, but 25 and three-fourths miles (60 to a degree), that is 8 and two thirds nautical leagues, or 24,500 toises, because the dimensions of the Gulf of San Blas (also called Ensenada de Mandinga because of II.305 the small river by this name that empties there). This gulf cuts into the land for 17 fewer miles than had been assumed in 1805, when they surveyed the

1. See my *Essai politique sur le royaume de la Nouvelle-Espagne*, Vol. II, p. 862. By comparing the two maps in the *Depósito hidrográfico de Madrid*, entitled *Carta esférica del Mar de las Antillas y de las Costas de Tierra Firme desde la isla de la Trinidad hasta el golfo de Honduras*, 1806, and *Quarta Hoja que comprehende la provincia de Cartagena*, 1819, one can see how well founded were the doubts I articulated fifteen years ago about the relative position of the most important points on the isthmus's north and south coasts. Formerly (▾Don Jorge Juan, *Voyage dans l'Amérique Meridionale*, Vol. I, p. 99), Panama was believed to be 31′ in arc west of Portobelo. La Cruz (1775) and López (1785) followed this assumption, which is based on nothing more than a summary of directions made en route with a compass. Already in 1802, López (*Mapa del Reino de Tierra Firme y sus provincias de Veragua y Darién*) had placed Panama 17′ to the east of Portobelo. On the 1805 map, from the Depósito, this difference in meridians was reduced to 7′. Finally, in 1817, the map from the Depósito situates Panama 25′ to the *east* of Portobelo. Here are some other latitudes that pertain to the isthmus's size:

	1819 map	1817 map
Southern coast between the mouths of the Río Juan Díaz and Río Jucume east of Panamá, in the meridian of Punta San Blas	8° 54′	9° 2′ ½
Northern coast forming the bottom of the gulf of Mandinga, or San Blas, south of the *Islas Mulatas*	9° 9′	9° 27′ ¾
According to the 1805 map the latitudinal difference for the isthmus's minimum width is nearly 14,250 toises; according to the 1817 map it is nearly 24,463 toises. Punta San Blas, NW section of the gulf of Mandinga	9° 33′	9° 34′ ½

Because this Cape does not have the same landmass to the north as the bottom of the gulf does near the mouth of the Río Mandinga, it follows that the gulf's location is, according to the first map, at 24′; according to the second, at 7′. It is likely that the discrepancies in latitude measurements that result from Mr. Fidalgo's last expedition should be attributed to the lack of *artificial horizons* and the difficulty of observing the sun with instruments of reflection amidst an island group and above a sea whose horizon is not open. Farther west, the average width of the isthmus, between Castillo de Chagre, Panama, and Portobelo, is 14 nautical leagues. The *minimum* width (8 leagues) is two to three times less that of the width of the Suez Isthmus, which Mr. LePère puts at 59,000 toises.

Islas Mulatas archipelago. No matter how reliable the astronomical observations that Madrid's Royal Hydrographic Office used for its 1817 map of the isthmus, it should nevertheless be remembered that these measurements concern only the north coast, and that they appear never to have been calibrated with the south coast either through triangulation or through chronometric measurements (using time differences). In other words, the question of the width of the isthmus does not depend exclusively on determining latitudes.

II.306 Having recently received a few excellent barometers built to ▾Mr. Fortin's specifications, Colombia's government will be able to undertake barometric leveling—a highly precise operation in the Torrid Zone—before engaging in geodesic leveling, always a slow and costly process. I am satisfied that corresponding observations in these regions can be trusted within 4 or 5 toises because of the remarkable regularity of hourly variations. The areas that call for closer scrutiny are the following: the *Coatzacoalcos Isthmus* between the headwaters of the Río Chimalapa and the Río del Pasco; the *Nicaragua Isth-* II.307 *mus*[1] between the eponymous lake and the isolated Granada and Mombacho volcanoes; the *Panama Isthmus* between Venta de Cruces, or, rather, the Indian village of Gorgona, 3 leagues below Cruces, and Panama port between the Río Trinidad and the Río Caymito, between Mandinga Bay and the Río Juan Díaz, and between the Ensenada de Anachacuna (to the west of Cape Tiburón) and the San Miguel Gulf where the Río Chuchunque, or Tuyra, empties; the *Cúpica Isthmus* between the South Sea coast and the confluence of the Río Naipi with the Río Atrato; and finally, the *Chocó Isthmus* between the Río Quibdó, an upper tributary of the Atrato and the Río San Juan de II.308 Charambira. Individuals used to making accurate observations and equipped with nothing but barometers, reflective instruments, and clocks would be able to resolve in a matter of months problems that have occupied the merchant peoples of both worlds for centuries. When I was listing the regions

1. If it were only a question here of *primary and secondary waterways* fit to revitalize inland trade, I would also have cited the Verapaz and Honduras coastlines. The *Golfo Dulce* reaches more than 20 leagues inland at the Sonsonate meridian, so that the distance between Zacapa village (in Chiquimula province, near the south edge of the *Golfo Dulce*) and the Pacific coastline is no more than 21 leagues. Rivers from the north come close to the water that runs into the South Sea from the Izalco and Sacatepéquez Cordilleras. To the east of *Golfo Dulce*, in the Comayagua *partido*, there are the Río Grande de Motagua, or *Río de las bodegas de Gualán*, the Camalecón, the Ulúa, and the Leán, which are navigable by large pirogues 30 or 40 leagues inland. Quite likely, the cordillera that forms the ridge dividing the two seas here is cut by some transversal valleys. The interesting work that Mr. Juarros published in Guatemala states that the waters of the beautiful Chimaltenango valley run off to the southern and to the northern coasts. Steamboats will one day revive trade on the Río Motagua and the Río Polochic.

well suited for linking the two seas, I did not ignore the Chocó Isthmus—that is, the land of *platiniforous deposits* that stretches from the Río San Juan de Charambira to the Río Quibdó—because this is the only point where a connection between the Atlantic Ocean and the South Sea has existed since 1788. The small Raspadura Canal, which a monk, a priest of Novita, had the Indians of his parish dig in a ravine periodically filled by natural floods, facilitates inland navigation for 75 leagues between the mouths of the Río San Juan, below the Noanama, and that of the Atrato, which is also known as Río Grande del Darién, as Río Dabeiba, and as Río Chocó.[1] By this route, significant amounts of Guayaquil cacao reached Cartagena de Indias during the wars that preceded the revolution in Spanish America. The Raspadura Canal—I believe I was the first to inform Europe of its existence—offers passage only to small boats. But it could easily be expanded[2] if the streams known as

II.309

1. I could add the synonym San Juan (del Norte), if I did not fear confusing the Atrato with the Río San Juan (de Nicaragua) and the Río San Juan (de Charambira). The Río Dabeiba took its name from a female warrior who, according to the early writers of the conquest, ruled the mountainous region between the Atrato and the sources of the Río Sinú (Zenu), to the north of the city of Antioquia. According to Petrus Martyr d'Anghiera's work (*Oceanica*, p. 52), a local myth confused the woman with the female deity of the high mountains who threw lightning bolts. Today, we recognize the name Dabeiba in the names of Mount Abibe, or Avidi, given to the *Altos del Viento* at 7° 15′ latitude west of the Boca del Espíritu Santo or the Cauca's banks. What is the Ebojito Volcano that La Cruz and López place in the nearly empty region between the Río San Jorge, a tributary of the Cauca, and the sources of the Río Murri, a tributary of the Atrato? The existence of this volcano seems highly doubtful to me.

2. [Caballero y Góngora,] *Relación del estado del Nuevo Reyno de Granada que hace et Arzobispo Obispo de Cordova a su sucessor el Exc. Sr. Fray Don Francisco Gil y Lemos* 1789, fol. 68 (unpublished manuscript compiled by the secretary to the archbishop-viceroy, ▾Don Ignacio Cavero). [Pombo,] *Representación que dirigió Don José Ignacio Pombo al consulado de Cartagena en* 14 *de Mayo* 1807 *sobre el reconocimiento del Atrato, Zenú y San Juan*, fol. 38 (manuscript). The Raspadura ravine (or Bocachica) receives water only from the Quebradas de Quiadocito, the Platinita, and the Quibdó. According to data that I acquired (in Honda and Vilela, near Cali) from people employed in the Chocó gold dust trade (*rescate*), the Río Quibdó, which connects with the Mina de Raspadura Canal, joins with the Río Zitara and the Andagueda near Quibdó village (commonly called Zitara). But, according to an unpublished map that I have just received from Chocó, which shows the Raspadura Canal (lat. 5° 20′ ?) joining both the Río San Juan and the Río Quibdó a little above the Mina de las Animas, the village of Quibdó is situated at the confluence of the little river by this name and the Río Atrato, which, 3 leagues upriver, is joined by the Río Andagueda near Lloro. From its mouth (lat. 4° 6′) at the Sud de la Punta de Charambira, the great San Juan River (flowing toward the NNE), receives the Río Calima, the Río Nó (above Noanama village), the Río Tamana, which passes near Novita, the Río Iro, the Quebrada de San Pablo, and finally, near Tadó village, the Río de la Platina. Only the river

II.310–13 the Caño de las Animas, the Caliche, and Aguas Claras were joined. It is easy to establish reservoirs and *feeding trenches* in a region like Chocó, where it rains all year and thunder can be heard every day. Because Mr. Caldas's barometric observations have not yet been published, we do not know the elevation of the ridge between San Pablo and the Río Quibdó. We know only that there are some *gold washes* in this region, up to 360 and 400 toises above sea level and never below 50 toises. The canal's position in the interior of the continent, its considerable distance from either coast, and the frequent rapids and waterfalls (*raudalitos y choreras*) on the rivers between Charambira and the Gulf of Darien, which one must ascend or descend to arrive at one sea or the other, are obstacles far too difficult to overcome to build *an oceanic navigation route* through Chocó. Although this route does not offer passage to schooners of considerable tonnage, it still merits attention from a prudent administration: it will revive interior trade between Cartagena and the province of Quito, and between the port of Santa Marta and Peru. By way of conclusion, we will point out that the Madrid ministry never enjoined New Granada's viceroy to block the Raspadura ravine; nor did he inflict the death penalty on those who sought to reestablish a canal in Chocó, as has been alleged in a recent report.[1] It is true that this worrisome policy recalls the order to the viceroy of New Spain, during my stay in America, to uproot the vine stalks in the *provincias internas*. But hatred for the colonies' viticulture came from the influence of a few Cádiz merchants, eager to protect their perceived long-time monopoly, while a small ravine crossing the jungles of Chocó more easily escaped the ministry's vigilance and the metropole's envy.

After having examined the locations of various ridges according to the im-

---

valleys in Chocó province are inhabited. The province has three trade routes: to the north with Cartagena by means of the Atrato, whose banks are completely uninhabited from 6° 45′ latitude; to the south with Guayaquil, and, before 1786, with Valparaíso, by means of the Río San Juan; and to the east with Popayán province, by means of the Tambo de Calima and Cali. It takes one day to pass downriver from Tadó to Noanama on the Río San Juan. From Noanama, it takes four days to go to Tambo de Calima (lat. 4° 12′), and from there to Cali (lat. 3° 25′). It takes five days in the Cauca valley from Popayán, crossing the Río Dagua or the San Buenaventura, and the western Cordillera of the Popayán Andes. I have gone into these local details because maps confuse the Raspadura ravine, which serves as a canal, with Calima and San Pablo's *portages*. The San Pablo *Arastradero* also leads to the Río Quibdó, but several leagues above the mouth of the Raspadura Canal. Merchandise (*géneros*) sent from Popayán by means of Cali, Tambo de Calima, and Novita to *Chocó del Norte*, that is, to Quibdó (Restrepo, *Estado de Colombia* in 1823, p. 24), commonly takes the path of the San Pablo *Arastradero*. The geographer La Cruz calls the entire isthmus, between the sources of the Río Atrato and the Río San Juan, the *Arastradero del Toró* (on the elevation of the *gold zone*, see [Caldas,] *Semanario de S.-Fé*, Vol. I, p. 19).

1. Robinson, [*Memoirs of the Mexican Revolution*,] Vol. II, p. 266.

perfect information that I have been able to cull up to now, it remains to be proven, by means of analogy with what men have already built, whether the possibility of an oceanic link in the New World can be realized at the current level of our modern civilization. The more complicated a problem becomes and the more dependent it is at the same time on many intrinsically variable elements, the more difficult it becomes to determine the *maximum* effort of intelligence and physical prowess that people are able to exercise. For thousands of years, from the unknown ages when the pyramids at Giza were built to the construction of our gothic steeples and St. Peter's dome, men have not erected edifices more than 450 feet tall. But would one dare conclude from this that modern architecture cannot reach elevations scarcely forty times higher than termite hills? If it were merely a question of average sized canals no more than 3 to 6 feet deep and used only for inland navigation, I could mention canals built a long time ago that have crossed mountain ridges 300 to 580 feet high.[1] England alone, whose canals are 584 nautical leagues long, has nineteen that cross ridges between west- and east-coast rivers. For a long

II.314

II.315–16

1. Here is the partial data for ten canals, arranged in order of height at their points of highest elevation:

Names of the canals	Elevation of summit levels in Paris feet
Canal de Languedoc or *du Midi.* (Length, 122,480 toises; average depth, 6 feet 2 inches; number of locks, 62; construction costs, in the day of Louis XIV, nearly 16,280,000 francs; current value, 33 million francs). Principal Waterway	582
Leominster Canal. (Lgth. 37,745 toises; costs, 14 million francs). Secondary Waterway	465
Huddersfield Canal. (Lgth. 15,900 toises; costs, 6.5 million francs). Secondary Waterway	409
Leeds and Liverpool Canal. (Length, 106,700 toises; number of locks, 91; cost, 14,400,000 francs). Principal Waterway	404
Canal du Centre, between Saône and Loire. (Length, 58,300 toises; depth, 5 feet; number of locks, 80; cost, 11 million francs) Principal Waterway	403
Grand Trunk Canal, or *Trent and Mersey.* (Length, 272,000 toises; depth, 4 to 5 feet; number of locks, 75; cost, 9 ½ million francs). Principal Waterway	382
Grand-Junction Canal. (Length, 74,400 toises; depth, 4 feet 3 inches; number of locks, 101; cost, 48 million francs). Principal Waterway	370
Canal de Briare, constructed in 1642, the oldest canal with a summit level. (Length, 14,500 toises; depth, 4 feet; number of locks, 40; cost, 10 million francs). Principal Waterway	243
Forth and Clyde Canal. (Length, 34,000 toises; depth 7½ feet; number of locks, 39; cost, 10 million francs). Principal Waterway	155
Caledonia Canal. (Length, 18,500 toises; number of locks, 23; depth, 18 feet 9 inches; cost, 19 million francs). Principal Waterway	88

I added abbreviations for *Principal* and *Secondary Waterways* to distinguish canals that, following British usage, are thus classified. The locks in the first category are at least

time, engineers thought so little of 582 feet as the *maximum* elevation—that is, the height of the lock that connects the Naurouze pass with the Midi Canal—that one can reasonably expect from this kind of hydraulic work that the famous ▾Mr. Perronet considered the Burgundy Canal a very feasible project. II.317 This canal runs between the Yonne and the Saône and must pass over (near Pouilly) an elevation of 621 feet above the Yonne at low water. Combining inclined planes and railroads (*railways*) with navigation routes has enabled ships to pass through the Monmouthshire Canal at an elevation of a thousand feet. Yet, such structures, however important for the prosperity of a country's internal trade, can hardly be called *oceanic navigational canals*.

In the present discussion, we are considering transoceanic transport by ships whose shape and tonnage is appropriate for the India and China trade. The industry of the peoples of Europe already offers us two examples of transoceanic links on large scale: the Eyder or Holstein Canal [Germany] and the Caledonian Canal. The first of these structures, built between 1777 and 1784, connects the Baltic and North Seas between Kiel and Tönningen and has II.318 only 6 locks and a difference in elevation of 28 feet. It separates the continental part of Denmark from Germany and makes the rather dangerous Kattegat and Sund passages unnecessary for medium sized ships. This canal takes ships between 140 and 160 tons,[1] coming from Russian and Prussian ports and headed for England, the Mediterranean, Philadelphia, Havana, and even Africa's west coast. These ships have *drafts* of no more than eight to ten feet.[2] Typically built in Holland or the Baltic, they have very flat bottoms and therefore a large capacity relative to the small amount of water they displace. The II.319 Caledonian Canal—not the most useful but certainly the most magnificent hydraulic endeavor to this day—is an *oceanic canal* in the full sense of the term. It connects Scotland's eastern shore at Inverness to its western shore at fort Williams through a gap that nature itself seems to have provided for this purpose. The navigable portion is 17 leagues long (20 to a degree), of which only six and a half are man-made. The rest is formed by Loch Oich and Loch Lochy, which were once separated by a rocky sill. This canal was built in 16

---

64 feet long and 14 feet wide; the locks of the second are also 64 feet long but only 7 feet wide. The *Canal de Monsieur*'s highest point is 590 feet above the Rhine's water level.

1. From 75 to 90 last [German unit of measure]. The capacity of barges sailing England's heavy navigation canals is generally no more than 40 to 60 tons. On the Languedoc Canal, the largest boats are 120 tons. The majority of merchandise transported to England can be broken down into small quantities and take many forms, including coal, iron, and brick. This is not the case with French wine and oil casks.

2. Feet are still the old French unit measure, pieds *de roi*. Six feet equal 1.949 meters, unless otherwise stated.

years: it provides passage to 32-cannon frigates and cargo ships for overseas trade. Its average depth is eighteen feet eight inches (6.09 meters), and its greatest width at the bottom is 47 (15.2 meters). The locks, which number twenty-three, are 160 long by 37 feet wide.

In the practical applications I present at the end of this Section, I am being guided only by analogies with already implemented work. First, I will note that the width of the Cúpica and Nicaragua Isthmuses, whose ridges are of insignificant height, is nearly the same as the width of the land that the man-made section of the Caledonian Canal crosses. The Nicaragua isthmus is quite similar to the gap in the Scottish Highlands because its inland lake, and this lake's link to the Antillean Sea through the Río San Juan is like the River Ness: a natural connector between the mountain lakes and the Gulf of Murray. In Nicaragua, as in the Scottish Highlands, there would be only one ridge to pass over, for if the Río San Juan[1] were 30 to 40 feet deep for most of its course, as I have been assured it is, then one would need to engage only in partial *channeling* by means of weirs and lateral trenches. II.320 II.321

As for the depth of the oceanic canal planned in central America, I think that it could be even shallower than the Caledonian Canal. During the last fifteen years, new systems of trade and navigation have led to notable changes in the capacity, or cargo, of the ships most commonly used for trade with Calcutta and Canton. Attentive examination of the official list of ships from London and Liverpool engaged in the India and China trade over a two-year period (from July 1821 to June 1823) shows that, out of 216 ships, two *thirds* were II.322

1. This point, near the Campeche timberlands (*cortes de Madera*), had attracted the commercial world's attention long before the publication of Mr. Bryan Edwards's excellent work on Jamaica ([*The History, Civil and Commercial, of the British Colonies in the West Indies*], Vol. V, p. 213). See ▾La Bastide, *Mémoir sur un nouveau passage de la Mer du Sud à la Mer du Nord*, p. 7. There are three possibilities for a canal in Nicaragua (as I explained in my *Political Essay on the Kingdom of New Spain*): either from Lake Nicaragua to Papagayo Gulf, or from this same lake to Nicoya Gulf, or from Lake León, or Managua, to the mouth of the Río Tosta (and not from Lake León to Nicoya Gulf, as suggested by the otherwise quite knowledgeable editor of the *Biblioteca Americana*, 1823, August, p. 120). Is there a river that goes from Lake León to the Pacific Ocean? I doubt it, though old maps indicate connections between the lakes and the sea (*Essai politique sur le royaume de la Nouvelle-Espagne*, Vol. I, p. 15). The distance from the southeastern edge of Lake Nicaragua to the Gulf of Nicoya varies quite notably (between 25 and 48 miles) on Arrowsmith's map of South America and on the beautiful map from the Madrid Hydrographic Office with the title *Mar de las Antillas*, 1809. The width of the isthmus between the east bank of Lake Nicaragua and the Papagayo Gulf is 4 to 5 nautical leagues. The Río San Juan has three mouths, the smallest two called *Taure* and *Caño Colorado*. One of Lake Nicaragua's islands, Ometepe, has a volcano [Concepción] that is said to be still active.

below 600 tons, one fourth between 900 and 1,400 tons, and one seventh below 400 tons.[1] In France, the *average tonnage* of ships engaged in the India trade from the ports of Bordeaux, Nantes, and Le Havre is 350 tons. The nature of the operations undertaken in distant waters determines the capacity of the ships one employs. When one wishes to load indigo in Bengal, it may seem sufficient, perhaps even preferable at times, to use a ship between 150 and 200 tons. The system of small shipments is especially active in the United States, where one recognizes all the benefits of swift loading and unloading II.323 and speedy capital circulation. The average American ship bound for India around the Cape of Good Hope, or for Peru around Cape Horn, carries 400 tons. South Sea whalers have a capacity of only two or three hundred tons. Spanish America is stuck on old habits, using larger vessels in times of peace. For example, during my stay in Mexico, 120 to 130 ships with an average capacity of 500 tons came to Veracruz from Spain. Only during wartime do expeditions of 300-ton ships make their way for Cádiz.

In the current state of world trade, these data adequately prove that a canal such as the one planned between the Atlantic Ocean and the South Sea would be large enough, provided that the size of its *segments* and the capacity of its locks could accommodate ships between 300 and 400 tons. These are the minimum dimensions that the canal must have. In keeping with what we have II.324 indicated earlier, this *minimum* assumes a capacity almost double that of the Holstein Canal but less than that of the Caledonian Canal. The former accommodates ships between 150 and 180 tons, the latter 32-gun frigates and commercial vessels of over 500 tons. It is true that tonnage has only an indirect relationship with a ship's *draft*, because how a ship is built changes both II.325 how it handles and what it can carry. Admittedly, however,[2] an average depth

1. ▾Phipps, *East India Shipping, a Return to the Order of the House of Commons, London* 1823. I converted British into French tons, the latter being 10 percent lighter.

2. I assume that a foot and a half of water below a ship's rudder would be enough for a ship to sail through a well-dredged canal of perfectly calm water. Despite the many elements that can influence the *draft* of ships with the same capacity, one can allow for the following approximations:

Port				Water draft		
1200	to	1300	tonnage	19	at	20 feet
750		800		17		18
500		600		15½		17
300		400		14		16
200		250		11		12

It behooves me to recall the relevant data pertaining to the practical solution of this problem, given that this is a matter that interests all men willing to reflect on the fate of

of 15½ to 17½ feet would suffice for a canal designed for ships between 300 and 400 tons. This would be fifteen inches less than the Caledonian Canal, which was designed by the great builders ▾Mr. Rennie, Mr. Jessop, and Mr. Telford. It would be twice the depth of the Forth and Clyde canals.

Europe's gigantic structures, which we are invoking as examples here, cost no more than 4 million piasters and had only slight elevations to pass over—less than 90 to 100 feet. The canals that cross ridges of 400 to 600 feet are no more than 4 to 6 feet deep. Difficulties naturally increase with the ridges' elevations, with the depth of the excavations, and with the width of the locks but not their number. It is not simply a question of digging a canal. It is necessary to ensure that the amount of water taken from higher areas on the summit level will always be sufficient to fill the canal and replace what is lost to locks, evaporation, and filtration. We saw earlier that the height of the ridge would pose less of an obstacle for a transoceanic canal on the Cúpica and Coatzacoalcos Isthmuses than the state of the riverbeds (the Río Naipi and the Río del Pasco), which would need to be *channeled*, either by means of hydraulic machines operated by steam pumps or by means of weirs or lateral trenches. In the intendancy of Nicaragua, the great depth of the Río San Juan and especially of Lake Nicaragua (*Laguna de Granada*)—between 17 and 14 feet, according to Mr. Robinson, and 20 to 55 feet, according to Mr. Juarros—would make a similar undertaking, if not redundant, at least less difficult. The Panama mountains probably rise to an elevation equal to that of the lock chambers of the Canal du Centre (between Châlon-sur-Saône and Digoin) and of the Grand Junction Canal (between Brentford and Braunston). It is even possible that the isthmus's mountains are higher yet and that no transversal valley completely divides them from south to north. Such disadvantageous sites probably should not be chosen, but we should point out that a high sill does not irrevocably rule out the joining of the seas, unless there is also insufficient water at higher elevations for use at the summit level. The sequence of seven to eight locks that harness 64-to-70-foot waterfalls along the Briare and Languedoc canals[1] have long been acknowledged as extraordinary feats, despite the small size of their locks and their shallowness (between 5 and 6 feet). *Neptune's Staircase* in the Caledonian Canal offers a similar succession of conjoined locks but on a much grander scale: frigates can be lifted to a height of 60 feet in a very short amount of time. This work costs no more than 257,000 piasters, that is, five times less than three mine shafts in Valenciana in Mexico. Ten *Neptune's Staircases* would permit 500-ton ships to pass over a ridge of

II.326 II.327 II.328

humanity and the progress of civilization at large. The Crinan Canal, in Scotland, is also between 11 and 14 feet deep over a length of 3 leagues.

1. Near Rogny and Fonserane.

600 feet, a point higher than the chain of the Corbières between the Mediterranean and the Atlantic Ocean. We are discussing here only the possibility of structures that might be built, without saying that they must be built.

The amount of water necessary to fill a canal increases with filtration, with the frequency of traffic (which causes the loss of *lockage water*[1]), and with the size, but not the number, of chamber locks. In the tropics, the ease of collecting enormous amounts of rain water in reservoirs far surpasses anything that European engineers can imagine. When Louis XIV wanted to beautify the gardens of Versailles, ▾Colbert was led to hope that rain would provide 9 million cubic toises of water for 12,700 hectares of lawn connected to ponds and artificial lakes.[2] But only 19 to 20 inches of rain fall in the Paris area each year, while at least 100 to 112 inches fall in the New World's Torrid Zone, particularly in jungle areas.[3] This prodigious difference shows how a skilled engineer could make the most of central America's climate by connecting springs, feeding trenches, and well-established reservoirs. Despite the high air temperature, losses to evaporation will not outweigh the benefits of tropical rain gathered in deep basins. The fine experiments that ▾Mr. Prony conducted

II.329

II.330

1. The *lockage water* is the volume of water that a lock needs to raise or lower a ship.

2. Unfortunately, they could retain only one one-hundred-and-fiftieth, the rest being lost to filtering. They were forced to build a ▾Marly machine. Huerne de Pommeuse, *Sur les canaux navigables. Supplement*, p. 45.

3. See Vol. VII, p. 305; VIII, pp. 423–27, 399–403. Even in Kendal in the west of England, the annual average rainfall is 57 inches. Bombay has between 72 and 106 inches; Saint-Domingue gets 113 inches. Mr. Antonio Bernardino Pereira [do] Lago, an infantry colonel in the corps of engineers, insists that he measured 23 feet, 4 inches and 9.7 lines (English measurements), that is, about 260 French inches, of rain in San Luis de Maranhão (lat. 2° 29′ south.) in 1821 alone. Although I am inclined to doubt such a prodigious amount of water, I do have in my possession the barometer, thermometer, and rain gauge readings that Mr. Pereira insists he made *each day during three different periods*. These Brazilian observations were published in the sixteenth volume of the *Annaes das Ciências, das Artes e das Letras*, pp. 54–79. The observer, describing the instruments he used in the *resumo das observaçoes meteorologicas*, explicitly stated that the plate with which he collected the rainwater had the exact same diameter as the cylinder that contained the scale. This diameter was only 6 inches (English). I hope that this important observation can be verified in Maranhão and in other parts of the tropics where rain is abundant, for example, on the Río Negro, in Chocó, and on the Panama Isthmus. The amount that Mr. Pereira [do] Lago noted is 2.5 times greater than what was observed on average in Saint-Domingue. But the amount of water that falls on the west coast of England exceeds by three times what is annually recorded in Paris. There are vast differences between very close latitudes. Captain Roussin reported that 151 inches of rain fell on Cayenne during the month of February alone! (Arago in *Annales du Bureau des Longitudes*, 1824, p. 166; Prony, *Description hydrographique et historique des marais Pontins*, pp. 33, 110, 116.)

in the Pontine Marshes and that Mr. Pin and Mr. Clauzade[1] undertook on the Languedoc Canal showed 348 lines of evaporation each year at 41° and 43½° latitude. I did not conduct enough experiments in the tropics to arrive at a generalizable result. But, assuming that the weather is equally calm on the French Riviera and in the Torrid Zone, that the mean annual temperature is 15° and 27° C, and that the average relative humidity is 82° and 86° (according to a ▾hair hygrometer) respectively, I would conclude, with Mr. Gay-Lussac, that proportionate evaporation in the two zones is 1 to 1.6, while the amount of rain that each receives is 1 to 4. Furthermore, it must not be forgotten that canals not only lose water through evaporation but that they receive all the water that falls on the vast stretches of terrain that surround them. One must distinguish between two kinds of water volume that hydraulic structures require: the water in the entire canal, that is, its width and its length, and the water in the locks, that is, the *replenishment amount*[2] of each lock, or the amount of water that drops from a higher lock to a lower lock each time that a boat passes through. Both lose water to evaporation and to filtering, although losses due to filtration (very difficult to estimate) decrease over time. The length and depth necessary for an *oceanic canal* in the New World thus affects the amount of water needed to fill it at the beginning when the excavations have just been completed, or after necessary repairs have been made. But disregarding losses due to filtering and evaporation, the amount of water needed to fill a canal each year depends only on the volume and the number of the *locks*, that is, the volume of the *replenishment amount* and the level of traffic. I am emphasizing these technical considerations to dispel the fear that there might not be enough water to fill an oceanic canal of some length. If such a canal were also to serve small boats engaged in inland trade, smaller locks could be added to the larger ones to save water. This was done with the Grand Junction Canal, and there have been plans to do the same with the Caledonian Canal for some time now.[3] It seems rather likely that the

II.331 II.332 II.333 II.334

1. Ducros, *Mémoires sur les quantités d'eau*, 1800, no. 2, p. 41.

2. In segmented locks, the *flotation amount* [lockage water], or the volume of water in which the ship floats up or down during its passage from one lock to another, needs to be taken into account. More water is needed during the ascent than during the descent, and the distribution of chutes and the height of the locks influences greatly how much water a canal needs (Ducros, *Mémoires sur les quantités d'eau*, p. 39. Prony, in Mr. de Pommeuse's work [*Des Canaux navigables*], p. 23; ▾Girard, in *Annales de Physique et de Chimie*, 1823, Vol. XXIV, p. 137).

3. The capacity of the Languedoc Canal, or the amount of water necessary to fill the entire canal, is 7 million cubic meters, according to Mr. Clauzade's calculations. 14 million cubic meters of water are used annually for 960 passages in each direction. This level

II.335 province of Nicaragua will be chosen for this link between the two Oceans, in which case it will be easy to establish an unbroken navigable route. The isthmus to be crossed is only 5 to 6 nautical leagues. It is encumbered by some hills where it is narrowest, between the west bank of Lake Nicaragua and the Papagayo Gulf. But it otherwise consists of vast expanses of savannas and plains that have excellent thoroughfares for carts and carriages[1] (*camino carretero*) between the town of León and the Realejo coast. Lake Nicaragua is perched above the South Sea for the entire length of the Río San Juan, which runs for 30 leagues. The basin's elevation is so well known in the country that it was once seen as an insurmountable obstacle to the construction of a canal. People feared either a sudden flooding to the west or a lowering of the San Juan's waters where, during times of drought, rather dangerous rap- II.336 ids develop above the old Castillo de San Carlos.[2] Nowadays, the skill of engineers and builders has become sufficiently perfected that they do not fear such dangers. Lake Nicaragua might serve as the upper basin, much like Loch

---

of water use, caused by locks that are a little too big and by heavy traffic from small craft, stands in a 2 to 1 ratio with the canal's capacity. 3.5 million Cubic meters per year are required to replenish the water after shutting down the canal up to Fresquel River, and this water flows over 9 days from the upper basin or the artificial source (Andréossy, p. 256; Pommeuse, pp. 258 and 265). Total evaporation from the canal, the reservoirs, and the feeders during the 320 days of navigation is estimated at 1.9 million cubic meters (Ducros, *Mémoires sur les quantités d'eau*, *p.* 41). Comparing the Caledonian Canal to the Languedoc Canal, I find that their surface areas are 5 to 1; the length of each of the canal's dugout parts (excluding the route through Scotland's lochs) is 1 to 6.5. It follows from these comparisons that the capacities of the two canals—one accommodating flat-bottomed boats that can carry between 100 and 120 tons, the other accommodating 32-gun frigates—are nearly the same. The difference in how much water is used in a lock depends on *replenishing and flotation amounts*. The Caledonian Canal's floodgates are 37 feet wide and the locks are 160 feet long; the Languedoc Canal's locks are 31 feet wide in the middle, 20 feet at the gates, and 127 feet long. We saw above that the dimensions of an American interocean canal can be smaller than that of Scotland's great canal.

1. This is the principal route by which merchandise is sent from Guatemala to León. The merchandise is unloaded in Conchagua port in the Golfo de Fonseca, or Amapala.

2. ▾This small fort, captured by the British in 1665, is popularly known as El Castillo del Río San Juan. According to Mr. Juarros, it is situated 10 leagues away from the eastern edge of Lake Nicaragua. In 1667, another fortress, called *Presidio del Río de San Juan*, was built on a rock at the river's mouth. Already, during the sixteenth century, the *Desagüadero de las Lagunas* had attracted the Spanish government's attention. They ordered ▾Diego López Salcedo to found the town of Nueva San Jaén near the left bank of the *Desagüadero* or Río San Juan. But the town was soon abandoned, just like the town of Brussels (*Bruselas*) near the Nicoya Gulf. In their current undeveloped state, the banks of the Río San Juan are disease ridden.

Oich does for the Caledonian Canal, and regulating locks will allow only the amount of water that is required to fill them. As I have shown elsewhere, the slight discrepancy in sea level between the Antillean Sea and the Pacific Ocean is due only to the difference between high and low tides. There is a similar difference between the two seas that Scotland's great canal connects. Even if tides were six toises and constant, as they are between the Mediterranean and the Red Sea,[1] conditions would be no less favorable for building a canal. The wind is strong enough on Lake Nicaragua to render it unnecessary for ships to be towed by steamboats from one sea to the other. But steam engines' power would be of great use for traveling from Realejo and Panama to Guayaquil, where, during the months of August, September, and October, calm winds alternate with crosswinds. II.337

In my ideas for linking two seas, I have assumed only the simplest means for the implementation of such a vast project. Steam pumps that fill the lock chamber and underground tunnels, like those proposed for the mountainous part of the Panama Isthmus and those of the Saint Quentin Canal which are 2,900 toises long,[2] are preferable for routes of inland navigation. Suffice it for me to demonstrate the possibility of an oceanic canal in Central America. Estimated construction costs for the *earthworks* (excavations and landfill), for the locks, for the basins, and for the feeding canals depend on the choice of locale. The Caledonian Canal, the most impressive structure built to this day, cost nearly 3.9 million piasters. That is still 2.7 million piasters less than the Languedoc Canal,[3] adjusting the value of the silver mark down to current exchange rates. ▾During Bonaparte's Egyptian campaign, Mr. LePère projected that expenses for the Suez Canal would rise to 5 or 6 million piasters, a third of which would be spent on the secondary Cairo and Alexandria canals. The Suez Isthmus, including the part not bounded by the sea, is 59,000 toises II.338 II.339

1. Even the ancients overcame the difference in sea level between the Red Sea and the Pelusiac [former eastern] branch of the Nile, even though they did not know locks. They knew at most how to plug their *euripes* [Gr. waterways] with wooden beams.

2. This *tunnel* is 15 feet wide. According to ▾Mr. Laurent's plan, the underground canal would have been 7,000 toises long without interruption (nearly 3 leagues), 21 feet wide and 24 feet high. Its length would have surpassed by one-sixth the famous gallery of mines in Clausthal (the Georg Stollen) in the Harz Mountains. To recall the kinds of underground structures that men can build, let me also mention the two large drainage galleries in Freiberg's mine district in Saxony: one is 29,504 toises, the other 32,433 toises. If this latter had been dug in a straight line, it would be almost twice the width of Pas-de-Calais, the Straight of Dover.

3. Pommeuse, [*Des Canaux navigables*,] *p.* 308. Canal maintenance, moreover, cost 25.67 million francs between 1686 and 1791 (see General Andréossy's thoughtful work, *Histoire du Canal du Midi*, p. 345).

wide (more than 20 nautical leagues), and the projected canal with its 4 locks[1] II.340 would be able to accommodate ships with a draft between 12 and 15 feet during many months of the year—as long as the Nile's rainy season lasts. Even assuming that the cost of connecting the two oceans in the New World was as expensive as building the Languedoc, Scottish Highlands, and Suez Canals, I do not think that this consideration should delay the implementation of such a grand structure. Already the New World offers many examples of equally impressive works. New York State constructed a canal longer than 100 leagues between Lake Erie and the Hudson River in only 6 years. In a report addressed to the state legislature, costs were estimated at about 5 million piasters.[2] II.341–42 When one takes in all the gigantic but unexceptional structures built over the course of two centuries to lower the lakes' water in the valley of Mexico, it is apparent that the same amount of effort could cut through the Nicaragua and Coatzacoalcos Isthmuses, perhaps even the Panama Isthmus, between Gorgona (on the Río Chagre) and the South Sea coasts. In 1607, a subterranean canal 3,400 toises long and 12 feet deep was dug in northern Mexico on the back slope of Nochistongo hill. The viceroy, the ▾Marquis de Salinas, rode through half of it on horseback. The open trench (*tajo de Hue-*

1. *Déscription de l'Égypte* (*État moderne*), 1808, Vol. I, pp. 50, 60, 81, and 111. The ancient canal that connected the Red Sea to the Nile (*the canal of kings*), navigable, if not under the Ptolemaists, at least under the Khalifes, was only a detour from the Pelusiac branch near Bubastis; it was 25 leagues long. Its depth sufficed for heavy, seaworthy ships; it seems to have been at least 12 to 15 feet deep.

2. ▾Warden, *Description statistique, historique et politique des États-Unis de l'Amérique septentrionale*, Vol. II, p. 197; Morse, *New System of Modern Geography*, 1822, p. 122. This canal, 294,590 toises long, is only 4 feet deep (two-thirds the depth of the Languedoc Canal, which is half as long). Lake Erie is 88 toises above the Hudson River's average water level. First, the boats descend evenly through 25 locks from Buffalo on Lake Erie to Montezuma on the Seneca River (passing through Palmyra and Lyons), a length of 166 English miles and 30 toises of perpendicular descent. Then, they climb 8 toises from Montezuma to Rome, on the Mohawk, for 77 miles; finally, they descend once more, without interruption, 66 toises by means of 46 locks, over 113 miles from Rome to Albany on the Hudson River by way of Utica. The total descent is, therefore, 9 toises less than a boats' descent from the highest level on the Languedoc Canal to the Mediterranean. I recall here that this is the *maximum* slope that I ascended *on a naturally navigable pathway* on one of the greatest rivers in South America without cataracts or rapids. To get from Cartagena de Indias to Honda, one has to row down the Río Magdalena, negotiating a total drop of 135 toises. This drop is fifty percent greater than the descent from Lake Erie to the Hudson River, but the Río Magdalena's navigable course is a third longer. When one considers the river's gentle decline between Morales and its mouth, it becomes apparent that a boat travels 80 leagues down a naturally navigable route without locks to arrive at a plateau of 100 toises, which amounts to a 0.43 toises drop over a 1,000 toise waterway.

huetoca), which currently carries water out of the valley, is 10,600 toises long. A considerable part of it runs through a transportation hub. The trench has 140 to 180 feet of perpendicular depth [i.e., at a right angle] and is 250 to 330 feet wide near the top. Between 1607 and the time when I visited in January 1804, the cost of all hydraulic works[1] by the *Desagüe [drain] of Mexico* had risen to 6.2 million piasters. How can we worry about not raising enough money for an oceanic canal when we recall that only the family of the ▾Count of Valenciana had the courage to dig four mine shafts[2] in Guanajuato that cost 2.2 million piasters altogether? Even assuming that the annual cost of cutting through the isthmus reached seven or eight hundred thousand piasters for a certain number of years, stockholders could easily float this sum, as could the various American states that would reap invaluable commercial benefits from a new sea route to northern Peru, Quito's west coast, Guatemala, and Mexico, toward Nootka, the Philippine Islands, and China. II.343

Enlightened members of the new governments of equinoctial America recently consulted me on how to carry out this plan. I think that a company that issues shares should not be formed until the possibility has been put in place of building, between 7° and 18° northern latitude, an oceanic canal large enough to accommodate ships of three to four hundred tons and until its place and location have been fixed. I will refrain from commenting on whether this area "should constitute a separate republic under the name ▾*Junctiana*, under the jurisdiction of the federation of the United States," as was recently proposed in England by a man whose intentions are generally laudable and unbiased. Whichever government lays claim to the soil where the grand sea junction will be established, the benefits of this hydraulic work should belong to every nation of the two worlds that contributes to its construction by acquiring stock. Local governments in Spanish America can survey locations, level ridges, gauge distances, sound lakes and rivers, and estimate the river and rainwater necessary to fill the upper basin. These preliminary labors are not very costly, but they must be undertaken systematically on the Isthmuses of Tehuantepec or Goazacoalcos, Nicaragua, Panama, Cúpica or Darién, and Raspadura, or Chocó. Once maps and profiles of the five territories can be put before the public's eyes, more people on both continents will be persuaded that a link between the two oceans is indeed possible; this would help spur the formation of a public corporation. A free discussion will clarify the II.344 II.345

1. I gave the detailed history of these works, based on official documents, in my *Essai politique sur le royaume de la Nouvelle-Espagne*, Vol. I, pp. 204–35.

2. *Tiro Viejo, Santo Cristo de Burgos, Tiro de Guadalupe,* and *Tiro general*, whose depths are 697, 460, 1,061, and 1,582 feet, respectively (old French unit of measure).

advantages and disadvantages of each locale, and soon, one or two locations will emerge. The *canal corporation* will undertake a second, more rigorous examination of local conditions. Expenses will be estimated, and the implementation of this important work will be entrusted to engineers who have actually competed with each other over similar projects in Europe.

If an *oceanic canal* proves unworkable, one could dig *section canals* in some of the five places we have just named to facilitate inland trade. Because such projects would be highly profitable to stockholders, it would perhaps II.346 make sense to charge the initial survey to a corporation. A ship would bring engineers and instruments, one after the other, to the deltas of the Río Atrato, the Río Chagre, and Mandinga Bay, to the Río San Juan and Lake Nicaragua, and to the Coatzacoalcos (or Tehuantepec) isthmuses. Systematic surveying will accelerate these operations; so will an appreciation of the different advantages of the sites to be compared. After having established the best locale and gauged the dimensions of the structure according to the tonnage of the ships or boats to be used, the *first survey corporation* could appeal to the public for more funds and constitute itself as an *implementation corporation*, either for, hopefully, a *transoceanic canal* or for *secondary (inland) waterways*. The method of implementation that I have just outlined is appropriately prudent for an affair that concerns commerce between the two worlds. The *canal company* will find shareholders among those governments and citizens who, II.347 unswayed by the lure of profit and giving way to more noble instincts, will pride themselves at the thought of contributing to a project that is worthy of modern civilization. Of course, it is wise to recall here that the profit motive, the basis of all financial speculations, is inevitably part of this enterprise, which I enthusiastically support. The dividends from companies that obtained concessions to open canals in England prove the profitability of these enterprises for shareholders. With a transoceanic canal, cargo duties can be much higher for ships that wish to benefit from new passages to Guayaquil and Lima, the sperm whale hunting grounds, America's northwest coast, or to Canton to shorten their route and avoid the high southern latitudes, which are frequently dangerous during bad weather. Traffic on the canal will grow as trade further familiarizes itself with the new route from one Ocean to the other. Even if dividends are not that high and the capital invested in this enterprise does not yield returns equal to those that many government backed II.348 bonds offer, it would still be in the interest of Spanish America's large states, from the Mosquito Coast all the way to Europe's final outposts, to support this enterprise. To reduce the usefulness of canals and great trade routes to the revenue from passage tolls, without considering the overall stimulus to national industry and prosperity, is to forget centuries old lessons from experience and political economy.[1]

Careful study of the history of trade among peoples shows that the direction of the India trade changed not only because of growing geographical knowledge and advances in navigation but also, in great measure, because of a general displacement of world civilization. From the Phoenician period to the British Empire's ascendancy, trade has shifted steadily from east to west, from the Mediterranean's east coast to the western edge of Europe. If this movement continues west, as everything indicates it will, the debate over how best to round Africa's southern edge to reach India will not have the same importance. The Nicaragua Canal has advantages for ships leaving the mouth of the Mississippi distinct from those it offers to ships that load up on the banks of the Thames. When comparing the various routes around the Cape of Good Hope, around Cape Horn or across one of Central America's isthmuses, one must carefully distinguish among the goods and the different positions of the people involved. The problem trade routes pose is different for a British trader than for an Anglo-American merchant. Similarly, this important problem is resolved variously by those involved in direct trade with Chile, India, and China, and those whose speculations are directed either toward northern Peru and the west coasts of Guatemala and Mexico, or toward China by way of the American northwest, or toward the sperm whale hunting grounds in the Pacific Ocean. The creation of a canal most clearly favors these latter three objects of European and United States' maritime commerce. By way of the proposed Nicaraguan Canal, one travels[2] 2,100 nautical leagues from Boston to Nootka, the old center of the otter pelt trade in the American northwest. The same journey takes 5,200 leagues via Cape Horn—the still current trade route. For a vessel coming from London, these respective distances are 3,000 and 5,000 leagues. These data would mean a shorter route for U.S. Americans by 3,100 leagues; for the British, by 2,000 leagues. This is not to take into account either the possibility of crosswinds or the very different dan-

II.349 II.350 II.351

1. It is in light of this beneficial influence that one should appreciate the perhaps too costly construction of Languedoc Canal, which cost 33 million francs, and which yields only 800,000 francs annually, out of a gross revenue of 1.5 million francs. This is barely a 2.5 percent return on capital investment. This is also the Central Canal's net profit.

2. I based these distance estimates, calculated in collaboration with ▾Mr. Beautemps-Beaupré (chief engineer geographer of the royal navy), on more or less direct routes. This was sufficient for arriving at comparative figures. To obtain travel distances, it would be necessary to take into account crosswinds and currents and increase the duration of the travel by about one fourth or one fifth.

gers of the two trade routes that we are comparing here. When it comes to direct trade with India or China, the comparison comes out far less in favor of sailing across central America in terms of the relation between time and distance. Vessels that normally sail from London to Canton around the Cape of Good Hope, crossing the equator twice, travel 4,400 leagues; from Boston to Canton, 4,500 leagues. If the Nicaragua Canal were built, these respective trade routes would be 4,800 and 4,200 nautical leagues.[1] With the current advances in navigation, the typical duration of a trip from the United States or from England to China, around the edge of Africa, lasts between 120 and II.352 130 days.[2] Applying these figures to voyages from Boston and Liverpool to the Mosquito Coast, and from Acapulco to Manila,[3] one finds that it takes 105 to 115 days to sail from the United States or England to Canton staying entirely within the northern hemisphere without ever crossing the equator, that is, by taking advantage of a canal through Nicaragua and of the constant II.353 trade winds in the most peaceable part of the Great Ocean.[4] The difference in time would barely be one sixth. One cannot return by the same route, but the voyage out would be safer during every season. I think that a nation with fine settlements on the southern tip of Africa and on the Île de France would generally prefer the route from west to east. The canal's principal and true benefits would be faster passage to America's west coasts,[5] a faster trade route

1. 5,800 leagues from London to Canton around Cape Horn, or 1,400 more than around the Cape of Good Hope; 5,900 leagues from Boston to Canton around Cape Horn.

2. In Boston, there have been some rare examples of 98 days. Warden, *Description statistique, historique et politique des États-Unis de l'Amérique septentrionale*, Vol. V, p. 596.

3. For Galleons, add 40 to 60 days. See my *Essai politique sur le royaume de la Nouvelle-Espagne*, Vol. II, p. 720, and Tuckey, *Maritime Geography*, Vol. III, p. 497.

4. Steam power has not been taken into account in these time estimates. The French engineers who estimated the costs of the Suez Canal assumed that a trade route from the French territories in India through the projected canal would cut half the distance and one third or one fourth of the time from the Cape of Good Hope route. *Description de l'Égypte* (*État moderne*), Vol. I, p. 111. Precise calculations concerning the *average duration* of a trip from London to Calcutta and Canton, and from Liverpool to Buenos Aires and Lima (and *vice versa*), would be highly desirable. This information should be gathered over an adequately long period so that variations due to the influence of seasons, winds, currents, ships' constructions, and navigational error might be minimized in the overall averages. Duration of travel is one of the most important variables in the movements of commercial peoples—a vital movement that grows from century to century as the art of navigation is perfected.

5. Exceptions are the Peruvian coastline south of Lima and the Chilean coastline which are very difficult to navigate from north to south. It would be faster to go from Europe to Valparaiso and to Arica [a town in Chile] by way of Cape Horn than through a Nicaraguan Canal. The canal would not be advantageous for trade on the west coast south of Lima, un-

from Havana and from the United States to Manila, and speedier expeditions from England and Massachusetts to the fur coast (the northwest coast) or the islands in the Pacific Ocean, from where they then visit the markets of Canton and Macao. II.354

I add to these commercial considerations a few political thoughts about the effects that the projected transoceanic canal might have. Such is the state of modern civilization that world trade cannot experience large changes without those changes affecting social organization. If we succeed in cutting through the isthmus that joins the two Americas, East Asia, hitherto isolated and unattainable, will, despite itself, enter into the most intimate contact with the peoples of the European race who inhabit the Atlantic seaboards. One might say that the strip of land that blocks the equinoctial current was, for centuries, China's and Japan's bulwark of independence. It is imaginable that in the future there will be a clash between mighty peoples over the desire for exclusive access to the new trade route between the two worlds. I admit that it is neither my confidence in the moderation of monarchial and republican governments nor my sometimes somewhat shaken faith in the progress of enlightenment and in honest profits that assuages this fear. If I abstain from discussing such remote political developments, it is to avoid entertaining readers with the pleasure of an idea that exists only in the minds of a few men interested in the public good. II.355 II.356

Contrary to what is stated in some very recent works, Lake Nicaragua and the Río San Juan do not belong to New Granada's territory. The lake is separated from the Colombian territory of Veragua by the province of Costa Rica, the southernmost province in the former kingdom of Guatemala. Situated in a very thinly settled region, particularly on its east coast, almost on the border of two independent central and southern American states, the large structures that would join the oceans could count for their defense only on Porto-

---

less the ferrying were done by steam ships. These days, North America's trade with China is pursued in three ways: (1) ships from the United States loaded with piasters sail directly from New York or Boston to Canton around the cape of Good Hope, to buy tea, nankeen, silks, porcelains, etc., and come back the same way; (2) ships are sent around Cape Horn either to hunt seals and sperm whales in the Pacific Ocean, or to visit America's northwest coast. If they do not acquire enough fur, they take on sandalwood or ebony in Polynesia. They take these goods to Canton and return around the Cape of Good Hope; (3) other ships engage in shady trade over many years, visiting in turn Madeira, the Cape of Good Hope, and Île de France, or New South Wales, some ports in South America and the Pacific islands. During the voyage, they round either the Cape of Good Hope or Cape Horn. But because they always reach Canton at the end of this long voyage, they return to the United States by way of the southern edge of Africa. A canal would greatly affect the last two routes we have just charted.

belo and Cartagena, two fortresses downwind from the Castillo de San Juan in Nicaragua. There is probably also a land route from Guatemala to León, but it is more than 135 leagues away. As things stand now, it is less the fortified places than the poverty of the countryside, its lack of cultivation and the dense vegetation between Darién and the 10° or 11° northern parallel that II.357 would render fruitless any invasion of an enemy who might inopportunely land on the east coast. In the course of addressing this important question, I cannot rely on a more authoritative testimony than that of ▾General Don José Ezpeleta, New Granada's viceroy until 1796. In an unpublished report that I have in my possession and that is addressed to his successor, ▾Don Pedro de Mendinueta,[1] this seasoned soldier expresses his thoughts on the defense of the Panama Isthmus: "Your Excellency knows that the king, our lord, had the brigadier Cramer survey these vast American possessions. This famous engineer weighed the dangers that we still run and pointed out the fortifications that we should put up against the enemy. The Isthmus of Panama is an object of the highest military importance, which Your Excellency should not lose from His sight for a single moment. This importance is based on its geographical configuration and on its proximity to the South Sea. It offers three points of defense: toward the north, Portobelo, and the fort of San Lorenzo de II.358 Chagre; and toward the south, Panama City. The bluffs dominating Portobelo make for an effective fortification of this poor and sparsely populated city impossible. The batteries of San Fernando, Santiago, and San Gerónimo seem adequate to me for the port's defense. Fort Chagre at the mouth of the river of the same name is, in my opinion, the isthmus's principal point, always assuming that an attack would come from the north. However, neither the capture of Portobelo nor the fall of the fort of San Lorenzo de Chagre determines control of the Panama Isthmus. The country's true defense consists in the difficulty any sizable expedition would have in penetrating the interior. On the southern coasts, which are entirely uninhabited, this difficulty would exist even for two or three isolated travelers."

After having discussed the area of the united Provinces of Venezuela, their population, products, and trade both in their current state and in their projected future growth, it remains for me to discuss finances, or state revenue. II.359 This subject is of such political importance because it is one of the pillars of a government's existence. But, after ▾prolonged civil strife—a thirteen-year war that devastated both the agricultural sector and trade relations and dried up the principal sources of public revenue—we find ourselves with an en-

1. [Mendinueta,] *Relación del Gobierno*, Part four, Chapter III, fol. 118, 122, 123 (manuscript).

tirely transitional state of affairs that has little to do with the country's natural wealth. It is necessary to go back again to the period before the revolution to find a sounder point of departure for judging the state of affairs in a time of confidence and peace. Between 1793 and 1796, annual liquid revenue averaged 1,426,700 piasters a year, not including the tobacco monopoly. Adding to this number 586,300 piasters for the monopoly's net profit (the average for the same period), the revenue of the *Capitanía general de Caracas* was 2,013,000 piasters, excluding collection costs. In the last years of the eighteenth century and the first years of the nineteenth, this revenue shrank because of disruptions in the maritime trade. But from 1807 to 1810, it rose to more than 2.5 mil- II.360 lion piasters (1.2 million piasters from customs; 700,000 piasters from the tobacco monopoly; and 400,000 piasters from the land and sea ▾*alcabala*. All these revenues were offset by government expenditures. From time to time, there were surpluses (*sobrante líquido*) of 200,000 piasters that reverted to the Madrid treasury, but such payments were extremely rare. Once Caracas no longer received *situados* [subsidies] from New Spain, it was forced periodically to draw on Santa Fé's equally meager funds. According to my research, gross revenue for all the provinces that currently constitute the republic of Colombia rose to a *maximum* of 6.5 million piasters at the time of the revolution.[1] The metropole's government never drew more than one twelfth of these funds. In my *Political Essay on the Kingdom of New Spain*, I demonstrated that, when mining and trade were most active, the Spanish colonies in America *had gross revenues of thirty-six million piasters, of which the colonies'* II.361 *domestic administration absorbed nearly twenty-nine, and only seven to eight million piasters of which went into Madrid's royal treasury*. According to these data, which are derived from official documents and whose accuracy has not been called into doubt for fifteen years, it is surprising to see that, during serious discussions of political economy, the metropole's financial troubles are still so often attributed to the colonies' emancipation. Throughout America, taxes on imports and exports are the primary sources of public revenue. This revenue has become gradually more plentiful ever since the court stripped the ▾Guipúzcoa Company of its commercial monopoly in Venezuela. This was, in a *royal cédula*'s odd turn of phrase, a company "in which anyone can take part without compromising one's nobility and *without losing one's honor or reputation*." If one bears in mind that Havana customs alone brought in more than three million piasters during these past years, and if at the same time one considers Venezuela's territorial expanse and agricultural bounty, then II.362

1. In his report to the Bogotá congress (May 5, 1823), ▾Don José María del Castillo estimates the *rentas ordinaries* at only 5 million piasters.

the continual growth of public revenue in this beautiful part of the world appears assured. But the realization of this hope, along with all other hopes that we have just expressed, depends on the return of peace, wisdom, and institutional stability.

In this chapter, I have sketched the components that I had the opportunity to bring together in statistical form over the course of my travels and thanks to my continuing communications with Spanish-Americans. As a historian of the colonies, I have presented the facts in all their simplicity, for the attentive and accurate study of facts is the only way[1] in which to dispense with vague conjectures and vain declamations. Such circumspection is particularly indispensable when one fears giving in too easily to the lure of hope and the tug of old affections. Emerging societies have something of the charm of youth; they share youth's refreshing sentiments, its naïve trust, and even its credu- II.363 lity. They offer a more attractive spectacle to the imagination than the old peoples, who with their grouchiness and defiant austerity, seemed to have frittered away all happiness, hope, and faith in human perfectibility.

Venezuela's great struggle for independence lasted for more than twelve years. Like most periods of civil unrest, this period was filled with all the heroism, generosity, shame, and turmoil that ignited passions produce. The sense of common danger solidified connections between men of various races, who, whether spread over the Cumaná steppes or isolated on the Cundinamarca plateau, are as physically and morally diverse as the climate in which they live. At several points, the metropole regained control of certain districts; but these conquests were only ephemeral, as revolution always erupts again with ever greater violence when the underlying evils that caused them in the first place persist. To aid defense and make it more robust, power was central- II.364 ized and a vast state formed, from the Orinoco delta beyond the Riobamba Andes and the banks of the Amazon. The *Capitanía general* of Caracas was joined to the viceroyalty of New Granada, from which it had separated only in 1777. This reunion, which will always be indispensable for external security, this centralization of power in a state six times larger than Spain, was motivated by political alliances. The new government's smooth operation has vindicated the wisdom of these moves, and the congress will find fewer obstacles to the implementation of its projects to benefit national industry and civilization, provided that it allows greater freedom for the provinces and makes them feel the advantages of institutions that they signed into existence with their own blood. In every form of government that has existed to this day, in republics as well as in moderate monarchies, improvements need to be

1. [Chabrol,] *Recherches statistiques sur la ville de Paris*, 1823, Introduction, pp. I and V.

gradual if they are to be beneficial. New Andalucía, Caracas, Cundinamarca, Popayán, and Quito have not become federated states like Pennsylvania, Virginia, and Maryland. Without provincial juntas or legislatures, all parties are directly subject to Colombia's congress and government. According to the constitution (art. 152), the president of the republic appoints departmental and provincial ˅intendants and governors. Naturally, such dependence is not always to the liking of those who prefer to discuss local concerns freely among themselves, and it has also occasionally stirred up discussions that one might deem geographic in nature. The former kingdom of Quito, for example, is close to both Peru and New Granada in its customs and the language of its mountain people. If Quito had a provincial junta and paid the congress only those taxes necessary for Colombia's defense and general welfare, the feeling of an autonomous political existence would make inhabitants less interested in the central government's place or seat. The same reasoning applies to New Andalucía and Guyana, which are governed by intendants whom the president appoints. It can be said that, until now, the provinces find themselves in a position little different from those of the United States' *territories* with populations still below 60,000. Peculiar circumstances that are difficult to grasp at a distance have probably made it necessary to centralize a civil administration to such an extent; any change would be dangerous as long as there are external enemies. But what is useful for defense is not always what sufficiently fosters individual liberties and promotes public wealth once the struggle is over. History itself proves that, when these difficulties are not carefully and wisely resolved, they have more than once become shoals upon which the people's affection and enthusiasm foundered. Without breaking the ties that should bind Colombia's various territories (Venezuela, New Granada, and Quito), partial life could spread little by little throughout this large political body—not to fragment but to invigorate it.

II.365

II.366

The powerful North American union has remained isolated for a long time without associating with states that have similar institutions. Although, as we already mentioned above, its westward progress has slowed considerably on the right bank of the Mississippi, it will nevertheless continue its march until it reaches Mexico's *inland provinces*, where it will encounter a European people of another race, other customs, and a different religion. Can the sparse populations of these provinces, belonging, as they do, to another emerging federation, resist, or will they succumb to the same westward torrent that transformed the inhabitants of lower Louisiana into citizens of an Anglo-American state? The very near future will resolve this issue. Elsewhere, only Guatemala, an unusually fertile country that has become a Central American republic, separates Mexico from Colombia. The political divisions between Oaxaca,

II.367

Chiapas, Costa Rica, and Veragua are founded neither on natural borders nor on indigenous customs or languages; they are based solely on a habitual dependence on Spanish authorities located either in Mexico, Guatemala, or II.368 Santa Fé de Bogotá. It seems quite natural that Guatemala may one day join the Isthmuses of Veragua and Panama to that of Costa Rica. Quito connects New Granada with Peru, much like La Paz, Charcas, and Potosí connect Peru with Buenos Aires.[1] The intermediary parts that we have just named form a bridge from Chia to the Cordilleras of Upper Peru, from one political entity to another, like the transitory forms through which various groups of organisms in nature link each other. In neighboring monarchies, adjacent provinces have always led to those distinct demarcations that are the effects of greatly centralized power; in federated republics, the states on the edges of each system oscillate for some time before finding a stable balance. It almost does not matter to the provinces between the Arkansas and the Río del Norte [Rio Grande] whether they send their deputies to Mexico City or to Washington. If, one day, Spanish America showed a more collective predilection for the II.369 same federalism that the United States has already exhibited in several ways, a whole spectrum of federations would emerge from the interaction of so many systems and groups of states. I only touch upon the relationships born of this singular assemblage of colonies stretched over an uninterrupted line of 1,600 leagues. In the United States, we have seen an old Atlantic state split into two, each having distinct representation. The separation of Maine and Massachusetts in 1820 occurred in the most peaceable manner. Splits of this sort will probably take place with some frequency in the Spanish colonies. Given local customs and mores, it is, however, to be feared that they will be more turbulent. When a people of the European race naturally inclines toward provincial and municipal independence, and when the copper-colored natives have an equally pronounced taste for political fragmentation and for the freedom of small communities, then the best form of government is that which knows how to render such a national penchant less harmful to the general interest and to the unity of the body politic without attacking it directly. There is more still: II.370 the importance of geographical divisions in Spanish America, based on the relationships between both places and customs established over many centuries, have kept the metropole either from preventing or slowing the separation of the colonies by trying to install ˇSpanish Princes in the New World. To rule such vast possessions, six or seven centers of government would have been needed, and this multiplicity of centers (viceroyalties or Captaincies-general)

1. Vol. XI, p. 226.

would have been opposed to the founding of new dynasties at a time when they might have had some benefit for the metropole.

▾Bacon[1] said in one of his political aphorisms: "It were good therefore that men in their innovations would follow the example of time itself; which indeed innovateth greatly, but quietly, and by degrees scarce to be perceived." This good fortune is not vouchsafed in colonies when they arrive at the critical moment of their emancipation. This was even less the case in Spanish America, thrown into the struggle to extricate itself from foreign domination, not, at first, to gain total independence. May a lasting peace replace partisan strife! May the seeds of civil discord, sown for three centuries to assure the dominance of the metropole, be gradually smothered! And may the states of industrial and mercantile Europe learn once more that prolonging the New World's political unrest has hurt their economies by suppressing demand for their goods and by depriving themselves of a market that has already reached more than 70 million piasters a year! Current exports from Spanish America, the United States, France, and Great Britain can be represented by the following numbers: 100, 103, 140, and 375.[2] II.371

Many years will probably pass before 17 million inhabitants, spread out over an area one fifth larger than all of Europe, will achieve stability through self-government. The most critical moment comes when people who have been oppressed for a long time suddenly find themselves free to arrange their lives to pursue their own prosperity. It has been said time and again that Spanish Americans are not culturally advanced enough for free institutions. I remember that, not so long ago, the same reasoning was applied to other II.372 II.373

1. See the article on innovations in Bacon, *Essays Civil and Moral*, no. 25 (*Opera omnia*, 1730, Vol. III, p. 335).

2. I have shown elsewhere (*Essai politique sur le royaume de la Nouvelle-Espagne*, Vol. II, pp. 748–49) that, according to the most conservative estimates, Spanish America already in 1805 needed 59 million piasters worth of foreign imports. This is nearly three times more than the United States was offered eight years after Great Britain had recognized their independence. To give a sense of comparable figures, I mention imports and exports for the two greatest trade nations in the world: the British from both Europe and America. Between 1821 and 1823, annual imports from Great Britain rose to 30,203,000 pounds sterling; exports rose to 50,636,800 pounds sterling. In the United States, exports were 64,974,000 dollars in 1820; imports, 62,586,000 dollars. In an earlier period, from 1802 to 1804, exports were 68,461,000 dollars in an average year; imports were 75,306,000 dollars. It follows that imports from the United States and from Spanish America were equally impressive in the year immediately before civil unrest broke out in the latter region. It should not be forgotten that everything imported to Spanish America is completely consumed there, and not reexported. In 1821, exports and imports for France were 404,764,000 and 394,442,000 francs, respectively.

peoples who were said to be too civilized. Experience no doubt proves that with nations, as with individuals, talent and knowledge are often irrelevant to happiness. But without denying the need for a certain amount of enlightened and popular instruction for the stability of republics or constitutional monarchies, we believe that this stability depends much less on the level of intellectual culture than on the force of national character, a mixture of energy and calm, passion and patience, that sustains and perpetuates institutions. It also depends on the local circumstances of a people and, finally, on the relations that a state has with its neighbors.

If modern colonies at the moment of their independence manifest a more or less pronounced tendency in favor of republican forms, the cause of this II.374 phenomenon should not be exclusively attributed to the principle of imitation that works on crowds even more than it does on an isolated person. It is, above all, founded on the position in which societies find themselves when they are suddenly cut off from a world with a longer history of civilization, free of all external ties, composed of individuals who do not recognize any class as having a natural monopoly on power. Titles that the mother country has bestowed on a very small number of American families have not given rise to what Europe refers to as a titled aristocracy. Freedom can dissolve into anarchy with some audacious official's temporary usurpation of power, but the real elements of monarchy are nowhere present in the modern colonies. In Brazil, those elements were imported from abroad at a moment when this vast country enjoyed a profound peace and the metropole had fallen under a foreign yoke.

Thinking about the interrelatedness of human affairs, one can see how the condition of modern colonies, or, rather, the discovery of a half settled con- II.375 tinent in which a colonial system could develop in such extraordinary ways, must have revived republican forms of government on such a large scale and in such a widespread fashion. Famous writers have regarded the changes in the social order in most of Europe in our day as the belated effects of early sixteenth-century religious reformation. We must not forget that this memorable period, during which ardent passions and the taste for absolute dogmas were the rocks on which European politics foundered, was also the time of the conquests of Mexico, Peru, and Cundinamarca. This conquest, to borrow some noble formulations from the author of ▾*The Spirit of Laws*, left the metropole with an enormous debt to pay so it could acquit itself before humanity. Vast provinces, opened to colonists by Castilian valor, were unified by common ties of language, custom, and creed. This is how, by a strange coincidence of events, the reign of Charles V, Europe's most powerful and absolute monarch, prepared the struggle of the nineteenth century and laid the foundation for the political associations that, barely conceived, surprise us with their scale and the uniform tendency of their principles. If Spanish America's emancipation is consolidated, as everything now gives us hope to believe that it will, a body of water—the Atlantic—will have on its two shores forms of government that even though they are opposed are not necessarily enemies. The same institutions cannot be salutary to all people in both worlds. A republic's growing wealth is not an offense to monarchies when those are governed with wisdom and with respect for the laws and for the public liberties. II.376

POPULATION OVERVIEW ACCORDING TO RACE, LANGUAGE, AND RELIGION

II.377

The Antillean archipelago has a surface area of nearly 8,300 square leagues (20 to a degree). The four large islands of Cuba, Haiti, Jamaica, and Puerto Rico comprise 7,200 leagues, or nearly nine tenths. The *area* of equinoctial America's islands is therefore about equal to the Kingdom of Prussia and twice as large as the state of Pennsylvania. Its *population density* is nearly identical to that of Pennsylvania, and about a third of Scotland's.[1] Over many years, I have worked meticulously to learn the number of inhabitants who belong to different castes and colors and whom the deplorable development of the colonial economy has brought together in the Antilles. The strange coexistence of such diverse populations is a problem that touches so closely on the misfortunes the African race has experienced and on the dangers facing human civilization that I did not want to rely on the little that is already published elsewhere. Through an active correspondence, I have consulted respectable and enlightened men who showed an interest in my work and sought to improve it by correcting my first results. It is my pleasure to express here my profound gratitude to ▾Lord Holland, Mr. Charles Ellis, Mr. Wilmot, undersecretary of state in the colonial department, General Macauley, Sir Charles M'Carthy, last governor of Sierra Leone, the Chevalier Mackintosh, Mr. Clarkson, Mr. David Hodgson, and Mr. James Cropper of Liverpool.

II.378

1. See Vol. XI, pp. 57 and 58.

Population of the Antilles (End of 1823)

Names of the Islands	Total population	Slaves	Observations and variations
I. BRITISH ANTILLES	776,500	626,800	In 1788, the total population of the British Antilles was estimated at 528,302; 454,161 were slaves. Bryan Edwards in 1791: 455,684 slaves; 65,305 whites; 20,000 free people of color. Colquhoun in 1812: total 732,176, 634,096 of whom were slaves; 33,081 free people of color; 64,994 whites. Melish: total 673,070, 70,430 of whom were whites, and 607,640 slaves. Individuals belonging to the *Methodist* church in 1823 in the British Antilles: 23,127 blacks and people of color, and 8,476 whites. (*[Substance of the] Debate of 15 May* 1823, p. 180.)
a) JAMAICA	402,000	342,000	In 1734, 86,146 slaves; 7,644 whites; in 1746, 112,428 slaves; 10,000 whites; in 1768, 176,914 slaves; 17,947 whites; in 1775, 190,914 slaves; 18,500 whites; in 1787, 250,000 slaves; 28,000 whites; in 1791, 30,000 whites; 10,000 free people of color; 250,000 slaves; in 1800, 300,939 slaves; in 1810, 320,000 slaves; in 1812, 319,912 slaves; in 1815, 313,814 slaves; in 1816, 314,038 slaves; 45,000 free people; in 1817, 345,252 slaves. (For 1658 old reports count 1,400 slaves; 4,500 whites; for 1670, 8,000 slaves; 7,500 whites; for 1673, 9,504 slaves). Between 1770 and 1786 Jamaica imported: 610,000 black slaves, of whom one fifth were reexported to other islands; 488,000 remained on the island (Bryan Edwards, Vol. II, p. 64). Between 1787 and 1808, more than 188,785 slaves were imported; thus, all in all, 676,785 blacks were imported over the course of 108 years. Nevertheless, there are only half that number on Jamaica, fewer than 350,000. (Hatchard, *Review of Registry Laws,* p. 74. Cropper, *Letters to M. Wilberforce,* 1822, p. 19, 29, 40). Other estimates put Jamaica's import of Africans since the *conquest* as high as 850,000. (*East and West India Sugar,* 1823, p. 34. James Cropper, *Relief for West Indian Distress,* 1823, p. 13. Wilberforce, *Appeal to Religion, Justice and Humanity,* 1823, p. 49). The population of free people of color is generally estimated too low. Mr. Stewart, who has lived on this island for twenty years (until 1820), estimated it at 35,000 and the number of whites at 25,000. According to *official registries* that Mr. Wilmot kindly conveyed to me in 1817: 343,145 slaves; in 1820, 341,812 slaves. Out of a slave population of 342,000, barely 600 marriages (257 per year) have been legally contracted over the past fourteen years. (*Substance of the Debate of the House of Commons,* 1823, p. 164.)
b) BARBADOS	100,000	79,000	Already in 1786, Mr. Morse estimated the total population at 79,220; in 1805, 60,000 slaves; 17,130 free people; in 1811, according to a census widely reputed to be quite precise: 79,132 slaves; 2,613 free people of color; 15,794 whites. In 1823, there were probably 16,000 whites; free people of color, whose ranks are increasing quite a bit, 5,000; total population, perhaps 100,000. According to *official registries* from 1817, 77,493 slaves; in 1820, 78,345 slaves.
c) ANTIGUA	40,000	31,000	In 1815, 36,000 slaves; 4,000 frees; in 1823, probably 4,000 free people of color; 5,000 whites. According to *official registries* from 1817, 32,269 slaves; in 1820, 31,053 slaves.

d) ST. CHRISTOPHE OR ST.-KITTS.	23,000	19,500	In 1791, 20,435 slaves; 1,900 whites; in 1805, 26,000 slaves; 1,800 whites; perhaps 2,500 free people of color. According to *official registries,* 20,137 slaves; in 1820, 19,817 slaves.
e) NEVIS	11,000	9,000	In 1809, total: 9,300, 8,000 of whom were blacks (▾Chalmers); in 1812, total: 10,430, of whom 9,326 were slaves. *Official registries* from 1817: 9,603 slaves; from 1820, 9,261 slaves; nearly 1,000 free people of color; 450 whites.
f) GRENADA.	29,000	25,000	In 1791, according to Bryan Edwards: 23,926 slaves; 1,000 whites; in 1815, 29,381 slaves; 1,891 free people. *Official registries* from 1817, 28,024 slaves; from 1820, 25,677 slaves; free people of color, nearly 2,800 today; 900 whites.
g) SAINT VINCENT AND THE GRENADINES.	28,000	24,000	In 1791, 11,853 slaves; 1,450 whites; in 1812, total 27,455, 22,920 of whom were slaves; 1815 total: 23,493, including 2,130 frees. *Official registries* from 1817, 25,255 slaves; from 1820, 24,252 slaves.
h) DOMINICA	20,000	16,000	In 1791, 14,967 slaves; 1,236 whites; in 1805, 22,083 slaves; 4,416 frees; 1811 total: 25,031, 1,325 of whom were whites; 2,988 free people of color; 21,728 slaves. The relation of blacks or free people of mixed race to whites is here, as everywhere, quite uncertain; today, the former group is possibly twice the latter. *Official registries* from 1817, 17,959 slaves; from 1820, 16,554 slaves. From Dominica and the Bahama islands, slaves are often exported to Demerary, where the climate induces frighteningly high mortality rates, even among those people of color who are acclimated.
i) MONTSERRAT	8,000	6,500	In 1805, 9,500 slaves; 1,250 frees; in 1812, 6,534 slaves; 442 frees. (In 1825, according to sounder estimates: 1,500 free, barely one-fifth of whom were whites.) *Official registries* from 1817: 6,610 slaves; from 1820, 6,505 slaves. Mr. Morse's estimate puts the total population in 1822 at 10,750; but it is too high.
k) VIRGIN ISLANDS BRITISH ANEGADA, VIRGIN GORDA, AND TORTOLA.	8,500	6,000	Much uncertainty. Apparently, there were 6,000 slaves in 1820; 1,200-1,500 free people of color; 400 whites. Nonetheless, in 1788, it was thought that the slave population reached 9,000. (Melish put Tortola's total population at 10,500 in 1822, and Virgin Gorda's at 8,000!)
l) TOBAGO	16,000	14,000	In 1805, 14,883 slaves; 1,600 frees; in 1811, 16,897 slaves; 935 frees; 1815 total: 18,000. *Official registries* from 1817: 15,470 slaves; from 1820, 14,581 slaves (today there are probably 2,000 frees, 1,200 of whom are of color). For 1822, Mr. Morse (*Modern Geography*, p. 236) reckons a total of 16,483, 15,583 of whom were slaves and free people of color, and 900 whites.
m) ANGUILLA AND BARBUDA.	2,500	1,800	Uncertain.
n) TRINIDAD.	41,500	25,000	In 1805, 19,709 slaves; 5,536 frees (▾MacCallum). The 1811 census, reputed to be quite accurate: total 32,989, 2,617 of whom were whites; 7,493 free people of color; 1,736 free Indians; 21,143 slaves. *Official registries* from 1817, 25,941 slaves; from 1820, 23,537 slaves. Estimates of the always growing population on this island tend to run much too low. In 1822, Mr. Morse put the total population at 28,477; nonetheless, today it would not be surprising to find at least 14,000 free people of color, 4,006 whites, and almost 24,000 slaves.
o) SAINT LUCIA	17,000	13,000	In 1788, the total was put at 20,968, 17,221 of whom were slaves; 1810 total 17,485, 14,397 of whom were slaves, 1,878 free people of color, and 1,210 whites. *Official registries* from 1817: 15,893 slaves; from 1820, 13,050 slaves.

Names of the Islands	Total population	Slaves	Observations and variations
p) BAHAMA ISLANDS	15,500	11,000	In part already outside of the torrid zone. 1810 total: 16,718, 11,146 of whom were slaves. (Today, there are probably 11,000 slaves, 2,500–3,000 free people of color, and 1,500 whites.)
q) BERMUDA ISLANDS	14,500	5,000	A small archipelago situated in the temperate zone and far away from the rest of insular America. 1791 total: 10,780, 4,919 of whom were slaves; 1812 total: 9,900, 4,794 of whom were slaves.
II. FRENCH AND SPANISH HAITI.	820,000		For the *French part,* Mr. Necker allowed for total population of 288,803 in 1779, and a total of 520,000 in 1788, including 40,000 whites, 28,000 freedmen, and 452,000 slaves. In 1802, Mr. Page estimated the total population at only 375,000, 290,000 of whom were laborers. According to ▼General Pamphile de Lacroix's 1819 observations, there were 501,000 people in the *French part,* including 480,000 blacks, 20,000 people of mixed race, and 1,000 whites. In the *Spanish part,* there were 135,000 people, including 110,000 Blacks, and 25,000 whites. General Macaulay, whose research always exhibits his love of humanity and of the truth, thinks that Haiti's total population exceeds 750,000, including 600,000 blacks and people of mixed race and 4,000 whites in the *French part,* and 120,000 blacks and people of mixed race, and 26,000 white creoles in the *Spanish part.* In the *French part,* the number of mixed-blood people [sang-mêlés] adds up to 24,000. The last *official census* counted 935,335 people, including 99,408 in the district of Jacmel alone; 89,164 in Port-au-Prince; 63,536 in Cayes; 58,587 in Aguni; 55,662 in Leogane; 53,649 in Mirabalais; 44,478 in Nepper; 38,566 in Cap Haiti; 37,927 in Tiburón; 37,652 in Jeremie; 37,628 in Saint Marc; 35,372 in Grande Rivière; 33,542 in Gonaïves; 33,475 in Lembé; 32,852 in Marmelade; and 20,076 in Santo Domingo. (*New Monthly Magazine,* 1825, Feb., p. 69). Measures by the Haitian government to arrive at an accurate result are unknown. As I am committed in all of my works in political economy to err on the side of the *lowest* possible figures, I cut the result of the official census by one-ninth. Currently, the lower and upper limits are 800,000 and 940,000. Highly exaggerated claims that can be attributed to political interests have swelled Haiti's population to more than one million. It is certain that under the tutelage of far-sighted institutions this population is growing extremely rapidly.
III. SPANISH ANTILLES	943,000	281,000	
a) CUBA	700,000	256,000	According to an official document presented to the Cortes in Madrid in 1821, the total population equals 630,980, 290,021 of whom are white, 115,691 are free people of color, and 225,268 are slaves. [Zayas y Chacón,] *Reclamación hecha por los representantes de la Isla de Cuba, contra los aranceles,* p. 7. From 1817 to 1819, the number of imported slaves was between 15,000 and 26,000. [Jameson,] *Letters from the Havana to John Wilson Croker, Esq.,* 1821, pp. 18–36. These imports are frightening, because Rio de Janeiro itself does not receive that many slaves these days; in 1821, 20,852 slaves; in 1822, 17,008 slaves; in 1823, 20,610 slaves; [Chalmers,] *Official Correspondence with the British Commissioner,* 1823, B., pp. 109, 121. Alexander Caldcleugh, *Travels in South America,* 1825, Vol. II, p. 296. (Mr. Melish, in his *American Geography,* still accords to Cuba a population of only 435,000 for 1823.)

b) PUERTO RICO	225,000	25,000	In 1778, the total population was estimated at 80,650; in 1794, at 136,000, including 15,000 whites, 103,500 free people of color, and 17,500 slaves; but the more reliable official 1822 census puts the total population at 225,000, with 25,000 slaves (Poinsett, *Notes on Mexico,* Philadelphia, 1824, p. 5). If we assume that the number of whites did not increased above 22,000, this census would yield 178,000 free people of color, an estimate that strikes me as exaggerated in comparison with the free people of color on the entire island of Cuba.
c) MARGARITA	18,000	12,400	▾Mr. [Vargas] Ponce: 14,000, including 2,000 Indians.
IV. FRENCH ANTILLES	219,000	178,000	Probably more than 25,000 freedmen.
a) GUADELOUPE AND ITS DEPENDENCIES (LES SAINTES, MARIE-GALANTE, DESIRADE, AND A PART OF SAINT MARTIN).	120,000	100,000	In 1788, total population 101,971, including 13,466 whites, 3,044 free people of color, and 85,461 slaves. According to official data which Mr. Moreau de Jonnès kindly conveyed to me the total 1822 population has 120,000, including 13,000 whites, 7,000 free people of color, and 100,000 slaves. Other official data put Guadeloupe's 1821 total at 109,404, including 12,802 whites, 8,604 free people of color, and 87,998 slaves.
b) MARTINIQUE	99,000	78,000	In 1815, the total population was assumed to be 94,413, with 9,206 whites, 8,630 people of color, and 76,577 blacks. According to the official 1822 census, the total population has 98,125 with 9,660 whites, 10,173 free people of color, and 76,914 slaves.
V. DUTCH, DANISH, AND SWEDISH ANTILLES.	84,500	61,300	
a) SAINT EUSTACIA AND SABA.	18,000	12,000	No island is clouded by more uncertainty. Mr. Malte-Brun (*Géographie,* Vol. V, p. 748) estimates the total 1815 population at 6,400, including 5,000 whites, 600 free people of color, and 800 slaves; but this number of whites is hardly likely. Mr. J. van den Bosch (*Nederlandsche Overzeesche Bezittingen* 1818, Vol. II, p. 232) suggests 2,400; Mr. Morse's new *Geography,* however, which has been carefully compiled overall (*New System of Modern Geography,* 1822, p. 249) shows at 20,000.
b) SAINT MARTIN	6,000	4,000	Morse, *ibid.,* p. 248. One part is French, the other Dutch.
c) CURAÇAO	11,000	6,500	Melish: 8500. Hassel: 14,000. Van den Bosch (Vol. II, p. 227), for 1805, a total population 12,840. Dutch Antilles overall, 35,000, of which 22,500 slaves.
d) SAINT CROIX	32,000	27,000	In 1805: 2,223 whites; 1,664 freedmen; 25,452 slaves. Total: 29,339.
e) SAINT THOMAS	7,000	5,500	In 1815, 726 whites; 239 freedmen; 4,769 slaves. Total: 5734.
f) SAINT JOHN.	2,500	2,300	In 1815: total 2,120, including 102 whites; 1,992 slaves. Mr. Hassel estimates the total 1805 population of the Danish islands at 38,695; Mr. Colquhoun estimates it, in 1812, at 42,787, with 37,030 slaves.
g) SAINT BARTHOLOMEW	8,000	4,000	Morse, p. 249.

II.389 The observations I place beside the results that are currently the most accurate contain some historical notes on gradual population growth. These notes, of widely variable accuracy, are nevertheless only *variantes lectionum*: they are simply opinions about population figures during such and such a period. Most often, I have used as the basis of my calculations not these estimates but *official registries* from the past few years. When there are no *records*, only general considerations can guide us as to the value of statistical results. When considering topics that are debated violently and that touch on humanity's most cherished interests, one must avoid partisan exaggerations and take the middle ground between the views of landowners and those of groups whose goal is to alleviate slavery's miseries. Comparisons between records from different periods do not always give an exact sense of mortality rates among slaves in the colonies of various nations. There are countries in II.390 which surreptitiously imported slaves are given the names of deceased slaves. When exact numbers are unavailable, there is much to be gained from determining *lower numerical limits*, so as to be able to assert that there are at least 342,000 slaves in Jamaica, 79,000 in Barbados, and 100,000 in Guadeloupe. Numbers derived either from the census or from the registry of slaves (*slave registry returns*) show only these *lower limits* for a given period. Owners have an interest in keeping low the numbers of their slaves in the registry. In the registries, emancipation[1] is confused with deaths, and births are intentionally undercounted. Until today (1817 to 1824), registries have generally tend to prove that the black population is shrinking faster in the smaller islands of II.391 the British Antilles than in Jamaica and wherever colonists invest considerable capital in exploiting land that produces foodstuffs in abundance. Official registries for 1817 showed 617,799 slaves for the twelve islands of the British Antilles and 604,444 slaves for 1820—a loss of one sixteenth over three years. Jamaica lost only one two-hundred-and-fifty-seventh; on the small islands, losses fluctuate between one tenth and one sixtieth. I am not saying that these figures are true, merely that they come from the *registries*. The distinction between whites and the *free colored population* presents such large difficulties that, at the end of 1823, not even the *Colonial Office* had exact information on this important subject. But with the most laudable intentions in mind, the British government has recently adopted measures to resolve a problem more closely connected to public safety than any other issue. In Havana, free blacks

1. Adam Hodgson, *A letter to* ▼*Jean-Baptiste Say*, 1823, p. 37. [*Substance of the*] *Debate of the 15 May 1823*, p. 184. ▼[Bridges,] "Bridges on the Effects of Manumission," 1823, pp. 51 and 85.

amount to five thirteenths, or 38 percent, of the population. Generally, however, their number can be put at only two fifths. In some colonies, estimates of the free population are no less uncertain than estimates of the slave population. There are individuals who enjoy lives full of freedom without their freedom being legally recognized. II.392

In the registries that give population numbers for the islands, the words *blacks* and slaves are typically used synonymously. The slave population nevertheless includes small numbers of mixed-race African [mulâtres] and other mixed people. I believe that their numbers are one twentieth at most, and I have calculated the figure for enslaved blacks in my table on America's black population based on this assumption. Cuba's census shows more substantial ratios: one tenth to one twelfth for Havana. In 1810, there were 2,300 *pardos esclavos* and 26,400 *morenos esclavos* out of an enslaved population of 28,700. Concentrations of different mixed-race slaves [mulâtres et de races mixtes] are a defining trait of sizable cities in the Spanish Antilles.

As for the population of the island of Saint-Domingue (Haiti), I think that I have settled on a sufficiently low estimate. We have partial data from the official census, district by district. Simple considerations based on positive calculations make it conceivable that Haiti's population might currently be as high as 820,000. ˅Mr. Page still assumed a population of 500,000 for both the Spanish and French parts in 1802, after the colony's misfortunes. Assuming an annual growth rate (r) of only 0.016 (which means that the population would double every 44 years), I arrive at a population of 686,800 for 1822. If one assumes more rapid growth, similar to what the slave population in the southern United States is experiencing ($r = 0.026$, which would double the population every 27 years), then the population will be 835,500 in 1822. But there is no doubt that Mr. Page's numbers for 1802 are too low. For 1788, Necker assumed 520,000 for the French part and 620,000 for all of Saint-Domingue. Many years of peace and tranquility have since followed, interrupted by a few years of disorder and carnage. Even ˅Jamaica's maroon population has grown, and this does not include the runaway slaves who join it from time to time. It makes more sense to accept that, despite civil wars and emigration, the population remained stable at 600,000 for 14 years (1788 to 1802). Using this data with either one of the two hypotheses ($r = 0.016$ or $r = 0.026$), we arrive at 824,200 or 1,002,500, respectively. The Haitian government's last official census counted 935,300. Preferring more conservative figures, I settled on 820,000. II.393 II.394

Black population of continental and insular America.

1°	*Black slaves:*	
	Antilles, insular America	1,090,000
	United States	1,650,000
	Brazil	1,800,000
	Spanish colonies on the continent	307,000
	British, Dutch, and French Guyana	200,000
		5,047,000
2°	*Free blacks:*	
	Haiti and the other Antilles	870,000
	United States	270,000
	Brazil maybe	160,000
	Spanish colonies and from the continent	80,000
	British, Dutch, and French Guyana	6,000
		1,386,000

Recapitulation

Unmixed Blacks, thus excluding racial mixtures:

5,047,000	slaves.	79 percent
1,386,000	frees	21
6,433,000		

II.395 Being accustomed to living in countries where whites are as numerous as they are in the United States has singularly influenced people's ideas about how various races are distributed in different parts of the New World. The counted numbers of blacks and racially mixed people are artificially low. According to my tables, they number more than 12,861,000 or 37 percent, while the white population does not exceed 13.5 million or 38 percent. In 1822, ▾Mr. Morse still assumed America's total population to be 50 percent white, 33 percent Indian, 11 percent black, and 6 percent mixed [de races mixtes]. For the Antillean archipelago, ▾Mr. Carey and Mr. Lea posited a population of 2.050 million, with 450,000 whites and 1.6 million blacks and people of mixed race, meaning that whites would make up 22 percent of the population. We just saw II.396 that the proportion is still slightly less favorable, and that, out of a total population of 2,843,000 for the Antilles, 17 percent are white and 83 percent are people of color, both slaves and frees; thus, the ratio between white and colored population is 1 to 5.

Parts	Total Population	Black Slaves and some people of mixed race	Free People of Color people of mixed race and blacks	Whites
Spanish Antilles	943,000	281,400	319,500	342,100
Haiti	820,000		790,000	30,000
British Antilles.	776,500	626,800	78,350	71,350
French Antilles	219,000	178,000	18,000	23,000
Dutch, Danish, and Swedish Antilles	[84,500]	61,300	7,050	16,150
Total Antilles	2,843,000	1,147,500 (40 percent)	1,212,900 (43 percent)	482,600 (17 percent)

Breakdown of races in Spanish America

1° *Indigenous* (Indians, redskins; primitive or American copper-skinned race, without mixtures with either whites or blacks)

Mexico	3,700,000
Guatemala	880,000
Colombia	720,000
Peru and Chile	1,030,000
Buenos Aires with the Sierra provinces	1,200,000
	7,530,000

2° *Whites* (Europeans and descendants of Europeans, without black and Indian mixtures, the so-called Caucasian race) II.397

Mexico	1,230,000
Guatemala	280,000
Cuba and Puerto Rico	339,000
Colombia	642,000
Peru and Chile	465,000
Buenos Aires	320,000
	3,276,000

3° *Blacks* (African race, without mixtures of white or Indian blood, free blacks and slaves)

Cuba and Puerto Rico	389,000
Continent.	387,000
	776,000

4° *Mixed races of blacks, whites, and Indians* (Mulâtres, Mestizos, Zambos, and mixtures of mixtures)

Mexico	1,860,000
Guatemala	420,000
Colombia	1,256,000
Peru and Chile	853,000
Buenos Aires	742,000
Cuba and Puerto Rico.	197,000
	5,328,000

II.398

Recapitulation

according to the races' predominance

Indians.	7,530,000	or	45	percent
Mixed races	5,328,000		32	
Whites.	3,276,000		19	
Blacks, African race	776,000		4	
	16,910,000			

Breakdown of races in continental and insular America.

1° *Whites:*

Spanish America.	3,276,000
Antilles, without Cuba, Puerto Rico, and Margarita	140,000
Brazil	920,000
United States	8,575,000
Canada	550,000
British, Dutch, and French Guyana	10,000
	13,471,000

2° *Indians:*

Spanish America.	7,530,000
Brazil (from the Río Negro, Río Blanco, and the Amazon)	260,000
Autonomous Indians, east and west of the Rocky Mountains, on the frontier of New Mexico, Mosquitos, etc	400,000
Autonomous Indians in South America	420,000
	8,610,000

II.399

3° *Blacks:*

Antilles with Cuba and Puerto Rico	1,960,000
Continental Spanish America	387,000
Brazil	1,960,000
British, Dutch, and French Guyana	206,000
United States	1,920,000
	6,433,000

4° *Mixed races* [races mélangées]:

Spanish America	5,328,000
Antilles, without Cuba, Puerto Rico, and Margarita	190,000
Brazil and United States	890,000
British, Dutch, and French Guyana	20,000
	6,428,000

RECAPITULATION

Whites	13,471,000	or	38	percent
Indians	8,610,000		25	
Blacks	6,433,000		19	
Mixed race [races mixtes]	6,428,000		18	
	34,942,000			

Calculations based on the 1810 and 1820 censuses show (growth rate = .002611) at least 1.623 million slaves in the United States for the end of 1822 (Vol. IX, pp. 177 and 178, and [African Institution,] *Sixteenth Report of the African Institute*, p. 324) and at least 1,708,300 for the end of 1824. In 1820, free people of color came to more than 238,000. In 1811, in the two colonies of Demerary and Essequibo, there were already 71,180 slaves, 2,980 free people of color, and 2,871 whites, for a total of 77,031. Berbice's total population was 25,959, including 550 whites, 240 free people of color, and 25,169 black slaves. The combined total populations of Demerary, Essequibo, and Berbice in 1811 exceeded 103,000, of which more than 96,000 were slaves. According to ▼J. van den Bosch (Vol. II, p. 114), in 1814, there were 47,032 slaves in Demerary, 16,187 slaves in Essequibo, and 22,223 slaves in Berbice, a total of 85,442. General Macauley believed that Demerary's population in 1823 was 83,900, including 77,400 slaves, 3,000 free people of color, and 3,500 whites. He assumed 25,430 people for Berbice, of whom 23,180 were slaves, 1,500 free people of color, and 750 whites. The *official registries* that Mr. Wilmot sent me count Demerary's slave population at 77,867 for 1817, and 77,376 for 1820; the numbers of Berbice's slave population are 23,725 for 1817 and at 23,180 for 1820. It seems likely that Dutch, British, and French Guiana currently have 236,000 slaves. In 1821, French Guiana had a total population of 16,000, not including Indians: 12,000 slaves, 1,000 whites, and 3,000 free people of color. According to official documents, as of January 1, 1824, there were 1,035 whites, 1,923 free people of color, 701 Indians, and 13,656 slaves, totaling 17,315. The number of blacks spread out over the vast Spanish American continent is so small (below 390,000) that it fortunately forms only 2.5 percent of the continent's population. Salutary changes in the status of slaves are imminent. According to the laws governing the newly indepen-

II.400

II.401

dent states, slavery will gradually vanish; the republic of Colombia pioneered gradual emancipation. This measure, at once prudent and humane, can be credited to the selflessness of GENERAL BOLIVAR, whose name is memorable no less for his civic virtues and his restraint in victory than for his resounding military achievements.

BREAKDOWN OF AMERICA'S TOTAL POPULATION BY RELIGION

I. *Roman Catholic*			22,486,000
a. Continental Spanish America		15,985,000	
Whites	2,937,000		
Indians	7,530,000		
Mixed races and blacks	5,518,000		
	15,985,000		
b. Portuguese America		4,000,000	
c. United States, Lower Canada, and French Guyana		537,000	
d. Haiti, Cuba, Puerto Rico, and French Antilles		1,964,000	
		22,486,000	
II. *Protestants*			11,636,000
a. United States		10,295,000	
b. British Canada, Nova Scotia, Labrador		260,000	
c. British and Dutch Guyana.		220,000	
d. British Antilles		777,000	
e. Dutch, Danish, etc. Antilles		84,000	
		11,636,000	
III. *Autonomous non-Christian Indians*			820,000
			34,942,000

II.402

This table shows only the major Christian denominations. I believe that I have sufficiently accurate data[1] on the number of Roman Catholics compared to Protestants, but I will not explore in detail the divisions within protestant or evangelical churches. Some partial estimates, for example, of the number of Roman Catholics in Louisiana, Maryland, and British Lower Canada, are perhaps a bit too imprecise. But this imprecision concerns figures that have a negligible impact on the overall numbers. I think that the Protestant population throughout continental and insular America, from the southern reaches

II.403

1. An earlier form of these materials appeared [in my "Evaluation numérique de la population du Nouveau Continent"] in the *Revue protestante*, no. 3, p. 97 (see my letter to ▾Mr. Charles Coquerel). More precise data on Cuba, Haiti, and Puerto Rico's populations have led to some corrections of the partial data.

of Chile to Greenland, has a 1 to 2 relationship with the Roman Catholic population. On North America's western coast, several thousand individuals follow the Greek rite. I do not know how many Jews live in the United States or on some of the Antillean islands. Their numbers are small. Free Indians, not belonging to any Christian community, stand in a 1 to 42 relationship with the Christian population. Today, the Protestant population in the New World is growing much faster than the Catholic population. It is likely that, in less than half a century, the 1 to 2 ratio will change in favor of the Protestant community, despite the prosperity that will come to Spanish America, Brazil, and the island of Haiti, thanks to independence, the progress of reason, and free institutions. In Europe, assuming a total population of 198 million, there are about 103 million Roman Catholics, 38 million followers of the Greek rite, 52 million Protestants, and 5 million Muslims. The numerical ratio between Protestants and members of both the Roman Catholic and Greek Orthodox churches is therefore about 1 to 2.7. The ratio between Protestants and Roman Catholics is the same in Europe as it is America. The tables appended to the end of this chapter are closely connected, because, throughout these regions, differences in race, origin, singularity of language, and state of freedom powerfully influence men's dispositions toward one sect or another. II.404

Predominant languages on the new continent II.405

1°	*English language:*		
	United States		10,525,000
	Upper Canada, Nova Scotia, New Brunswick		260,000
	British Antilles and British Guyana		862,000
			11,647,000
2°	*Spanish:*		
	Spanish America;		
	that is:	Whites	3,276,000
		Indians	1,000,000
		Mixed races and blacks	6,104,000
	Spanish part of Haiti		124,000
			10,504,000
3°	*Indian languages:*		
	Span. and Port. America, including autonomous tribes		7,593,000
4°	*Portuguese:*		
	Brazil.		3,740,000
5°	*French:*		
	Haiti		696,000
	Antillean dependencies of France, Louisiana, and French Guyana		256,000
	Lower Canada and a few autonomous Indian tribes		290,000
			1,242,000

6°	*Dutch, Danish, Swedish, and Russian:*	
	Antilles	84,000
	Guyana	117,000
	Russia on the NW coast	15,000
		216,000

II.406

Recapitulation

Language	English	11,647,000
	Spanish	10,504,000
	Indian	7,593,000
	Portuguese	3,740,000
	French	1,242,000
	Dutch, Danish, and Swedish	216,000
		34,942,000

Romance Languages	15,486,000	European languages	27,349,000
Germanic Languages	11,863,000		
		Indian languages	7,593,000

I did not make separate mention of German, Gallic (Irish), and Basque, because the rather large number of individuals who retain these languages as their mother tongues also know either English or Castilian. Speakers who typically use indigenous languages currently have a 1 to 3.4 ratio with speakers of European languages. Due to the more rapid growth of United States, II.407 Germanic languages will imperceptibly gain on Romance languages in the total numerical proportion. But because of the growing Spanish and Portuguese presence in Indian villages, where barely one twentieth of the population speaks even a few words of Castilian or Portuguese, Romance languages will also spread. I believe that there are more than seven and a half million indigenous people throughout America who still speak their own languages and practically do not know any European idioms. Such is also the opinion of Mexico's archbishop and several equally respectable clerics who have lived in Upper Peru for a long time and with whom I was able to confer on this topic. The small numbers of Indians (perhaps a million) who have entirely forgotten their native languages live in large cities and in the populous villages that surround them. Among those who speak French on the New Continent are 700,000 blacks of African descent. Despite the laudable efforts of the Haitian government to promote public education, such a circumstance is not conducive to maintaining the language's purity. It can generally be assumed that, in both II.408 continental and insular America out of 6,433,000 blacks, more than 45 percent speak English, more than 50 percent speak Portuguese, more than 14 percent speak French, and more than 12 percent speak Spanish.

These population tableaus, considered from the perspective of the relations among different races, languages, and creeds, contain variable elements that represent the current state of American society in numerical approximations. A study of this type deals only with very large quantities; partial estimates may gain more rigorous precision over time. The language of numbers, the only hieroglyphics preserved amidst the signs that represent thought, does not require interpretation. There is something gravely prophetic in these inventories of the human species: the New World's entire future seems inscribed within them.

ANNOTATIONS

Unless context made it more sensible to do otherwise, we have annotated a reference or allusion at its first occurrence. Boldface signals related annotations, and readers should consult the index for page numbers.

Weights and Measures

ARPENT: a unit either of length or of land area used in France from the sixteenth to the eighteenth century; also used in Québec and Louisiana.

ARROBAS: old unit of weight; a Spanish arroba equaled 25 pounds, whereas a Portuguese arroba equaled about 32 pounds.

BARREL: a unit of volume that varies depending on the liquid in question: 31½ gallons, or 119.24 liters, of wine; 36 gallons, or 136.27 liters, of beer; or nearly 178⅔ pints of Paris.

BOCOYES: also *bocois* (casks or barrels); storage bins for molasses. A bocoy holds between 1,200 to 1,500 gallons.

BOUCAUT: hogshead, that is, a large barrel for shipping beef, olive oil, wine, vinegar, brandy, salt, and sugar in the French colonial period.

BURGOS FEET (*PIES DE BURGOS*): ca. 10.95 inches (27.83 cm); also called *vara de Burgos* and *vara de Castilla*. In 1536, the official length of the *vara* in Spain was 32.8748 inches. Burgos feet were commonly used after 1750.

CWT: hundredweight; a unit of weight measurement created by U.S. merchants in the late 1800s; it equals 100 pounds. The "C" represents the Roman numeral for 100, and the letters "W" and "T" abbreviate "weight."

CABALLERÍA: also *cavallería* or *caballería de tierra*; a unit of land area in Spanish-speaking countries in Europe and in the Americas, including Texas and the Caribbean, from the sixteenth to the twentieth century. Typically 64 *manzanas*, each of 10,000 square *varas*.

FANEGA: a unit either of length or of weight. Humboldt uses it as the latter. Values differ according to region: in Venezuela, for instance, it equals 3.334 U.S. bushels.

GOLD CASTELLANOS: the earliest gold coins in Spain. They varied in weight and were originally known as Alfonso's *doblas*, after Alfonso XI (1311–50) who first had them minted.

GOLD MARKS (*MARCS D'OR*): a currency used in France until the late eighteenth century. Not to be confused with the *Goldmark* that became the German currency in 1871.

HOGSHEAD (HHD): a unit of volume with a range of possible values, typically equivalent to 63 gallons or 238 liters in the United States.

KING'S FEET (*PIEDS DU ROI*): the most widely used unit of measurement in France.

One king's foot equaled 324.83938497 millimeters, or 12.789 inches or 12 *pouces* or 1728 *lignes*. Legend has it that the measurement was based on the actual size of Charlemagne's foot.

LAST (German: BURDEN): a German unit of measure; 1 *last* equals 2 tons.

LEAGUE: any of a number of European units of measurement ranging from 2.4 to 4.6 statute miles (3.9 to 7.4 km). The *land league* in English-speaking countries is generally 3 statute miles (4.83 km), although variations range from 7,500 feet to 15,000 feet (2.29 to 4.57 km). The Normans introduced this ancient Gaul unit to England, and the Romans estimated it at 1,500 paces (a pace, or *passus*, was nearly 5 feet, or 1.5 meters). The Spanish used land leagues of about 2.63 miles (4.23 km). At times, leagues were also used as a unit of area measurement. The *square leagues* of old California surveys equal 4,439 acres or 1,796 hectares. In the late eighteenth century, leagues also referred to the distance from which cannons' shot could be fired at menacing ships offshore, which resulted in the 3-mile offshore territorial limit. *Nautical leagues* or geographical leagues come in a variety of different lengths. In the eighteenth and nineteenth centuries, one nautical league equaled 3 nautical miles. The *international nautical league* equals 5.556 kilometers, and the *British nautical league* equals 18,240 feet, that is, about 5.559552 kilometers.

MINUTE OF ARC: also *arcminute* or *MOA*; a unit of angular distance equal to one-sixtieth of a degree.

MYRIAMETER: a metric measure of length containing 10,000 meters; equal to 6.2137 miles.

PALM: about 9 inches in Spain, a little longer in Italy.

PESO ENSAYADO: a common unit of accounting in Peru until about 1735 but not an actual coin. It equaled 450 *maravedis*.

PIASTER: a unit of currency originally equal to 1 silver dollar or peso. Original French word for the U.S. dollar. Modern French uses *dollar* for this unit of currency as well. Slang for U.S. dollars in the Francophone Caribbean, especially Haiti.

PIASTRAS FUERTES: a *piece-of-eight* or *ryall-of-eight*; the Spanish dollar (also *piece of eight*) is a coin worth eight *reales*. Minted in the Spanish Empire after a currency reform in 1497. Until an act of the U.S. Congress discontinued the practice in 1857, it was legal in the United States. Used widely in Europe, the Americas, and the Far East, it became the first world currency by the late eighteenth century.

PINTS OF PARIS. *See* BARREL.

PICUL: various units of weights in Southeast Asia and China. It equals a little over 60 kg or 133 pounds.

PIPA: a traditional Portuguese unit of volume similar in size to the English pipe, the latter typically equaling 2 hogsheads. The *pipa* has become a metric unit of 500 liters, that is, 0.5 cubic meter, 132.085 U.S. gallons, or 109.996 British imperial gallons. In the sixteenth and seventeenth centuries, a *pipa* equaled between 525 and 625 liters.

QUINTAL: also *centner*; a unit of weight that equals 100 kg in the metric system. Variations: in Germany, the *Zentner* refers to 50kg; in Switzerland and Austria, it equals 100 kg.

REAL DE PLATA: Spanish silver coin first introduced by Pedro I of Castile (1334–69); created in 1642 alongside the *real de vellón* made of billon; exchange rate was 1 to 2.

SOLES: past and present currency in several Latin American countries.

SQUARE DEGREE: two-dimensional equivalent of a degree, used to measure solid angles.

SQUARE MILE: an area of 640 acres.

TAREA: a unit of land measurement in Spanish America with historically greatly varying lengths. Still in use, for example, in the Dominican Republic, where it is equal to 628.86 square meters.

TERCIO: an old measure of weight in parts of Central America and the Caribbean, equal to about 150 pounds.

TOISES: old French unit of measurement equal to 1.949 meters, or about 6.3946 French feet.

VARA: *stick* or *pole*; a traditional Spanish and Portuguese unit of distance whose length varied: in Spanish Latin America, it typically equaled about 33 inches. Often used in land measurement in Texas as equal to 33⅓ inches or 84.667 centimeters; 33 inches or 83.82 centimeters in California, but only 32.993 inches or 83.802 centimeters in Mexico. About 34 inches, or 86.4 centimeters, in the Southern Cone. The Spanish *vara* equaled only 32.908 inches, or 83.587 centimeters, whereas the Portuguese *vara* equaled 5 *palmos* (palms), about 110 centimeters, or 43.3 inches.

ZURRÓN: translated as *seroon* in the nineteenth century; not a unit but a type of packaging. It is a bale wrapped in animal hide; also a hamper or basket used to pack indigo, raisons, figs, tobacco, tea, and so on.

VIIn. **Jabbo Oltmanns** coauthored the first two volumes of the *RECUEIL*, composing it on the basis of Humboldt's notes and travel journals. Their collaboration illustrates how Humboldt could produce such a large amount of solidly researched publications in what we would now consider rather separate fields of study. Specifically, Humboldt worked in partnership with Oltmanns on analyzing his astronomical observations. He also planned to team up with **Jean-Baptiste Biot** for a volume on magnetic observations. See also Willdenow.

IX. In his *Essai politique sur le royaume de la Nouvelle-Espagne* (1:10), Humboldt defines CHRONOMETRIC LINES as "a series of points whose longitude is based only on the movement of time but whose end points coincide with the results of the *absolute methods*." For lack of reliable geometrical information, most eighteenth and early nineteenth-century navigators had to establish their own measurements to know the exact positions of their routes. For this to happen, some of the most modern ships were equipped with a marine chronometer, a precise and portable timekeeper that needed to be set to a fixed geographical position; the exact time of this location was then compared to the actual position of the voyage. The time difference between the chronometer time and the ship's local time, which was established by stellar observations, determined the exact longitude. By pointing to his method of establishing a "single system of coordinates" with his own instruments, Humboldt indicates both the exactness of his American measurements and the modernity of his methods. Humboldt's chronometer, which he describes at the beginning of his *Relation historique*, was built by Pierre-Louis Berthoud, nephew of the famous French chronometer-maker Ferdinand Berthoud (1727–1807). It was among the most advanced chronometers of its time. As Humboldt enthusiastically points out, his chronometer had once belonged to the "great [Jean-Charles] **Borda**" (1733–99), the famous French mathematician and naval general with whom Humboldt specified the magnetic inclination of the Paris Observatory in 1798. Magnetic inclination (or magnetic dip), which Robert Norman (c. 1550–1600) first described in 1581, is the angle that a compass needle makes with the horizon at any point on the earth's surface. See also absolute longitudes.

x. In 1822, **Francisco Antonio Zea** approached **Georges Cuvier** about a scientific mission to Colombia; the mission came about thanks to Humboldt and **François Arago**'s intervention. It included four young men—**Boussingault**, **Rivero y Ustariz**, Justin Goudot (died c. 1850), and François Désiré ROULIN (1796–1874). Roulin was a medical doctor, physiologist, and painter who collaborated with Humboldt on a map on the surroundings of Bahía Honda, Mariquita, and the mines of Santana.

xiii. The Spanish Admiralty's cartographic publishing office, the DEPÓSITO HIDROGRÁFICO [Hydrographic Office], was established in Madrid in 1797. See also Navarrete.

Ventura de BARCAÍZTEGUI (1762–c. 1815) was a Basque geographer and surveyor who lived in Cuba from 1790 to 1794. The Spanish naval officer and cartographer José Antonio DEL RÍO Cosa (d. 1804) sailed to Cuba in 1776 and created a map of Havana in 1798. Humboldt notes that Barcaíztegui and Del Río corrected existing maps of Cuba's southern coastline in 1795 and 1804; their findings were published in 1821. The map itself, however, states that Barcaíztegui had the revision made in 1793 rather than in 1795.

Felipe BAUZÁ y Canas's (1764–1834) role in **Malaspina**'s second expedition exceeded that of a scientist. The Spanish sketcher and naval officer was in charge of the maps, hydrographic charts, and landscape drawings that were to be produced during the projected voyage and for which Malaspina had retained several painters. Bauzá kept Humboldt well informed about the details of Malaspina's expedition; his letters were an important resource as the Prussian naturalist prepared for his own voyage to America. Bauzá is frequently mentioned in the *Political Essay of the Island of Cuba* in connection with hydrographic measurements, which both scientists exchanged over years. Bauzá shared data both from the Malaspina expedition and from his work as director of the **Hydrographic Office** in Madrid.

While in the service of Spain, the Italian naval captain Alessandro MALASPINA (1754–c. 1809) organized two expeditions to circumnavigate the world via the South American costal lines (1786–88 and 1789–94). He traveled together with the Spanish naval officer José de Bustamante y Guerra (1759–1825), the naturalists and botanists Felipe Bauzá, Antonio de Pineda y Ramírez (1751–92), **Thaddeus Haenke**, and **Louis Née,** and the hydrographer Cayetano Valdés y Flores, also known as Cayetano Valdés y Bazán (1767–1835). Malaspina's mission was to perfect the cartographic knowledge of the American continent, broaden the scope of botanical investigations, and establish new trade routes. The vast exploits of his expeditions were to be published in Spain in seven volumes, but Malaspina fell victim to an intrigue against him and was sent to prison in La Coruña for ten years. After Napoleon intervened, Malaspina was banished to Italy where he died. The *Personal Narrative* of his voyages was not published until 1885.

XIV. The mathematician Pedro de MEDINA (1493–1567) was Spain's royal cosmographer during the sixteenth century. His influential *Arte de navigar* (1545) was the first manual for compass navigation; it was translated into many languages, as was his *Regimiento de navegación* [Navigation manual] from 1563. Medina was in charge of training naval pilots for the West Indies.

XIVn. Joaquín Velázquez CARDENAS de León (1732–86) was a trained mathematician and jurist from New Spain. A self-taught astronomer, he made accurate observations of eclipses, the passing of Venus around the sun in 1769 from Baja California, and determined the longitude and latitude of Mexico (1773). Velázquez requested that the crown create both a School of Mines and a Mining Board, of which he was the first director. The oldest technical education institution in America, the School of Mines was approved in 1783 and inaugurated in 1792. Joaquín Velázquez Cardenas de León is not to be confused with Joaquín Velázquez de León (1803–82), a Mexican astronomer and mathematician. The Mexican astronomer and archeologist Antonio LEÓN Y GAMA (1735–1802) published the first exact observation of the longitude of Mexico, which the astronomer **Lalande**, one of Humboldt's close contacts, brought to wider attention. When Fausto D'Elhuyar (1755–1833) opened the Mining School in Mexico City, Cardenas de León appointed Gama professor of mechanics, pyrotechnics, and aerometry. Gama wrote about experimental physics, medicine, mathematics, and Mexican antiquities; his essay about the Aztec Calendar Stone was vital for Humboldt's own work on the Aztec calendar in *Vues des Cordillères*.

Historian and hydrographer Martín Fernández de NAVARRETE (1765–1844) joined the Spanish navy in 1780. He moved on to the naval department in 1796 and eventually established its **Hydrographic Office**. By 1824, he directed the Spanish Royal Academy of History. Navarrete collected five volumes of fifteenth- and sixteenth-century narratives of Spanish voyages of discovery, which were published between 1825 and 1837. Navarrete's remark in **Zach**'s *Correspondence astronomique*—a note he added to his translation of an article by Luis de Salazar (1758–1838)—reads as follows:

> A foreigner who traveled in our America a few years ago accused the Spanish of ignorance and oversight, priding himself to have been the first to have positioned—thanks to astronomical observations—Mexico City in its true longitude

> and latitude. But already the Jesuit P. *Juan Sánchez* had observed a lunar eclipse in said capital on November 17, 1584. *Ferdinand de los Ríos* made similar observations there in 1610 in order to derive from them the longitude. Disregarding a few others, both earlier and later, there are, among others, excellent observations made in 1791 by *Don Denis Galliano*, which can be found on page 79 of the second publication.

The foreigner in question is Humboldt, whose own comments are a rebuke to Navarrete.

XV. The *QUARTERÓN* was a famous eighteenth-century Spanish map based on latitude and longitude measurements by Bartolomé de la Rosa, a naval major of the Royal Spanish Armada vessel *Barvolento*. The map was created in the seventeenth century better to protect Spain's American colonies and coastal lines from foreign attacks.

XVI. Jacques CASSINI (1677–1756), also known as Cassini II, was part of the Cassini dynasty of geographers with close ties to the centralized State (another such dynasty were the **Buaches**). Jacques Cassini was the son of the famous Italian astronomer Giovanni Domenico (or Jean Dominique) Cassini (Cassini I, 1625–1712), one of the founders of the Paris Observatory. Jacques was actually born at the observatory and eventually succeeded his father as its director, as his son and grandson would in their turn. Cassini II is best known for his two major treatises from 1740, the *Elements of Astronomy* and the *Astronomical Tables of the Sun, Moon, Planets, Fixed Stars, and Satellites of Jupiter and Saturn.* A member of the Académie des sciences and the Royal Academy of Sciences in London, Cassini II is also believed to have proved that the degrees of the terrestrial meridian grow smaller from the equator towards the pole, a conviction that appeared to have been supported by the geodesic measurement that Pierre Bouguer and **La Condamine** made in Peru in 1735. Cassini held fast to his erroneous notion for the rest of his life, even in the face of Newton's gravitational theory and other evidence to the contrary. See also Chappe d'Auteroche.

Marcos Antonio Riaño (de) GAMBOA (1672–1729) was a Cuban astronomer and mathematician.

The German mathematician and astronomer Jabbo OLTMANNS (1783–1833) played a major role in assembling and revising the astronomical and barometrical results of Humboldt's measurements in the Americas. He held a professorship in Berlin beginning in 1824 and worked closely with Humboldt on the ***Receuil** d'observations astronomiques* [Compendium of astronomical observations].

Franz TRIESNECKER von Paula (1745–1817) was a Jesuit astronomer, mathematician, and surveyor in Vienna, Austria, where he headed the Observatory. Astronomer and physicist Johann Tobias BÜRG (1766–1834) was university professor in Vienna from 1792 to 1813. Together, Triesnecker and Bürg edited and published the journal *Ephemerides astronomicae*. The "Tabulae Mercurii, Martis, Veneris, Solares" and observations of the sun, moon, planets, and positions of stars were printed in this periodical between 1787 and 1806.

Humboldt frequently discusses geodesic measurements by distinguishing *relative* from *ABSOLUTE* LONGITUDES. When determining the absolute longitude of a specific location, the geologist calculates the distance between his current position and a predetermined fixed position with the help of a chronometer. To verify one's own position, the time of the chronometer is compared to the time of the current location. In nineteenth-century travels, this location was usually established by observing Jupiter's moons, for which Humboldt used the achromatic telescope by Dollond (London). With the current location as a new anchor point, one can then measure longitudes for surrounding locations that are relative to the first established, absolute, longitude. It was not until 1884 that the Royal Greenwich Observatory became the international standard for this chronometric method. In Humboldt's day, exact measurements were much harder to produce and created much discussion in the scientific community. See also chronometric lines.

José ESPINOSA y Tello (1763–1815) was the first director of the **Hydrographic Office** in Madrid; he was succeeded by **Felipe Bauzá**, a former shipmate from the second **Malaspina** expedition. Humboldt wrote in a letter to Heinrich Berghaus (1797–1884) in February 1828: "May I ask you please to return the Espinosa to me in 8 to 14 days, because I use this very book often. You know that it is old and needs to be handled with great care" (*Briefwechsel Alexander von Humboldt's mit Heinrich Berghaus* 1:122). The book in question is Espinosa's 1809 *Memorias sobre las observaciones astronómicas hechas por los navegantes españoles en distintos lugares del globo* [Account of the astronomical observations made by Spanish mariners in different parts of the world], which Berghaus had borrowed for several months. According to Berghaus, Humboldt consulted this work, then a rarity in German libraries, especially while he was writing his "Reasoned Analysis of the Map of the Island of Cuba."

XVII. The scientist, merchant, and traveler Francisco SEIXAS Y LOBERA (also Seijas y Lovera) (1650–c. 1705) published several books, among them a fourteen-volume history of the Spanish kingdom (1702–4). Seixas y Lobera was a former district governor of Tacuba and a distinguished member of the Spanish gentry. Both **Espinosa y Tello** and **Fernández de Navarrete** mentioned him in their publications on the history of navigation.

In 1760, the French Académie des sciences sent one of its members, the astronomer Abbé Jean CHAPPE d'Auteroche (1722–69), to Tobolsk to observe the famous passage of Venus across the sun. In 1771, D'Auteroche published *Voyage en Sibérie fait en 1761.* Several years later, he was chosen for another observatory journey, this one to California, where he succumbed to dysentery. His *Voyage en Californie pour l'observation du passage de Venus* [Voyage to California to observe the transit of Venus] in 1772 was published posthumously by César-François Cassini (Cassini III, 1714–84). See also Jacques Cassini.

Vicente DOZ y Funes (1734–81) was the captain of the *San Agustín.* He was part of an expedition that José de Carvajal y Lancaster (1698–1754), Spain's foreign minister, organized to establish the borders between the Spanish and the Portuguese possessions in South America (1754–61).

José de San Martín Suárez was a Spanish naval officer who created the *Mapa y plano del seno mexicano* [Map and plan of the Gulf of Mexico] (1787) and two additional maps. All three are available in digital form at the Library of Congress website.

xviii. As **Espinosa** points out in his *Memorias sobre las observaciones astronómicas* [Account of astronomical observations], Captain Juan Henrique (or Enrique) de la Rigada explored the Old Bahama channel in 1792. He contributed to a 1799 map published through the **Hydrographic Office** in Madrid. Ridaga's new course of the Old Bahama Channel is attached to the March 12, 1793 correspondence of the postmaster of La Coruña, José Zorrilla y Monroy, with the general directors of revenue. Rigada was then a captain of a mail boat. The documents are preserved in the *Correspondencia y expedientes de la Admón. de La Coruña* (1796/98) in the Archivo General de Indias in Seville, Spain.

Cosme Damián Churruca y Elorza (1761–1805) was a military sailor, who, together with **Galiano**, led a geographical expedition to map the Straits of Magellan. He spent some time at the Cádiz observatory and, in 1792, signed on to **Fidalgo**'s expedition to North America and the Antilles, where he drew valuable maps (published in 1801 and 1802). By the time the Franco-Spanish war interrupted his work, Churruca had completed twenty-four charts, including charts of the coasts of Cuba, Haiti, and Puerto Rico. He died in the Battle of Trafalgar.

xviiin. Spanish naval officer Mariano Isasbiribil (or Isasbirivil) mapped the coastlines of Guatemala and Peru between 1800 and 1808. He died on his return voyage to Spain in 1811.

xx. Humboldt's map of the interior of Cuba is indeed limited, since he never visited Cuba's interior and western regions. The west was later explored by the Guadeloupian Jean-Baptiste Rosemond de Beauvallon (1819–1903). This "continuer of Humboldt" published his account of Cuba in 1844 as *L'Île de Cuba*; the book's Spanish translation was reprinted in Cuba in 2002.

xxi. José Joaquín Ferrer y Cafranga (1763–1818) was a Spanish astronomer. After his ship's capture in the 1770s, he was brought to Britain where, thanks to his family's influence, he was able to use his time for studies, discovering his aptitude for mathematics and astronomy. He took astronomical measurements in America, which earned him the recognition of the American Philosophical Society (United States). His astronomical observations included those on the island of Cuba. He lived in the United States in the early nineteenth century and returned to Europe in 1814. At the observatory of Cádiz, he subsequently prepared geographical statistics on Spain.

Cuban librarian, editor, and astronomer Antonio Robredo (died c. 1830) was among the founding editors of Havana's first newspaper, the *Papel periódico* (October 1790); in 1800, he started the journal *Aurora: Correo político-económico de la Habana*. Robredo, who met Humboldt in Havana, was also the first librarian of the Havana public library (1792), which improved considerably when he donated three hundred books and moved it to his house. In that same year, he edited the *Calendario manual y guía de forasteros de la isla de Cuba* [Almanac and visitors' guide to the island of Cuba], which had begun in 1791. In 1796 and 1806, he made metrological and astronomical observations in Cuba.

XXII. French hydrographer and cartographer Alexandre-Pierre GIVRY (1785–1867) participated in the 1817–18 Bayadère expedition led by captain **Roussin**, which produced new and important cartographical data for coastlines of West and South Africa, the Caribbean, and especially Brazil.

XXIII. James Wallace MONTEATH (1793–1854) contributed to establishing longitudes for Grand Cayman Island, which were reported in **Purdy**'s *Columbian Navigator* (1823).

XXIV. Pedro de SILVA was a mathematician from Cuba who aided topographers, cartographers, and military engineers with geodesic calculations. Silva collaborated in the development of the first map of the most populated part of Cuba, published in 1809 as *Mapa topográfico de la parte más poblada de la isla de Cuba, levantado recientemente por acuerdo y a expensas del real consulado de la misma isla establecido en la Havana y con aprobación del superior Gobierno. Año 1805* [Topographic map of the most populated part of the island of Cuba, recently published with the consent and at the expense of the Royal consulate of this island founded in Havana with approval of the government].

XXV. Naval officer Antoine Hyacinth Anne de Chastenet, COUNT OF PUYSÉGUR (1752–1809), was the second of three sons from an illustrious family of French nobility. He suffered from dry asthma and sought the help of Anton Frederick Mesmer (1723–1815), who had explored the idea of the physiological influences of the planets in an attempt to harmonize astronomy and medicine. The count himself started to experiment with animal magnetism, introducing it aboard the ship under his command, the *Frédéric-Guillaume*. He founded a magnetic society in **Saint-Domingue** in 1784.

XXVI. Ciriaco CEVALLOS (or CEBALLOS) (1764–1816) was a Spanish cartographer who participated in the **Malaspina** expedition. Zeballos on North Vancouver Island is named after him.

XXVII. José de LUYANDO (1773–1835), who in the *Almanaque náutico* (1807) simplified Juan Escalante de Mendoza's (born c. 1545) calculations of longitudes for observing lunar distances. According to **Fernández de Navarrete**, Escalante de Mendoza's *Itinerario de la navegación de los mares y tierras occidentales* [Navigation routes for the western oceans and lands] (1575) can be considered that era's premier reference book of marine knowledge.

XXIX. Anthony de MAYNE is mentioned repeatedly in **Purdy**'s *Columbian Navigator* (1823). Based mainly on his survey, a general chart of the West Indies and gulf of Mexico was drawn. It described the gulf and Windward passages; the coasts of Florida, Louisiana, and Mexico; the bay of Honduras and Mosquito Coast; and the coast of the Spanish main up to the Orinoco delta. The map is from 1824 and was corrected in 1832.

XXX. Francisco María (or Mathias) CELI was a captain in the Spanish Royal Fleet. In 1757, he explored parts of what is now Florida, carefully drawing what was probably the earliest map of Tampa Bay. He also sketched the *Plano de la Bahía de Jagua* [Plan of the Bay of Jagua], now Cienfuegos Bay. He collaborated with Thomas JEFFERYS (c. 1695/1710–71), a British cartographer and engraver, on the *Plan of the Town and Harbour of San Juan de Puerto Rico* (1768).

In more recent history, CAYO CONFITES is known as the place from which, in October 1947, Dominicans in Cuba prepared to launch an invasion of the Dominican Republic to depose dictator Rafael Trujillo (1891–1961). Trujillo was eventually assassinated.

XXXI. The HOUR ANGLE is an astrometric measurement that indicates the angle between the celestial meridian of an observer and the hour circle of a celestial object measured to the west of the meridian. The celestial meridian is set by the global position of the observer, which can be determined by longitude measurements. The hour circle of a celestial object, that is, a visible star, and the hour circle of the observer's position result in the hour angle, which determines the time elapsed since the celestial body last crossed the observer's meridian, expressed in hours and minutes. The hour angle can also be expressed in degrees, 15° of arc being equal to one hour. See also absolute longitude.

XXXII. John PURDY (1773–1843) of Laurie & Whittle in London was the foremost hydrographic authority of his time. Although he himself did not participate in any hydrographic expeditions, he compiled charts and wrote navigational aids based on reports from others. He first published the *Columbian Navigator* in 1817.

Captain Andrew LIVINGSTON published his "Remarks on Cape Antonio, the S.W. of Cuba, and the Isle of Pine" in the *Columbian Navigator*'s 1823 edition. Livingston prepared a manuscript translation of the 1820 *Derrotero de las Antillas* [Navigation manual for the Antilles], to which he added his own notes, allowing Purdy to present a fuller description of the Mexican Sea and the coasts of Cuba and Haiti.

The French naval officer and diplomat Baron Albin Reine ROUSSIN (1781–1854) surveyed the Brazilian coastline in 1819. Humboldt mentions **Givry** alongside Roussin. They collaborated (with **Desaulses de Freycinet**) on the *Carte de la province de Rio de Janeiro* [Map of the province of Rio de Janeiro] (1824).

XXXVI. Cartographer Pierre LAPIE (c. 1777–1850) was a captain first-class of the imperial corps of engineer-geographers. See his *A Map of South America* (1814).

XLII table. Charles MASON (1728–86) was a British astronomer and surveyor. In 1756, Mason worked at the Royal Greenwich Observatory under James Bradley (1693–1762). In his *Lunar Tables in Longitude and Latitude according to the Newtonian Laws of Gravity* (1778), Mason published, for the first time, 1,200 observations by Bradley. Mason's own observations of the transits of Venus across the sun from the Cape of Good Hope (in 1761 and 1769) became the basis for the *Nautical Almanac* (1773). Mason is famous for working with Jeremiah Dixon (1733–79), who also accompanied him to the Cape of Good Hope, on the survey that defined the Pennsylvania-Maryland border, which is still known as the Mason-Dixon Line. Mason and Dixon were the first British scientists to measure a meridian arc; they also observed the first gravity measurements taken in North America. After Mason's death, the British Commission of Longitude published his revisions to lunar observations as *Mayer's Lunar Tables, improved by Mr. Charles Mason* (1787).

Don José Antonio (de) TISCAR was a Spanish general and governor of Barinas province in Venezuela around 1813. In 1819, Ferdinand VII (1784–1833) sent troops from Spain in support of the viceroy of Peru; Tiscar was among them. He left Cádiz for Lima on the vessel *Alexandro 1* together with a number of other ships, at least one of them bound for Antarctica. Tiscar's ship sprang a leak and had to return to Spain.

Clemente NOGUERA coauthored various maps of the Americas published through the **Depósito Hidrográfico** in Madrid between 1810 and 1828.

In 1801, Jesuit-trained French astronomer Joseph-Jérôme Lefrançais de LALANDE (1732–1807) published the *Histoire céleste française*, which contained a catalogue of over forty-seven thousand stars; one of them is Lalande 21185. His work was reputedly based on observations made by his nephew Michel Lalande (1766–1839). In 1760, Lalande became professor of astronomy in the Collège de France, a post he held for forty-six years, and in 1768 was appointed director of the Paris Observatory. Among his disciples were Pierre Méchain and Jean Baptiste Joseph Delambre (1749–1822). Humboldt knew all of them well from his Paris years. See also León y Gama.

Together with Delambre, with whom Humboldt was in contact during his entire voyage, astronomer and hydrographer Pierre François André MÉCHAIN (1744–1804) had measured the meridian arc from Dunkirk (France) to Barcelona in the mid-1790s. Their aim was to establish a basis for the units of length in the metric system for the French national legislature. Méchain also discovered eleven comets and calculated their orbits. He became a member of the French Royal Academy of Sciences in 1782.

German astronomer Johann Friedrich WURM (1760–1833) published his *Geschichte des neuen Planeten Uranus* [History of the new planet Uranus] in 1791. His son Johann Friedrich (d. 1839) wrote about theology, mathematics, and philology. Another son was Christian Friedrich (1803–59), a professor in Hamburg, who devoted his time to commercial and political work.

XLIII table. French astronomer and Catholic priest Alexandre Guy PINGRÉ (1711–96) undertook several long voyages in the 1760s and 1770s. In 1769, he accompanied the Comte Charles Pierre Claret de Fleurieu (1783–1810) on the *Isis* to study the behavior of marine **chronometers**, using the journeys for astronomical observations. He was a member of the French Academy of Sciences.

Alexander MACFARLANE (d. 1755) was a wealthy Jamaican plantation owner from Scotland and a member of the Legislative Assembly who had a pronounced interest in astronomy. He was elected to the Royal Society in 1747, but there is no record that he ever appeared for formal admission. He observed an eclipse of the moon and a transit of Mercury. He bequeathed his collection of astronomical instruments to the University of Glasgow in Scotland, where he had graduated in 1728 before moving to Jamaica. The instruments are now housed in the MacFarlane Observatory (built in 1757).

Oltmanns mentions in the **Receuil** *d'observations astronomiques* (1810) that one of five observations that helped calculate Jamaica's longitude was made when Captain Bartholomew CANDLER (d. 1722) observed a lunar eclipse at Port Royal in 1722. The British Admiralty had sent Candler to ascertain the latitudes and longitudes of the British islands in the West Indies in 1717.

Mathematician Nevil MASKELYNE (1732–1811) contributed a valuable almanac of astronomy to the science of navigation. A close friend of **Joseph Banks**, he was admitted to the Royal Society of London in 1758, which sent him to observe a transit of Venus on the island of St. Helena three years later.

French naval officer and hydrographer Louis Isidore DUPERREY (1786–1865) accompanied **Freycinet** on the expedition around the world from 1817 to 20. From 1822 to 1825, Duperrey, who was interested in magnetism, commanded an expedition to Oceania and South America. Both Duperrey and Freycinet contributed information to several maps.

Captain FOUQUE commanded the Royal vessel *L'Aigrette* that visited South America in 1821. He determined differences in longitude from Montevideo to Martinique.

Joseph LARTIGUE (1791–1876) was with the French Navy. He was involved in publishing maps of South America during the 1820s.

XLIV table. The French naval officer Louis Claude Desaulses de FREYCINET (1779–1842) explored Australian waters from 1801 to 1804, together with Captain **Baudin** and Freycinet's brother, Louis Henri (1777–1840). The maps and journals from the journey were published in 1805. In 1817, Freycinet returned to Australia and also visited the Marianas, Hawaii, and various other Pacific islands. Although he lost his ship off the Falkland Islands, the records of his expedition were saved. His *Voyage autour du monde* was published in thirteen volumes and four atlases (1824–44.) See also Duperrey.

Scottish naval officer Basil HALL (1788–1844) traveled to the Orient, South America, and the United States. His father was the geologist and chemist Sir James Hall (1761–1832). Having joined the navy in 1802, Basil Hall accompanied British ambassador William Pitt (1773–1857) to China in 1815. Hall's *Account of a Voyage of Discovery to the West Coast of Corea, and the Great Loo-Choo Island* was published in 1818. Two years later, he set sail for South America, relating his experiences in *Extracts from a Journal Written on the Coasts of Chili, Peru and Mexico* (1824). During that voyage with naval officer Henry Foster, he conducted a series of geophysical pendulum experiments, which were published in the 1823 *Philosophical Transactions of the Royal Society*; they also appeared as a separate publication in 1824. Hall subsequently left the navy, moved to the United States, and eventually returned to England, where he published his *Travels in North America in the Years 1827 and 1828* (1829), a piece of frank criticism that was not appreciated in the land of the free. Hall spent the last two years of his life as a mental patient in a hospital.

Henry FOSTER (bap. 1796–1831) was a British geophysicist. In 1823, he was asked to accompany **Sabine**'s voyage to Greenland and Norway, after which Foster sailed with Sir William Edward Parry (1790–1855) in search of a northwest passage (1824–25) and, again in 1827, toward the North Pole. He received the Copley Medal for his reports on the series of geomagnetic and astronomical observations he published in the *Philosophical Transactions of the Royal Society* (1826). In early 1831, he drowned on a canoe return trip down the River Chagres after having measured differences in longitude across the Isthmus of Panama by means of rockets.

A member of the religious order of the Minimes, French astronomer, mathematician, and botanist Louis FEUILLÉE or Feuillet (1660–1732) traveled across the Atlantic to make scientific observations in South America (1708–11). His travel journal was published in 1725. His legacy includes a drawing of a monster he claimed to have seen in Buenos Aires in 1708. Louis XIV (1638–1715) awarded Feuillée a pension in recognition of his services.

The ATREVIDA and the DESCUBIERTA were ships of the **Malaspina** expedition.

XLV table. Louis GODIN (1704–60) was a French astronomer who traveled to the new continent (1735–42) with **La Condamine** and the French geo-physicist Pierre Bouguer (1698–1758). Godin's first cousin Jean Godin des Odonais (1713–92) was a French cartographer who traveled the Amazon. In 1770, Jean Godin's wife, Isabel de Casa Mayor (1728–92), embarked on her own adventurous Amazon journey in search of her husband. They had met in Quito.

XLVI table. A HYPSOMETER may be used instead of a barometer to determine differences in barometric pressure between two locations. It is used to determine the boiling point of water with precision. Hypsometry is the relation between area and altitude.

1.2. Reputedly the first Mexican place that the Spanish explorers visited in 1517, CABO CATOCHE is located in the northeastern part of the Yucatán Peninsula about 150 miles from western Cuba.

1.3. Charles V of Spain (1500–58) adopted as part of his emblem the PILLARS OF HERCULES decorated with scrolls and the imperial motto "Plus Ultra," signaling his break with medieval notions of military caution.

1.3-4. SANTO DOMINGO, FLORIDA, AND NEW SPAIN cut ties with the Spanish metropole around 1820. In 1492, **Columbus** had landed on the island of Santo Domingo, which he named "Hispaniola" (Little Spain). **Saint-Domingue** (later Haiti) was a portion in the west of the island that was temporarily under French rule. In 1804, Haiti became the world's first free black republic. Haiti invaded Santo Domingo in the early 1820s and named the new state the "Independent State of Spanish Haiti." The occupation lasted until 1844, when the Dominicans declared their sovereignty. After a brief return to Spain, the Dominican Republic finally achieved independence in 1865. FLORIDA was lost to Great Britain in 1763, returned to Spain in 1783 with the British defeat in the Colonies, and finally ceded to the United States in 1819; it became the twenty-seventh U.S. state in 1845. The *Virreinato de Nueva España* [VICEROYALTY OF NEW SPAIN] was established in 1535, comprising the land north of the Isthmus of Panama to which were later added Upper and Lower California and the eastern territory along the Gulf of Mexico to Florida. While Cuba remained under Spanish rule until 1898, Mexico became independent on August 24, 1821, with the Treaty of Córdoba; Mexico's wars of independence had begun in 1810. To Mexican conservatives, independence from Spain was a way of sparing New Spain albeit temporary liberal changes in the religious, economic, and social realms. The First Mexican Empire, as it was named, lasted only until 1823, when Mexico became an independent republic. Concerned that recognizing the independence of the Latin American colonies might risk war with Spain and its allies, President James Monroe (1758–1831) did not officially recognize the countries of Argentina, Peru, Colombia, Chile, and Mexico until March 1822.

The transatlantic **slave trade** had been abolished in Great Britain and in the United States in 1807; Spain did not follow suit until 1820, and even then continued to engage in the slave trade until at least 1830. The STRAITS OF FLORIDA were the dividing line between Florida, which had been ceded to the United States in 1819, and Cuba; it was thus the dividing line between a part of the Americas that had outlawed the international slave trade and another region where the trade persisted.

1.7. The revolution in the French colony of SAINT-DOMINGUE (1791–1804), the only successful slave rebellion in history, created the second independent state in the Americas and the first black republic in the world. The catalyst for this revolution is believed to have been a vodun service performed at Bois Caïman by the Jamaican Dutty (Zamba) Boukman (d. 1791) in August 1791. Within a few weeks, at least one hundred thousand slaves joined the revolt, and the violence escalated. Over the next few years, about one hundred thousand blacks and twenty-four thousand whites were killed, and most sugar plantations were destroyed. In 1801, Toussaint L'Ouverture (c. 1743–1803), a self-educated domestic slave who had emerged as a major revolutionary leader, issued a constitution for an independent country with himself as perpetual head. Napoleon Bonaparte (1769–1821) retaliated by sending troops, and, in 1802, Toussaint was betrayed into surrendering. He was shipped off to France as a prisoner and died in the Jura Mountains. After only a few months of Napoleonic rule in Saint-Domingue, Jean-Jacques Dessalines (c. 1758–1806), a former ally of Toussaint's, initiated a rebellion that would lead to the defeat of the French troops in 1803. On January 1, 1804, Dessalines formally declared the sovereignty of the state he called "Haiti," after its Arawak name. France recognized Haiti in 1825, the United States not until

1862. The Haitian Revolution (as we now know it) had probably even more of an impact on Cuba than had the U.S. Revolution. Since Saint-Domingue was ruined economically by the early 1790s, the demand for Cuban sugar soared, and so did Cuba's demand for slaves. The revolution also instilled profound and lasting fears in the possibility of another large-scale slave insurrection in Cuba, known as the "Africanization of Cuba" scare. There were indeed several slave conspiracies in Cuba that followed in the wake of the Saint-Domingue Revolution, among them the conspiracy by José Antonio Aponte (c. 1760–1812) in 1812 and the conspiracy of La Escalera [The Ladder] in 1844. Both were defeated.

HUMBOLDT AND **BONPLAND**'S HOSTS IN CUBA included the **Count O'Reilly** and the family of **Luis de la Cuesta**. During their stay, the two voyagers met prominent planters and members of the **Patriotic Society**, such as **Arango** (who later commented on Humboldt's writings on Cuba), **Nicolás de la Puerta Calvo y O'Farrill**, the Count of **Jaruco y Mopox**, and the Marqués de Real Socorro (1774–1811), on whose plantations Humboldt was also a guest. They exchanged information with members of the local scientific community, such as the physician Tomás Romay (who had worked on yellow fever), the chemist **Ramírez**, the astronomer **Robredo**, and the botanist Antonio de la Ossa. They also met with local authorities, notably Salvador de Muro y Salazar, the island's Captain-general, the Intendant José Pablo Valiente y Bravo (1740–1817), and **Valle Hernández**, the honorary secretary of the merchant guild. They also encountered Sebastián de la Puerta (1749–1820), Luis María Peñalver de Cárdenas (1749–1810), the Count José de Bayona y Chacón (1676–1757), the Count Gabriel Peñalver y Cárdenas of Santa María de Loreto (1737–1806), **Ignacio O'Farrill y Herrera**, the philosophers José Cipriano de la Luz y Caballero (1800–1862) and José Agustín Caballero (1762–1835), and the poet Manuel de Zequeira (1760–1846). Luz y Caballero, who visited Humboldt in both Paris and Berlin, awarded his friend the epithet "second discoverer of Cuba." Humboldt notes humbly that he and Bonpland received the same kind of welcome from Havana's elite as the princes of the House of **Orléans** had only a few years earlier.

1.7. The botanist and physician Aimé Goujaud BONPLAND (1773–1858) accompanied Humboldt during his entire American journey. Bonpland is named as coauthor of Humboldt's thirty-volume *Voyage aux régions équinoxiales du Nouveau Continent*. In 1804, Bonpland had initially returned to Paris with Humboldt, who succeeded in having him named chief gardener of Malmaison, Napoleon's residence. Humboldt convinced Napoleon to grant Bonpland a lifelong pension. But despite both his recognition in Paris and the pending classification of New World plants that Humboldt insisted he complete, Bonpland decided to return to South America in 1816. He emigrated first to Argentina but eventually moved to Uruguay and Paraguay, where he owned a cattle ranch and fields of cotton, sugar, and yerba mate. In Paraguay, he was initially imprisoned under (false) charges of being a spy. Upon his release in 1831, he devoted himself to his collections and herbariums, keeping in constant touch with the European scientific centers. Until the end of his life, he lived between Corrientes (Santa Ana), where he was appointed director of the Natural History Museum, and in San Borja.

1.11. Actual TOBACCO FACTORIES did not exist in Cuba before 1770. Up to that point, the pickers either rolled cigars on the spot or the tobacco leaf was sent to Seville, Spain.

Only five years prior to Humboldt's landing in Cuba, Havana was provided with a POLICE (*comisario*) for each of its eight *cuarteles* as a first measure to assure public safety in the city. In the 1820s, Governor Nicolás de Mahy y Romo (1757–1822) set up a larger group of sixty policemen to strengthen their presence in the streets.

1.12. Joseph PERUANI (also José or Giuseppe Perovani) (d.1835), a Venetian painter and architect, advertised in Philadelphia in 1795 that he had come to the United States "to try to satisfy respectable citizens of America." The following year, he exhibited the statue of a goddess contemplating a bust of George Washington. He also painted a full-length portrait of Washington in 1796, which was sent to Spain, where it remained in obscurity until 1918. After having been invited to Havana to decorate the cathedral with frescos, Peruani came to influence Cuban art.

Felipe de Fondesviela or Fons de Viela y Ondeano (1722–84), also known as MARQUIS DE LA TORRE, was governor of Cuba from 1771 to 1777. Besides building the Paseo and improving the city streets, he gave Havana its first whale oil street lights.

1.13. Luis de LAS CASAS y Aragorri (1745–1800) was governor of Cuba from 1790 to 1795. He was the first president of the famous **Patriotic Society** (Sociedad Económica de Amigos del País), where Humboldt was invited, during his second stay in 1804, to present the results of his mineralogical research conducted in the hills of Regla and Guanabacoa. There is an oil painting of Luis de las Casas by Juan del Río (1748–1819) in the Museum of the City of Havana.

Juan Procopio Bassecourt y Bryas, COUNT OF SANTA CLARA (1740–1820), was governor of the Island of Cuba from 1796 to 1799. He continued the expansion not only of Havana's *Paseo*, as Humboldt points out, but also of the city's fortifications, which had been initiated by his predecessor, **Luis de las Casas**.

The *BARRACONES* were one of the many symbols for the inhumane treatment that Cuba's slave population suffered. In contrast to the more humane *bohíos*, individual slave cabins that accommodated slaves in open household units, the Cuban barracks were enclosed masonry structures, in which the living areas of men and women were separated into prisonlike cells. On some plantations, these barracks were even surrounded by wooden or stone walls.

CHARLES III (1716–88) was King of Naples and Sicily from 1735 to 1759, and King of Spain from 1759 to 1788. His support for France in the Seven Years' War (1756–63) produced an immediate reaction from Great Britain, which sent a large fleet across the Atlantic to take control of Havana. The **British invasion** lasted from March to August of 1762 and represented a major symbolic and military loss for colonial Spain.

Christopher COLUMBUS (1451–1506), the famous Genovese explorer, initiated the European colonial era with his four voyages to America between 1492 and 1504. On the first of his four journeys, Columbus arrived at the island of Cuba on October 28, 1492, after having visited five islands in the Bahamas; their exact locations still remain under dispute. Humboldt's achievements in the scientific arena are, in many ways, related to the accomplishments as the first "discoverer" of the Americas. Named by **Bolívar** the "second discoverer" of the Americas, Humboldt (whose mother's maiden name was *Colomb*) discusses Columbus throughout his writings. Humboldt read the Genovese captain's travel diaries while working on vol. 3 of his *Relation historique*, and he emphasized in his own texts the importance of the *intellectual* dimensions of Columbus's "discovery" without denying the atrocities that the Spanish *Conquista* had brought to the New World. The parallels that Humboldt saw between Columbus and himself in this respect are particularly evident in his *Examen critique de l'histoire de la géographie du Nouveau Continent* (1836–39) [A critical analysis of the history of geography of the New Continent], which focused on the fifteenth and sixteenth centuries. Humboldt describes Columbus as the prototype of the European discoverer who—unlike a mere adventurer—followed an elaborate preconceived plan (though replete with errors) and was equipped with profound geographical knowledge and cartographical skills. Humboldt encouraged Washington Irving to write his biography of Columbus, *A History of the Life and Voyages of Christopher Columbus* (1828).

Fernando CORTÉS de Monroy y Pizarro (1485–1547) was a Spanish *hidalgo* (low gentry) and adventurer. Sent in 1519 by the colonizer and first governor of Cuba, **Velázquez de Cuéllar**, to pillage the eastern coasts of Mexico, Cortés instead organized a small expedition of six hundred men to conquer and colonize the new land. With the help of his native interpreter and mistress Malintzín (also known as Malinche, c. 1501–50) and the allied Tlaxcaltecas, Cortés and his men rapidly gained control of the Mexican heartland and, in 1521, overthrew the Aztec capital of Tenochtitlan, where they captured and finally killed **Moctezuma** as well as his successor, Cauhtemoc (c. 1495–1522), the last kings of the Aztec Empire.

1.14n. Like Humboldt himself, the German botanist Karl Sigismund KUNTH (1788–1850) was a student of **Willdenow**'s. Kunth was Humboldt's assistant in Paris from 1813 to 1819, classifying the plants that Humboldt and **Bonpland** had collected during their journey through the Americas.

1.16. On April 25, 1802, a MAJOR FIRE broke out in the suburb of Jesús María outside of Havana's city. One hundred ninety-four houses were burned to the ground, as a result of which more than ten thousand of the mostly poor inhabitants sought shelter in the adjacent neighborhood of La Salud. Some of them were resettled to the districts of Mariel and Matanzas with the help of the **Count of Jaruco y Mopox**.

1.18. Between March and August of 1762, British naval troops put Havana under siege, a response to the Spanish alliance with France in the Seven Years' War (1756–63). Humboldt, who stresses the importance of Havana's defenses, was very aware of the symbolic value of the BRITISH TAKING THE *MORRO*, a heavily armed fortification of Havana's port entry, which resulted in the surrender of the Spanish troops two weeks later. It was the first time that Havana had fallen to foreign invaders. Cuba remained in

British hands for nearly eleven months, during which a host of merchants descended on the island. Exchanging Havana for Florida in the **Treaty of Paris** (February 10, 1763—there were several over the centuries), Britain withdrew its troops from the island; but trade relations with the British remained intact. This brief moment of British presence on the island had a major impact not only on Cuba's society and economy but also on administrative and other reforms throughout the Spanish colonies. Putting an end to the colonial trade restraints enforced by the metropole, the British reforms under occupation aimed at opening up the trade routes, removing agricultural monopolies, strengthening Cuba's Creole elites, and centralizing political power in Havana. Many of these reforms were incorporated into a master plan by the Count of Ricla, Ambrosio Funes de Villalpando y Abarca de Bolea (1720–80), who ruled as governor of Cuba from 1763 to 1765. Cuba's new society and economy were to be based on the large-scale import of slaves, whose labor would support an agricultural infrastructure that relied almost completely on sugarcane, tobacco, and coffee.

1.26. *Mestizos, zambos,* and PARDOS are among the many terms that existed in Spanish America to describe degrees of racial mixture. *Mestizos* are persons of mixed European and Native American ancestry. *Zambo*, a word to which Cuban anthropologist Fernando Ortiz (1881–1969) attributes a West African origin, was a designation used in Cuba and other parts of Spanish America for persons of mixed African and Amerindian descent. In the United States by contrast, *zambo* refers to individuals of African descent, specifically descendents of a Negro/a and a Mulatto/a. The Spanish use of the word may have influenced the English use of *sambo* as a racial stereotype. *Pardo* (brown) refers to the first-generation descendants of African and Spanish parents; it is a term used almost interchangeably with *mulato*.

1.36. The CUESTA, O'REILLY, and SANTA MARÍA FAMILIES were among the richest and most prominent aristocrats, merchants, and sugar mill *hacendados* in Havana. Humboldt stayed in both families' residences. While Luis de la Cuesta offered Humboldt to take care of one of the three of his identical herbarium collections, Count Pedro Pablo O'Reilly y de las Casas (1768–1832) offered Humboldt his in-town palace for the safe storage of his **Nairne** thermometer and other precious instruments. See also Juan Gonzáles.

1.37. Dionisio Alcalá GALIANO (or GALEANO) (1762–1805) was a Spanish naval officer, cartographer, and explorer whom Humboldt met in Havana. An island in British Columbia, Canada, is named after him; he came to that area in 1792, together with Cayetano Valdés y Flores, who had been part of the **Malaspina** expedition in 1789.

For BORDA'S COMPASS, see chronometric lines.

SEXAGESIMAL, a numeral system with sixty as the base, originated with the ancient Sumerians in the 2000s BCE, was transmitted to the Babylonians, and is still used—in modified form—for measuring time, angles, and geographic coordinates.

1.38. British astronomer Edward SABINE (1788–1883) experimented with determining the shape of the Earth and studied the planet's magnetic field. As the astronomer of the expeditions of Sir John Ross (1818, 1777–1856) and Sir William Edward Parry (1819) in search of the Northwest Passage, Sabine carried out pendulum measurements in the Arctic. His wife, Elizabeth Sabine (1807–79), translated Humboldt's *Cosmos* and other works into English. See also Henry Foster.

1.39n. In 1794, the naval officer and hydrographer Élisabeth Paul Édouard ROSSEL (1765–1829) became commander of a French maritime expedition in search of a lost expedition to the Pacific. Bruni d'Entrecasteaux's (1737–93) expedition (1791–96) had two ships: the *Espérance*, under the command of Captain Jean-Michel Huon de Kermadec (1748–93), and the *Récherche*, with first and second lieutenants Dauribeau and Rossel. In "Practical Treatise on finding the Latitude and Longitude at Sea" (1815), a translation of which appeared in Griffiths's *Monthly Review* in 1817, Rossel mentions his friend Jean-Baptiste BIOT (1774–1862) as having praised his work on nautical astronomy. Biot's *Traité élémentaire d'astronomie physique* was first published in 1805 and in a revised version, with Rossel's input, in 1810. Biot was a French physicist who joined **Gay-Lussac** on a first scientific balloon excursion in 1804 and also worked with Humboldt's friend **François Arago** on geodetic operations and triangulations.

Christopher HANSTEEN (1784–1873), a Norwegian astronomer and physicist, was noted for his research in geomagnetism. He took measurements in London, Paris, Finland, and, from 1828 to 1830, in Siberia, although not as part of Humboldt's expedition there, at roughly the same time. Carl Friedrich Gauss (1777–1855) ultimately disproved Hansteen's theories of geomagnetism.

1.41. Baron Bernhard August von LINDENAU (c. 1780–1854) was the first minister of the interior of the German state of Saxony. Starting in 1807, the astronomer edited the journal *Monatliche Correspondenz zur Beförderung der Erd- und Himmels-Kunde*. The journal had been founded by **Zach** in 1800 and was published until 1813.

1.43. José Sánchez CERQUERO (1784–1850) joined the Spanish Navy in 1778 and, in 1812, became director of the Marine Guards Academy of Cartagena, which had been founded in 1776 to provide future naval officers with the necessary scientific training in navigation. He left this post in 1816, when he was appointed to the astronomical observatory of San Fernando. From 1825 until 1846, Cerquero served as the observatory's director and used his post to reorganize the institution, visting for that purpose the observatory at Greenwich and observatories in France and Belgium. He also improved the *Nautical Almanac*. A member of the Royal Astronomical Society and the Royal Meteorological Society, he was fluent in English, French, Italian, and German and published a wealth of scientific works. Among other topics, Cerquero wrote about methods for obtaining latitude at sea by observing altitudes and about calculating eclipses.

Spaniards of French descent, the engineers Francisco LEMAUR y de la Muraire (1770–1841) and Félix LEMAUR y de la Muraire (1767–1841) were two of the four sons of Carlos Lemaur y Burriel (c. 1720–85), designer of the Guadarrama-Rozas canal in Spain (1785). He committed suicide later that year, and his sons completed this project. They were also the unsuccessful planners of the old Güines canal in Cuba and technical advisors to the Railroad Commission that was formed on the island in 1830.

1.44. ALDEBARAN (or Alpha Tauri) is the brightest star in the constellation Taurus and one of the brightest stars in the nighttime sky. NASA's Pioneer 10 spacecraft will reach and pass by Aldebaran in about two million years.

GEOGNOSY (i.e., the orderly succession of geological formations, or the superposition of geological strata), which literally means "knowledge of the earth," has now been replaced by its near equivalent, *geology*, but it was originally far more specific than that, insisting on five basic rocks formations and on the so-called Neptunian theory of geology: all the rocks on the earth's crust had been deposited there as the all-encompassing ocean receded. Abraham Gottlob Werner (1750–1817) was mining inspector and a celebrated instructor at the Freiberg School of Mines in the German state of Saxony, where he taught the study of rocks and minerals and geognosy. Werner's students, including Humboldt (since 1791), cherished their teacher even as they set about refuting theories that, in some circles, had hardened into dogma. See also von Buch and volcanism.

1.47. The LOMAS DE SAN JUAN (or SAN JUAN HILL) are a mountain ridge in Cuba. In the United States, San Juan Hill, the highest point of the San Juan Ridge, became famous in 1898 during the Spanish-Cuban-Philippine-American War, a power struggle between the young and aspiring United States—best personified in the figure of Theodore Roosevelt (1858–1919)—and old Spain over Cuba and the Philippines. On the U.S. side, the war has been celebrated as "the splendid little war," whereas the Spanish filed it away as "the disaster." Roosevelt resigned a post as assistant secretary of the navy to become the second in command of a colorful and unorthodox regiment called the Rough Riders (1st Volunteer Cavalry). On July 1, 1898, the Rough Riders joined in the capture of Kettle Hill, after which they moved on to assist in the taking of San Juan Hill. Roosevelt came to personify ideal U.S. masculinity at the turn of the twentieth century and ultimately ascended to the presidency after William McKinley (1843–1901) was assassinated in 1901.

1.49. The Cuban chemist Francisco RAMÍREZ was a passionate mineralogist. When Humboldt and **Bonpland** visited Cuba for the second time (from March 19 to April 29, 1804), they gathered materials that they had left with Ramírez. Ramírez was a student of Joseph Louis PROUST's (1754–1826), a French chemist who held professorships in different places in Spain but early and late in his career worked—in family tradition—as a pharmacist in France. An analytical chemist, Proust is best known for his law of definite proportions. He was exclusively concerned with inorganic binary compounds, such as metallic oxides, sulfides, and sulfates. Ramírez probably studied under Proust in Spain before Humboldt traveled to the New World.

1.50n. José Martín Félix de ARRATE y Acosta (1701–64), one of the first historians of Havana, completed his *Llave del Nuevo Mundo, antemural de las Indias Occidentales:*

La Habana descripta [Key to the New World, fortress of the West Indies: Havana described] in 1761; it was first published in its entirety in 1827 (portions of it had been printed in the *Patriota Americano* in 1812). Alongside his defense of Havana's Creole oligarchy, Arrate laid out economic statistics and policies for Cuba, which were inspired by modern economic theorists.

1.56. Adelbert von CHAMISSO (1781–1838) is best remembered today as the author of *Peter Schlemihl's Remarkable Story* (German, 1814). As a child in the throngs of the French Revolution, he had been moved from France to Prussia, eventually abandoning his French mother tongue for German. Chamisso was also a noted scientist. As a naturalist, he joined an exploration journey from the South Sea to the Bering Straits under the command of Otto von Kotzebue (1787–1846) from 1815 to 1818. The travel account Humboldt quotes was published in 1821. From 1804 to 1806, Chamisso coedited the *Berliner Musenalmanach* with Humboldt's friend Karl August Varnhagen von Ense (1785–1858), whose correspondence with Humboldt is available in published form. Chamisso is a good example of how science today has become so specialized that it is difficult to conceive how one person might be at once a literary storyteller and an acclaimed scientist, a description that also fits Humboldt.

Joseph Paul GAIMARD (c. 1790–1858) was a French naturalist who embarked on various journeys of exploration. In 1817–20, he traveled to the Mariana and Hawaiian Island groups. He wrote about these journeys, and at least two species are named in his honor. Humboldt likely had in mind Gaimard's 1823 (or 1825) paper "Mémoire sur l'accroisement des polypes litophytes" (coauthored with Jean-René Quoy [1790–1869]), which discussed geological aspects of coral formation.

1.56n. Alexandre Moreau DE JONNÈS (1778–1870) was a soldier and an adventurer who traveled in Asia and in the Antilles, notably in St. Vincent, Saint-Domingue, Martinique, and Dominica, but also in Venezuela and on the Orinoco. He was Aide-de-camp for numerous admirals and generals, often in charge of maps, and was known for his risky missions. He crossed the Atlantic ten times and participated in numerous expeditions. Upon his return to France in 1814, the government placed him in charge of statistical and topographical work and, in 1819, he was awarded the first prize in statistics from the Académie royale des sciences in Paris.

1.58n. Abbé PIETRO MARASCHINI (1774–1825) was an Italian geologist educated at a seminary in Vicenza and at Padua, Italy. He served as inspector of mines during the Napoleonic government, traveling extensively through Italy, including the Vicentine prealps, to study rock formations. He authored several essays on lithostratigraphy between 1810 and 1823.

1.60n. During his second visit to Cuba, Salvador de Muro y Salazar, the MARQUIS OF SOMERUELOS (1754–1813), governor and captain-general of the island and president of the **Patriotic Society**, asked Humboldt to undertake a mineralogical investigation in the hills of Guanabacoa and Regla near the city of Havana. To the disappointment of the Cuban authorities, Humboldt could not find proof of significant silver or gold deposits. He presented his results before the *Society* on April 13, 1804; the actual publication of the slightly revised "NOTICIA MINERALÓGICA" in the Cuban journal *El Patriota Américano* did not come until 1812 (see the HiE website for the digital facsimiles and a translation).

1.62n. In 1508, **Nicolás de Ovando** charged Sebastián de OCAMPO (c. 1450–c. 1525) with the task of circumnavigating Cuba. Ocampo visited the port of Havana, calling it "Puerto Carenas"; this is where he sealed two ships using the natural pitch (*chapopote*) that he found in the area. See also repartimiento.

1.65n. A graduate of Edinburgh University, the physician and geologist John MACCULLOCH (1773–1835) taught chemistry at the Royal Military Academy (1803–21) and at the East India Company (1814–35). MacCulloch's main contribution to medicine was the compilation and analysis of data on fevers. He published two works on the subject: *Malaria* (1827) and *An Essay on the Remittent Diseases* (1828). He joined the Royal Geological Society of London in 1808 and served as its president from 1816–18. After years of fieldwork in Scotland he published detailed descriptions of his findings on Scottish rocks in the *Transactions of the Geological Society of London*. A keen observer, analyst of crystalline rocks and erosion, he published *A Description of the Western Islands of Scotland* (1819) and *A Geological Classification of Rocks* (1821). He also worked on a Scottish geological map (four miles to an inch) published in 1836. His posthumous *Proofs and Illustrations of the Attributes of God* (1837) argues that geological evidence revealed the work of God in creation.

A mentor to Charles Robert Darwin (1809–82), Sir Charles LYELL (1797–1875) taught geology at King's College in London. In 1826–27, he published his first essays in the *Quarterly Review*, the official journal of the Geological Society of London. Lyell's seminal work, *Principles of Geology* (1830–33), shifted away from **Cuvier**'s view that unexplained catastrophes caused earth's changes and popularized the theory that modern causes were adequate to explain the past (this is known as actualism). His data source was a compilation of the physical and topographical changes recorded within human history published by **von Hoff** between 1822 and 1824. Lyell served as president of the Royal Geological Society for two terms (1835–37 and 1849–51) and received the society's highest award in 1866 (the **Wollaston** Medal). Among other books, he also published *Elements of Geology* (1838–41), the aim of which was to explain his theory of geology to a wider audience. He was knighted in 1848.

1.66. Edme Jean Antoine d'Orval DUPUGET, Comte Du Puget d'Orval (1742–1801), was a French general and naturalist who served as military inspector of the French colonies and traveled through the West Indies and Saint-Domingue from 1784–86. Once back in France, he was appointed to several high ranking military positions. During the **Reign of Terror** he was imprisoned, later released, and finally retired to write about natural history. In 1796, Dupuget was elected to the National Institute of

Arts and Sciences (Section of Natural History and Mineralogy) and published several articles about mineralogy in the *Journal des mines* in Paris.

1.68n. Giorgio GALLESIO (1772–1839) was an Italian botanist who, although trained in law at the University of Pavia in Italy, had a passion for pomology. He published several books on the topic, including *Teoria della riproduzione vegetale* [Theory of plant reproduction] (1816) and his famous *Pomona italiana ossia trattato degli alberi fruttiferi* [Italian fruits, or treatise about Italian fruit trees] (1817–39), an important collection of images and description of fruits and fruit trees from Italy. The Gallesia, a flowery plant species, is named after him.

1.69. CALORIC means "of or pertaining to heat." In the eighteenth century, a widely accepted caloric theory explained phenomena of heat and combustion with the help of a hypothetical weightless fluid known as caloric. Only by the mid-nineteenth century did the caloric theory cease to be influential and was replaced by the recognition that heat is a form of energy transfer. "Caloric updraft," then, is an ancestor to the theory of radiation cooling, which describes the nighttime cooling when the ground, lacking incoming solar energy, continues to radiate away heat, causing temperatures to drop. See also Wells and Wilson.

The *ZANJA DE ANTONELLI*, also known as the Zanja Real, is the first aqueduct the Spanish built in the Americas. The RÍO CAUTO is the longest river in Cuba.

1.72. Walter OUDNEY (1790–1824) was a British explorer and medical doctor from Scotland. He traveled in Africa and died in Katagum, Nigeria. Bornu is an old kingdom in what is now Nigeria; some sources say it is an old name for Nigeria.

1.74. Dominique François Jean ARAGO (1786–1853) was a French physicist and astronomer with whom Humboldt had close contact during his entire scientific career. On one occasion, Arago admonished Humboldt that he did not know how to write books.

1.75. William Charles WELLS (c. 1757–1817), born in South Carolina of Scottish descent, prepared for a medical career in Scotland under authorities who were naturalists rather than only medical doctors. In 1814, he wrote an essay on dew, for which the Royal Society of London awarded him the Rumford Medal. He also formulated evolutionary theories long before Charles Darwin did, trying to understand how different races of humankind might have come into existence. In the introduction to his paper on dew, Wells mentioned Patrick WILSON (1743–1811), professor of practical astronomy at the University of Glasgow, who wrote a paper on hoarfrost, published in 1788 in the first volume of the *Transactions of the Royal Society of Edinburgh*. Wilson had first published on the subject in 1780 and 1781. It is likely that Humboldt was referring to the 1815 essay, in which Wells pointed out that he and Wilson had had the same idea independently of one another.

1.76n. The chemist Thomas Luc Augustin HAPEL (DE) LACHENAIE (1760–1808) was chief apothecary of the Military Hospitals of Guadaloupe. He embarked for Guadeloupe in 1784 and remained there for the rest of his life. In 1798 and 1802 he published "A New Process for Claying Sugars" in the *Journal of Natural Philosophy, Chemistry, and the Arts* edited by Charles Nicholson (1753–1815), a British engineer who invented various scientific instruments. André Pierre LE DRU (1761–1825), a French naturalist who was ordained to priesthood early during the French Revolution, traveled as an official botanist with an expedition under Captain **Baudin** to the Canary and West India islands in 1796. He owned a fine herbarium and botanical garden and, in 1809, published the "Memoires sur les ceremonies religieuses et vocabulaire des Guanches" [Account of the religious ceremonies and the vocabulary of the Guanches] (in *Memoires de l'Académie Celtique*), followed in 1810 by the *Voyage aux isles de Tenerif, la Trinite, St. Thomas, Ste. Croix, et Porto Rico* [Travels to the islands of Tenerife, Trinidad, St. Thomas, St. Criox, and Puerto Rico].

1.77. Bento SÁNCHEZ (or Sánches) DORTA (1739–95) was a Portuguese astronomer and a member of the Portuguese Academy of Sciences. His *Memorias da Academia real das sciencias de Lisboa* [Reports of the Royal Academy of Sciences of Lisbon] were based on nearly two thousand meteorological and astronomical observations conducted seven days a week between 1781 and 1788. The *Memorias* were published only partially, in 1797, 1799, and 1812.

A French priest of the congregation of Saint Lazare, Jean François RICHENET (1759–1836) lived and worked in Macao, China, as a missionary between 1801 and 1815. In Macao, Richenet undertook atmospheric measurements for three years, working—as did Humboldt—with Six's **maximum-minimum thermometer.**

1.79n. Baron Christian Leopold VON BUCH (1774–1853) was a Prussian geologist and geographer who studied under Abraham G. Werner at the Freiberg School of Mines in Saxony from 1790 to 1793, where he must have met Humboldt. Starting in 1797, Buch investigated the Alps, moved on to Italy, and visited the Auvergne Mountains in 1802. His volcanic theories contributed to revising Werner's theories. He also traveled to Scandinavia and the Canary Islands. His extensive wanderings and writings significantly influenced the development of geology. Humboldt expressed his high regard for Buch in *Kosmos* and in letters to Varnhagen von Ense. At the point of his death, Humboldt owned nearly forty items authored by Buch. See also geognosy.

Francisco ESCOLAR (d.1826) kept a register in Santa Cruz, Tenerife, from 1808 to 1810, on which Buch commented in his *Physicalische Beschreibung der Canarischen Inseln* [Physical description of the Canary Islands]. Apparently, Escolar prepared statistics on the Canary Islands between 1793 and 1806. He studied law in Spain in the 1790s and also took chemistry and botany at Göttingen. He translated the 1801 *Principes d'économie politique* (1804) by Nicolas François Canard (1754–1833).

1.83. The Scottish botanist Robert BROWN (1773–1858) was part of a British expedition under the command of Matthew Flinders (1774–1814) along the northern and southern coasts of Australia from 1801 to 1805. In 1810, he took charge of **Joseph Banks**'s library, which he inherited after Banks's death. Brown donated the library to the British Museum when he became keeper of botany there. Humboldt esteemed Brown, whom he had met in Paris in 1816, so highly that he procured an annual pension for him through Sir Robert Peel (1788–1850).

1.83n. Joannes or Johannes de LAET (Latinized as Ioannes Latius) (1581–1649) was a Dutch geographer and director of the Dutch West India Company. Laet's history of the New World, entitled *Novus orbis*, has been called the finest description of the Americas published in the seventeenth century and one of the foundation maps of Canada. Laet was the first to print maps with the names Manhattan, New Amsterdam (New York), and Massachusetts. He was embroiled in a controversy with Hugo Grotius (1583–1645) about the origin of the American peoples. It is much less known that he also worked on an early Anglo-Saxon dictionary that has been lost.

1.85. Olof Peter SWARTZ (1760–1818) was a Swedish botanist.

1.87. The physicist Edward NAIRNE (1726–1806) was one of the most prominent British manufacturers of scientific equipment. Humboldt, who was known for using only the finest instruments, worked with this and other thermometers made by Nicholas Paul (Geneva), Jesse Ramsden (London, 1735–1800), Pierre Bernard Megnié (Madrid, 1751–1807), and **Fortin** (London). They were mainly used for determining air temperature.

1.89. RÍO BLANCO PLANTATION was owned by the Count of **Jaruco y Mopox**.

1.91n. Bryan EDWARDS (1743–1800) was a British politician and historian who became a member of parliament in 1796. After his father's death in 1756, his maternal uncle Zachary Bayly (1721–69), a wealthy Jamaican merchant, took charge of him, and Edwards moved to Jamaica in 1759. Edwards inherited two of Bayly's estates. A leading member of the Colonial Assembly in Jamaica, Edwards returned to England in 1782. He published a history of the British Colonies in the West Indies in 1793. Edwards was also a secretary of the African Association, wrote poems on the history of the West Indies, and composed a brief sketch of his life which was published in 1801. He engaged in scientific debates about the impact of the New World environment on plants, animals, peoples, and cultures. Representing a moderate planter position, Edwards opposed the abolition of the slave trade. See also Beckford and MacFarlane.

1.92. Humboldt's teacher in Freiberg, Abraham Gottlob Werner (see geognosy), had minimized the significance of the earth's movement, but Humboldt and others increasingly recognized its role in shaping the earth's surface. By calling for more data on what was called VOLCANISM, or vulcanism, Humboldt points toward this new direction in geological study. While traveling in the Andes, Humboldt, once a convinced neptunist, realized that volcanic activity—evidence of HEAT AT THE EARTH'S

CORE—played an important role in creating mountains. Another of Werner's students, **Leopold von Buch**, studied extinct volcanoes in central France. Even though he dedicated his studies on the Alps (1802–9) to Werner, he confirmed that massive earth movements had actually created these mountains.

1.97-98. Pánfilo de NARVÁEZ (1470–1528) was a Spanish conquistador and lieutenant under Cuba's first governor, **Velázquez de Cuéllar**. With a special authorization by the Spanish king to found new colonies and begin trade, Velázquez saw an opportunity to prevent his rival, **Cortés**, from conquering the Mexican mainland; he sent out Narváez to stop Cortés' expedition and arrest him. Together with the Spanish captain Alvar Núñez Cabeza de Vaca (c. 1490–c. 1557), Narváez undertook, in 1528, an expedition through the Gulf of Mexico from Tampa Bay to Galveston, which took a disastrous turn. After landing in Tampa Bay, Narváez had decided to split up his troops of four hundred men and eighty horses into two expeditions, one to sail north along the western coast, the other to follow the same route on land. But the tropical jungles of the Florida peninsula and the steady attacks by Apalaches obstructed the land route, and the two expeditions never saw each other again. Two years and several shipwrecks later, few expeditioners were still alive, among them Cabeza de Vaca, who continued his odyssey along the Texan Coast and the Rio Grande towards the heartland of New Spain (Mexico) in search of Spanish settlements until 1536. We know of these first European pioneers to the West from de Cabeza deVacas' famous *Naufrágios y Comentarios* [Shipwrecks and commentaries]. The most complete English version of this narrative is the three-volume *Álvar Núñez Cabeza de Vaca: His Account, His Life, and the Expedition of Pánfilo de Narváez* by Rolena Adorno and Patrick Pautz (1999).

1.98. Brigardier Tomás de UGARTE y Liaño (1754–1804) was one of several officers of the Spanish Royal Navy whom Humboldt met in Lima and Callao. In 1799, Ugarte was in charge of designing the ports of the South Sea, from Chiloe to the north coast of Veraguas Province. In 1802, he became chief of the entire Spanish fleet but retired a year later. Humboldt was familiar with Ugarte's maps, several of which can be found in the Karpinski Collection in the Library of Congress.

1.100. BARACOA was believed to be the first place where **Columbus** landed in Cuba and where, according to legend, he placed a cross at the later site of the Baracoa harbor. **Velázquez de Cuéllar**, the island's first governor, made Baracoa Cuba's first capital in 1511. In 1518, Baracoa, which is located in today's Cuban province of Guantánamo, officially became a city, and the first Cuban bishop was appointed there.

1.108. Franz Xaver Freiherr von ZACH (1754–1832) was a German-Hungarian astronomer under the patronage of Ernst II (1745–1804), Duke of Saxe-Gotha-Altenburg. Zach built an observatory on the Seeberg near Gotha and directed the observatory—one of the most important of the time—from 1791, when it was completed, until 1806. During this period, Zach enlisted twenty-four astronomers from across Europe in a systematic search for new comets and for the planet between Mars and Saturn, expected on the basis of Johann Elert Bode's (1747–1826) law (the Titius-Bode law). The main result of this effort was the discovery of several minor planets (commonly called asteroids). Zach's most lasting achievement was editing three scientific journals from 1798 to 1826.

Established in the kingdoms of late-medieval Spain as a court to administer royal justice, the *AUDIENCIA* was one of the most important governmental institutions of Spanish colonial America. During the sixteenth century, Audiencias were established in the various administrative districts (viceroyalties, captaincies general) of Spanish America. The first Audiencia in the New World was that of Santo Domingo, set up in 1511 with jurisdiction over the Caribbean islands. In 1817, the Spanish Crown officially agreed to transfer the Royal Audiencia to the Cuban city of Santa María del Puerto Príncipe, today's Camagüey. Since the 1970s, Cuba has been organized into fourteen administrative districts or provinces and one special municipality (Isla de la Juventud, formerly the Isle of Pines): (1) Pinar del Río (formerly Nueva Filipina); (2) La Habana, which before 1976 included the province and the city of Havana; (3) Cuidad de la Habana; (4) Matanzas; (5) Ciénfuegos (part of Las Villas before 1940); (6) Villa Clara (part of Las Villas, or Cuatro Villas, before 1940); (7) Sancti Spíritus (part of Las Villas before 1940); (8) Ciego de Avila; (9) Camagüey (Puerto Príncipe prior to 1899; also included Diego de Avilas); (10) Las Tunas; (11) Granma; (12) Holguín; (13) Santiago de Cuba (Oriente, which included the provinces of Las Tunas, Granma, Holguín, and Guantánamo); (14) Guantánamo.

1.109. Juan de Witte Hoos (UBITE) was actually appointed Bishop of Santiago de Cuba in 1517 and resigned in 1525. It is possibly that Humboldt meant to write 1518 instead of 1528.

1.110. Today, the Cuban province of HOLGUÍN, along with the neighboring province of Guantánamo, is home to the Parque Nacional Alejandro de Humboldt [Alexander von Humboldt National Park], which is a World Heritage site.

1.113. In Cuba's toponymy, ABAJO and ARRIBA [down and up] signify western and eastern. This use originates in ships' navigation. It is used in other countries as well, notably in Nicaragua.

By some accounts, Pedro VALDÉS Balnueva was governor of Cuba from 1602 to 1607 or 1608. Valdés would have become captain-general in 1607, the year when **Philip III** created the captaincy-general of Cuba as a new administrative unit. At the time, the island was divided into two governorships (Havana and Santiago de Cuba); the governor of Havana was also the captain-general of Cuba. Humboldt's date, 1601, may have been a printer's error (1 instead of 7). By most counts, Cuba had nineteen, not sixteen, governors prior to Valdés.

Conquistador Diego VELÁZQUEZ DE CUÉLLAR (c. 1465–1524) was the first Spanish governor of Cuba. He accompanied **Columbus** on his second voyage in 1493. Together with **Cortés**, Velázquez embarked for Cuba in 1511, after Columbus's son

Diego had put him in charge of conquering that island and named him *adelantado* (governor). Velázquez founded Baracoa, Bayamó, Santiago de Cuba, and Havana and encouraged settlement. He also organized explorative journeys and sent Cortés to conquer the mainland of Mexico. Velázquez twice unsuccessfully attempted to stop Cortés, but even his commander, **Pánfilo de Narváez**, sided with Cortés. The courts instructed Velázquez to ignore Cortés, since they had begun receiving Aztec treasures from Mexico.

DIEGO COLUMBUS (c. 1479–1526), Columbus' eldest son, was a viceroy of the Indies for fifteen years and spent most of his life trying to secure his father's claims through legal channels. Diego arrived in Santo Domingo in 1509, three years after his father's death, to succeed Governor **Ovando**.

I.122. The Cuban SUGAR PLANTATIONS that were established and expanded in the wake of the **Saint-Domingue** revolution operated more like industrial factories than feudal estates, working the slave around the clock. See also *barracones*.

I.128. Abbé Guillaume Thomas François RAYNAL (1713–96) was a French Jesuit and historian who contributed significantly to preparing the intellectual climate for the French Revolution. His most important work was the *Histoire philosophique et politique des deux Indes*, translated as *History of the East and West Indies*, which appeared in several successive and expanded editions between 1770 and 1789. A very popular work, the *Histoire* was one of the centerpieces of a discourse on the New World as lacking civilization and cultural vitality, which Humboldt fervently opposed in his own writings. Because of its radicalism—the encyclopaedist Denis Diderot (1713–84) is credited with having authored its more radical passages—the *Histoire* was placed on the Roman Catholic Church's *Index of Forbidden Books* in 1774, and Raynal was banished from Paris in 1781. His property having been confiscated, he died in poverty. See also Nuix.

I.129n. Andrés CAVO (1739–1803) was a Jesuit historian from Nueva Galicia (present-day Mexico). Before the Jesuits' expulsion, he was a professor at the Ignacio de Puebla Seminary and missionary in Nayarit. He traveled to Italy with Father José Julián Parreño (1728–85), a Cuban Jesuit and former rector of the Colegio de San Idelfonso in Mexico City, whose biography he completed in 1792.

I.132. In 1811, when progressive Cubans were increasingly concerned about the slave trade and the growing numbers of slaves on the island, José Miguel Guridi y ALCOCER (1763–1828), the Mexican deputy to the Courts of Cádiz, and Agustín de ARGÜELLES y Álvarez González (1776–1843), a liberal Peninsular lawyer, politican, and diplomat, petitioned the Spanish Courts to put a stop to the transatlantic slave trade and to end slavery in the colonies. Later that same year, Arango y Parreño similarly appealed to the courts on behalf of Havana's merchant guild and the **Patriotic Society**. For the text of the petition see Arango, "Representación de la Ciudad de la

Habana a las Cortes el 20 de julio de 1811" [The representation of the city of Havana in the courts on July 20, 1811].

Lawyer, politican, and planter Francisco de ARANGO y Parreño (1765–1837) was one of the most prominent figures in late eighteenth and early nineteenth-century Cuba. Arango shaped many of the economic discussions about a more efficient handling of the Cuban trade and about agricultural infrastructure, especially the development of the big sugarmills with large numbers of slaves. After the revolution in **Saint-Domingue**, Arango moderated his position on slavery, discussing posibilities for the abolition of the slave trade and promoting the immigration of white European population. Fighting against the tobacco monopoly and other agricultural and commercial problems for Cuba, Arango became one of the foremost advocates of economic liberalization in nineteenth-century Cuba. During the early 1790s, Arango traveled to England to see how merchants there ran the slave trade and how manufacturers ran their factories. Upon his return in 1792, he founded the Sociedad Económica de Amigos del País (**Patriotic Society**), a deliberative body that also gathered statistics and economical infromation. It also inspired the founding of Cuba's first newspaper, *El Papel periódico*, in 1793. Arango met with Humboldt during both of his stays and showed him his own modern *ingenio*, "Las Ninfas," and other sugar estates in the region. Humboldt relied on Arango for many economical and statistical data and maintained close relations with him for several years after his departure from Cuba. Most notably, Arango felt the need to comment in writing on Humboldt's *Essai politique*, which led to the publication of the *Observaciones sobre el ensayo político de la isla de Cuba por el baron de Humboldt* [Observations about the *Political Essay on the Island of Cuba* by the Baron von Humboldt] in 1827. These observations had a strong impact on later Spanish editions of Humboldt's text, in which Arango's notes have become a paratextual part of Humboldt's *Essai* itself. For this edition, which aims to restore the original character of Humboldt's text, Arango's *Observaciones* have been added to the secondary apparatus and can be consulted at www.press.uchicago.edu/books/humboldt/.

1.133. The influential Sociedad Económica de Amigos del País—literally, economic society of the friends of the country—was founded in Havana in 1793, when governor **Las Casas** approved a petition from twenty-seven prominent Cubans. Foremost among them was **Arango**. Later on, this society was also known as the Sociedad Patriótica de Amigos del País, which is why Humboldt refers to it as the PATRIOTIC SOCIETY.

1.136. José Pascual de ZAYAS Y CHACÓN (1772–1827) was a Cuban-born Spanish military officer. In 1789, he sailed to Spain and began a successful military career, during which he was distinguished for his participation in the Peninsular War (1807–14). In 1820, he became a representative of Havana to the Spanish Courts.

1.139n. The term CREOLE is a problematic rendering of the Spanish *criollo*, which dates to the seventeenth century and refers primarily to place of birth, that is, to the American colonies of the Spanish Empire. *Criollos* were effectively second-generation

Americans whose social standing was below that of the first generation, Spanish-born colonists. *Criollo*, then, refers to social class rather than color, though some argue that it also implies whiteness as a default. Humboldt does not use *criollo* (or the French *créole*) as racial terms but specifies "white creoles"and "black creoles" to express racial distinctions. The English word *creole* as it has been used in the United States since the nineteenth century has different racial, or ethnic, connotations: it typcially refers to a racial mixture that includes African ancestry.

1.140. Irish writer John Wilson CROKER (1780–1857) served as the first secretary of the Admiralty from 1809 to 1830. While lacking ambition, the politician Croker was an excellent secretary valued highly for his eloquence.

Joaquín Antonio Vigil de QUIÑONES was mayor of San Juan de los Remedios in 1833 and 1836. José de AGUILAR, a governor of Nueva Filipina (now Pinar del Río) in the early 1800s, carried out the first population census of Nueva Filipina in 1819.

1.142. Comte Gilbert Joseph Gaspard de CHABROL DE VOLVIC (1773–1843) was a French military engineer educated at the École Polytechinque in Paris, France. A senior official under Napoleon, he joined the expedition to Egypt as a scientist. As the prefect of the Seine Departement and chief administrator of Paris (1812–30), he produced the *Recherches statistiques sur la ville de Paris et le département de la Seine* [Statistical research about the city of Paris and the department of La Seine] in collaboration with Jean-Baptiste-Joseph Fourier (1768–1830). The *Recherches*, four large volumes published in 1821, 1823, 1826, and 1829, was the first collection of demographic, economic, and meteorological statistics on Paris. One of the landmark accomplishments in the field of urbanism, it was the most significant source of raw data on population in France in early nineteenth century.

Jacques NECKER (1732–1804) from Geneva was a banker who served as the general director of finance under King Louis XVI of France (1754–93) between the 1770s and 1790 (he had moved to Paris in 1750). His daughter was Madame de Staël (1766–1817). Both returned to Switzerland in 1790, during the French Revolution. See also Lafayette.

1.144. So-called white letters were official certifications of "white" (that is, Iberian) ancestry that light-skinned persons of African descent could obtain from an *audiencia*, usually for a price. It was an accepted form of WHITENING or *blanqueamiento*.

1.147. A slave's ability to purchase his or her freedom was facilitated by a legal custom that had developed in the island since the sixteenth century: *coartación*. *Coartación* allowed slaves to agree with their masters on a fixed price for their freedom and to make payments toward it over time, in installments. Such agreements were legally binding and restricted the master's capacity to sell or mortgage a slave who had already paid a portion of the price. For the slaves' right to go into the service of another master (BUSCAR AMO, literally to seek a master), see Humboldt's explanation in a footnote in p. 150 of this text.

I.148. In his 1827 notes on Humboldt's text, **Arango** remarked that there were still remnants of CUBA'S INDIGENOUS POPULATION in the Taíno town of Jiguaní near Bayamo (in what is now the Cuban province of Granma).

I.149. The GUANCHES were the first-known inhabitants of the Canary Islands; they migrated to the archipelago sometime between 1000 BCE and 100 BCE and are now extinct as a distinct people (only mummies survive after the resisting Guanches finally succumbed to disease). The Spanish conquest of the Canary Islands, dating back to 1402, was complete by the end of that century.

I.150. For *ENCOMENDERO* see encomienda in Las Casas.

Spanish historian Francisco López de GÓMARA (1511–60) was **Cortés**' secretary. With his *Historia de las Indias* [History of the Indies], which Humboldt cites frequently, Gómara became one of the prominent chroniclers of the New World. Gómara's version of the Spanish conquests, especially his reports on Cortés's invasion of Mexico (1519–21) in his *Historia de la conquista de México* [History of the conquest of Mexico] (1522), was countered by Bernal Díaz del Castillo's (1492–1585) dramatic and colorful *Historia verdadera de la conquista de la Nueva España* [True history of the conquest of New Spain], which narrates Cortés's invasion through the eyes of an ordinary soldier, who—unlike Gómara and many other chroniclers of the time—had actually been in the New World and participated in the conquest. The *Historia verdadera* was published in 1632; the complete text was not recovered until 1904.

I.151n. Gilbert FARQUHAR MATHISON (d. 1828), the son of a Jamaican planter, was a writer of travel narratives about South America and the Caribbean.

The German geographer and statistician Johann Georg Heinrich HASSEL (1770–1829) published various statistical and geographical writings about Austria, Prussia, the German Confederacy, Helvetia (a part of Switzerland), Italy, the Ionian Islands (a group of islands in Greece), Great Britain, Spain, Portugal, Denmark, Sweden with Norway, Russia, Poland, the Asiatic kingdoms, the two East Indian peninsulas, Japan, British and Russian North America, the United States of North America, Mexico, Guatemala, and Australia. Humboldt owned at least seven of Hassel's books. Hassel had published statistics on Russia as early as 1805–7. Hassel also collaborated with others, among them Adam Christian Gaspari (1752–1830).

I.152n. ALEXANDER VI (1431–1503), born Rodrigo Borgia y Doms in Spain, was pope from 1492 to 1503. His neglect of spiritual matters contributed to the emergence of the Protestant Reformation. Alexander issued the *Inter Caetera*, a Papal Bull of May 4, 1493, in the wake of **Columbus**'s discoveries, which granted Spain the exclusive right to explore and claim the seas and lands west of a north-south demarcation line about 320 miles west of the Cape Verde Islands. Portugal was granted similar rights to the east of that line. Spain and Portugal amended this papal order in the Treaty of Tordesillas (1494), even though no other European power recognized the treaty. The Latin text of Humboldt's quotation is as follows: "Certas insulas remotissimas et etiam terras firmas invenerunt, in quibus quamplurimae gentes, *pacifice viventes,* nudae incedentes, *nec carnibus vescentes*, inhabitant, et, ut nuntii vestry possunt opinari, gentes ipsae *credunt unum Deum creatorem in coelis esse.*"

ZEMÍS (or *cemís*) are material representations of the spirits of gods and ancestors, which the Taínos revered in their public rituals and ceremonies. They usually have the shape of small animals or human faces and were carved of wood, clay, stone, shell, and bone.

1.153n. GARCILASO DE LA VEGA, El Inca (1539–1616), was a relative of the Inca Atahuallpa (c. 1502–33). Born Gómez Suárez de Figueroa, he was the son of the Spanish conquistador Sebastián de la Vega Vargas and the Inca ñusta [princess] Chimpu Ocllo (d.1571), baptized Isabel Suárez. He adopted the epithet "El Inca" to signify pride in his indigenous ancestry. Garcilaso lived in Spain from 1561 until his death; he was the first American to write in Spanish. He authored an extensive history of the Incas, the *Comentarios Reales de los Incas* [*Royal Commentaries of the Incas*], which was first published in Lisbon in 1609 and 1617. The *Comentarios* were banned in Peru in 1780 at the outset of the rebellion against colonial dominance led by José Gabriel Condorcanqui, better known as Tupac Amaru II (1742–81).

Juan NUIX de Perpiñá (1740–83) was a Catalán who lived in Italy after the expulsion of Jesuits from Spain. He published his *Riflessioni imparziali sopra l'umanità degli Spagnuoli nell'Indie contro i pretesi filosofi e politici: Per servire di lume alle storie de' Signori Raynal e Robertson* [Impartial reflections on the humanity of the Spanish in the Indies against the alleged philosophers and politicians: To illustrate the histories of Raynal and Robertson] in Venice in 1780. Humboldt quotes from one of two Spanish translations, the one from 1782 by Pedro VARELA Y ULLOA (b. 1737), secretary to the royal council. The second translation by Nuix's brother Josef from 1783 was little known. **Varela y Ulloa**'s translation was sponsored by the crown and included an essay by him criticizing colonialism. In portraying Spanish colonialism as unique and benign, he captured the essence of Nuix's book, which was written as a defense against **Raynal**'s and William Robertson's (1721–93) attacks on colonialism (in *The History of America*, 1777). Nuix also argued that **Pope Alexander VI**'s 1493 bull had not meant that the conquistadors should rule Native Americans but only convert them to Christianity.

PHILIP III OF SPAIN (1578–1621) reigned from 1598 to 1621. His favorite minister, Francisco Gómez de Sandoval y Rojas, the Duke of Lerma (c. 1552–1625), in cohuts with the Archbishop of Valencia, Juan de Ribera (1532–1611), succeeded in convincing the king to expel the *moriscos* [Moors] from Spain; Philip III signed an edict to this effect on April 9, 1609. What finally persuaded him was that he could confiscate Moorish assets and properties, which was a dramatic boost to the royal coffers. Ribera's proposal to enslave the *moriscos* for work in galleys and mines, however, was rejected.

After having been uprooted by the French Revolution, Joseph Marie (Count de) MAISTRE (1753–1821) became an opponent of the European Enlightenment. From 1803 to 1817, he served as ambassador to the Russian court at St. Petersburg. Humboldt quotes from his *Soirées de St. Petersbourg* (1821), one of Maistre's major works. Maistre critically analyzed the French Revolution, held a providential view of history, and supported the idea of the supremacy of Christianity and the absolute rule of sovereign and pope. Jack Lively (1930–98) translated and edited Maistre's writings in 1965.

I.154. During his fight against the Spanish invaders, the Taíno ruler HATÜEY (d. 1512) was captured and ordered burned alive by **Velázquez de Cuéllar**, even though **Las Casas** tried to intervene on the cacique's behalf. (The word *cacique*, meaning chief or leader, is Taíno in origin; the Spanish imported it from the Antilles to the rest of their colonies.) Las Casas subsequently recounted the story of Hatüey's execution in his *History of the Indies* and his *Brief Account of the Destruction of the Indies* to expose of Spanish cruelty toward the Indians—a cause to which Humboldt would contribute significantly in his *Political Essay on the Kingdom of New Spain*. Cubans still celebrate Hatüey as their first national hero.

Pedro de BARBA (d. 1521) was a Spanish soldier who served as first lieutenant and governor of Havana under **Velázquez de Cuéllar**. While **Cortés** was preparing his Mexican conquest from the newly founded city of Veracruz, Velázquez had ordered Barba—as he did Narváez—to detain Cortés. Barba, like Narváez, disobeyed, integrating himself into the conquest troops instead.

Hernando DE SOTO (c. 1500–42) was a conquistador in Nicaragua and Peru (with Pizarro) who became governor of Cuba. After the failed expedition to Florida by **Pánfilo de Narváez**, the more experienced de Soto was sent on a second voyage to the tropical peninsula (1539–43). On his famous route through Florida, along the Alabama River and across the Mississippi, de Soto searched for the mythical "Seven Cities" and a pathway to the Pacific—both without success. After his death in 1542, his expedition continued south across the Mississippi and along the coast to Tampico, where the remaining crew dispersed.

Spanish conqueror Diego de MAJARIEGOS founded San Cristóbal de las Casas—he named it Villa Real de Chiapa—in Mexico's southwest region of Chiapas in 1528. He was governor of Cuba from 1556 to 1565 (slightly different from how Humboldt recalls it) and governor of the Province of Venezuela from 1570 to 1576.

I.155n. Among his many treasures, Columbus had brought back to Spain a **Taíno** ornament from pre-Hispanic Cuba; it turned out to consist of an alloy of copper, silver, and some gold, which the locals called *GUANÍN*. In his inventory, Columbus referred to guanín as "base gold" and was surprised that the Taínos valued it more highly than they did pure gold. Guanín is of a reddish color and has a distinctive odor; archeologists contend that its iridescent qualities facilitated associations with the heavens, the divine, and thus with power. Even though Cuba did have copper ore, guanín appears not to have been made on the island but imported from the the mainland, likely the **Guianas**.

There is significant evidence that the ancient Americas were not isolated from the old worlds, including Africa, as historians used to believe. Muslims were probably one of the most important contact people before **Columbus**'s voyage. Evidence leading to the presence of Muslims in the ancient Americas comes from a number of sculptures, oral traditions, eyewitness reports, artifacts, and inscriptions. In 1513, the Spanish explorer Vasco Núñez de Balboa (1475–1519) encountered African people in the Isthmus of Darien. As late as the mid-nineteenth century, a number of Manding place names still survived in Panamá. There is additional proof of the presence of Ethiopians in pre-Columbian America. Father Roman, one of the first Catholic missionaries to arrive in the New World, recorded that a tribe of black men came from the south and landed in Haiti, that they were armed with darts of guanine, and that they were known as black Guaninis. It is possible that these were the Blacks of QUARECA mentioned by **Peter Martyr d'Anghiera**. They could also have belonged to other ancient Black Nations of the Americas, such as the black Carabees of St. Vincent or the Jamassi of Florida.

The CARIB people once lived in the Lesser Antilles; the Spanish called them "Caribis," by which they meant "Canibales" [cannibals], a likely corruption of the Carib and/or Taíno words "Canibis" or "Caniba" [literally, the brave ones], by which the Carib peoples referred to themselves. Their reputation for ferocity stems from their countless wars with the Arawaks (Taínos). The Spanish did not fight the Caribs but the British and the French did in the seventeenth century, when they became interested in the Lesser Antilles. In a peace treaty, the Caribs were given full possession of Dominica and St. Vincent. Today, a few of them survive on Dominica; their original type has been much altered by past mixing with runaway slaves from other parts of the Caribbean. See also Maroon.

I.158n. Humboldt's doubts about this claim are indeed well-founded: the first FRANCISCAN convent in Baracoa, Cuba, dates back to 1516; it was moved to Santiago de Cuba in 1522. The first missionary activities, however, date back to 1511 (Franciscans) and 1515 (Dominicans). It was not until the Spanish sailor **Sebastián de Ocampo** had circumnavigated the island in 1508 (other sources mention 1509–10) that Cuba was proven to be an island. This affirmed the assumptions made by the cartographer Juan de la Cosa (c. 1460–c. 1510) in his famous map of the Americas (1500) and finally refuted **Columbus**'s initial belief that he had reached the East Indian continent. It took another four years to conquer and "pacify" the island under the rule of **Velázquez de Cuéllar**, during whose military campaign seven major settlements were founded: Asunción (today's Baracoa, 1511), San Salvador (Bayamo, 1513), La Santísima Trinidad (Trinidad, 1514), Sancti Spíritus (1514), San Cristobál de La Habana (1514), Santa María del Puerto del Príncipe (Puerto Príncipe, 1515), and Santiago de Cuba (1515), the island's first capital.

I.163n. Antonio del VALLE HERNÁNDEZ was a member of the Havana merchant guild [Consulado] in 1795 and also Cuba's first demographer. In 1800, he recorded the situation of Cuba's sugar mills, describing the large plantations as industrial rather than agricultural enterprises, even though many mills (much to his dismay) continued to follow older practices. In some cases, however, the spirit of progress and inquiry had resulted in improvements in the grinding mills and kettle trains and the occasional installation of water-powered and steam-powered mills.

1.166. Physician and botanist Joel Roberts POINSETT (1779–1851) was a U.S. Representative from South Carolina from 1821 to 1825 and from 1830 to 1831. In 1809, President James Madison (1751–1836), one of Humboldt's correspondents, had sent Pointsett to South America to investigate the prospects of the revolutionists in their struggle for independence from Spain. Poinsett was minister to Mexico from 1825 to 1829 and, from 1837 to 1841, served as secretary of war in the cabinet of President Martin Van Buren (1782–1862). Humboldt quotes from Poinsett's 1824 *Notes on Mexico*.

1.167. The term *INGENIO* refers to a sugar plantation in its entirety, not just to the mill. It was used until 1880, when the term *central*, which referred to large sugar factories, began to replace it and *ingenio* was used mainly to refer to the mill itself, rather than to the entire operation.

1.168. Pedro Gómez REYNEL from Portugal was the first to be granted a contract to sell 4,250 slaves annually to Spain's American colonies for nine years, starting on May 1, 1595. In 1601, the concession was withdrawn and awarded to the Portuguese Juan Rodriguez Coutinho, merchant and governor of Loango (Congo). Slave trade had already occurred several years prior to this monopolization. The first black slaves arrived in the New World in 1510, but they had been purchased in Lisbon, not in Africa.

In 1615, Lisbon banker Antônio Rodríguez DE ELVAS (or Delvas) was given the right to import African slaves to the New World for a period of eight years. During the early years of colonization, this right—called ***asiento***—was awarded as a monopoly over a restricted period of time. The contractor was to transport three thousand five hundred slaves a year to the Indies in return for an annual fee of one hundred fifteen thousand ducats paid to the crown. The special advantage that the royal treasury had from the increase in import taxes was made up to contractors, such as Rodríguez de Elvas, whom the crown relieved of taxes on merchandise he might export either from Spain or from the colonies. See also slave trade.

During the WAR OF SUCCESSION (1701–14), several European powers joined forces to stop a possible unification of the kingdoms of Spain and France under a single Bourbon monarch, which would have upset the balance of power in Europe. The 1713 **Treaty of Utrecht** helped end the war, which had been fought mainly in Europe but also in the American colonies. As a result of the treaty, Spain's European empire was divided. Spain ceded Gibraltar and Minorca to Great Britain and also agreed to give the British a thirty-year ***asiento*** for importing slaves to the Spanish colonies. The British South Sea Company was in charge of this trade.

1.169. In 1778, Jerónimo de ENRILE y Guersi (or Garci) (d.1813) was named Marquis de Casa Enrile in Havana, where he worked as the representative of the Compañía Gaditana del Real Asiento de Negros.

Until 1789, Spain had regulated the transatlantic SLAVE TRADE mainly by awarding contractual monopolies for the direct importation of a fixed number of African slaves into the Spanish colonies either to individuals or to companies or agencies. Previously, all such imports had to be cleared through Seville. These contracts were known as *ASIENTOS*. Because they limited the numbers of imported slaves, they also encouraged significant illegal trade. The first *asiento* was granted to the Portuguese **Gómez Reynel** in 1595. In 1789, **Arango** successfully fought the renewal of a contract with a British firm of ship builders and slave merchants, Peter Baker (d. 1796) and John Dawson of Liverpool, arguable the wealthiest slave traders of their times. The abandonment of the *asiento* system was the beginning of the free trade in slaves in the Spanish Empire, which made it possible for far larger numbers of African slaves to be shipped to Cuba and to Puerto Rico than had been the case before, when the majority of the transatlantic trade had supplied the continental colonies. Contraband trade in slaves continued after the cessation of the transatlatic trade in 1820.

I.176. The Quaker merchant James CROPPER (1773–1840) was a leading supporter of abolition in Britain, even though he himself imported slave-grown cotton. Cropper traded with Ireland, North America, India, and China, and the wealth he amassed enabled him to become active in religious and philanthropic affairs. He was associated with **William Wilberforce** and with the more radical wing of the British antislavery movement, which demanded the immediate and unconditional extinction of slavery. Cropper also established mills in Ireland to provide employment for impoverished Irish peasantry and was a founder of the pioneering Liverpool and Manchester Railway. He dabbled in education by establishing an agricultural school for boys and built a school and orphan house on his estate. During the last years of his life, he published pamphlets on the condition of the West Indies.

I.184. The PEACE OF VERSAILLES refers to the Treaty of Versailles (also called Treaty of Paris—one of many) signed on September 3, 1783. It ended the U.S. American War of Independence. The treaty was an impressive feat of diplomatic skill on the part of Benjamin Franklin (1706–90), John Adams (1735–1826), and John Jay (1745–1829).

I.187. Alejandro RAMÍREZ (1777–1821) was intendant of Havana from 1816 to 1821. His administration, which promoted the development of Cuba's commerce, agriculture, and industry, took a census of the population and the resources of the island and abolished the tobacco monopoly. In Havana, he established a botanical garden, an anatomical museum, a free academy of drawing, and numerous public schools. He served as president of the **Patriotic Society**.

I.189. SUGARCANE, which grows in tropical and subtropical climates, came originally from New Guinea. It was one of the earliest goods to be traded between Europe and the East. Portugal and Spain introduced sugarcane to the Canaries and other Atlantic islands. On his second voyage, **Columbus** brought sugarcane to the Americas, where cultivation began around 1517. The Cuban sugar industry did not begin to develop until 1595, when Philip II of Spain (1527–98) contracted with **Gómez Reynel** to bring African slaves to the Antilles.

1.189n. Gonzalo FERNÁNDEZ DE OVIEDO y Valdés (1478–1557) was a natural historian and became the official chronicler for Charles V in 1532. Oviedo crossed the Atlantic for the first time in 1514 as part of an expedition to the Isthmus of Darien. He lived most of his life in America, serving the crown as inspector of gold mines in Tierra Firme, governor of Cartagena, and councilman in perpetuity and governor of Santo Domingo on Hispaniola. The firsthand accounts of Central America's flora, fauna, and peoples in his *Sumario de la natural historia de las Indias* (1526) became the basis for his seminal *Historia General y Natural de las Indias* (1535). One of the most important chroniclers of the region, Oviedo also wrote the unpublished "Libro de blasón" [Book of heraldry] (c. 1528), a chivalric novel best know as *Don Claribalte* (1519), in addition to poetry and other literary works.

1.193n. The **Count of Jaruco y Mopox** commissioned Antonio LÓPEZ GÓMEZ to write a *Historia natural y política de la isla de Cuba* [Natural and political history of the island of Cuba].

1.195n. The British mathematician and astronomer Patrick KELLY (1756–1842), the master of the private Mercantile School in London, was appointed mathematical examiner at the Trinity House and in 1813 was honored with an LL.D. degree from the University of Glasgow. He was interested in weights and measures. Louis Benjamin FRANCŒUR (1773–1849) was a French mathematician with far-reaching interests; he pursued both an academic and a military career and published books on mechanics, mathematics, and astronomy. The French chemist Jean René Denis RIFFAULT des Hêtres (c. 1752–1826) developed a method for testing saltpeter in 1787. He was called to Paris to become general administrator of powders and saltpeter. He translated into French the third edition of the 1807 *A System of Chemistry in Five Volumes* by Thomas Thomson (1773–1852), a book from which Humboldt quotes.

1.196. Scottish economist and jurist Patrick COLQUHOUN (1745–1820) was a municipal administrator in London. He is considered the architect of modern policing, applying business principles to police administration in his 1796 *Treatise on the Police of the Metropolis* (second ed.). Humboldt cites his 1814 *Treatise on the Wealth, Power, and Resources of the British Empire*.

1.197. Charles Rose ELLIS (1771–1845) was made Lord Seaford in 1826. His family owned property in Jamaica and had ties to Britain. Ellis inherited Montpellier plantation in Jamaica in 1781.

1.198. The TREATY OF PARIS (not to be confused with the 1783 **Peace of Versailles** that is sometimes called the Treaty of Paris) was signed in Paris on February 10, 1763. It brought an end to the Franco-British wars, that is, the Seven Years' War in Europe and the French and Indian War in North America. The treaty was signed by representatives of Britain and Hanover on the one hand (they were under the same monarch then) and France and Spain on the other. Portugal was also included. The terms of the treaty stated that France would cede to Britain all of North America east of the Mississippi, excluding New Orleans and environs; the West Indian islands of Grenada, St. Vincent, Dominica, and Tobago; and all of France's conquests in the East Indies since 1749. France received from Britain the West Indian islands of Guadeloupe, Martinique, Marie-Galante, and Désirade; the islands of St. Pierre and Miquelon off Newfoundland; the West African colony of Gorée (Senegal); and Belle-Île-en-Mer off Brittany. Britain also relinquished St. Lucia to France. Spain recovered Havana and Manila, ceded east and west Florida to the British, and received Louisiana from the French, who retreated from Hanover, Hesse, and Brunswick.

I.202. SUGAR EXPORTS from **Saint-Domingue** dropped precipitously during the time of the revolution, since most of the country's sugar mills were destroyed and many of the sugar technicians fled either to Louisiana or to Cuba.

I.204. Isaac THURET (1771–1852), the Netherlands' consul-general at Paris, came from a French Protestant family that took refuge in Holland around the time of the Edict of Fontainebleau (1685). Thuret was the first family member to return to France, where his son, Gustave Adolphe Thuret (1817–75), a famous botanist, was born. Thuret is not to be confused with his namesake Isaac Thuret (1630–1706), the Royal clock maker in Paris.

I.210n. Humboldt's reference here is probably to Juan José RODRÍGUEZ, a student at the Real Seminario de Minería [Royal school of mining] in Mexico City who helped Humboldt with his geological maps.

Military commissioner José Ignacio Peñalver y Cárdenas, MARQUIS DE ARCOS (1736–1804), was the royal treasurer of Havana and a member of the **Patriotic Society** [Sociedad Económica de Amigos del País]. For Rafael O'FARRILL see O'Farrill y Herrera.

I.212. A *TRAPICHE* is the mill that extracts the syrup (*guarapo*) from the sugarcane. Since the late eighteenth century, the term has also been used to refer to small-scale sugar farms. Sets of wood-fired boilers were also known as *tren* (train), a designation that later came to be employed for vaccum-evaporation techniques. It was not until the 1850s that centrifuges were used to separate molasses from crystallized sugar.

French engineer Antonio BAILLY was a refuge from the revolution in **Saint-Domingue**. In Cuba, his projects included working on the Güines River drainage for crop irrigation (in 1800) and paving Havana's street with stones (in 1804). He was also involved in planning the Güines railroad.

The Count Nicolás de la Puerta CALVO Y O'FARRILL (c. 1758–c. 1800) was one of the many representatives of Havana's aristocracy who hosted Humboldt in Cuba. Renowned as the chair of theology of the Real y Pontífica Universidad de La Habana,

Calvo, himself propietor of two ***ingenios*** in the Valley of Güines, also stimulated Cuba's sugar industry. A practically minded reformer, he introduced new processes for sugar refining and for artificial irrigation. Together with the **Count O'Reilly** and his cousin **Arango**, Calvo accompanied Humboldt on his excursions to sugar plantations near Havana and in Güines, including his own, "La Holanda" (Humboldt visited "La Holanda" twice in 1801). A typical representative of a modern scientist with broad knowledge in the many areas of natural sciences, Calvo supplied Humboldt with important information from his own studies.

1.215. Sugar MAGMA is a viscous mixture of sugar syrup and sugar crystals, which is produced during the refining process.

William BECKFORD (1744–99), whom Humboldt consistenly calls "Bockford," was a Jamaican sugar planter, an Oxford-educated historian, and a patron of the arts. In 1756, Beckford inherited considerable estates in Britain and in Jamaica, including 910 slaves. Among his published works is *Descriptive Account of the Island of Jamaica* (1790), in which he defends the slave trade but urges amelioration. Humboldt mentions him twice without giving a specific reference. Beckford came from an influential and illustrious family: his uncle, William Beckford (bap. 1709, d. 1770), was a lord mayor of London, and his first cousin, William Thomas Beckford (1760–1844), was the author of *Vathek*. The Scottish botanist and physician William ROXBURGH (c. 1751–1815) studied medicine at the University of Edinburgh. In his late teens, he was appointed surgeon's mate and sailed twice to the East, eventually settling in Madras. He started submitting papers on natural history to the Royal Society and sometimes also sent seeds. In 1781, he moved on to Samulcottah where, among other things, he occupied himself with improving the manufacture of sugar. In 1793, he was appointed superintendent of the new Botanical Garden in Calutta. He wrote several papers on the Hindu method of cultivating the sugarcane. See also Edwards and *Indian Recreations*.

1.218. French physicist and chemist Louis-Joseph GAY-LUSSAC (1778–1850) coauthored several papers on eudiometric analysis and magnetic observations with Humboldt, his dear friend. Together with the young geological engineer Franz August O'Etzel (1784–1850), they traveled Switzerland, Italy, and Germany in 1805, visiting Wilhelm von Humboldt (1767–1835) in Rome. Louis Jacques THÉNARD (1777–1857), another French chemist, also collaborated with Humboldt, Gay-Lussac, and Louis Claude Berthollet (1748–1822)—Gay-Lussac had been Berthollet's student—on chemical experiments during Humboldt's Parisian years (1804–27). Humboldt used to call his three colleagues "Soda" (Thénard), "Potash" (Gay-Lussac), and "Ammonia" (Berthollet).

Jacques-François DUTRÔNE la Couture (1749–1814) wrote about sugar refining in **Saint-Domingue** around the turn of the nineteenth century. In 1791, he defended before the French National Assembly his idea to organize research locally through royal administration, in this case through the newly revived royal society of agriculture.

Henri Jacques Guillaume CLARKE (1765–1818), the duke of Feltre and marshall of France, was general governor of Prussia during the Napoleonic era. Humboldt frequently appealed to Clarke's goodwill to ease tense German-French relations during the French occupation (1806–13).

The Irish chemist Bry (Bryan) HIGGINS (c. 1737–1820) experimented with sugar refining and rum production. He detailed his research in a 1797 publication entitled *Observations and Advices for the Improvement of the Manufacture of Muscovado Sugar and Rum.*

The British chemist John Frederic DANIELL (1790–1845) is best known for his invention of the Daniell battery cell; he also invented a dew point hygrometer and a register pyrometer. British meteorologist Luke HOWARD (1772–1864) invented a nomenclature for different cloud formations (cirrus, cumlus, and stratus clouds) in 1802. His meteorological studies greatly influenced Humboldt and Johann Wolfgang von Goethe (1749–1832).

Henri BRACONNOT (1780–1855) was named director of the botanical garden of Nancy in 1807; he worked mainly on plant chemistry.

Jean-François DEROSNE (1774–1855), a Paris pharmacist, discovered a crystalline component of opium ("sel de Derosne" or Derosne salt) in 1803; it was later renamed narcotine. Derosne's brother Louis-Charles (1780–1846) focused on improving sugar production. In the 1830s, he invented a new vacuum boiler, which he marketed in Cuba with the help of the sugar chemist José Luis Casaseca (1800–1869), one of **Arango**'s protégés. Casaseca also translated Derosne's manual into Spanish. In 1841, Derosne installed his new machine at "La Mella," the plantation of Wenceslao Villaurrutia y Puente (1816–62) in Matanzas. The results of what came to be known as the "Derosne train" were dramatic.The family name is mispelled as "Desrones" in various publications, including Humboldt's text.

I.219. Humboldt describes the different types of sugarcane in his *Aspects of Nature in Different Lands and Different Climates* (translated by Elizabeth Sabine in 1849; pages 4–5).

I.220n. "*INDIAN RECREATIONS* (CALCUTTA, 1810)" may actually refer to two different works rather than one. William Tennant (1758–1813), a chaplain in the service of the East India Company, quotes amply from **Roxburgh** in his 1804 *Indian Recreations*, specifically in "Sect. I. V. The Hindoo Method of Cultivating the Sugar Cane." There are Edinburgh and London editions of the book, but none from Calcutta. John Fleming's (1747–1829) *Catalogue of Indian Medicinal Plants and Drugs*, however, was published in Calcutta in 1810 (by the Hindustani Press); a British edition appeared in London two years later. Fleming had managed the Royal Botanical Garden, Calcutta, for a few months in 1793 until Roxburgh was appointed later that year. He stepped in again when Roxburgh was on sick leave in 1805–7. Humboldt lists several references to Roxburgh's work on Indian sugarcane, but he is entirely silent on **Beckford**'s writings, even though he mentions both men twice in connection. Ironically, the footnote is actually attached to Beckford's rather than to Roxburgh's name.

1.221. John STEWART resided in Jamaica for a long time before returning to Scotland. In 1823 he published *A View of the Past and Present State of the Island of Jamaica*, which Humboldt quotes. Stewart thought of the abolition of the slave trade as a means to improving the treatment of the slaves who were already in the West Indies.

1.223. BEET SUGAR from Europe began to enter the global sugar market in the early nineteenth century. The commercial production of beet sugar, however, gained importance only after the 1820s.

1.223n. Jean-Pierre BARRUEL (1780–1838), director of the Chemical Laboratory at the Paris Medical School, perfected the process for extracting sugar from beet root, together with the French perfume dealer Maximin Isnard (1758–1825), who played a minor role in French revolutionary affairs. Barruel was the first to bring **beet sugar** production to a commercial level. His findings were published in *Mémoire sur l'extraction en grand du sucre des Betteraves, et quelques considérations sur leurs culture* [Report on the large-scale extraction of sugar from beets, and some considerations about beet cultivation] (1811) and also in an article in the *Moniteur* in March of that same year; Humboldt quotes from the latter. Barruel is sometimes confused with his more famous contemporary namesake, the Abbé Augustin Barruel (1741–1820), a Jesuit priest and publicist. See also Beaujeu and Chaptal.

1.226n. Donatien Marie Joseph de Vimeur, the VISCOUNT OF ROCHAMBEAU (1750–1813), commanded a division of Charles LECLERC's (1772–1802) army in France's unsuccessful 1802 expedition to **Saint-Domingue** to put down the revolution and defeat the military leader Toussaint L'Ouverture. Leclerc was the brother-in-law of Napoleon Bonaparte (later Napoleon I). In November 1803, Rochambeau was forced to surrender, and by 1809 the French had withdrawn from the island. Rochambeau died in Leipzig, Saxony, in the Battle of Nations, one of Napoleon's major defeats.

1.226. When the slave revolution broke out in **Saint-Domingue**, French planters were forced to flee CAP-FRANÇAIS (now Cap-Haïtien), the capital of the French colony. Some of the refugees went to New Orleans, Virginia, Baltimore, Philadelphia, and New York. Many went to Louisiana and to eastern Cuba, notably the port city of Santiago de Cuba. They brought with them their technical knowledge about sugar refining, which was more advanced at the time than that of the Cuban planters. One notable innovation they brought was the "Jamaican train," a long train of copper cauldrons that could be heated over a single fire at the same temperature.

1.227. Joaquín Santa Cruz de Cárdenas (1769–1807), the COUNT OF SAN JUAN DE JARUCO Y SANTA CRUZ DE MOPOX, was subinspector general of Cuba's military and owner of the county of San Juan. Jaruco y Mopox was the father of the famous María de las Mercedes Santa Cruz y Montalvo, Countess of Merlin (1789–1852). Mopox played a key role in the resettlement of the population of the Jesús María suburb after the **fire of 1802**. Humboldt visited him at his Río Blanco sugar plantation, where the Prussian experimented with improving the sugar processing installations, notably the *casa de caldera* (boiling room), where high temperatures created very difficult work conditions for the slaves. Mopox was the first to experiment with a steam engine to power a *trapiche* in 1797. The count accompanied Humboldt on his trip to the port of Batabanó, on the southern coast of Cuba, from where Humboldt and **Bonpland** departed from **New Granada** on March 5, 1801, on rough seas. Recent studies have cast doubt on the historical meeting of the two men, arguing that Humboldt met the Jaruco family but not the count himself, who was probably in Spain or on a royal expedition to Guantánamo, both in 1800–1801 and in 1804.

I.233n. Diego José de SEDANO (born c. 1761) was a lawyer and colonial administrator from Spain. A graduate in civil law from the University of Osuna in Seville, he joined the Royal Law Academy in Madrid in 1782 and was authorized to practice law anywhere in the Indies in 1785. He traveled to New Spain, where he joined the local College of Attorneys. In 1787, he began practicing law in Havana and became the interim legal advisor for the treasury and the intendant. Sedano became virtually indispensible in Havana, holding several key positions until his retirement in 1817, including that of fiscal of the Royal Treasury of Cuba and manager of the royal lottery.

I.235. *GUARAPO* is a Cuban word for pressed-out sugarcane juice; it is also a fermented drink made from this juice. While some have attributed the word to Quechua origins, Fernando Ortiz calls it an *afronegrismo* [Afrocubanism] with an etymology that moves from Arabic to Spanish to Portuguese. The name of the drink itself is related to the name for a beverage made in Congo and Angola from fermented corn or yucca. See also *trapiche*.

I. 237. HALURGY means working with salt. The word was not used in English until the mid-nineteenth century. Humboldt studied at the University of Göttingen from 1789 to 1790, during which time he became especially interested in mineralogy and geology. He moved on to the School of Mines in Freiberg, Saxony, where he spent his mornings underground in the mines, attended classes in the afternoon, and looked for plants in the evening. He left in 1792 without taking a degree and shortly thereafter was appointed to the Mining Department of the Prussian government. His post took him to the remote Fichtel Mountains in the Margraviate of Ansbach-Bayreuth, a recent addition to Prussia. Humboldt traveled from mine to mine, reorganizing them and improving conditions as he went along. His first diplomatic mission was a result of this work and took him to the salt-mining regions of Bavaria, Austria, and Galicia and into northern Italy and Switzerland. He resigned in 1797.

I.238n. José Ignacio ECHEGOYEN was the sugar master of the ***ingenio*** "La Ninfa" and a friend of its owner, **Francisco de Arango y Parreño**. Even though Humboldt admits that Echegoyen's numbers seem imprecise, he still uses them for lack of better data. In one of his handwritten annotations to his copy of this text, Arango corrects Echegoyen's analysis (which had actually been written by Arango himself), calling it outdated.

1.242n. The GALLIFFETS were old French aristocracy dating back at least to the fourteenth century. The Gallifets owned five sugar and coffee plantations in **Saint-Domingue**. In the early eighteenth century, the phrase "heureux comme nègre à Gallifet" (happy as a Gallifet Negro) was coined. Count Louis-François de Gallifet (1695–1761) was a lover of the wife of the French revolutionary, the Count of Mirabeau (1749–91). Gaston Alexandre Auguste, Marquis of Gallifet (1830–1909), a general who harshly suppressed revolts in the Paris Commune in 1871, was a minister of war in 1899 and involved in the infamous Dreyfus Affair.

William W. WHITMORE was a prominent parliamentarian who, alongside **Wilberforce**, was a member of the Society for Mitigating and Gradually Abolishing the State of Slavery throughout the British Dominions (the Anti-Slavery Society, for short) founded in 1823. **Zachary Macauley**'s *Anti-Slavery Reporter* was the society's journal.

1.245n. The chemist and mineralogist Alexander CALDCLEUGH (d. 1858) served as a private secretary to a British diplomat in Rio de Janeiro and traveled extensively in Brazil, Chile, and parts of present-day Argentina. In 1825, he published the travel narrative from which Humboldt quotes here.

1.248. John CRAWFURD (1783–1868), a Scottish Orientalist employed by the East India Company, successfully combined scholarship with diplomacy. Having received medical education in Edinburgh, Crawfurd, aged 20, was appointed to the northwestern provinces of India. He held various posts in Singapore, Thailand, and Indonesia. After the first Anglo-Burmese war, he served as the first British resident at the court of Ava in 1827. Crawfurd retired to England in the 1830s and unsuccessfully tried for Parliament. He published a number of books based on his far-ranging experiences, including a three-volume *History of the Indian Archipelago* (1820); Humboldt quotes from it. Crawfurd's *Descriptive Dictionary of the Indian Islands and Adjacent Countries* (1856) remains an important contribution to the study of early nineteenth-century maritime Southeast Asia.

1.249n. Louis François BENOISTON DE CHATEAUNEUF (1776–1856) was a former military surgeon who played a major role in the development of modern epidemiology. He also became France's leading economist and demographer in the 1820s and 1830s.

1.252. The *FACTORÍA DE TABACOS*, also known as the Royal Trade Company of Havana (Real Compañía de Comercio de la Habana), was created in 1717, when King Philip V of Spain (1683–1746) ordered that an *estanco* (agency or trade post) should be in charge of buying the Cuban tobacco crop and the cigars, leaving a specified amount for local consumption. It was effectively a way for the Spanish Crown to monopolize revenue from the tobacco trade, which caused significant discontent among the tobacco farmers. In 1817, **Arango**, along with others, succeeded in getting the *estanco* revoked per royal decree.

1.258n. The Spanish colonist Gabriel de CABRERA was likely traveling with **Columbus**. Humboldt paraphrases **Herrera y Tordesillas** in his reference to Cabrera. Herrera gives an account of Friar Cabrera speaking with native Cubans in the early sixteenth century. Cabreara was one of the first ten Spanish settlers in today's Guatemala, founding the settlements of Santa María de Jalapa and San Lucas Cabrera. He became the general proctor of the province of Guatemala.

COXCOX'S RAFT refers to an Aztec myth in which Coxcox, who also bears the name Teo-Cipactli, is the only male survivor of a big flood, which is why he has been likened to Noah. The myth also has elements of the story of the Tower of Babel: only one man and one woman, Coxcox and his wife Xochequetzal, escaped the flood in the hollow stem of a cypress and landed on the peak of Mount Colhuacan. Their children were born deaf and mute; a dove eventually gave them their tongues and innumerable languages. The Toltecs, the Aztecs, and the Acolhua were said to have descended from fifteen of Coxcox's children.

The EXPLORATORY BIRD famously appears in William Wordsworth's (1770–1850) poem of the same title (Sonnet X, 1841).

1.259. Carl Ludwig WILLDENOW (1765–1812) classified many of the plants that Humboldt and **Bonpland** had brought back from their journey and that Humboldt sent his friend from Havana. Willdenow can be considered one of the most influential of Humboldt's teachers, since he stimulated in Alexander the "infinite thirst for the examination of unknown objects" (as Humboldt noted in 1801). See also Juan Gonzáles and Kunth.

French entomologist Pierre André LATREILLE (1762–1833) was adopted by the mineralogist Abbé René-Just Haüy (1743–1822) in 1778. Latreille was arrested in Bordeaux during the French Revolution and supposedly released when he discovered a new kind of beetle in his prison cell. He came to direct the Entomology Department of the Museum of Natural History in 1799 and was elected a member of the French Academy of Sciences in 1814. In 1829, he succeeded Jean-Baptiste Lamarck (1744–1829) as professor of entomology.

1.278. COCHINEAL is an intensely red dye obtained from the crushed dried bodies of female cochineal insects (Dactylopuis coccus). It is used for food coloring and as fabric dye.

1.281n. Bertrand HUBER (1780–1846) was an attaché in the foreign affairs ministry, probably in France. He translated *Letters from the Havana* by Robert Francis JAMESON (1774–1854), adding, as Humboldt himself observed, "much important information on trade and Cuba's customs system," such as statistics by the Spanish economist and naturalist **de la Sagra y Pérez**.

Ignacio Benito NÚÑEZ (1792–1846) was an Argentine writer and journalist. He entered the military service in 1806 and held a number of public positions; he was promoted to the rank of captain in 1829. In addition to being involved in rebuffing the British in 1806 and 1807, Núñez supported the *movimiento de mayo* that led to Buenos Aires' declaration of independence from Spain. Núñez was one of the founding members of the Sociedad Literaria de Buenos Aires and editor of its periodical, *El Argos* (1821–25). The Literary Society also published *La Abeja Argentina*, considered one of the foremost magazines in Argentina. Núñez also contributed to *El Centinela*, *El Nacional*, and the *Revista Europea*.

1.283. William Davis ROBINSON (b. 1774) was a U.S. merchant who had traded with the Spanish authorities in Venezuela since 1799. He became openly critical of Spain and eventually made the acquaintance of insurgents during the Mexican Revolution. In 1816, he went to New Spain with a passport issued by the Secretary of State James Monroe. He was eventually captured and imprisoned in Oaxaca but escaped in 1819. Upon his return, he wrote his *Memoirs of the Mexican Revolution*.

1.284. Jurist Timothy PITKIN (1766–1847) was a U.S. Representative from Connecticut. He graduated from Yale in 1785, where he had studied mathematics, natural philosophy, and astronomy. He was a member of the House of Representatives repeatedly between 1790 and 1805 and served in the U.S. Congress from 1805 to 1819. In 1818, he became a delegate to the convention that framed a new state constitution. He also wrote literary works. In 1816, he published *A Statistical View of the Commerce of the United States of America,* from which Humboldt quotes.

1.285. Matheo PEDROSO (1719–1800) was one of Cuba's most prominent merchants and *regidor perpetuo* (alderman for life) of Havana. His immense economic success was due not only to lending money to his planter cousins and friends but also to buying lands himself; at his death, they were valued at $2 million. His own capital was likely family money made either by his ancestors, who came to Cuba as lay officials for the Inquisition, or by his father, who was municipal treasurer for many years.

1.286n. Luis Juan Lorenzo de CLOUET y Piettre (1776–1848) was lieutenant colonel of the infantry from Louisiana who was attached to the general staff in Havana. He made a contract with the government to obtain land for forty families. By 1823, about 845 settlers had arrived from Louisiana, Bordeaux, Philadelphia, Baltimore, Santo Domingo, and elsewhere. Clouet was the first mayor of the city of Bahía de Jagua in the south of Cuba.

1.287. Cuban statesman Claudio Martínez de PINILLOS Y CEBALLOS, the Count of Villanueva (1782–1853), who had fought in Spain against Napoleon, was intendant of Havana from 1825 to 1852.

1.289. For Francisco and Félix LEMAUR, see Francisco Lemaur.

Cuba had the first RAILROAD in Latin America and in the Caribbean: between Havana and Bejucal (1837) and between Havana and Güines (1838).

1.289n. Brigadier General Juan de O'FARRILL (d. 1819) was granted a special concession to import the first steamboat to Cuba at roughly the same time that the steam-powered sugarmill was introduced and long before these new techniques were available in Spain. He modernized dramatically the transportation methods available to the sugarcane planters, of which the O'Farrill family was one of the most prominent representatives. As early as 1823, three steamboats plied regularly between Havana, Matanzas, Cárdenas, San Juan de los Remedios, and Bahía Honda.

I. 291n. Peter BAUDUY was a planter from Saint-Domingue who had moved his residence to Wilmington, Delaware, by 1801. In addition to being the architect responsible for building the Wilmington Town Hall, Bauduy was also involved in canal construction in Delaware. In 1811, he joined the French manufacturer Éleuthère Irénée du Pont (1771–1834), who had been the first assistant of the renowned French chemist **Lavoisier** in 1791, in establishing a woolen mill on the Brandywine River, along with a powder manufacturing business.

I.294. For DECREES OF FREE TRADE see slave trade.

I.300 table. The *ALMOJARIFAZGO* was a tariff that had to be paid on all merchandize that entered or departed the Spanish empire, including goods in transit that were shipped through the various peninsular and American ports. This tax or tariff was first created by Alfonso X the Wise (1221–84), together with the ***alcabala***. Initially, the *almojarifazgo* was a percentage of the value of all imports; it was calculated on the basis of the value that the merchandize would have in the Americas, not in the port of embarcation. Alfonso XI, eventual successor to Alfonso the Wise, attempted to replace the Jewish tax inspectors of old (*almojarifes*) with Christian officials, which the courts of Alcalá authorized in 1348. The *almojarifazgo* was abolished in 1783.

I.301. Juan de MICOLAETA BARRUTIA (d, 1803) was the administrative head of the Real Factoría de Tabaco [Royal Tobacco Company] in 1774 and in 1778. He wrote the "Instrucción para el mejor gobierno y dirección que deben observar los labradores de tabaco en la jurisdicción de La Habana" [Manual for the better regulations and rules that the tobacco workforce in the jurisdiction of Havana must heed].

I.305. HERE, I END THE *POLITICAL ESSAY*: Humboldt's translator John Sidney Thrasher (1817–79) took this remark as an excuse to excise pages 305 to 336 from his 1856 English version. For further materials on Thrasher see www.press.uchicago.edu/books/humboldt/.

I.307n. Henry BOLINGBROKE (1785–1855) was an English clerk who lived in Demerary from 1799 to 1805. In 1807, he published *A Voyage to the Demerary* based on the letters to his family. It is an important source of information on that region and time period.

I.314n. The French aristocrat Marie Joseph Paul Yves Roch Gilbert du Motier, MARQUIS DE LAFAYETTE (1757–1834), played important roles both during the U.S.

American and French Revolutions. He became an outspoken advocate of religious toleration and the abolition of the **slave trade**. Having first arrived in America in 1777, he considered slaves "property" at least until 1781. His friend John Laurens (1754–82), an associate to George Washington (1732–99) and the proponent of a plan to offer slaves their freedom in exchange for service in the Continental Army, likely influenced him to change his mind. In a letter from February 5, 1783, Lafayette asked for Washington's collaboration in an experiment to emancipate slaves and employ them as tenant farmers. When Washington was reluctant to put the idea into action, Lafayette turned elsewhere for support, including to Armand Charles Augustin de la Croix (1756–1842), the young Duke of Castries, whose father, Marshal Charles Eugène Gabriel de la Croix, MARQUIS OF CASTRIES (1727–1801), was minister of the navy. The marshal brought the idea to the attention of Louis XVI, the last king of France (1774–92) in the Bourbon line before the French Revolution of 1789. The king approved, and the experiment started in the French colony of present-day French Guiana in 1786 with the gradual emancipation of nearly seventy slaves on land along the Oyapok River. Lafayette's agents in Cayenne saw to it that the slaves were educated and received wages for their labor; their sale was explicitly forbidden. Lafayette hoped to show that slavery would prove unnecessary under such conditions. Through Marie Jean Antoine Nicolas de Caritat, Marquis of Condorcet (1743–94), a French philosopher, mathematician, and early political scientist, Lafayette was introduced to JEAN-FRANÇOIS HENRY DE RICHEPREY (1751–87), land registrar in Upper Guyana (1779–86). Richeprey gave up his position in order to carry out Lafayette's scheme but died of yellow fever soon thereafter. For a lack of successors, the project was abandoned. Lafayette last visited the United States during a fourteen-month farewell tour in 1824–25. On that occasion, he repeatedly expressed his interest in the welfare of African Americans and had frank private discussions about slavery with Thomas Jefferson (1743–1826) and James Madison. Lafayette also supported a number of prominent women writers and reformers, such as Germaine de Staël (better known as Madame de Staël; see Necker). On his last trip to the United States, he was accompanied by Frances Wright (1795–1852), who established a utopian community in Nashoba, Tennessee, in 1826, the first experiment in the United States with a truly egalitarian and interracial society.

1.315. When wars of independence began to erupt in Spanish America in 1810, military recruiters turned to slaves, and thousands responded to this opportunity to achieve personal freedom. Slaves initially remained staunchly royalist, as for example in Buenos Aires where the Spanish Crown had ordered blacks to be incorporated into militia units and offered freedom to those who defended the city. Many revolutionaries, by contrast, displayed little sympathy for slaves. In VENEZUELA, racial unrest was particularly devastating. Slaves there joined the royalist forces in considerable numbers and threatened a race war like the one on **Saint-Domingue**. During the first republic in Venezuela (1811–12), Francisco Miranda (1750–1816) offered slaves freedom after ten years of military service, but his offer alienated Creole slaveholders and did not attract many slaves. **Bolívar** eventually realized that he could not defeat royalist forces without either the support or the neutrality of the slaves.

1.316. Claims about the aloofness of the COPPER-COLORED INDIGENOUS RACE to which Humboldt refers here should be attributed to centuries of abusive treatment of native populations in the Americas at the hands of the Spanish. Despite the fact that the new laws Spain had enacted in 1542 comprehensively abolished the enslavement of Indians, widespread abuses of the native populations, who worked, for instance, in mining, persisted. Already during his first encounter with Amerindians on the Venezuelan coast in 1799, Humboldt referred to the indigenous peoples as "copper-colored, half-naked Indians and mestizos." Later, a Guaiquery Indian, Carlos del Pino, accompanied him and **Bonpland** for sixteen months of travels in the Americas. But while Humboldt regarded native cultures as "closer to nature" and thus as culturally and racially different from Europeans, he was also fascinated by indigenous cultures. He wrote about these cultures at length, in *Views of the Cordilleras and Monuments of the Indigenous Peoples of the Americas* and the *Political Essay on the Kingdom of New Spain*, repeatedly blaming Spanish governors and missionaries for native peoples' cultural, political, and social decay. Humboldt insisted in his travel diary that the destiny of the Europeans and their descendants was inseparable from that of America's indigenous peoples and that both had to share all rights and privileges that the progress of civilization had to offer.

1.317. As early as in 1538, PORTUGAL started to ship African slaves to Brazil as a labor force for the colony's huge sugar plantations and cattle ranches first established during the captaincy-general of Martim Afonso de Sousa (1500–64). Over the next three centuries, Brazil imported between three and four million Africans—about one-third of the total number of slaves brought to the Americas. The threat of slave uprisings persisted throughout the colonial period. Brazil acquired the status of kingdom in 1815 under the rule of the Portuguese prince and later King João VI (1767–1826), who fled to his American colony in the wake of the Napoleonic wars to establish his court in Rio de Janeiro in 1807. Returning to Portugal in 1821, he left the new kingdom in the hands of his son Pedro I of Brazil (1798–1834), who declared Brazilian independence in 1822. By that time, the country counted roughly four million inhabitants, probably half of them slaves of African birth or descent. In 1824, the first constitution was proclaimed, and the new monarchy was recognized by the United States and other countries. The landowner aristocracy, which was in firm control of the government throughout the first decades of the new regency, resisted any criticisms of slavery. Even though the state's legal system slowly began according Brazilian slaves a few rights, fear of violent uprising on the part of the black population remained a large factor in Brazilian white Creole society. Indeed, in 1835, a few years after the publication of Humboldt's *Political Essay on the Island of Cuba*, a major slave revolt took place in Bahia, one of nine uprisings there since 1800. Slavery was not abolished in Brazil until 1888, two years after Cuba, making Brazil the last country in the Western Hemisphere to do so.

1.318n. The CONGRESS OF VIENNA (September 1814 to June 1815; that is, shortly before Napoleon's final defeat) reorganized Europe after the Napoleonic Wars. It was the most comprehensive treaty that Europe had ever seen. Two treaties signed at Paris in 1814 and 1815 ended the Napoleonic Wars. The 1814 treaty contained an article enjoining the upcoming Congress of Vienna to decree the abolition of the **slave trade**. The Congress indeed produced an agreed declaration that unanimously condemned the slave trade. Its abolition was, however, not prescribed. As regards THE SLAVE TRADE'S CESSATION in Washington and London, in March 1807, President Thomas Jefferson signed into law an act to "prohibit the importation of slaves into any port or place within the jurisdiction of the United States." The British House of Lords passed an Act for the Abolition of the Slave Trade a few weeks later. The Massachussetts bill entitled "An Act to prevent the importation of slaves into this Province" was first read in the House of Representatives in March of 1767 rather than in 1769, as Humboldt notes. It was read again soon thereafter under the title "An Act for preventing THE UNNATURAL AND UNWARRANTABLE CUSTOM OF ENSLAVING MANKIND in this Province, and the importation of slaves into the same." It was read twice and, finally, a committee was formed to write a bill for imposing a duty on imported slaves as a legislative discouragement of slavery.

Educated at Georgetown College and St. Mary's College in Baltimore, with an honorary LL.D. from Harvard University, Federalist jurist Robert WALSH, JR. (1784–1859) held a professorship of general literature at the University of Pennsylvania (1819–28). Walsh edited the *American Review of History and Politics* (1811–12), the Philadelphia *National Gazette and Literary Register* (1820–35), and the *American Quarterly Review* (1827–35). In addition to the *Appeal* to which Humbold refers here, Walsh wrote various papers that are collected in *Didactics, Social, Literary, and Political* (1836). Edgar Allan Poe's eulogy of him as a thinker, scholar, and writer appeared in the *Southern Literary Messenger* in 1836. From 1837–44, Walsh was consul-general of the United States in Paris, where he remained until his death.

Diego de AVENDAÑO (1594–1688), a Jesuit jurist and theological author from Segovia, became one of the most eminent defenders of the indigenous and enslaved populations of the Spanish colonial Americas. Living most of his life in Peru as Jesuit theologist, university rector, and spiritural leader of various Jesuit communities, Avendaño became one of the key advocates for granting the indigenous populations the legal status of *personae miserabiles* (poor people, who require the protective mercy of the Spanish Crown). A zealous nationalist and convinced of the supremacy of the Spanish monarchy, Avendaño exposed in his famous juridical and theological *Thesaurus Indicus* the desolate living conditions of the Peruvian population, but without abandoning the typical judgments of his time, portraying the *indio* as simple yet weak, obedient yet lazy. His moralist opposition to the common practice in the Spanish colonies especially targeted forced labor of the native population in mines and the African **slave trade**, which he condemned as "injusticia malvada" (a wicked injustice). By referring to Avendaño's interventions in favor of the indigenous people, Humboldt alludes to a discussion that was very present among Spanish scholars, theologists, and jurists of the sixteen and seventeenth century, among them the famous **Bartolomé de Las Casas**.

The Scottish historian Mountstuart ELPHINSTONE (1779–1859) was a colonial administrator in India, where he worked for the East India Company (1796–1827). As a civil servant deeply interested in Indian literature and philosophy and the Persian classics, he wrote about his travels through northwestern India and Afghanistan. The governor of the Bombay Presidency (former province of British India, 1819–27), he compiled a set of regulations commonly known as the Elphinstone Code. The code was in force from 1827 to 1860 and laid the basis for the system of state education in India, with the establishment of colleges at Poona (Pune) and Nasik. Significantly, the curriculum at these institutions included classical Indian sciences and literatures. Once back in England, Elphinstone helped create the Royal Geographical Society of London and published *History of India: The Hindu and Mohametan Periods* (1841).

I.319. In 1819, the question of admitting Missouri into the union was discussed in the United States, which meant, above all, a dispute between southern and northern states as to whether the new state would be a slave state. A Missouri bill was introduced in the U.S. House of Representatives amended by a New York proposition to prohibit slavery in the new state, that is, to prohibit the further introduction of slaves into Missouri and to free slaves when they reached the age of twenty-five. Heated debates ensued. The so-called Missouri Compromise finally broke the deadlock: Maine was admitted as a free state, Missouri was admitted as a slave state, and the rest of the former Louisiana Territory was to be free. Humboldt's indignation about THE ADOPTION OF AN IMPRUDENT AND HARMFUL LAW is aroused by the fact that Missouri did enter the union as a slave state. See also Missouri Territory.

I.319n. Rufus KING (1755–1827) was a U.S. senator, lawyer, and diplomat. A member of the Continental Congress from Massachusetts (1784–87), a delegate to the Federal Constitutional Convention at Philadelphia in 1787 and to the State Convention in 1788, King was one of the signers of the U.S. Constitution. A four-term senator from New York, King vigorously opposed the admission of **Missouri** as a slave state. After an unsuccessful bid for the U.S. presidency, he returned to being minister to Great Britain. He was chairman of the Committee on Roads and Canals and the Committee on Foreign Relations.

I.320 The GUIANAS consisted of British, Dutch, French, and Portuguese Guiana. British Guiana existed from 1831, when Berbice, Essequibo, and Demerary were united under British authority. British Guiana, which became Guyana upon its independence in 1970, changed hands with perplexing frequency during the wars (mostly between the British and the French) from 1780 to 1815. During a brief French occupation, Longchamps, later called Georgetown, was established at the mouth of the Demerara; the Dutch renamed it Stabroek and continued to develop it. The British took over in 1796 and remained in possession, except for short intervals, until 1814, when they purchased Demerara, Berbice, and Essequibo, which were united as the colony of British Guiana in 1831. Dutch Guiana, which is now known as Suriname and gained full independence from the Netherlands in 1973, became a Dutch colony in 1667, after the Dutch defeated the British in that area. The colony of French Guiana (Guyane) was settled by the French in the seventeenth century; it became an overseas département of France in 1946. Portuguese settlers from Madeira, the Cape Verde islands, and Brazil were brought to British Guiana as alternative labor after the abolition of slavery in the British colonies in 1833. Spanish Guyana was an administrative region in Venezuela, and Portuguese Guiana, to the extent that such an area actually existed, was part of Brazil.

1.321. Robert NORRIS (d. 1791) was a British trader in West Africa from the 1750s to the 1780s. He represented Liverpool slave traders before governmental investigative committees in the 1780s. In 1789, he published *Memoirs of the reign of Bossa Ahádee, King of Dahomy an inland country of Guiney*, to which Humboldt seems to refer without naming the book. Humboldt often uses this strategy of omission when he culls the work of proslavery authors for statistics and other information; a case in point is **Beckford**. Norris defended the slave trade in his *Memoirs*, thus preventing heavy regulation of the slave trade in the early 1790s.

1.322. Many Africans died during the MIDDLE PASSAGE. Captains of some transatlantic slave ships bound for the Americas not infrequently decided to jettison their "cargo" in order to collect the insurance. The most notorious of the cases was that of the *Zong* from 1781. Humboldt was likely familiar with this case, which was widely publicized. On his way from Africa to Jamaica, the *Zong*'s Captain, Luke Collingwood, commandeered his crew to throw overboard 133 slaves. The underwriters brought the case to court in London in 1783 (*Gregson v. Gilbert*) for discrepancies in the claims that the ship's owner had made. Although murder was not at issue, the case made the public aware of the atrocities of the Middle Passage. Among those who came out against the murder of the African slaves was Olaudah Equiano (Gustavus Vassa, 1745–97), author of an important early slave autobiography. The case was instrumental in influencing the formation of the first British antislavery society four years later. It also inspired J. M. W. Turner (1775–1851) to paint the picture entitled *Slavers Throwing Overboard the Dead and Dying—Typhoon Coming On*, known as *The Slaveship* (1840).

The 1814 ANGLO-SPANISH TREATY limited slave trade in Spain to its own colonies. The trade north of the equator was prohibited. The original treaty was signed in Madrid on July 5 (not 15, as Humboldt has it), and additional articles on August 28, 1814. The TREATY BETWEEN GREAT BRITAIN AND PORTUGAL was signed in Vienna on January 22, 1815. It discussed provisions in legal slave trade at length, limiting the slave trade to the Portuguese possessions south of the equator. Substantial financial compensations were involved for both Spain and Portugal. See also Congress of Vienna.

1.322n. Achille Léonce Victor Charles, the third DUKE OF BROGLIE (1785–1870), was a French statesman and diplomat. He was born at Paris, the son of Charles Louis Victor, Prince of Broglie (1756–94), and grandson of Victor-François, second Duke of Broglie (1718–1804). While his grandfather emigrated during the **Reign of Terror**, his parents were imprisoned.

1.323. Humboldt is referring to the Maroon War in JAMAICA IN 1795, when the Trewlany Town Maroons declared war against the British for a second time, mainly due to their dissatisfaction with a 1739 treaty that had been brokered by Cudjoe (c.1680–1744) to end the first Maroon War, started in 1731, or, some argue, in 1655. The four other Maroon towns did not join the rebellion, and their treaty with the British remained in force until Jamaica gained its independence in 1962.

1.323n. After his studies of law at the University of Salamanca, Spain, Manuel Ignacio de ABAD Y QUEIPO (c. 1775–c. 1824) went to Guatemala, where he worked as a lawyer (*promotor fiscal*) for five years before relocating to Valladolid (present-day Morelia, Michoacán) to serve under the Bishop Antonio de San Miguel Iglesias (1726–1804). In 1809, Abad y Queipo was appointed bishop of Valladolid but never confirmed in that position. He was a fervent social critic. Hoping to assuage social inequalities between Spaniards, Indians, and other people of color in New Spain to prevent an insurrection against the monarchy, he proposed to abolish caste distinctions, eliminate Indian tributes and the mercantile tax, and redistribute some privately owned lands. Despite his reformist views, Abad was the first to excommunicate the revolutionary Miguel Hidalgo y Costilla (1753–1811) on September 24, 1810. Abad left Vallalodid when the Hidalgo Rebellion broke out; he had warned of it as imminent in his famous *Carta a la regencia española* of May 30, 1810. Before returning to Spain in 1815, he published *Colección de los escritos más importantes que en diferentes épocas dirigió al govierno* [Collection of the most important writings that were directed at the government in various epochs] (1813). Once in Spain, the Inquisition tried him for Jansenism and confined him to the Rosario Convent in Madrid. Ferdinand VII (1784–1833) pardoned him and even appointed him minister of justice in 1816. In 1824, Abad was once more arrested and sentenced to six years confinement in the Jeronymite Monastery of Santa María de la Sisla (near Toledo, Madrid), where he died.

1.324n. José María ZAMORA y Coronado (1785–1852) was a Costa Rican administrador who had studied at León (Nicaragua) before practicing law in Guatemala. While he had intended to sail to Spain in 1809, his ship was captured and he was forced to remain in Cuba. From 1811–26 he lived in Puerto Príncipe, and then returned to Havana to hold a number of administrative positions, among them the high office of regent of the audiencia of Havana. In 1849, he secured a position in Madrid, where he retired and died.

The Spanish word *CIMARRÓN* originally referred to runaway dogs or sheep of a herd, but it came mainly to describe slaves who sought refuge in the mountains, where they established fortified settlements called *palenques*. The rebellious slaves who lived in them were also called *apalencados*. There have been records of fugitive slaves in Cuba since the first shipments of African slaves in the early sixteenth century. The main *palenques* in Cuba—they did exist elsewhere in the Americas, notably in Jamaica and Brazil—were located in the mountains of the province Oriente, in Pinar del Río, and in the swamps of Zapata. In the eastern part of Cuba, the hills of Cuzco were preferred refuges. The most prominent example of a Cuba *cimarrón* is Esteban Montejo (d. 1973), who was born in 1860 on a sugarmill in Flor de Sagua in Villa Clara. Montejo's story was recounted from interviews in *Biografía de un cimarrón* (1966) by Miguel Barnet (1940–). Translated both as *Biography of a Runaway Slave* and *Autobiography of a Runaway Slave*, this narrative is considered one of the most influential examples of testimonial literature in Latin America.

One of the leading members of Havana's saccharocracy, the BARRETO family is of Portuguese origin; its Cuban presence dates back to the mid-seventeenth century. The first Count of Barreto was Jacinto Tomás Barreto y Pedroso (1718–91), who was known for his cruelty. He died under mysterious circumstances, with his corpse being removed from the public wake. His father was Antonio Barreto, infrantry captain of the Morro fort and lifelong alderman of Havana. As the city's provincial mayor, he was in command of a special police force whose goal was, as Humboldt vividly describes, to capture and immure fugitive slaves with the help of specially trained bloodhounds. When Jacinto Tomás took the place of his father, he converted the prison into a lucrative source of revenue, detaining not only fugitives but also any slave unable to document his status. Jacinto Tomás forced the slaves to such hard labor that only a few managed to escape from prison alive.

I.327n. The *CODE NOIR* [Black Code] was a decree passed by King Louis XIV in 1685. It was the harshest of the slave codes. Giving slaves and free persons of color virtually no rights, it severely restricted their activities, forbade the exercise of any religion other than Roman Catholicism, and ordered all Jews out of the French colonies. It also passed the condition of slavery on through the mother, a practice that was later adopted by other slave codes, notably in the United States. By contrast, the main code regulating slavery in Castile was the thirteenth-century code *Las Siete Partidas* of Alfonso X the Wise. This remarkable legal code contained not only specific regulations about various aspects of slaves' lives but also legal and moral principles that limited masters' rights. In the event of murder, for example, the law did not distinguish between free persons and slaves as victims. Some rights of slaves were recognized and protected, including the ownership of goods and the right to self-purchase. The right to marriage could even be exercised even against the master's wishes. But when the crown attempted to publish its relatively benign 1789 black code in Cuba, the planters mobilized to impede the promulgation of the new law on the basis that it would undermine slave discipline and morale. Some traditional rights were restricted, such as the possibility of interracial marriages, which required special permission by the authorities after 1805.

I.331. Nicolás de OVANDO y Cáceres (1460–1511) sailed to the New World in 1502 with a large fleet of thirty ships, which also included **Pizarro**. As governor of Hispaniola (1502–9), Ovando laid the foundations for colonial economics and the centralized bureaucratic control of Spain's colonies. He recommended a system of forced labor for the native populations, which the crown approved in 1503 (see *encomienda* in **Las Casas**). The introduction of this system accelerated the native popluation's almost complete extinction in the Caribbean. When the Spanish had come to the island in 1492, the native population had been an estimated 400,000; by 1508, it was down to 60,000. Because the new colonists needed more workers for the gold mines and farms, they brought in Indians from the Lucayas (now the Bahamas) and African slaves from Spain (*Ladinos*, that is, of African origin but born in Spain); those were the first African slaves brought to America. In 1509, Ovando was recalled to Spain. He was succeeded by **Diego Columbus** but was permitted to retain his property.

1.334. The FRENCH REIGN OF TERROR was a particularly violent period during the French Revolution. It lasted from September 5, 1793 to July 27, 1794 (9 Thermidor, year II of the revolutionary calendar). At least three hundred thousand suspects were arrested, seventeen thousand were officially executed, and many died in prison. The radical Jacobin leader, Maximilien François Marie Isidore de Robespierre (1758–94), was eventually executed himself.

1.334n. Pompée-Valentin Baron de VASTEY (1781–1820) was a Haitian historian, educator, and politician. At the age of fifteen, he joined Toussaint L'Ouverture's army and later fought under Dessalines. A leading statesman and advisor to King Henri Christophe (1767–1820), he was decorated with the national honor of Knight of the Royal and Military Order of St. Henry. Upholding the newly independent state, Vastey wrote extensively against European views of superiority and the colonial system.

1.335. MISSOLONGHI, or Messonlonghi, was a major rebel stronghold in the Greek War of Independence. It is notable both for the siege, which it sustained in 1822 and 1823 against the Turkish army, and for the heroic defense of 1825–26. The poet Lord George Gordon Byron (1788–1824), who supported the Greek struggle for independence, died there in 1824 and is commemorated by a cenotaph and a statue. The CHIOS Massacre refers to the slaughter of over eighty thousand Greeks on the island of Chios by Ottoman troops in 1822. Psara (Humboldt calls it IPARA) joined the Greek War of Independence on April 10, 1821. When Egyptian forces invaded the island in 1824, part of the population managed to flee; others were either sold into slavery or slaughtered. More than fifteen thousand Greeks were massacred.

1.336. What follows up to the supplement corresponds to Humboldt's *Relation historique*, vol. 3, pages 445–501 (quarto edition).

Gabriel de ARISTIZÁBAL y Espinoza (1743–1805) was a Spanish admiral who conceived and carried out the idea of transferring to Havana the remains of **Christopher Columbus** in 1795–96 in the warship *San Lorenzo*. These remains had been lodged in the cathedral of the city of Santo Domingo, on the island of Santo Domingo, since 1536. Doubts have arisen about the genuineness of these remains when, in 1877, the purported true bones of Columbus were discovered *in the same cathedral*: a lead box filled with bones and engraved with the words "D. de la A. P.er A.te" [Descubridor de la América Primer Admirante]. The 1877 discovers argued that Aristizábal had erroneously removed the remains of Diego, Columbus's son. When the Spanish withdrew from Cuba in 1898, the disputed bones were transferred from Havana's cathedral to the cathedral of Sevilla, Spain, where they still are. After comparing Columbus's DNA with that of his brother, a team of genecists found that the remains at Sevilla do indeed belong to Columbus. The 1877 remains, which might be part of Columbus's actual skeleton, are currently at the Faro a Colón in Santo Domingo, Dominican Republic.

Thomas Nicolas BAUDIN (1750–1803) was the captain of a French expedition to circumnavigate and determine the Australian (New Holland) costal line. Despite the significant complaints Humboldt had heard about Baudin's trustworthiness, he nevertheless had planned to meet up with Baudin in Guayaquil, where the captain was supposed to stop over on his expedition. It was because of these plans that Humboldt had altered his original route from Havana to Mexico in 1801, taking the landway via the Andes to Quito. In Havana, Humboldt learned that Baudin had changed his travel route to sail on the African route to Australia.

I. 338. Sir Joseph BANKS (1743–1820) was a famous British explorer and naturalist. Having inherited a considerable fortune in 1761, Banks traveled extensively in Newfoundland and Labrador (1766) and in Iceland (1772), collecting plants and other specimens. He also accompanied Captain James Cook (1728–79) on his first voyage around the world (1768–71), notably to the "South Sea Paradise" of Tahiti, where the crew of the *Endeavor* stayed for three months, and to New Zealand and Australia. As longtime president of the Royal Society, Banks vigorously promoted science, making his own residence a meeting place for the exchange of ideas. His herbarium and library are now part of the British Museum. Humboldt met Banks in London through his friend Georg Forster (1754–94), who had been part of Cook's second journey (1772–75) before becoming a leader of the German Jacobins. Banks gave Humboldt the use of his extensive private library, and the two men stayed in touch until Banks's death. See also Robert Brown.

While in Havana, **Bonpland** had divided the herbarium into three nearly equal parts to protect the precious collection from being stolen or from being destroyed during the long voyage still ahead of the tropical travelers. The "English collection" was sent to the Scottish botanist **John Fraser**, who transferred the valuable freight to **Willdenow**. One was stored in Havana at the residences of **de la Cuesta** and **Francisco Ramírez** and picked up by Humboldt and Bonpland during their second stay in Cuba in 1804. The third collection was supposed to be taken back to France by the young Franciscan friar JUAN GONZÁLES, who had accompanied the European travelers on their route through Venezuela. But, as Humboldt notes briefly here, Gonzáles died in a tragic shipwreck and the entire collection was lost at sea. Notice did not reach Humboldt until two years later, when he was traveling in Mexico.

I.341. General Gonzalo O'FARRILL Y HERRERA (1754–1831), a cousin of **Nicolás de la Puerta Calvo y O'Farrill**, was one of the most prominent figures of the powerful O'Farrill clan. In addition to his service to the Spanish Crown as minister in Prussia, O'Farrill served as lieutenant general of the Royal Spanish Army under King Charles IV (1748–1819). Ignacio O'FARRILL Y HERRERA (d. 1884), the second Marquis of Almendares, was Gonzalo's brother and captain of Havana's cavalry. The O'Farrill brothers and the Prussian baron had set up an agreement to exchange their pensions for the time of Humboldt's American travels. Humboldt received the money for his expedition expenses from Ignacio O'Farrill, while Gonzalo O'Farrill received his payments as ambassador in Berlin from the Prussian State. Another sibling, Rafael O'Farrill y Herrera (1769–1845), was a military officer in Cuba who was very active in the **Patriotic Society** (Sociedad Económica de Amigos del País), serving as its secretary and spokesperson on repreated occasions between 1793 and 1832.

I.342n. Louis-Philippe D'ORLÉANS (1773–1850), the "Citizen King of the French" from 1830 to 1848, became the Duke of Orléans after his father had been executed in 1793 during the French Revolution. Together with his two younger brothers, Louis Antoine Philippe (1775–1807, the Duke of Montpensier) and Louis-Charles (1779–1808, the Count of Beaujolais), he spent two years in exile in the United States, traveling across the country and even visiting Nashville, Tennessee. In 1797, the three brothers were registered as merchants in Philadelphia. In 1798, they sailed from New Orleans to Cuba, where they spent about a year as hosts of local elites before making their way back to Europe.

I.343. Captain de Mendoza y Ríos's famous brother is the astronomer and mathematician Juan José de MENDOZA y Ríos (1761–1816). After graduating from the Royal College of San Isidro in Madrid, Juan José joined the army and was transferred to the navy in 1776. While in the navy, he wrote and published *Tratado de Navegación* [Treatise on navigation] (1787). In 1792, after he had left Spain for France, he was elected foreign correspondent to the French Academy of Sciences. In that same year, he moved to London where, in 1793, he was elected fellow of the Royal Society for his contributions to nautical astronomy. In addition to several essays on nautical astronomy and cartography, he published a collection of mathematical tables as aids for nautical astronomy. These tables, entitled *Colección de tablas para varios usos de la navegación* [Collection of tables for various navigational uses] (1800), were widely reprinted and translated into several languages.

I.344n. Jacques Nicolas BELLIN (1703–72) was a French hydrographer and geographer who drew maps of the Americas, particularly of Canada, but also a 1762 map of Cuba. His maps were to serve the interests of those in power, but he may also be seen as a representative of the Atlantic enlightenment. José de SAN MARTÍN SUÁREZ drew a map of the Gulf of Mexico in 1787.

Aaron ARROWSMITH (1750–1823) was a British geographer and cartographer, who established his reputation with a large map of the world in 1790; it was based on the best sources available in his day. After Arrowsmith's death, his business was carried on by his sons and a nephew. What is noteworthy about Humboldt's listing these three cartographers in one breath is that, in doing so, he combined representatives of the French, the Spanish, and the British Empires, who successively drew their visions of parts of the Americas. There is an interesting subplot regarding Arrowsmith. Humboldt had stated in his December 20, 1811 letter to Thomas Jefferson that "Mr.

Arrowsmith in London has stolen my large map of Mexico." Actually, Humboldt's map had already been swiped in Washington in 1804: while Humboldt had lent the map to Albert Gallatin for copying, General James Wilkinson (1757–1825), a double agent for the Spanish, seems to have made an unauthorized copy,which he passed on to **Zebulon Pike**. As a result, versions of this map appeared in print before Humboldt himself could publish it in 1811.

1.346. Georges Baron de CUVIER (1769–1832) was the leading French zoologist and anatomist of his time. Humboldt for a time endorsed his theories on global catastrophe, formulated as a corrective to the concept of evolution, which Cuvier vehemently rejected, believing that species remained unchanged once created. Cuvier's bone comparisons made him one of the founders of modern paleontology. Humboldt frequently consulted Cuvier while working on his American narratives.

1.346n. Étienne Geoffroy SAINT-HILAIRE (1772–1844), a French naturalist, first studied medicine but switched to law and joined the church, through which he obtained a position in the *Jardin des plants* (formerly *Jardin du roi*). **Lacépède** having left France in the midst of the revolution, Geoffroy Saint-Hilaire, in 1793, found himself a professor of zoology at the newly created National Museum of Natural History, where he quickly established a notable reputation. From 1795 to 1797, he took the young **Cuvier** as a student, but their professional opinions would come to differ significantly in later years. In 1898, Bonaparte appointed Saint-Hilaire to his **Egyptian campaign**. This appointment had a profound impact on Saint-Hilaire's life and work. He briefly meddled in politics but renounced them after Napoleon's fall, focusing exclusively on his scientific work. He lost his sight in 1838. At the National Museum of Natural History, he was succeeded by his son Isidore Geoffroy Saint-Hilaire (1805–61) in 1824.

1.347. Josephus Nicolaus LAURENTI (1735–1805) was an Austrian physician and naturalist. His greatest contribution to herpetology and zoology is his *Specimen medicum* (1768), which names and describes—for the first time—many reptiles, their toxicity, and the function of their poison. Laurenti also was the first to describe a cave organism and thus is responsible for the early beginnings of biospeleogy.

French naturalist Georges Louis Leclerc COUNT OF BUFFON (1707–88) enjoyed pointing out that he had inherited his intelligence from his mother, a woman of spirit and learning. In 1753, Buffon delivered a famous discourse on style, arguing that "the style is the man himself." He did not contribute to the French *Encyclopédie*, even though he was a friend of Denis Diderot and Jean Le Rond d'Alembert (1717–83). Buffon's son was guillotined during the French Revolution in 1794. Buffon, who was director of the Jardin du Roi, is particularly famous for the first attempt at presenting systematically existing knowledge in natural history, geology, and anthropology in one multivolume book, the *Histoire naturelle, générale et particulière* (1749–1804, 42 volumes). Buffon studied mountains and glaciers and was one of the first to point out that climate, species, and the position of the continents were not fixed but ever changing, arguing that separate regions with a similar environment have distinct faunas (this became known as Buffon's law). After his death, the work was continued by Bernard Germain Étienne Médard de la Ville-sur-Illon, the **Count of Lacépède**.

1.348n. Michel Étienne DESCOURTILZ (1775–c. 1835) was a French naturalist who, like many planters, traveled to **Saint Domingue** between 1799 and 1803, hoping to recover lost fortunes. Descourtilz described the climate as responsible for corrupting the Creole way of life and observed the use of plants for abortion by slave women who wanted to spite their masters. He noted his impressions of **Leclerc**'s expedition in 1801 to put down the insurgency and was himself captured by black troops. His three-volume *Voyages d'un naturaliste* [Travels of a naturalist] was published in 1809. Given Descourtilz's perspectives on the Saint-Domingue Revolution, it is noteworthy that Humboldt merely comments on his knowledge of crocodile behavior.

1.351. William DAMPIER (1651–1715) was a British pirate and a pioneer in scientific exploration. An orphan at age sixteen, he sailed off to Newfoundland, the East Indies, and the Gulf of Mexico. From 1678 to 1691, he engaged in piracy. By 1699, he traveled to Australia as an explorer for the British Admiralty, but his ship, the *Roebuck*, had a dangerous leak and had to be abandoned. The crew was stuck on a South Atlantic island but was eventually rescued by homeward-bound warships. Dampier returned to pirating after that. His *New Voyage Round the World* (1697) was very popular. **Wafer** (c. 1640–1705) sailed with Dampier for some time, and Wafer's travel account was, at some point, attached to Dampier's.

1.352n. Albertus SEBA (1665–1736) was an apothecary and collector in Amsterdam. He assembled two impressive collections of exotic natural history objects, mainly by purchasing curiosities from travelers. These collections included specimens from around the world: shells, stones, stuffed mammals, hundreds of snakes, insects, plants, two-headed fetuses, and other rarities. Impressed with the first of these collections, Peter the Great (1672–1725) bought it in 1717. Seba quickly started anew with the proceeds from the sale. In 1731, he commissioned thirteen artists to illustrate his second collection for publication. He himself would only see two of the four volumes of his *Cabinet of Natural Curiosities* (1734–65), one of the first great illustrated natural history books of the eighteenth century. The Sebaea, a flowering plant, is named after him.

1.357n. With eight years of experience in the book publishing business, Awnsham CHURCHILL (1658–1728) became the publisher, book dealer, and the financial agent of John Locke (1632–1704). Locke encouraged and advised Churchill in the publication of the multivolume *Collection of Voyages and Travels* (1704), which Humboldt quotes. A supporter of the English Revolution of 1688, he was arrested for publishing, in 1687, an open letter by Gaspar Fagel (1634–88) criticizing the king's religious policy.

Juan Bautista MUÑOZ y Ferrandis (1745–99) was a Spanish historian. In 1770, Charles III appointed him cosmographer of the Indies, with the task of writing a comprehensive history of America. Muñoz could only publish the first volume of this enormous undertaking, *Historia del Nuevo Mundo* [History of the New World] (1793). In the process of his research, he ordered the Spanish archives, which served as the basis for an address to the Real Academia de la Historia in 1794 about the apparitions of Our Lady of Guadalupe in Mexico.

1.358. For Pierre-Louis BERTHOUD (1754–1813), see chronometric lines.

1.361. According to the *Leyenda Negra* [Black Legend], which continues to hold sway over many interpretations of the history of Spanish colonialism, no indigenous culture survived the conquest. This is only partly correct. CUBA'S NATIVE PEOPLES, who appear to have lived on the island since at least 1000 BCE, were the Guanajatabey, the name they applied to themselves. Archeologists discovered that the now extinct population of which the Guanajatabeys were a vestige had once extended across all of the West Indies. Hunters, gatherers, and farmers, these native Cubans cultivated tobacco, a crop that would become very important to the island's economy. **Columbus** had come in contact with remnants of this original population in the western part of Cuba. They are sometimes referred to as the Ciboney, who were actually part of the Taíno (Arawak) community. Unlike the Taínos, the Guanajatabey did not worship **zemís**; nor did they have the warlike behavior for which the Caribs were known. As **Arango** pointed out to Humboldt, a few native villages survived at the time (in 1826) in the unfrequented east of Cuba.

1.362. Petrus (Pietro or Peter) MARTYR D'Anghiera (1457–1526) was a renowned historian of the Spanish conquests. A chaplain at the court of King **Ferdinand II** of Aragón and Queen **Isabella I** of Castile, he became a member of Emperor CHARLES V's Council of the Indies in 1518. His ambitious work, *De Orbe Novo* [*On the New World*] (1530) was published over the course of more than twenty years. It is based on unidentified documents from several explorers, including **Columbus**. His 812 collected letters are another valuable source of information about his period. Charles V was Holy Roman Emperor (1519–56), king of Spain (as Charles I, 1516–56), and archduke of Austria (as Charles I, 1519–21). The Spanish and Habsburg Empire he inherited comprised large parts of Europe (Spain and the Netherlands, Austria and the kingdom of Naples) and the Spanish colonies overseas.

1.363. Woodes ROGERS (c. 1679–1732) was a British privateer who became governor of the Bahamas. During his privateering expeditions around the world (1708–11), he rescued Alexander Selkirk (1676–1721) from a Pacific island. Daniel Defoe's (1660–1731) *Robinson Crusoe* (1719) is based on Selkirk's story. After his appointment as governor of the Bahamas in 1717, Rogers helped suppress piracy in the Caribbean. He published his *Cruising Voyage Round the World* in 1712. The Frenchman Philibert COMMERSON (1727–73) was the official naturalist on Louis de Bougainville's (1729–1811) voyage around the world from 1766 to 1769; the expedition went to Tahiti the year before James Cook did. Denis Diderot wrote a fictional account of Bougainville's stay in Tahiti, which was published posthumously in 1798 (*Supplément au Voyage de Bougainville*). Before embarking in 1766, Commerson prepared a will in which he dedicated his body to science. A dolphin (*Cephalorhynchus commersonii*) and the Spanish mackerel (*Scomberomorus commerson*) are named after him.

I.363n. Bernard Germain Étienne Médard de la Ville-sur-Illon (1756–1825), the COUNT OF LACÉPÈDE, had considerable talents in both music and physics. In addition to attracting the attention of several renowned composers, notably Christoph Willibald Ritter von Glück (1714–87), Lacépède also won favor with **Buffon**, who secured him a position at the Jardin du Roi. He continued Buffon's *Natural History of Animals* after the latter's death. Lacépède's best-known works deal with the oviparous quadrupeds, reptiles, fishes, and cetaceans; the most important of these is his *Natural History of Fishes*. Active in French politics, Lacépède was exiled during the **Reign of Terror**.

II.9n. Nils Abraham BRUNCRONA (1763–1846) volunteered to the Swedish army's dragoon regiment (mounted infrantrymen) at the age of thirteen and later entered and had a successful career in the Swedish navy. In 1820–21, as director of the Swedish Pilot Service, Bruncrona lead a team to examine the water measurements on the Baltic Sea's west coast. The results, a table showing a drop in the sea level on the Swedish coast for a period of about four decades, was published in *Transactions of the Royal Swedish Academy* (1823) and then in the *Annalen der Physik* (1824). After retiring from the navy, Bruncrona designed private gardens and public parks. His fellow countryman Carl Peter HÄLLSTRÖM (1774–1836) graduated from the Academy of Turku (later the University of Helsinki). Working with Bruncrona, he made observations of the water level in the Gulf of Bothnia from 1770 to 1795, publishing his findings with Bruncrona's in *Transactions.* A talented geographer and cartographer, Hällström published scores of maps of Sweden and Finland. In 1804, he was elected member of the Royal Swedish Academy of Sciences.

The Gotha-born German geologist and diplomat Karl Ernst Adolf von HOFF (1771–1837) was best known for his extensive work on earthquakes. Like Humboldt himself, Hoff subscribed to the idea of dynamic geology. He was also the major German proponent of **Lyell**'s theories of actualism in which the present is regarded as the key to the past.

II.11. Nikolaus Joseph Freiherr von JACQUIN (1727–1817), a Dutch physician, botanist, chemist, and mineralogist, was sent to the Antilles and Central America in 1755 by Francis I (founder of the Hapsburg dynasty, 1708–65) to collect plants for the Schönbrunn Palace. He amassed a large collection of animal, plant, and mineral samples and wrote about his travels. In 1762, Jacquin was appointed professor of minerals and mining at the Mining Academy in Schemnitz (Hungary, now Slovakia). In 1768, he became professor of botany and chemistry and director of the botanical gardens of the University of Vienna. One of his students at Vienna was **Thaddeus Haenke**.

II.12. Humboldt's allegorical use of the encounter of birds and humans recalls the use of BIRDS as omens in Homer's *Iliad*.

II.13n. Written in 1503 during **Columbus**'s fourth voyage, The *LETTERA RARISSIMA* was originally in Spanish. It was circulated in a 1505 Italian translation, which was translated into other languages. A Spanish retranslation appeared in 1825. Humboldt owned an 1810 Italian version of the letter that, he believed, had previously belonged to **Malte-Brun**. The letter was addressed to FERDINAND II (1452–1516), also known as Ferdinand the Catholic, king of Aragón and Castile (Ferdinand V) from 1474 to 1504, in joint reign with his wife, QUEEN ISABELLA I (1451–1504). Ferdinand united the Spanish kingdoms and set Spain on the course of modern imperial expansion.

The Peruvian historian Antonio DE LEÓN PINELO (1589–1675), who became the official chronicler of the Indies in 1637, was a tireless compiler of laws, ordinances, and other publications by and about the Spanish empire. His *Epitome de la Biblioteca Oriental i Occidental, nautica i geográfica* [Nautical and geographical compendium of the oriental and occidental library] from 1629 is one of the greatest early bibliographies about Latin America.

II.16. MOCTEZUMA II, also known as Motecuhzoma Xocoyotzin (Montezuma in English) (c. 1466/80–1520), was the ninth ruler of Tenochtitlan, who expanded the Aztec empire to its maximum through warfare. The Spanish Conquest began during his reign (1502–20).

II.17n. STRABO (c. 64 BCE–c. 23 CE) was a Greek geographer and historian. His *Geography* is the only surviving work that discusses the peoples and countries known to Greeks and Romans at the time. Seldom used in ancient times, it was better known in Byzantium. It first appeared in Western Europe in a Latin translation in Rome around 1469. Book 16 (of seventeen) discusses Assyria, Babylonia, Syria, and Arabia. Humboldt owned individual books from various Latin editions (among others, from 1587, bought in 1832), in French (from 1819), and in German (from 1831–34). It is not clear which edition of Book 16 he would have used in the mid-1820s.

II.22–23. A Spanish missionary and Dominican priest, Bartolomé de LAS CASAS (1484–1566) came to "La Isla Española" (today's Haiti and Dominican Republic) in 1502, with his father Pedro de Las Casas, who was part of **Columbus**'s second journey. In 1511, Las Casas joined the conquest of Cuba at the hands of **Velázquez de Cuéllar**, who offered Las Casas an *encomienda*, a special form of *REPARTIMIENTO*, originally an allotment of lands either won or yet to be won. The colonial *encomienda* system, however, included no allocation of land or rents, as it originally had when used on the Spanish Peninsula. It was purely a forced labor system, in which the state temporarily assigned Indians to individual Spaniards who, in turn, agreed to take care of them. The *encomienda* system was a common practice among Spanish conquerors until 1542, when it was prohibited by Spanish law. Like most Dominican clerics an adversary of the *encomienda*, Las Casas finally abandoned his life as an *encomendero* in Cuba and tried to establish a better form of coexistence with the natives in his attempt of autonomous Dominican colonization and missionary work near Cumaná. But because of repeated attacks by Spanish slave hunters, the project soon failed. Shocked by the brutality with which the Spanish treated the Indians, Las Casas became an eager defender of the indigenous cause, first as head of the Dominican mission Verapaz, Guatemala, and then as bishop of the Mexican province of Chiapas from 1544 to 1546. Frustrated by his experiences, Las Casas returned to Spain in 1547, where his political efforts to denounce the systematic violence against the native population culminated in the famous "Dispute of Valladolid" in 1550. In this public controversy, the Dominican cleric became the *Indian advocate* against Juan Ginés de Sepúlveda (1494–1573), who was intent on legitimating violent colonization and the enslavement of the native populations as a natural right and part of a just war against the barbaric pagans. In the end, Las Casas won the argument, which had a positive impact on the colonial legislation of the sixteenth century, known as the "Leyes de Indias." But Las Casas's writings were even more influential. The *Historia de las Indias* is one of the most prominent early chronicles of the Spanish discoveries and a highly authoritative biography of **Columbus**. The *Brevísima relación de la destrucción de las Indias,* where Las Casas presents a shocking report of the Spanish violence against the Indians, became a key document in the propagandist efforts against Spain by rivalizing colonial competitors—Britain, France, and Holland. Both the *Historia* and the *Brevísima relación* are crucial to understanding the history of the Americas and constitute an important backdrop for the controversies surrounding the abolition of slavery in later centuries. Because of his proposal to replace indigenous with African slaves (which he later revoked), Las Casas's personal and political integrity has often been called into question. Humboldt, however, typically emphasizes Las Casas's humanitarian goals.

II.23. Spanish historian Antonio de HERRERA y Tordesillas (1559–1625) was born in Cuéllar, in the province of Segovia. Philip II of Spain appointed Herrera as the first official chronicler of the Indies. Later in his life, Herrera was appointed secretary of state by Philip IV (1605–65). Humboldt quotes from one of his most famous work, the *Historia General de los Hechos de los Castellanos en las islas y tierra firme del Mar Oceano* [General history of the deeds of the Spanish on the Tierra Firme of the ocean seas] (1601–15), better known as *Historia de las Indias occidentales* [History of the west Indies], which is the title Humboldt uses throughout.

II.28n. John FRASER was the Scottish botanist with whom Humboldt sent part of his collection to his friend **Willdenow**. Humboldt may have mixed up three generations of Frasers here and was probably referring to the Cuba trip of John and John Fraser Jr. at the time of his own American travels. There are three generations of Frasers who are of interest here. The older John Fraser (1750–1811) was a Scottish botanist, who worked as a botanical collector for the Russian Czar later in life. Having started out as a businessman in London, he eventually became a plant collector and crossed the Atlantic repeatedly for that purpose. John Fraser explored the Southern Appalachians and introduced shrubs from North America and Cuba to Europe. The Fraser magnolia, the Fraser fir (also known as the she-balsam), and Fraser sedge are named after him. John Fraser had two sons; the older son, John FRASER JR. (b. 1780), joined his father on tours to North America, notably in 1800–1801, when they visited Thomas Jefferson at Monticello and also went to Cuba. After his father's death, John Jr. returned to America, where he gathered a considerable collection of plants.

II.29n. William Hyde WOLLASTON (1766–1828) earned a medical degree from Cambridge in 1793 and practiced medicine in England until, in 1799, his older brother George gave him a considerable sum of money that allowed him to devote himself to natural scientific inquiries, especially to chemistry and metallurgy. Wollaston made considerable profits after finding a way of converting cheap granular platinum ore, which had been smuggled out of **New Granada**, into a very pure platinum powder. A lifelong bachelor, Wollaston also made significant contributions to botany, mechanics, electrochemistry, astronomy, crystallography, physiology, optics, and scientific instrumentation. He became a member of the Royal Society in 1793, alongside **Joseph Banks**. Later in life, he recommended to the House of Commons Select Committee of Weights and Measures the adoption of the British imperial gallon and served on the government's Board of Longitude from 1818 to 1828. Shortly before his death, Wollaston submitted a paper, "On the Water of the Mediterranean," which was published in the *Philosophical Transactions of the Royal Society of London*.

II.31. SPICA (Latin "Head of Grain") or Alpha VIRGINIS is the brightest star in the constellation Virgo; the fifteenth brightest star in the sky. It is about 250 light years away from Earth. Centaurus is a bright triple-star constellation that may only be seen from south of about 40° northern latitude. It is about 4.2 light years away. Crux, commonly known as the Southern Cross, is at about 12 hours 30 minutes right ascension (that is, the coordinate on the celestial sphere analogous to longitude on the Earth) and 60° south declination (or the angular distance south of the celestial equator). It is visible only from south of about 30° northern latitude. First described as a constellation by the astronomer Augustine Royer in 1679, it has been mentioned since antiquity. The constellation's five bright stars form a somewhat irregular cross.

II.32. *DECLINATION* is astronomical terminology represented with a lowercase Greek letter δ (delta). It describes the angular distance of a body north or south of the celestial equator. Together with the east-west coordinate (right ascension), declination defines the position of an object in the sky. The Italian monk Giuseppe PIAZZI (1746–1826) became a professor of mathematics in Palermo, Sicily, where he founded an astronomical observatory and compiled a catalogue of the position of 7,646 stars. He showed that most stars are in motion relative to the sun.

Spanish cartographer Antonio de ULLOA (1716–95) participated in a French expedition to Ecuador from 1736 to 1742 and, in 1758, became governor of Huancavelica in Peru. See also Jorge Juan y Santacilia.

II.34. At the time of Humboldt's voyage, Venezuela was one of the seven provinces of the captaincy-general of Caracas known then commonly as TIERRA FIRME. Venezuela means "Little Venice," a name that dates back to the sixteenth century, when the Spanish navigators and conquerors Juan de la Cosa, Alonso de Ojeda (c. 1465–1515), and Amerigo Vespucci (1451–1512) named this coast in memory of the famous Italian metropolis. Humboldt frequently refers to the entire *Capitanía general* as "Venezuela."

II.35. Infantry Captain Luis Alejandro de BASSECOURT, a nephew of the **Count of Santa Clara,** then governor of Cuba, was deputy governor of Trinidad, Cuba, from 1797 to 1799. Humboldt describes Bassecourt in his travel diaries as "talented and lissome," a man from an influential family and commited to the city's defense; in 1797, he valiantly (and successfully) defended Trinidad against an attack by British warships. Bassecourt went on to become a military leader in his Spanish motherland in the independence wars against the Napoleonic occupation.

A wealthy landowner in Trinidad, Antonio PADRÓN (1757–1823) was Humboldt's host during his brief stay in that city. The eighteenth-century house is today the Arquaeological Museum of Guamuhaya, exhibiting mostly pre-Columbian Cuban heritage.

II.42. Thomas TOOKE (1774–1858) was a British financier and economist who advocated free trade. He explained low prices with underlying cyclic conditions in his publications on *High and Low Prices* (1823), from which Humboldt quotes, and on *Considerations on the State of the Currency* (1826). He had started out his adult life in business in St. Petersburg and retired as governor of the Royal Exchange Assurance Corporation in 1852.

II.46. William IRVING (1771–1853) was one of many Scots employed in the British civil service. Like his father Thomas Irving (c. 1738–1800) before him, William was inspector-general of imports and exports at the London Customs House from about 1811 until 1851. He must not be confused with the merchant-politician and author William Irving (1766–1821) of New York City, older brother of Washington Irving (1783–1859).

II.51. Jules Paul Benjamin DELESSERT (1773–1847) was a French banker, philanthropist, collector, and botanist who generously sponsored scientific research, botanical expeditions, and individual collectors. He had a notable botanical library and was a member of the French Academy of Sciences. He founded the first cotton factory in Passy in 1801 and the first sugar factory in 1802.

There were at least two brothers DES BASSAYNS. Humboldt is referring to the oldest brother, Philippe Panon des Bassayns Richemont (1774–1840), who had conducted diplomatic missions since 1814 and was connected with the Isle Bourbon (Réunion). Another brother was Charles des Bassayns (1782–1863). In his private journals, Matthew Flinders (1774–1814), the British navigator who charted much of the Australian coast, made frequent references to visits from and correspondence with the des Bassayns.

II.52. Sir Robert Townsend FARQUHAR (1776–1830) was an administrator with the East India Company (1793–1823) in various posts at Madras (now Chennai, India), Amboyna (now Ambon, Indonesia), and Penang (Malaysia). In 1807, he published *Suggestions for counteracting any injurious effects upon the population of the West India colonies from the abolition of the slave trade,* encouraging Chinese labourers to migrate to the West Indies. From 1810–17 and 1820–23, Farquhar served as the English governor and commander-in-chief in Mauritius. In 1817, he concluded a treaty abolishing the slave trade in Madagascar with the ruling Merina. Once in England, he adopted the additional surname Townsend, was elected to parliament (1825), and directed the East India Company and the Alliance Insurance Company.

II.53. The THREE PRESIDENCIES of British India were Bengal, Madras, and Bombay.
The British politician George Gerard de Hochepied LARPENT (1786–1855) was part of the East India house of Cockerell & Larpent, chairman of the Oriental and China Association, and deputy-chair of the St. Katherine's Docks Company. He made several unsuccessful runs for Parliament in the 1840s. In addition to writing a wide-circulated pamphlet on the protection of West Indian sugar in 1823, he edited, in 1853, the *Private Journals of Francis Seymour Larpent*, his half-brother (1776–1845) and, in 1854, *The History of Turkey* by his grandfather, the British diplomat Sir James Porter (1720–86).

II.56n. The British politician and philanthropist William WILBERFORCE (1759–1833) became the acknowledged leader in the struggle to abolish slave trade and, subsequently, slavery in the British overseas possessions. His activism was in part connected to his conversion to evangelical Christianity in 1784–85. In 1787, he was involved in founding the Anti-Slavery Society, together with **Clarkson** and **Macaulay**. See also Cropper, Whitmore, and Mackintosh.

II.57. William HUSKISSON (1770–1830), a British statesman who championed free trade, was secretary of the Treasury (1804–5), president of the Board of Trade (1823), secretary of the colonies, and leader of the House of Commons (1827). On March 8, 1824, he gave a speech on "Mr. Hume's Motion for the Reduction of the Sugar Duties." A hesitant man who is said to have had "a peculiar aptitude for accidents," Huskisson died from injuries incurred at the opening of the Liverpool and Manchester Railway, when he lost his balance and fell onto the rails in front of an approaching train.

II.58. The French writer and statistician Jacques PEUCHET (1758–1830) published a dictionary of commercial geography (1799–1800) and a topographical and statistical description of France (1805); Humboldt quotes from the latter. Peuchet's 1827 *Les secrets de la police de Louis XIV à Louis-Philippe* [The secrets of the police from Louis

XIV to Louis-Philippe] inspired both Honoré de Balzac (1799–1850) and Alexandre Dumas (1802–70).

In 1800, Jean Antoine Claude CHAPTAL, the Count of Chanteloup (1756–1832), was minister of domestic affairs under the rule of Napoleon I (1769–1881). A chemist as well as a politician, Chaptal published an account of beet sugar in 1821.

II.60. In 1825, Count Pierre Laurent Barthélemy de SAINT-CRICQ published documents relating to trade with the new states in America through the *Bureau de commerce et des colonies*. He is also remembered for terminating the love affair between his young daughter Caroline (c. 1872) and the Hungarian composer and pianist Franz Liszt (1811–86) in 1828. Caroline eventually agreed to an arranged marriage with Bertrand d'Artigaux; as a result, Liszt suffered a nervous breakdown and succumbed to religious mania. He never forgot Caroline and even remembered her in his will, although she died fourteen years before he did.

II.61. Diplomat Albert GALLATIN (1761–1849) was the fourth secretary of the U.S. Treasury and representative and senator-elect from Pennsylvania. He was born in Geneva, Switzerland, graduated from the University of Geneva in 1779, and promptly ran away to Boston, Massachusetts. Having served in the Revolutionary army, he became an instructor of French at Harvard in 1782 but moved to Virginia in 1785. He was elected to the U.S. Senate and took the oath of office in 1793, but a petition was filed claiming that he did not meet the citizenship requirement. Early in 1794, the Senate declared his election void. Gallatin served in Congress from 1795 to 1801, when he was appointed secretary of the Treasury by President Thomas Jefferson. Reappointed under James Madison, he served until 1814, moving on to negotiate the Treaty of Ghent. From 1815 to 1823, he served as U.S. envoy extraordinary and minister plenipotentiary to France, and from 1826 to 1827 as minister plenipotentiary to Great Britain. Upon his return, he became president of the National Bank of New York. Humboldt owned some of his later works; his reference to Gallatin in this text, however, appears to be based on the fairly extensive correspondence they had rather than on publications.

Adam SEYBERT (1773–1825) was a representative from Philadelphia who studied at the University of Pennsylvania and in Europe. Interested in chemistry and mineralogy, he befriended the young geologist **Isaac Lea** in Philadelphia. In 1797, Seybert was elected a member of the U.S. American Philosophical Society. Humboldt quotes here from his 1818 *Statistical Annals of the United States*.

II.69. Johann Daniel Georg MEMMINGER (1773–c. 1840) edited and authored contributions for the *Beschreibung des Königreichs Württemberg* [Description of the Kingdom of Württemberg] in the 1820s and 1830s. He had started publishing descriptive accounts of that region in 1812. A notable family from that particular German state, the Memmingers are also of considerable interest to U.S. history: Christopher Gustavus Memminger (1803–88) was a German-born Confederate secretary of the Treasury, who has generally been held responsible for the collapse of his government's credit during the U.S. Civil War. He was born in Württemberg but immigrated to Charleston, South Carolina, after his father's death in 1819. He became a successful lawyer and helped draft the provisional constitution of the Confederate states. After the Civil War, he was granted a presidential pardon that allowed him to go back to practicing law in South Carolina. He also made a career in chemical manufacturing.

II.70. From the sixteenth to the nineteenth century, the Europeans used the term BARBARY COAST (derived from Berber) to refer to the middle and western coastal regions of North Africa (today's Morocco, Algeria, Tunisia, and Libya).

II.72. Antoine Laurent LAVOISIER (1743–94) was a leading French chemist, who was executed for tax fraud during the **Terreur** of the French Revolution. Although trained in law, he preferred to follow his scientific inclinations and, in 1768, became a member of the Academy of Sciences in Paris. His wife, Anne-Marie Paulze (1758–1836), collaborated in his research.

II.73. Du Temple de BEAUJEU was a self-educated Frenchman born in the traditional French province of Maine and orphaned at the age of fifteen, after which nature had served as his teacher. Between 1806 and 1811, he traveled to Italy, Spain, Germany, Denmark, Flanders, Holland, Sweden, Poland, and Russia. In 1809, he became interested in agriculture. In 1823, Adolphe Jules César Dureau de la Malle (1777–1857) published an account of Beaujeu's contributions to science, including a lengthy reproduction of Beaujeu's descriptions of sugar production from sugar beets, connecting the scientific discoveries with Beaujeu's life. Beaujeu described a method of extracting **beet sugar** as an alternative to the method based on sugarcane that was used in the West Indies. While Humboldt quotes at length from Beaujeu's report to the Academy of Sciences in Paris from August 1826, he mentions the decreasing number of beet sugar factories in France much earlier. See also Chaptal.

II.77. After studying in Europe during his youth, Spanish economist and naturalist Ramón DE LA SAGRA y Pérez (1798–1871) went to Cuba in 1823, where he became director of the Botanical Gardens and the Agronomic Institute. A member of scientific organizations in Madrid, Paris, London, New York, Philadelphia, Geneva, and Moscow, he published writings on Cuba, the United States (which he visited in 1835), Belgium, Holland, and Spain. His *Historia económico-política y estadística de la isla de Cuba* appeared in 1831. Many other volumes on Cuba, including maps, followed in the next three decades.

II.79. French naturalist Horace Bénédict de SAUSSURE (1740–99) invented many of the scientific instruments of his time, among them the electrometer, the cyanometer and the **hair hygrometer**. In the opening chapter of his own *Relation historique*, Humboldt acknowledges the importance of Saussure's travelogue of his fourteen alpine expeditions, *Voyage dans les Alpes* (1769–96), for his own writings. The example set by Saussure—he was also one of the first to climb the Mont Blanc—served Humboldt as a model for a type of scientific-personal narrative that combined the advances in scientific exploration with the emotions and impressions of the traveler, interweaving meteorological observations with comments on the habits of the local population.

This subsection of the *Political Essay on the Island of Cuba*, which was reprinted as a separate article in 1826, is devoted to discussing **Atkinson**'s claim that Humboldt had erred in establishing the equator's mean TEMPERATURE.

II.80. The British astronomer and mathematician Henry ATKINSON was a highly successful teacher of mathematics with wide-ranging interests (as was typical for his day). He actively pursued research in mathematics and astronomy and read many papers to the Newcastle Literary and Philosophical Society. Two of his papers, including the one about Humboldt, were published by the Astronomical Society of London. Atkinson also delivered papers to the Literary and Philosophical Society on economics, engineering, and metaphysics. The paper in question here was a response to some of Humboldt's findings as recounted in *On Isothermal Lines* (1817; English, 1820/21). As Humboldt notes, David Brewster, who had published an English translation of *On Isothermal Lines* in one of his Edinburgh-based scholarly scientific journals, had published a note regarding Atkinson's claim. Humboldt provides a lengthy reasoning of his findings about the equator's mean temperature here, concluding that he is not convinced of the criticism being indeed valid. Humboldt does not name Atkinson again in this text. He refers to Atkinson's publication in French as "Refraction astronomiques" [Astromical refractions] (see vol. II, p. 91), even though the original publication was in English. After the publication of Humboldt's *Essai*, **Brewster** provided further evidence that suggested that Humboldt had not erred.

Richard KIRWAN (1733–1812) was an Irish chemist who also contributed to other areas of science and scholarship. In 1787, he published *An Estimate of the Temperature of Different Latitudes* in London. In his *Relation historique*, Humboldt remarks that he was unable to use Kirwan's "excellent work," because the latter did distinguish sufficiently between theoretical results and actual experiments. Although Kirwan's book on temperature is not his best-known work today, this topic appears to have preoccupied him later in life, for he developed a pathological fear of catching a cold. Visitors would find him in spring wrapped up in a cloak, shawl, and slouch hat, with a huge fire ablaze in the hearth.

II.81. Sir David BREWSTER (1781–1868) was a Scottish physicist and the editor of scholarly scientific journals based in Scotland. Particularly interested in optics and polarized light, he invented the kaleidoscope. In addition to writing a widely read biography of Sir Isaac Newton (1643–1727) in 1831, Brewster published an English translation of Humboldt's "Des lignes isothermes et de la distribution de la chaleur sur le globe" (1817) as "On Isothermal Lines, and the Distribution of Heat over the Globe" in five installations in his *Edinburgh Philosophical Journal* (1820–21). In his reference to Brewster, Humboldt mixed up the publication date, giving it as 1829 rather than 1826. A brief entry in an 1826 *Edinburgh Journal of Science* article, entitled "Meteorology" (vol. IV, no. VII, p. 180), observed that **Atkinson** had challenged Humboldt's findings on the equator's mean temperature. After the publication of Humboldt's *Political Essay on the Island of Cuba*, Brewster proclaimed Atkinson wrong in the same publication's 1827 issue (vol. VI, no. XI, pp. 118–19): "When Humboldt, in his admirable paper on Isothermal Lines, fixed the mean temperature of the equator at 81 ½°, he naturally gave a preference to observations made in the old world, where the distribution of temperature did not exhibit the same anomalies which occur in the New World. He accordingly used the mean temperatures of Senegambia, Madras, Batavia, and Manilla, whereas Mr. Atkinson, neglecting entirely the temperatures of the Old World, deduces his results solely from the American observations. Mr. Atkinson is therefore not correct in stating, 'that it appears, from data furnished by himself, that Humboldt has fallen into an error when he asserted that the mean temperature of the equator cannot be fixed beyond 81½°.'"

Baron Dirckinck von HOLMFELDT made observations on the water temperature on the coast of Peru, which he communicated to Humboldt in 1826. Von Holmfeldt conducted these experiments in 1825, as Humboldt notes in "On the Principal Causes of the Difference of Temperature on the Globe," published in the *Edinburgh New Philosophical Journal* (1827–28).

II.82. PERRINS, QUEVEDO, DAVY, and RODMAN are mentioned in various contemporary journals in one breath with Humboldt and in connection with latitudinal, longitudinal, and temperature measurements.

R. PERRINS was a surgeon in the British East India Company. He sailed to Bombay in 1800. Perrins's temperature table, which Humboldt cites, was published in 1804 in William Nicholson's *Journal* and in the following year in *Annalen der Physik* by Ludwig Wilhelm Gilbert (1769–1824). The measurements had been requested by Anthony Carlisle (1768–1840), who was interested in knowing "whether fishes possess any other temperature than that of the water in which they live."

JOSÉ DE QUEVEDO Y CHEZA (d. 1835) was an officer of the Spanish Royal Navy in charge of the vessel *Santa Rufina*. Humboldt engaged Quevedo to observe, during his voyage from Lima back to Spain around Cape Horn, two thermometers, one exposed to the air, the other submerged in the ocean waters at a shallow depth. Quevedo's readings were also published in the 1836 *Annalen der Physik und Chemie* by Johann Christian Poggendorff (1796–1877). Quevedo is famous for being the captain of the *San Leandro*, a battleship at Trafalgar in 1805. Because he was a skilled administrator, he was appointed military governor of Veracruz in 1812. He returned to Cádiz in 1815 and organized two expeditions to Puerto Rico and to Havana.

JOHN DAVY (1790–1868) was an Edinburgh-trained British army surgeon and inspector general of hospitals. He was the youngest brother and biographer of the famous British chemist Sir Humphry Davy (1778–1829), who founded the Geological Society of London in 1807. In 1812, John Davy discovered phosgene, the compound of carbonic oxide and chloride, while working to contest the conclusions of John Murray (1778–1820) on the composition of chlorine. While in the army, which he had joined in 1815, Davy traveled to various British colonies, including Ceylon (currently

Sri Lanka), the West Indies, and India. Humboldt quotes from Davy's *Observations on the Temperature of the Ocean and Atmosphere, and on the Density of Sea-Water, Made during a Voyage to Ceylon*. Davy used Fahrenheit, which Humboldt, untypically, convertes to centrigrade.

Humboldt mentions RODMAN's voyage from Philadelphia to Batavia in his *Relation historique*. Thomas Horsfield (1773–1859), a naturalist who studied medicine at the University of Pennsylvania, published Samuel Rodman's table on "Comparative Temperature of the Air and of the Water of the Ocean" alongside his own *Account of a Voyage to Batavia, in the Year 1800*. In 1799, Horsfield was appointed surgeon on the merchant ship *China*, sailing from Philadelphia to Indonesia in 1800. Rodman made his observations during that same trip to Jakarta (then Batavia) on the *China*. Rodman was likely related to the New England whaling merchant and Quaker leader Samuel Rodman (1753–1835); his name is mentioned in connection with the *Logbook of the Ship South America* (1823–25) by Edmund Gardner (1784–1874).

Humboldt quotes Perrins, Quevedo, Rodman, Davy, and **Churruca** in his "Observations on the Mean Temperature of the Equatorial Regions" in the *Edinburgh Journal of Science*, written in Paris in September of 1826. This article is also a follow-up on **Brewster**'s discussion of **Atkinson**'s claim that Humboldt had erred with regard to the equator's mean temperature.

II.84. Humboldt portrayed his isothermal work as a continuation of a project outlined by the German astronomer and cosmographer Johann Tobias MAYER (1723–62) to the Göttingen Societät der Wissenschaften [Goettingen Academy of Sciences] in 1755 and published by Georg Lichtenberg (1742–99) in 1775. Mayer developed a theory of heat distribution on the earth's surface in terms of trigonometrical equations with indeterminate coefficients. Humboldt held up Mayer's astronomical method as exemplary for his own enterprise, not because it enshrined gravitational equilibration as a universal heuristic, but because it placed precise measurement at the center of any attempt to reveal the laws of temperature or magnetism. No amount of theory could supply the basic terms of the equations that describe these lines. These laws, Humboldt emphasized, were essentially empirical and increased in accuracy with more precise and more frequent measurements.

II.87. Antonio Bernardino PEREIRA LAGO (c. 1777–1847) was a Portuguese military engineer. A member of the Royal Corps of Engineers, he served in the Brazilian northeastern region of Pernambuco. His *Estatística historica-geographica da provincial do Maranhão* [Historical and geographical statistic of the province of Maranhao] (1822), with seventeen maps of the region, was the first work on the subject printed in Portugal. Pereira do Lago also authored *Cinco annos de emigração na Inglaterra, na Belgica e na França* [Five years of travel in England, Belgium, and France] (1834, 2 vols.) in the form of letters to his wife, the first of which is dated November 8, 1828. In addition to authoring maps, surveys, and meteorological observations of the region, he wrote *Memoria sobre o forte do mar em Pernambuco* [Report about the fortress in Pernambuco] (1862) and *Itinerario da provincia do Maranhão, começado em Janeiro de 1820* [Travels in the province of Maranhao begun in January 1820] (1872).

II.88. Joseph RITCHIE (c. 1788–1819) was a "gentleman of scientific attainments" and a surgeon in London who knew the poet John Keats (1795–1821). Together with the traveler and navigator Captain George Francis LYON (c. 1795–c. 1832), Ritchie embarked on a journey to explore Northern Africa in 1819. Ritchie died of an illness in Morzouk, the capital of Fezzan (southwestern region of modern Libya), in November of that year. During his African trip with Ritchie, Lyon dressed like a Muslim, kept his head shaved, and let his beard grow. In1821, he published his African journal, which included scientific data to which Humboldt is likely referring here.

Different systematic approaches to measuring air temperature were used in the eighteenth and nineteenth century, among them the scales inventend by Daniel Fahrenheit (1690–1740) and Anders Celsius (1701–44). The scale of the RÉAUMUR THERMOMETER, created by the French physicist and naturalist René Antoine Ferchault de Réaumur (1683–1757) in 1730, contained 80 degrees and worked with ethanol instead of mercury.

II.89. The MAXIMUM-MINIMUM THERMOMETER, invented by James SIX (1731–93) in 1778, was one of the first instruments to measure reliably water temperature at great depth; it could withstand water pressure. All major marine expeditions of the nineteenth century used this instrument, which was capable of determining two temperatures at a time, the maximum and minimum, within a time frame of twenty-four hours.

II.91. French missionary Father Claude DE BÈZE (1657–95) was one of fourteen Jesuit priests to establish a Christian mission in the southeast-Asian kingdom of Siam (today's Thailand). De Bèze had been instructed to study the flora and fauna and later wrote his memoirs on the fall of the Prasat Thong Dynasty (*Revolution in Siam*, 1688). Father de Bèze also worked as a missionary in the Malaysian state of Malacca, where he observed a comet in 1689, and in Batavia, where he took multiple barometric measurements and discovered, in 1690, that the barometer remains unaffected by temperature changes throughout the equinoctial zone.

II.93. Humboldt took this part of the supplement from his *Relation historique,* vol. IX, pages 136–419 in the octavo edition, with some changes.

II.95. Girolamo BENZONI (born c. 1519) traveled to the Americas in 1541, where he visited the Antilles, the Isthmus of Darien, Guatemala, and the west coast of South America. His main purpose in the New World appears to have been commerce, which he often carried on with difficulty, because the Spanish did not look favorably upon foreign traders. Benzoni conceived an inveterate hatred of Spaniards and the Spanish government. Benzoni's *Historia del Mondo Nuovo* was published at Venice in 1565, at a time of the dispute over the treatment of the native populations. Several editions were published in rapid succession.

II.105. British political economist and demographer Thomas Robert MALTHUS (1766–1834) developed a theory on population growth, arguing that populations will always outgrow food supplies; therefore, reproduction should be restricted. He anonymously published his *Essay on the Principle of Population* in 1798; by 1826, it was available in a sixth edition. His line of thinking is often called Malthusianism.

II.106. The 1920 census put the U.S. POPULATION at about 105 million; the 1930 census put it at 122 million. Although Humboldt's predictions fall short, he could hardly have anticipated the continental expansion of the United States.

Was the United States situated entirely in the TEMPERATE ZONE by 1826? At the time, twenty-four states had entered the union. Including territories that had not yet become states, the settled areas at the time stretched from Maine to Florida on the East coast and extended West of the Mississippi. The **Missouri territory** comprised unorganized regions of what is now Iowa, Nebraska, the Dakotas, most of Kansas, Wyoming, and Montana. It bordered Spanish Texas and New Mexico to the South. The twenty-four states that were part of the union included (in chronological order): Delaware, Pennsylvania, New Jersey, Georgia, Connecticut, Massachusetts, Maryland, South Carolina, New Hampshire, Virginia, New York, North Carolina, Rhode Island, Vermont, Kentucky, Tennessee, Ohio, Louisiana, Indiana, Mississippi, Illinois, Alabama, Maine, and Missouri.

Mostly nondenominational missions had been established before the American Revolution (United States). Only some of them were modeled after the British voluntary societies that had led to the creation of METHODISM. After the War of 1812 (see Pike), various new missionary societies blossomed, including the denominational Methodist Missionary Society. They focused on frontier missions and Native Americans.

II.112. Claude-Louis MATHIEU (1783–1875) was a French astronomer and mathematician who worked with the Bureau des Longitudes in Paris. This scientific institution, which still exists, was founded in 1795. It had authority over the Paris Observatory and was charged with the improvement of nautical navigation, the standardization of time-keeping around the world, geodesy, and astronomical observation. Among its ten original founders were **Jean Dominique Cassini**, **Borda**, **Buache**, and Bougainville.

II.117. Fernando NAVARRO Y NORIEGA, auditor general of ways and means for New Spain, was highly regarded, among others by historian Hubert Howe Bancroft (1832–1918). Humboldt's "Rispuesta de un Mexicano al no. 200, del Universal" refers to Navarro's book *Memoria sobre la población del reino de la Nueva España* (1820). Navarro's other famous work, which Humboldt also cites, is the *Catálogo de los curatos y misiones que tiene la Nueva España en cada una de sus diocesis* (1813).

II.118. Pedro José de FONTE y Hernández Miravete (1777–1839) was archbishop of Mexico City from 1815 to 1837, the last Spaniard to occupy that seat in the Mexican capital. His opposition to the independence movements forced him to flee to Spain in 1821. When asked to return to his office, de Fonte decided to retire from his position.

José María del BARRIO was member of a Spanish aristocratic family from Guatemala and the Guatemalan minister in Mexico until the mid-1830s. Del Barrio probably met Humboldt in Paris in 1823.

Juan Domingo JUARROS y Montúfar (1753–1821) was one of the first Guatemalan historians. He wrote the *Compendio de la historia de la ciudad de Guatemala* [Compendium of the history of the city of Guatemala], of which the first volume was printed in 1808. It was published in its entirety as *Compendio de la historia del reino de Guatemala (Chiapas, Guatemala, San Salvador, Honduras, Nicaragua, Costa Rica 1500–1800)* [Compendium of the history of Guatemala] in 1856–57.

II.120. Humboldt's acquaintance Manuel de NAVARRETE had worked in the royal treasury at Cumaná since 1777. In April 1799, he wrote a lengthy letter to Humboldt, providing him with geographical and other details of the area.

II.123. Charles BROWN was among the thousands of volunteers from England who traveled to South American to join **Bolívar**. Brown enlisted in the artillery brigade and arrived in Angostura (today Cuidad Bolívar) in 1818. With a promise of a significant salary, a lump sum on landing on American soil, a grant of land after the war had ended, and quick promotion within the army's ranks, many from England, Scotland, and Ireland readily volunteered but were desilusioned upon arrival. Joining others in writing with bitterness about their volunteer experiences in Bolívar's army, Brown published *Narrative of the Expedition to South America.*

François Raymond Joseph DE PONS (1751–1812), an agent of the French Government at Caracas, traveled through the **Tierra Firme** region of South America between 1801 and 1804 and published his travel descriptions *Together With a View of the Manners and Customs of the Spaniards, and the Savage as Well as Civilized Indians* (from the title of the English translation, London 1806). This American travelogue was reviewed widely in Europe.

II.125. Conrad MALTE-BRUN, originally Malte Conrad Bruun (1775–1826), was the founder of the first modern geographic society. He was exiled from Denmark for his verses and pamphlets in support of the French revolution and moved to Paris. Among his works are the first six volumes of the *Précis de la géographie universelle* [Short essay on universal geography] (1810–26). See also La Bastide.

II.125. Venezuelan poet, philologist, jurist, and politician Andrés BELLO López (1781–1865) was one of the most prominent voices of the emerging nations of independent South America. As a young man of only eighteen years, "Bellito" (as Humboldt calls him in his travel diary) is believed to have met Humboldt and **Bonpland** in Caracas in 1799. Bello joined the Prussian's excursions through the Valley of Caracas and together they climbed the "Silla de Caracas" mountain, even though Bello—as so many others—did not manage to keep up and had to turn back halfway. During his London years (1810–29), Bello translated into Spanish parts of Humboldt's *Plantes équinoxiales* [Equinoctial Plants] and substantial excerpts of chapters 6 and 7 of his *Relation historique.* In 1812, Humboldt supposedly met Luis LÓPEZ Méndez (1758–1831),

Manuel PALACIO FAJARDO (also Faxardo) (1784–1819), and Bello (who was then Fajardo's secretary) in London, where all three were working with **Bolívar** as representatives of the *Junta de Caracas,* asking the British government for political support of their new independent nation. This is where they allegedly supplied Humboldt with data on slave and other populations of Venezuela. Fajardo, a Venezuelan physician, lawyer, chemist, and author of the famous *Outline of the Revolution in Spanish America* (1817), was an important part of Humboldt's broad international network of scientific colleagues and politicians. In promoting the cause of independence, Fajardo appealed to and met with President Madison and Secretary of State Monroe in 1812, as well as with Napoleon and Pope Pius VII (1742–1823) during the following year. Fajardo was a participant in the Angostura Congress of 1819 and appointed secretary of state by **Bolívar**. With the help of Fajardo's expertise, Humboldt perfected his map of **New Granada**, which appears as "Carte du cours de Rio Meta et d'une partie de la Chaîne Orientale des montagnes de la Nouvelle Grenade" [Map of the course of the Rio Meta and of a part of the eastern range of the mountains of New Granada] in his *Atlas géographique et physique du nouveau continent* [Geographical and physical atlas of the new continent] (1814).

II.126. Francisco José de CALDAS (1768–1816), also known as El Sabio Caldas, or Caldas the Wise, was a Colombian geographer, astronomist, and cartographer who taught mathematics at the University of Bogotá. He edited the weekly *Semanario de la Nueva Granada*. A leading participant in Colombia's independence movement, he was executed in 1816. See also Pombo and Mutis.

II.131. José Hipólito UNANUE y Pavón (1755–1833) edited the *Mercurio Peruano* (1791–95) and contributed to the Sociedad de Amantes del País. A year after publishing *Observaciones sobre el clima de Lima* (1806), in which he discussed endemic, epidemic, and sporadic diseases in relation to the region's physical geography and natural history, Unanue was named the colony's chief medical officer. An intellectual during the transition period from the colonial to the independence era, he acted as adviser to several viceroys; after independence, he was appointed Peru's first minister of finance. He became president of Peru in 1825.

Antonio José de IRISARRI Alonso (1786–1868) was a Guatemalan statesman, journalist, and interim supreme director of Chile (1814). After marrying a Chilean heiress, Irisarri took up residence in that country and joined the movement for independence in 1810. In charge of important public offices during the struggle for liberty, he was appointed minister of government and foreign affairs in 1818. After a political scandal, in which he was accused of high treason, Irisarri returned to his homeland and became minister of Guatemala and El Salvador from 1839 to 1848. He moved to the United States in 1850, where he spent the rest of his life.

II.132. Caesar Augustus RODNEY (1772–1824), a U.S. lawyer and politician from Wilmington, Delaware, was the nephew of Caesar Rodney (1728–1724), the signer of the Declaration of Independence who is depicted on the Delaware state quarter. In 1817,

Rodney was appointed by President James Monroe to lead a commission to investigate whether the newly formed South American republics should be officially recognized. Rodney strongly advocated such recognition and, with John Graham (1774–1820), published his findings in 1819 as *Reports on the Present State of the United Provinces of South America*. This report is thought to have contributed to the Monroe Doctrine. It also resulted in Rodney's 1823 appointment as United States minister plenipotentiary to the United Provinces of the River Plate, now Argentina, the last of his numerous political offices.

II.133. Henry Marie BRACKENRIDGE (1786–1871) wrote a travel account on South America in 1819. He was a son of the writer and Judge Hugh Henry Brackenridge (formerly Hugh Montgomery Breckenridge, 1748–1816), author of *Modern Chivalry* (1792), the first novel portraying frontier life in the United States.

Thomas POWNALL (1722–1805) was a governor of colonial Massachusetts (1757–60) and member of the English House of Commons. He was promoted to the governorship of South Carolina in 1760 but decided instead to accept a position in Europe. In 1764, he published *The Administration of the Colonies*, which is probably the work to which Humboldt is referring here. In the 1770s, Pownall also contributed to the journal *Archaeologia: or Miscellaneous Tracts Relating to Antiquity. Published by the Society of Antiquaries of London*.

II.134. The George HARVEY whom Humbolt mentions here should not be confused with the Scottish landscape painter Sir George Harvey (1806–76). We have been unable to find any informantion about the former other than his publications (see bibliography).

Adrien BALBI (1782–1848) was a geographer and statistician from Venice, who started out teaching mathematics, physics, and geography. In 1819, he left Italy and spent about two years in Portugal before moving to France for fourteen years, where, in 1822, he published the *Essai statistique sur le royaume de Portugal et d'Algarve* [Statistical essay on the kingdom of Portugal and the Algarve region]. He also published works on geography.

Polymath José Francisco CORREIA da Serra (1750–1823) was a Portuguese Abbot, philosopher, diplomat, and scientist. One of the founders of the Academy of Science in Lisbon, Correia had to leave Portugal after several conflicts over his progressive writings. In 1813, he departed for the New World, arriving first in New York. He was appointed Portuguese minister plenipotentiary at Washington, D.C., in 1816.

Augustin François César Prouvençal (also known as Auguste de) SAINT-HILAIRE (c. 1779–1853) was a French botanist who traveled in South America—mainly in south and central Brazil—between 1816 and 1822. He collected data in all fields of natural history.

II.139. All the sides of a WELL-CONDITIONED TRIANGLE are of equal length, or nearly so, thus reducing to a minimum any possible error in the measurement of an angle.

II.141n. Captain Joseph VARELA y Ulloa saw the ring of Saturn disappear once, in Cádiz in October 1773. He is mentioned in Cook's voyages.

II.146n. Geographer and cartographer Jean Baptiste Bourguignon D'ANVILLE (1697–1782) was the first geographer of the king of France. He greatly improved the standards of mapmaking by frequently revising his 1743 atlas. Humboldt owned D'Anville's *Nouvel Atlas de la Chine* [New atlas of China] (1737), his *Carte de la Province de Quito au Perou* [A map of the province of Quito in Peru], based on the journals of **La Condamine**, and his *Orbis veteribus notus* [New general atlas of modern geography] (1763).

II.152. Spanish cartographer Andrés BALEATO (1714–1817) was ordered by the viceroy of Peru, Francisco Gil de Taboada Lemos y Vallamarín (1736–1809), to undertake the necessary measurements to draw a map of the territory. This map later turned into one of the focal points for the polemic and controversial discussions of the border lines between the Andean countries of Peru, Bolivia, and Chile. Particularly at issue was Bolivia's right to have access to the Pacific Ocean.

II.153n. Humboldt's year is incorrect. The CONGRESS OF VENEZUELA, a product of the revolutionary movement, first met in March 1811. It drafted the constitution of Venezuela's so-called First Republic, which was officially publicized in December of that year. In a much debated move, this constitution granted legal equality to all men regardless of their race.

II.155. Zebulon Montgomery PIKE (1779–1813) was a U.S. army officer and explorer. Pike's Peak in Colorado, which he tried to climb and failed, is named after him. In 1806, in the wake of the Louisiana Purchase, Pike led an exploration party that traveled two thousand miles by boat on the Mississippi River, from St. Louis, Missouri, to Minnesota. In 1806, Jefferson commissioned him to explore the Arkansas and Red Rivers and to acquire information about the neighboring Spanish territory. His report noted military weaknesses and pointed out the possibilities of overland trade with Mexico, thus feeding the dream of westward expansion into Texas. In 1810, Pike published a Mexico map plagiarized from Humboldt; it was based on the illict copy that Wilkinson had given him (see also Arrowsmith). Pike died in battle during the War of 1812 (a war that, in spite of its name, actually lasted until 1815).

II.156n. José Cecilio DEL VALLE (c. 1776–1834) was born in Honduras and educated in Guatemala, where he rose in politics in 1820. He corresponded with luminaries in Europe. Having been elected to the presidency of Central America, he died before he could take office in 1834. The Federal Republic of Central America, or United Central Provinces of America, did not survive him for long: it existed from 1823 to about 1838, comprising the territories of the former captaincy-general of Guatemala.

II.158n. Humboldt owned a 1798 *Carte de la Guiane* [Map of Guiana] by Jean-Nicolas BUACHE de la Neuville (1741–1825), a nephew of the famous cartographer Philippe

Buache (1700–1773) who had contributed notably to the theory of physical geography and pioneered the use of contour lines to represent relief on maps. Although Jean-Nicolas Buache's fame was more due to his uncle's renown than his own achievements, he was still considered the premier geographer both during the ancien régime and in Revolutonary France. He was a member of the Academy of Sciences and the Bureau des Longitudes.

James Kingston TUCKEY (1776–1816) was a naval officer and explorer. After having served in the Caribbean, India, and the Far East, he was dispatched to Australia in 1802. Captured by the French on the way home, he was detained in France until the peace of 1814. Besides marrying a fellow prisoner's daughter, Tuckey used his time to write a four-volume treatise on maritime geography (1815). In poor health, he died during an expedition to explore the Congo River. The journals of his voyage were published posthumously.

II.162. In 1791, when the Botanical Expedition in **New Granada** headed by **Mutis** settled in Bogotá, Francisco Antonio ZEA (1770–1822) joined them as a scientific associate. Zea worked in botany under Mutis until 1795, when Zea's name came up in a scandal and he was sent to prison in Spain; he was eventually declared innocent. His involuntary journey to Europe proved felicitous, giving him the opportunity to study chemistry in Paris. Until 1804, he served as assistant director of the Royal Botanical Garden in Madrid. Zea also had a political career: in 1819, he became **Bolívar**'s vice president of the new Republic of Gran Colombia, which comprised the captaincy-general of Venezuela, the kingdom of New Granada, and the audiencia of Quito. Shortly after Zea's death in England, a publication entitled *Colombia from Humboldt and Other Recent Authorities* appeared in Spanish, presumably under Zea's auspices. Its authors were Robert Madie Neele and Frank Howard. For Zea's scientific significance, see also Roulin; for his political role, see Torres.

II.163. The engineer Francisco REQUENA was a governor of Maynas. In addition to publishing a description of Guayaquil in 1774, he also authored a history of Maynas and a narrative of his Amazonas explorations (1782).

II.170. The *Mapa Geográfico de América Meridional de 1775* [Geographical map of South America in 1775] by the Spanish cartographer, painter, and engraver Juan de la CRUZ Cano y Olmedilla (1734–90) was a landmark publication. Humboldt owned the original, which he had purchased in Paris.

Antonio CAULÍN (1719–1802) was a Franciscan missionary in the region that is now Venezuela. From 1754 to 1761, he was part of the Spanish-Portuguese border-survey expedition, which he described in his book *Historia coro-gráphica natural y evangélica de la nueva Andalucía* [History of natural and religious/evangelist topography in New Andalucia]. The book, written in 1759 and published after his death, also contains a history of the Franziscan missions in eastern Venezuela.

CUMANÁ is the first city on the American continent founded by the Spanish, in 1521, under the command of Gonzalo de Ocampo, who subsequently "pacified" the Cumaná coast to make it safe for pearl harvesting.

II.171. The JULY 5TH REVOLUTION refers to Venezuela's declaring its independence from Spain on July 5, 1811.

II.172. Friedrich von BUCHENRÖDER (1758–1824) published a map about Essequibo and Demerary in Amsterdam in 1798. A merchant and farmer who also had a military career, he settled in South Africa in 1803.

II.173. *KYK-OVER-AL* is Dutch and means "look all around." This may be the smallest fort ever constructed by the Dutch overseas. Kyk-Over-Al is the historic name of a small island, about 1.5 acres in size, located at the junction of the Mazaruni and Cuyuni Rivers. In 1940, *Kyk-Over-Al* became the name of a literary magazine published in Guyana and edited by the poet Arthur James Seymour (1914–89) and, after its revival in 1984, also by the poet and novelist Ian McDonald (b. 1933).

II.176. Tomás López de VARGAS Machuca (1730–1802) was one of Spain's most important geographers and cartographers in the eighteenth century. After studying grammar, rhetoric, and painting in Madrid, he began his career as a geographer in 1752, when he went to Paris with a group of other geographers, including **Antonio de Ulloa**. In Paris, he studied with **D'Anville**, and in 1755, he collaborated with **Cano y Olmedilla** on a nautical map of Mexico and the Antilles. Upon his return to Spain a few years later, Machuca was put in charge of the newly created Cabinet of Georgraphy and of creating maps of the entirety of Spain. His sons, notably Juan López de Vargas Machuca (fl. 1780–1812), carried on the family tradition.

II.177. The VICEROYALTY OF NEW GRANADA was created in 1717–23 and reconstituted in 1740. The uprising in Bogotá on July 20, 1810 is commemorated as Independence Day in Colombia, although these new governments swore allegiance to Ferdinand VII and did not begin to declare independence until 1811. In 1819, **Bolívar** invaded Colombia and decisively defeated the Spanish forces. There followed the decisive Battle of Carabobo, Venezuela, in 1821 and that of Pichincha, Ecuador, in 1822. Mopping-up operations were completed in 1823, while Bolívar led his forces on to Peru. The Congress of Angostura laid the foundation for the formation of the Republic of Colombia (1819–30), which was generally known as Gran Colombia because it included what are now the separate countries of Colombia, Panamá, Venezuela, and Ecuador.

II.179n. Thaddeus Peregrinus HAENKE (1761–1816) was a South American naturalist from Bohemia, who studied under **Jacquin** in Vienna. He joined the **Malaspina** expedition in Chile, accompanying them along the American coast as far as Vancouver Island. He traveled across Mexico, returned to Chile, and ended up in Cochabamba, Bolivia, in 1796, where he worked in a silver mine on the estate he had purchased. He accidentally poisoned himself in 1817. He bequeathed his botanical collection to the National Museum in Prague, but only a part of the collection reached its destination.

Haenke did not publish an account of his explorations but left notes and manuscripts that other naturalists consulted. Most of his manuscripts are now at the National Archives in Buenos Aires and at the Botanical Institute in Madrid.

II.181. MAITA CAPAC was the fourth emperor of the Inca Kingdom of Cuzco in Peru; he reigned from 1171 to 1211.

II.185. ENTRE-RÍOS is part of Argentina today. As a province of the viceroyalty of La Plata, colonial Uruguay was known as the BANDA ORIENTAL (Eastern Strip), referring to its location east of the Río Uruguay. The inhabitants called themselves *Orientales* (Easterners), a term they still use to refer to themselves. The Uruguayans' road to independence was much longer than those of other countries in the Americas; in was finally achieved in 1828. Early efforts at attaining independence focused on overthrow of Spanish rule, a process begun by **Artigas** in 1811, when he led his forces to victory against the Spanish in the battle of Las Piedras. In 1816, Portuguese troops invaded present-day Uruguay, which led to its eventual annexation by Brazil in 1821 under the name Provincia Cisplatina.

II.189. The FRENCH INVASION OF SPAIN (1808–14) was one episode of the Napoleonic Wars that led to the reorganization of Europe under the **Congress of Vienna**. As news reached BUENOS AIRES that Spain was being run over by Napoleon and his troops, and that Napoleon had installed his brother Joseph Bonaparte (1768–1844) as the new king, the city declared independence. This became known as the MAY REVOLUTION OF 1810 and eventually led to independence. The "revolution" was a series of events in the city of Buenos Aires that installed the first local government not designated by the Spanish Crown in the viceroyalty of the Río de la Plata (at the time, it comprised Argentina, Bolivia, Paraguay, and Uruguay). Although the May Revolution took place only in Buenos Aires, one of the consequences was that the head of the viceroyalty, Viceroy Baltasar Hidalgo de Cisneros y la Torre (c. 1755–1829), was ousted from office. Involved in these acts were mostly middle-high to high-class citizens of Buenos Aires. There was little violence; the term "revolution" has been loosely applied by Argentine tradition to highlight the changing of their governmental system.

II.190. General José Gervasio ARTIGAS Aznar (1764–1850) is sometimes called "the father of Uruguayan independence." In 1820, a military conflict broke out in the United Provinces of Río de la Plata between Artigas, the governor of the Eastern Province, and the commander of the Liga Federal and Artigas's subordinate governor of the **Entre-Ríos** Province, Francisco RAMÍREZ (1786–1821). Ramírez triumphed, and Artigas was forced into exile in Paraguay.

II.190n. Joseph James Thomas REDHEAD (1767–1847) was commissioned by the British government to study the natural sciences; he arrived in Buenos Aires in 1803.

II.191n. Martin Dobrizhoffer (1717–1791) was an Austrian Roman Catholic missionary who lived in Paraguay (1748–67), first among the Guaranis, and then among the Abipones in the Gran Chaco region (a lowland plain extending from southern Bolivia through Paraguay to northern Argentina). Returning to Europe on the expulsion of the Jesuits from South America, he settled at Vienna, obtained the friendship of Maria Theresa, survived the suppression of his order, and wrote the history of his mission.

II.193. The Treaty of Utrecht was a series of individual peace treaties signed in the Dutch city of Utrecht in March and April 1713 between the representatives of Louis XIV of France and Philip V of Spain, on the one hand, and representatives of Queen Anne of Great Britain (1665–1714), the Duke of Savoy (Victor Amadeus II, 1666–1732), and the United Provinces on the other. These treaties helped end the **War of Succession**. The treaty recognized Portugal's sovereignty over the lands between the Amazon and Oyapock rivers in Brazil. In Article 8 of the Treaty of Utrecht of April 11, 1713, the Vincent Pinçon River was designated as the border between Brazil and French Guyana in the region around the village of Counani, founded in 1788 by the Jesuits. But the eighth article of the treaty mentions "Japoc," which the Brazilians took to mean Oyapock, while the French argued that it was only the indigenous word for "river." According to the treaty, the rivers Oyapoc and Araguary could both be identified as the Yapoc. See also Congress of Vienna and La Condamine.

II.193n. Louis Née (c. 1737–c. 1807) was the French-born Spanish botanist of the **Malaspina** expedition besides the two naturalists Antonio de Pineda and **Thaddaeus Haenke**. Little is known about him. Prior to the expedition, he worked in the Royal Botanical Garden in Madrid. It seems that he had gone on earlier expeditions as well. Pineda described Née as a botanist with broad knowledge about both the theoretical and the applied aspects of botany. The botanist Antonio Josef Cavanilles (1745–1804) tried to discredit Née in a letter to **Mutis** from April 28, 1795. Mutis, however, was friends with Née and had a better opinion of him. Cavanilles seems to have changed his mind by 1797. Indeed, when Née's postexpedition funding was terminated, Cavanilles lobbied his influential friend **Muñoz y Ferrandis** to help ensure that the work could be continued. Née published little but collected thousands of plants, only few of which survive today in Madrid, Spain. During the expedition he supervised a few artists, including José Guio y Sánchez, José del Pozo (c. 1757–1821), Francisco Lindo, and Francisco Pulgar, whose task was to sketch plants.

José Joaquim Vitório da Costa, a Portuguese engineer and mathematician, was member of the joint boundary commission Spain-Portugal created in the 1777 Preliminary Treaty of Saint Ildefonso to determine lines of demarcation in the Amazon region. From 1806 to 1818, Vitório da Costa was governor of the former Capitania de São José do Rio Negro. In 1808, he founded the village of São João de Príncipe on the Rio Japurá. He drafted the maps of the Río Negro and the Solimões, on which the general map of South America was based.

Like his collegue Vitório da Costa, José Simões de Carvalho (born c. 1805), a

Portuguese geographer and mathematician, was a member of the joint boundary commission. He was appointed governor of São José do Rio Negro in 1805 but died of indigestion from seagull eggs before taking office.

Francisco José de LACERDA e Almeida (c. 1753–98) was born in Brazil and, in the 1770s, moved to Portugal for his education. In 1777, he was named one of the astronomers to demarcate the borders of Brazil. Another member of the joint commission was Pedro Alexandrino Pinto de SOUZA, a Portuguese engineer and naval officer, who was governor of French Guiana from 1809 to 1812. Antônio Pires DA SILVA Pontes Leme (c. 1750–1805) was born in Mariana, Minas Gerais, Brazil. In the early 1790s, he was appointed professor at the Naval Academy of Lisbon and later promoted to frigate-captain. He was member of the Royal Academy of Sciences. In 1798, he published *Construction and Analysis of the Geometrical Propositions and the Practical Experiments which serve as the Basis to Naval Architecture.* From 1800 to 1804, he was governor of the capitancy of Espirito Santo in Brazil.

French mathematician and explorer Charles-Marie de LA CONDAMINE (1701–74) led a French expedition in Ecuador from 1736 to 1742, one of the few that the Spanish allowed foreigners at the time. About La Condamine's remark concerning the river Vincent Pinçon: there were a river and a bay by that name in the nineteenth century. It was named after Vincent Yanez Pinçon, who had commanded one of **Columbus**'s ships and who was thought, by some, to have discovered Brazil in 1500. Another name for this river is Yapoc. See also Treaty of Utrecht.

II.196. This CONVENTION is known as the London Convention, the Anglo-American Convention of 1818, the Convention of 1818, or simply the Treaty of 1818. More precisely, it is the "Convention respecting Fisheries, Boundary, and the Restoration of Slaves," which the United Kingdom and the United States signed in London on October 20, 1818, to set the boundary and allocate property and fishing rights between them. This treaty was the first to recognize the three-mile limit of territorial seas, which dominated maritime law until replaced with the twelve-nautical-mile limit of the United Nations Convention on the Law of the Sea. Article 3 of this bilateral international agreement, which was extended indefinitely in 1827, provided for the joint occupancy for a decade of a Pacific Northwest region (48° northern latitude and 54°40′ southern latitude) known as the Oregon Country or Columbia District: part of present-day British Columbia, Montana, Wyoming, and all of Oregon, Washington, and Idaho. The northwestern boundary of both parties was permanently set at the 49th parallel of northern latitude in the "Treaty between Her Majesty and the United States of America, for the Settlement of the Oregon Boundary" of 1846. The treaty was negotiated for the United States by **Albert Gallatin**, ambassador to France, and Richard Rush (1780–1859), ambassador to Britain; for Britain by Frederick John Robinson (1782–1859), treasurer of the Royal Navy, and Henry Goulburn (1784–1856), an undersecretary of state. Ratifications were exchanged on January 30, 1819. Along with the Rush-Bagot Treaty of 1817, the Convention of 1818 marked the beginning of improved relations between the British Empire and its former colonies; it also paved the way for better relations between the United States and Canada.

The Adams-Onís Treaty, also known as the Purchase of FLORIDA or Transcontinental Treaty of 1819, settled a border dispute between the United States and Spain. In addition to ceding Florida to the United States, the treaty settled a boundary dispute along the Sabine River in Texas and firmly established the boundary of U.S. territory and claims through the Rocky Mountains and west to the Pacific Ocean. The treaty was ratified in 1821.

II.197. John MELISH (1771–1822) was a Scottish geographer and merchant who moved to Philadelphia in 1809. He traveled extensively in the United States and published accounts and maps based on his experiences and observations. The cartographer and statistical geographer Henry Schenck Tanner (1786–1858) did extensive engravings for him. In a letter to Melish dated January 13, 1813, Thomas Jefferson noted, "[t]o return to the merits of your work: I consider it so lively a picture of the real state of our country, that if I can poosibly obtain opportunities of conveyance, I propose to send a copy to a friend in France, and another one in Italy, who, I know, will translate and circulate it as an antidote to the misrepresentations of former travelers."

The MISSOURI TERRITORY was an initially unorganized territory known as the Louisiana Territory. To avoid confusion, it was renamed in 1812, when the new state of Louisiana joined the Union. In 1819, the Arkansas territory was separated from the Missouri Territory, and the state of Missouri joined the union. See also Missouri Bill.

II.198. Pierre TARDIEU (1784–1822), a French cartographer from a famous family of engravers, frequently worked for Humboldt and for **Leopold von Buch**. Étienne-Robert (or Adrian Hubert) BRUÉ (1796–1858) was a French cartographer and geographer, who worked for Humboldt during his years in Paris (1804–27).

II.200n. In 1797, Isidoro de ANTILLÓN y Marzo (1778–1814) graduated as doctor of law from the University of Valencia, where his interest and research in geography began. He worked as professor of geography, chronology, and history at the Real Seminario de Nobles (1800–1808) in Madrid. Opposed to slavery in the Spanish colonies, he was also an active member of a provincial junta during the French occupation of Spain, and he founded and edited several journals, such as *Semanario patriótico*, *Gaceta del gobierno,* and *La Aurora patriótica mallorquina*.

II.207n. Charles-Etienne COQUEBERT DE MONTBRET (1755–1831) was a diplomat who also headed the French empire's statistical department, taught physical geography, and founded the *Journal des mines*. He collaborated with Jean-Baptiste Omalius d'Halloy (1783–1875) on what they called a "mineralogical-agricultural map" of France, which they presented to the academy on February 19, 1821. The Ministry of the Interior had collected the information for the map in 1808–9.

II.210n. The Great SLAVE LAKE in the Northwest Territories (Canada) was named after a native people; the name has nothing to do with African slavery. It is the second largest lake in Canada and was first explored by the British fur trader Samuel Hearne

(1745–92) in 1771. The Lesser Slave Lake in Alberta, Canada, north of Edmonton, was first recorded in 1799 by David Thompson (1770–1857). It was not named until half a century later.

II.216n. In 1499, one year after **Columbus** had discovered the Orinoco Delta, when the explorer Alonso de Ojeda (c. 1465–1515) sailed into the area around Lake Maracaíbo, the Warao (also known as Waroa, Guarauno, GUARAO, and Warrau, which literally means "Boat People") lived in huts on stilts with thatched roofs but without walls. Ojeda named the area Venezuela, "Little Venice," because of the huts perched on stilts.

II.220n (table 103). Edme François JOMARD (1777–1862) accompanied the French **Egyptian campaign** in 1798 as a topographical engineer. He contributed several volumes to the monumental *Description to l'Égype*, a collaborative work that took decades to complete. A member of the French Academy of Inscriptions and Belles Lettres since 1818, Jomard was one of the founders of the Geographical Society in 1821. In 1828, he was appointed to the position of curator of the maps and charts preserved in the Royal Library of Paris, of which he was to become the main librarian.

II.225. In 1742, Carl Linnaeus (1707–78) named the CINCHONA (or chinchona) tree in honor of the Countess of Chinchón, the wife of a viceroy of Peru. The medicinal properties of the tree's bark had been known since the mid-sixteenth century: it could be processed to obtain quinine to treat malaria as well as fever and pain more generally, and quinidine, which is used for cardiac rhythmic disorders. In the 1850s and 1860s, increasing demand for quinine from Europeans living in the tropics led naturalists to smuggle *cinchona* seeds from South America to plantations in Asia.

II.226. IPECACUANHA is a flowering plant, whose root is most commonly used to make syrup of ipecac, a powerful emetic. Its name comes from the Tupi i-pe-kaa-guéne, translated as "road-side sick-making plant." It is native to Brazil. CORTEX ANGOSTURAE (angostura bark) is another medicinal herb, used to ease digestive ailments. It is also used as a tonic and to make bitters.

II.228. Pedro de URQUINAONA y Pardo, a wealthy Creole from Quito, came to know Humboldt while the Prussian stayed in the mansion of the Duke of Selvalegre, Juan Pio Aguirre y Montúfar (1756–1818). The latter's son, Carlos Aguirre y Montúfar (1780–1816), was to become one of Humboldt's most trusted American companions. Urqinaona, who later wrote about this in his *Memorias* (1820), accompanied Humboldt on his several attempts at climbing Mount Pichincha, from where the Prussian naturalist and painter took many of his inspirations for the sketches of his famous Andean panoramas.

II.230. Diego María de GARDOQUI y Arriquibar (1735–98) was a minister to the Spanish Crown, empowered to look after Spanish interests in North America during the revolutionary war that consolidated the independence of the United States. He continued to serve as the Spanish envoy to the United States until his death. The Spanish were then in control of the Louisiana Territory (see Missouri Territory).

II.234n. Humboldt's PLANT geography is a precursor to geobotany (also known as phytogeography), a branch of biogeography concerned with the geographic distribution of plant species. Humboldt is often called the "father of phytogeography" for advocating the quantitative approach that has come to define modern plant geography. Research in plant geography has also been directed to understanding the patterns of adaptation of species to the environment. See Humboldt's *Essay on Plant Geography* (1805–7).

II.235. The SIERRA NEVADA DE MÉRIDA is the highest mountain range in the largest massif in Venezuela, the Cordillera de Mérida, which, in its turn, is part of the northern extent of the Andes. It includes Venezuela's highest peak, the Pico Bolívar, which has an elevation of 16,342 feet. This range also includes the Pico Humboldt (16,207 feet).

II.239. Manuel TORRES (c. 1767–1822), a nephew of archbishop-viceroy Antonio Caballero y Góngora (1723–96), was forced to flee his native Venezuela in 1796. He settled in Philadelphia. On February 20, 1821, Torres conveyed an official note to the U.S. secretary of state, John Quincy Adams (1767–1848), requesting the acknowledgment of the independence of the Republic of Colombia, a union of the Republics of **New Granada** and Venezuela. The document also bore the names of **Zea** as president of Congress and of **Bolívar** as president of the Republic. Torres had been named chargé of Venezuela in 1819, with full powers to negotiate.

II.240. Humboldt is most likely referring to the numbers of wheat tillers here. Tillers are branches of the main stem that serve to increase the numbers of heads per acre in the wheat crop. Tiller count is an indicator of yield.

II.242n. Oxford-educated British botanist Aylmer Bourke LAMBERT (1761–1842) is best known for his *Description of the Genus Pinus* (1803). This and his earlier work on the cinchona benefited much from the advice of **Joseph Banks**.

II.249. GEM SALT is the old name for colored, highly mineralised ROCK SALT, back before iodized salt was manufactured. Rock salt (halite) is crystalline sodium chloride that has a great solubility in water. It can be found in rocks from all geologic periods. The terms are used interchangeably.

II.253. Jean-Baptiste Joseph Dieudonné BOUSSINGAULT (1802–87) was known for his work in agricultural chemistry and for the development of the first chrome steels. In 1821, **Bolívar**, desirous to found an institute for training engineers in Colombia, engaged Boussingault. Boussingault attempted to reach the peak of Chimborazo and described his feat in "Versuch einer Ersteigung des Chimborazo: Unternommen am 16. December 1831 (c.1835)" [Attempt to ascend Chimborazo on December 16, 1831], which prompted Humboldt to publish a detailed account of his own climb. Boussingault published *Viajes científicos á los Andes ecuatoriales; ó, Colección de memorias sobre física, química é historia natural de la Nueva Granada, Ecuador y Venezuela* [Scientific travels to the equatorial Andes, or collection of reports on the physics, chemistry, and natural history of New Granada, Ecuador, and Venezuela] (1849).

Mariano Eduardo de Rivero y USTARIZ (1798–1857) was a prominent Peruvian geologist, mineralogist, chemist, archeologist, politician, and diplomat. His publications about his discovery of Humboldtine (an iron-oxalate), demonstrating the existence of organic minerals, about deposits of copper and sodium nitrate near Tarapacá in the Atacama desert (today Chile), and about bird-guano and coal in Peru made him a pioneer of mining education in South America and the most notable Peruvian scientist of the nineteenth century. In 1822, the minister of Gran Colombia in Paris, **Francisco Antonio Zea**, asked Rivero y Ustariz, also upon Humboldt's recommendation, to found and run a mining school in Bogotá.

II.255. Juan José (or Jean-Francois) DAUXION-LAVAYSSE (1775–1826) came to the Carribean as a young man to visit his uncle in Guadeloupe. After his uncle's sudden death and the outbreak of the French Revolution, Dauxion-Lavaysse fled to Trinidad, where he became a planter. He is author of the *Voyage aux îles de Trinidad, de Tabago, de la Marguerite, et dans diverses parties de Vénézuéla, dans l'Amérique Méridionale* [Travels in the islands of Trinidad, Tobago, Margarita, and in various parts of Venezuela in South America]. In addition to the political and commercial resources of Venezuela and the islands in its vicinity, he also described the flora and fauna of the region, along with social manners and customs. Humboldt might have read with particular interest his remarks about the slave trade and the cruel treatment of slaves.

II.261n. South American military leader and statesman Simón BOLÍVAR (1783–1830) led the revolutions against Spanish rule in the viceroyalty of **New Granada** against the Spanish commander Pablo Morillo (1775–1836). The armistice negotiated between the two leaders in 1820 established South American continental independence from the Spanish metropole. Bolívar became president of Gran Colombia (1819–30), dictator of Peru (1823–26), and president of Bolivia (1825–26). As a young man, Bolívar, the son of a Venezuelan aristocrat of Spanish descent, had traveled to Europe, where he had met Humboldt on two occasions, in 1804 in Paris and in 1805 in Rome. Humboldt is supposed to have given Bolívar the revolutionary inspiration to liberate South America from Spanish dominion, but this episode is more fiction than fact. Although, even in his early travel diaries, Humboldt adopted a critical stance toward Spanish colonialism, he never endorsed a militant approachy to solving the social, economic, and political problems of the American colonies. As late as 1808, when the first confrontations between Spanish and Creole troops began, Humboldt was convinced that the Creole elites were too conservative to start a revolution. It seems higly unlikely that he would have influenced Bolívar so directly at such an early date.

II.270. José María QUIRÓS (c. 1750–c. 1824) was a writer and journalist who served as secretary of the merchant guild [Consulado] of Veracruz and as captain in the Spanish army in Mexico.

II.280n. There is no "recently" in the original Alexander WALKER quotation, although Humboldt includes it in his French version as if to stress the point that, so many years later, people still refer to his observations. Walker merely refers to Humboldt, noting that "[t]he river Cassiquiari [Casiquiare], long conjectured to be a strong branch of the Orinoco, but not known to be an arm of the Negro, communicates also with the Marañon by means of the Negro; its streams having been visited by M. de Humboldt, who encountered great perils in the undertaking, by the force of the current and other obstacles. The whole country for 300 miles was a complete desert, in which the ants and mosquitoes were so extremely troublesome as almost to deter the traveler from proceeding. He entered the Orinoco by the Cassiquiari in 3° 30′ north latitude, and mounted the current of the great river as far as Esmeraldas, the last Spanish settlement in that quarter."

II.286n. For Miguel del CORRAL, see Cramer.

II.287. Antonio María de BUCARELY Y Ursúa (1717–79) was the forty-sixth viceroy of New Spain from 1771 to 1779. Before that, he had served as governor and captain-general of Cuba (1766–71). His early career in Europe had been in the military—he participated in campaigns in Italy and Portugal, became a lieutenant general, and also served as inspector of coastal fortifications in Granada. As viceroy, he sent expedition to explore and settle coastal California, which was then a Spanish possession, and he took steps to prevent Russian invasions of North America. See also Cramer.

The COUNT Juan Francisco de Güemes y Horcasitas REVILLAGIGEDO (c. 1682–1766) was a Spanish colonial administrator and viceroy of New Spain (1746–55). His son, Juan Vicente Güemes Pacheco de Padilla, Count of Revillagigedo (1740–99), was born in Havana and also served as viceroy of New Spain (1789–94). Humboldt is likely referring to the son, who was rather effective. He improved living conditions in Mexico City and introduced the system of intendancies. Moreover, he founded the General Archives and, in 1793, the Museum of National History.

II.288. In 1774, on orders of **Bucarely**, the viceroy of New Spain, Agustín CRAMER surveyed the Isthmus of Tehuantepec for the possibility of an interoceanic waterway. Cramer was an engineer in command of the Fort of San Juan de Ulúa in Veracruz. His report is preserved in the *Archivo de Indias* in Seville, Spain. It is not clear whether Augustín Cramer is the Agustín Cramer Mañecas of Navarre who came to Cuba in 1763 as a brigadier of the army engineering corps to repair the damages done to Havana's forts by the British. It seems possible, considering that Bucarely, before being appointed viceroy, had been governor of Cuba since 1766. He might well have called upon his Cuban contacts when he pondered the option of an interoceanic waterway. Cramer Mañecas helped construct a warehouse for the Cuban Royal Tobacco Trading Post in 1770. In 1777, he wrote a study on commerce and navigation in **Guiana**. He also published an important treatise on Cuba's commercial activity. Miguel del Corral was Cramer's assistant.

II.290n. Count Antoine François ANDRÉOSSY (1761–1828) was a hydographer, diplomat, and general of the French Empire, who was part of Napoleon's **Egyptian campaign**. After his return from that expedition, Andréossy assisted Napoleon in the coup d'état of the 18th Brumaire (1799). He served as ambassador to England, Austria, and the Ottoman Empire and as military governor of Vienna. After the Battle of Waterloo, Andréossy was one of the commissioners sent to negotiate an armistice with the allied powers. A member of the Academy of Sciences, Andréossy authored works on artillery, military history, geography, and hydrography. In 1818, he published *Voyage à l'embouchure de la Mer Noire* [Voyage to the mouth of the Black Sea].

Michel Louis François HUERNE DE POMMEUSE (1765–1840) (mentioned in the text on p. 000, note 1), whose *Des Canaux navigables considérés d'une manière générale* [About navigable canals, considered a general fashion] (1822) Humboldt quotes, described Prony as a man of rare modesty and commendable talent. Huerne de Pommeuse also mentioned that Prony was inspector-general of bridges and highways and a director of the École nationale des ponts et chaussées, the engineering school founded by **Jean-Rodolphe Perronet** in 1747.

Like many French scientists of his day, the mathematician Pierre Charles François DUPIN (1784–1873) was educated at the École polytechnique in Paris. Perhaps his major contribution to geometry was the Dupin indicatrix of curved surfaces, from *Dévelopements de géométrie pour faire suite à la géométrie pratique de [Gaspard] Monge* [Developments in geometry for application to Monge's practical geometry] (1813). His *Mémoires* were the product of serveral visits he made to Great Britain in his capacity as a naval engineer to study civil and military infrastructures. A professor at the Conservatory of Arts and Professions in Paris since 1819, he was an advocate for workers' education and popularized mathematics by publishing several textbooks on mechanics and geometry in the 1820s. After becoming a baron in 1824, he began a political career and was later appointed senator by Napoleon III (1808–73). An ethical economist, Dupin helped pass the first child labor legislation in France in 1841.

Joseph DUTENS (1765–1848) was a French physician, engineer, and geometrician.

II.294n. According to some historical anecdotes, the name Perú derived from the name of Chief BIRÚ. **Garcilaso de la Vega, El Inca** recounts his version of an encounter of a Spanish sailor with a native in his *Royal Commentaries of the Incas*. The Spaniard asked what this land was called, and the Indian replied by giving his own name: "Berú." In other versions, the Indian gives the name of his chief.

II.294n. Vicente TALLEDO y Rivera (b. 1760) was deputy colonel of the Royal Corps of Engineers in New Granada around 1800. He drew a map of **New Granada**—*Mapa corográfico del nuevo reino de Granada* [Topographic map of the new kingdom of

Granada]—and corrected Humboldt's map of the Río Magdalena on orders of the viceroy.

II.296. Born in Popayán near Cartagena de Indias, José Ignacio POMBO (1761–c. 1815) was a philanthropist and supporter of progress, educational causes, and, above all, scientific experimentation. A friend and avid correspondent of José Celestino Mutis, Pombo was particularly known as a generous supporter of **Caldas**, giving him books and instruments, paying for his scientific excursions, and contributing to his journal. An unflagging patriot, Pombo used the entirety of his considerable fortune to aid the struggle for independence, in which two of his sons fought and died. Humboldt, who met Pombo in New Granada, was also interested in the latter's studies of interoceanic navigation in connection with the Atrato River.

Botanist José Celestino Bruno MUTIS y Bosio (1732–1808) headed the Royal Botanical Expedition in New Granada (that is, the modern countries of Colombia, Panama, Venezuela, and Ecuador) in 1782, which has been interpreted as the advent of the Enlightenment in the Spanish Empire. Mutis had ambitious plans, such as cataloguing new plants, building an astronomical observatory, maintaining correspondence, training painters, educating young natural philosophers, and supporting the development of agriculture, commerce, and culture in the colony. Humboldt stayed with Mutis in Bogotá from July to September 1801. See also Zea.

II.298n. José Manuel RESTREPO (c. 1781–c. 1863) was a naturalist and a historian of Colombian independence. A disciple of **Caldas**'s, he surveyed and triangulated his native province Antioquia in 1807, drawing a map and writing an account in 1809. See also Zea and Mutis.

II.301n. Colleges of *propaganda fide* were established by Observant Franciscan missionaries in New Spain. The SACRED CONGREGATION *DE PROPAGANDA FIDE* (sacra congregatio christiano nomini propaganda) is a department of the pontifical administration, whose purpose is the spread of Catholicism and the regulation of ecclesiastical affairs in non-Catholic countries. It was established in Rome in the seventeenth century.

II.302n. For Lionel WAFER, see Dampier.

II.303. Joaquín Francisco FIDALGO (d. 1820), a naval officer in Cartagena de Indias and commander of Venezuela's coast guard, was in charge of mapping the coastal areas of Venezuela and of the Antilles. Humboldt established contact with the Fidalgo expedition while in Cartagena, and **Caldas** acknowledged their contributions to the progress of the study of geography in New Granada. Fidalgo is named as a coauthor of an 1802 map of the Antilles, together with Churruca and Bauzá.

II.304n. The Spanish mathematician JORGE JUAN y Santacilia (c. 1712–c. 1773) accompanied **La Condamine** and **Antonio de Ulloa** on their expedition to Peru. Santacilia, using a baromter, was singularly responsible for measuring the altitude of the Peruvian mountains successfully.

II.306. London manufacturer Nicolas FORTIN (1750–1831) was one of many to equip Humboldt with some of the finest instruments of his time. Humboldt used the barometers during his American voyage mostly to determine altitude and changes in the atmospheric pressure. See also Nairne.

II.309n. Ignacio Cavero y CÁRDENAS (c. 1756–1834), who was in charge of tobacco exports and a customs officer in Cartagena de Indias for almost twenty years, was active in the struggle for New Granada's independence in 1810. As a result, he was forced to live in exile in Jamaica from 1815 to 1821, when he returned to Colombia.

II.316. Jean-Rodolphe PERRONET (1708–94) was a French engineer best known for his stone-arch bridges. In 1747, he founded the world's first engineering school, the École nationale des ponts et chaussées [National School of Bridges and Highways] and was appointed its director (one of his successors would be **Prony**). In 1791, in spite of the French Revolution, he finished work on the bridge Pont de la Concorde, which was originally called Pont Louis XV. He joined the Academy of Sciences in 1765. Perronet was also involved in the construction of the Burgundy Canal, a project that lasted—with interruptions during the French Revolution—from 1765 until 1832. The canal has a tunnel at its summit, through which one barge can pass at a time.

II.320n. Martin de LA BASTIDE was the secretary of the **Duke of Broglie**. In his *Universal Geography*, **Malte-Brun** observed that Lake Nicaragua, because of its tides, would probably not be much of an option for the vague project of a transoceanic canal, "which everyone has been able to dream of, but which it was reserved for M. Martin de la Bastide to publish, under the triple form of a pamphlet, a fan, and a snuff-box!" (vol. 5, p. 283).

II.322n. John PHIPPS was British colonial administrator until he retired in the late 1820s. Starting in 1794, Phipps lived in Calcutta for about thirty-five years, as the clerk to the British master-attendant, an officer in the dockyards appointed to tend to vessels of war. While in this position, he compiled vast amounts of information about trade and transportation. On the basis of those accounts, he published *A Series of Treatises on the Principal Products of Bengal; No. I. Indigo* (1832), *A Practical Treatise on the China and Eastern Trade* (1835), and *A Collection of Papers, Relative to Ship Building in India* (1840).

II.325. Scottish engineer John RENNIE (1761–1821) built and improved canals, docks, harbors, and bridges across Britain. He is best known for his London bridges. Of his nine children, two sons carried on his business, with the elder George (1791–1866) looking after the mechanical engineering and Sir John Rennie (1794–1874) after the civil engineering side. In 1798, Rennie became a fellow of the Royal Society. William

JESSOP (1745–1814) was a British engineer involved in the construction of canals, harbors, and early railways. Of Jessop's seven sons, four followed him into the engineering profession. A man of simple tastes, he supported younger engineers without envy and showed concern for the workers on his projects. The Scottish engineer and writer Thomas TELFORD (1757–1834) was partly responsible for the Caledonian Canal, which was constructed in response to his extensive 1801/02 recommendations for improving the highlands. He was also a consultant for the Göta Canal in Sweden, built roads and bridges, and devoted his energy to improving and building canals in light of threatening railway competition. He published poetry and wrote an autobiography that appeared posthumously in 1838.

II.328. Under King Louis XIV of France, Jean Baptiste COLBERT (1619–83) was comptroller general of finance (from 1665) and secretary of state for the navy (from 1668). His program of economic reconstruction helped turn France into the dominant power in Europe. In 1664, he was appointed superintendant of the king's buildings, in which capacity he was also in charge of the reconstruction of the Palace of Versaille, whose gardens were to glorify the king in every detail. **Ustariz**'s *Teórica y practica de comercio y de marina* [Theory and practice of trade and ports] (1724), the center piece of eighteenth-century Spanish mercantilism published in 1742, had high praise for Colbert.

II.328n. The Machine of MARLY was built on an arm of the Seine river. It consists of fourteen wheels of 30 feet in diameter, the axles of which have two cranks, one that moves the pistons, which push the water into the pipes and raise it up to the first reservoir; the other moves a system of crossarm levers, which run the length of the mountain up to the highest reservoir. These crossarms power the pumps that are in the reservoirs and pump the water from the lower resevoir to the upper and from there to the top of the tower, which is at the summit of the mountain, from where it runs over a large aqueduct that feeds different pipes to furnish all the waters of Versailles and Marly.

II.330–31. Gaspar Clair François Marie RICHE DE PRONY, Baron Riche of Prony (1755–1839), was a French engineer. As a land registrar since 1791, he oversaw the introduction of new trigonometric tables. A member of the Academy of Sciences since 1795, he also measured the speed of sound with great precision and was very much involved in public works.

Jean-Pierre CLAUZADE (1751–1832) was a member of the French Royal Academy of Sciences and chief engineer of the Royal Corps of Bridges and Highways. He was involved in the construction of the Canal du Midi (or Languedoc Canal) already at an early age, since his father was also an engineer. In late 1801, the younger Clausade succeeded Gilles PIN (Père) as chief engineer upon the latter's death. Clausade dealt with the question of water loss caused by boats passing through a lock and by evaporation. Clausade and Pin were codirectors of the Languedoc Canal project from 1767 to 1773; after that, Pin served as its sole director.

The HAIR HYGROMETER that Humboldt used was manufactured by Paul (Geneva). It helped determine air humidity by variations in the length of a hair or baleen applied to the apparatus. The hair hygrometer, invented in 1783 by the French naturalist **Saussure**, was a fairly simple mechanism: the degreased and strained hair had first to be exposed to "dry air"—usually produced artificially—and later to the environment in which the air humidity was to be determined. The vapor in the air would cause the hair to gain weight and therefore stretch. Humboldt, who took measurements wherever he was, needed this important meteorological tool to double-check and correct barometrical data (air pressure and altitude).

I.332n. Pierre Simon GIRARD (1765–1836) was a French engineer, founding member of the Institute d'Egypt, Napoleon's minister of the interior of Egypt, and president of the Académie des sciences of Paris (1830, member since 1815). Trained at École royale des ponts et chaussées (now École des ponts Paris tech), he won a prize in engineering for improving locks from the Académie des sciences in 1792. He was the chief engineer for the construction of the l'Ourcq Canal in Paris and wrote extensively on navigable canals including *Mémoires sur le canal de l'Ourcq* [Reports on the L'Ourcq Canal] (1831). In addition to several other works on the motion of fluids, he is the author of *Traité analytique de la résistance des solides et des solides d'égale résistance* [Analytical treatise on the resistance of solids and of solids of equal resistance] (1798), the first comprehensive modern treatise on the strength of different materials.

II.335n. This FORT was captured by the Welsh privateer Sir Henry Morgan (1635–88), who sailed up the San Juan River in 1665 to sack Granada (Nicaragua).

Diego López de SALCEDO y Rodríguez (d. 1547) was governor of Honduras from 1526 to 1530 and governor of Nicaragua in 1526 and 1527. The Río San Juan or Desagüadero is an outlet of Lake Nicaragua, which flows from the lake's southeastern end at San Carlos north to the border of Nicaragua and Costa Rica into the Caribbean Sea at San Juan del Norte. The river has been the cause of many a dispute between the two countries regarding rights of use. Migration within the United States from the East Coast to California, from 1850 to 1870, went by way of San Juan del Norte and across Lake Nicaragua. It is of little surprise, then, that the shipping and railroad tycoon "Commodore" Cornelius Vanderbilt (1794–1877) became interested in Nicaraguan opportunities at the time. The possibility of a transoceanic canal in Nicaragua, which Humboldt discusses here, also played a role.

II.338n. The French engineer Pierre-Joseph LAURENT (1713–73) planned and began the construction of the Saint-Quentin Canal in France. Defeating a two-tunnel project, he proposed an unprecedented engineering feat of a single underground tunnel of about nine miles (c. 6.5 m in width). A state-financed project, excavations of the single tunnel started in 1768 under Laurent's direction. In 1774, after his death, Laurent de Lionne, his nephew and assistant, took over the project, which was halted later that year. A commission composed of members of the Académie Royale des Sciences recommended not to continue the tunnel until it reopened in 1802. The project was finally finished as a two-tunnel project (not Laurent's) under Napoleon Bonaparte who inaugurated it on April 28, 1810.

II.339. Napoleon's EGYPTIAN CAMPAIGN (begun in 1798) led to a relatively brief French occupation of Egypt. Even though it was a military campaign in the war against Britain, designed to damage British trade, its scientific-scholarly motivations and results were significant. For instance, Napoleon brought with him the naturalist **Saint-Hilaire**. He also had surveyors look for possible routes for a canal between the Mediterranean and the Red Sea. Moreover, the French uncovered the Rosetta Stone, with its trilingual inscription, which made it possible to decipher hieroglyphics and thus establish Egyptology as a discipline. Reports on the numerous findings were published in the multivolume *Description de l'Égypte* [Description of Egypt] (1809–28), from which Humboldt quotes. In the *Description de l'Égypte* and in **Strabo**'s work, Humboldt found information about an ancient canal that connected the Red Sea to the Nile; he refers to it as *canal des rois* (canal of kings, see vol. II, p. 339n), a term also found in François-Marie Arouet Voltaire's (1694–1778) writings. This canal, likely the ancient Eastern Canal of Egypt, ran near the defunct Pelusiac branch of the Nile. Ancient Egyptians used canals for various purposes, including the transport of obelisks, which could weigh up to fifty tons, large sculptures, stones for construction, and for defense against Asian and other invasions. Strabo observed that canals in Egypt could fill up easily and therefore would have to be maintained carefully. Humboldt also mentions the modern Suez Canal project that connects the Red Sea with the Mediterranean. Today's Suez Canal does not have any locks. It was completed in 1869 under the leadership of Ferdinand de Lesseps (1805–94). Old inscriptions point to an unfinished ancient Suez Canal at the time of Darius the Great, King of Persia (522–486 BCE), which was completed under the Ptolemies around 250 BCE. The completion of a Suez Canal was delayed, in Darius's time and again in Napoleon's day, because of fears that the Red Sea, located at a higher altitude, would disasterously empty into the Mediterranean.

A member of the corps of engineers in the Napoleonic Egypt Expedition, the French engineer and architect Jean-Baptiste LEPÈRE, also spelled Le Père or Le Peyre (1761–1844), accompanied Napoleon from Cairo to Suez on December 1798. On Napoleon's order, LePère, in January 1799, carried out a complete survey of the Isthmus of Suez to support the construction of the canal; the construction was halted because of Napoleon's defeat. Together with Jacques Gondouin (1737–1818), LePère built the Vendôme Column in Paris (1806–11), with spiral bas-reliefs based on Trajan's Column in Rome, to commemorate Napoleon's campaigns at Austerlitz (1805 and 1806). LePère also designed and started the Saint-Vincent-de-Paul Church in Paris in 1824; the project was finished by his son-in-law, Jacques-Ignace Hittorff (or Jakob Ignaz, 1792–1867), almost twenty years later.

II.340n. David Bailie WARDEN (1778–1845), a member of the French Academy, graduated from New York medical college and was distinguished for his scientific achievements and broad knowledge. Among many other things, he published an "Inquiry concerning the Intellectual and Moral Faculties and Literature of the Negroes" (1810).

II.341. Luis de Velasco, the MARQUIS OF SALINAS (1535–1617), a son of the second viceroy of New Spain, was involved in constructing a royal canal, or DESAGÜE [drain] of Mexico. He served as viceroy in Peru from 1590 to 1595, and again in Mexico City from 1607 to 1611. **Philip III** made him the Marquis of Salinas in 1609. In 1611, the Marquis was recalled to Madrid to become president of the Royal Council of the Indies.

II.343. Antonio Obregón y Alcocer (d. 1786) was granted the title of COUNT OF VALENCIANA in 1780, after he struck the vein of silver that turned Guanajuato into the richest mining city in the Americas.

II.344. Humboldt is alluding to the British philosopher Jeremy Bentham (1748–1832), who was an early promoter of utilitarianism. In June 1822, Bentham sketched out a plan for an interoceanic canal in Nicaragua; the manuscript was entitled "JUNCTIANA Proposal." The projected Nicaraguan canal, built on land ceded by Mexico, would have Mexico as its northern and Colombia as its southern neighbor. The canal area was to be called "Junctiana," which Bentham envisioned as a new state within his "Anglo-American United States." The canal would link the Atlantic Ocean to the San Juan River, to Lake Nicaragua, and to the Pacific. Bentham noted in his manuscript that neither Humboldt nor anyone else had yet explored the idea that Nicaragua would be the best choice for such a canal. See also La Bastide.

II.350n. Charles de BEAUTEMPS-BEAUPRÉ (1766–1854) is known as the father of modern hydrography. As chief hyrographer of the French navy, he was also in charge of maps. From 1816 to 1838, he went on a mission to map all of the French coastlines. His geography teacher had been **Jean-Nicholas Buache**. As a young man, Beautemps-Beaupré had accompanied Antoine Raymond Joseph de Bruni d'Entrecasteaux on his last expedition (1791–92). See also Rossel.

II.357. José de EZPELETA y Veire de Gadeano (1740–1823) was governor of Cuba from 1785 to 1789 and viceroy of **New Granada** from 1789 to 1797, when he was replaced by Pedro de MENDINUETA y Múzquiz (1736–1825), a Spanish military officer and administrator, who had served as governor of New Mexico prior to this assignment. He received Humboldt and Bonpland in Bogotá with great fanfare. A supporter of important public works, such as the construction of the astronomical observatory and the improvement of hospital services, Mendinueta is also remembered as a backer of **Mutis**'s Botanical Expedition and the creation of the Sociedad Patriótica de Amigos del País.

II.359. The PROLONGED CIVIL STRIFE to which Humboldt refers here was the war of independence, which lasted effectively from 1811, when Francisco de Miranda first declared Venezuela's independence, to at least 1823, when **Bolívar** founded Gran Colombia.

II.360. To raise revenue, the Spanish Crown introduced the ALCABALA (an excise tax) to the American colonies in the late sixteenth century. This indirect tax was initially 2 percent; it was later raised to as much as 6 percent. The members of the clergy and many towns were exempted from this tax, and nobles sometimes collected the tax for themselves instead of passing it on to the crown. In order to increase American revenue, the viceroy of New Granada gave instructions to remove the *alcabala* and the brandy monopoly. Unsurprisingly, the *alcabala* was a steady source of friction between Spain and its colonies. In 1765, it became a trigger for the Quito Revolt. See also *almojarifazgo*.

II.360n. José María Eusebio Carlos del Rosario del CASTILLO y Rada (1776–1835) was acting president of the United Provinces of **New Granada** in 1814. He also served as vice-president of the Republic of Colombia in 1821.

II.361. The GUIPÚZCOA COMPANY, named for the Basque province where it was headquartered, was chartered by the Spanish Crown in 1728 and given a monopoly on trade between Spain and Venezuela. It was the only financially successful company for colonial trade that the crown created during the eighteenth century. The company was granted extensive commercial privileges to promote officially sanctioned trade and thus to prevent smuggling. While the company fostered the production of tobacco, indigo, cotton, and cacao in Venezuela, its business practices provoked a revolt among colonists in 1749. In 1778, it was abolished by law.

II.365. Provincial INTENDANTS, who came to replace the earlier district magistrates during the mid-eighteenth century, symbolized Spain's innovative practice of appointing salaried military and civilian career officials as colonial administrators. Intendants were effectively financial commissioners whose main function was to collect revenue for the Spanish Crown.

II.370. Both Peru and Mexico petitioned to have Bourbon or Hapsburg PRINCES placed on their thrones. No one volunteered, and they were forced to adopt presidential systems that proved far more unstable than Brazil's monarchy.

Baron Francis Bacon Viscount Saint Alban, better known as Sir Francis BACON (1561–1626), was lord chancellor of Britain from 1618 to 1621. This lawyer, statesman, philosopher, and master of the English tongue is best remembered today for his incisive worldly wisdom and for his power as a political orator.

II.375. *THE SPIRIT OF LAWS* [*Esprit des lois*] by Charles Louis de Secondat, Baron of La Brede and of Montesquieu (1689–1755), was a major contribution to political theory; it was first published in French in 1748 and in English two years later. The book was highly controversial and placed on the *Index Librorum Prohibitorum* in

1751. Already a year earlier, Montesquieu had written a subtle and good-humored *Défense de l'esprit des lois* [In defense of the spirit of laws]. A modest, affable, and absentminded man, he contributed an essay on taste to the French *Encyclopédie*.

II.378. Most of the individuals listed here are more or less directly linked to British efforts to abolish the transatlantic slave trade and slavery in the British colonies. Humboldt's LORD HOLLAND was Henry Richard Vassall Fox, the third Baron Holland (1773–1840). He was a British Whig politician who helped secure the abolition of the slave trade in the British colonies, even though his wife, Elizabeth Vassall (c. 1771–1845), had property in the West Indies. To ensure that his own children would benefit from the West Indies possessions, Baron Holland assumed the name of Vassall after his wife's former husband, the much older Sir Godfrey Webster (bap. 1749–1800), had committed suicide. Holland was close to his uncle Charles James Fox (1749–1806), a politician of such notoriety that he was the subject of more caricatures than any other person in the late eighteenth century.

Sir Robert WILMOT Horton, born as Robert Wilmot (1784–1841), was a British politician, public servant, and pamphleteer. In 1806, he married Anne Beatrix (d. 1871), eldest daughter and coheiress of Eusebius Horton (d. 1823) of Catton, Derbyshire; in 1823, Wilmot assumed the surname of Horton by royal license in compliance with his father-in-law's will. In late 1821, Horton was appointed parliamentary undersecretary in the Colonial Department, which he reformed. He was knighted in 1831 and appointed governor of Ceylon.

Zachary MACAULAY (1768–1838) was an active force in abolitionist circles and close to **Wilberforce**. His son, Thomas Babington Macaulay (Lord Macaulay, 1800–1859) was a historian. Zachary's brother, the Reverend Aulay Macaulay (1759–1819), was a scholar and antiquarian. Their brother Colin was known as General Macaulay (1760–1836). The general served in India for about thirty years, a few years of which he spent imprisoned in an underground dungeon; he returned from India in 1811.

Sir Charles M'CARTHY (1764–1824) was a colonial governor and army officer of Irish extraction. M'Carthy served with the military in the West Indies in the 1790s. By 1811, he was lieutenant-colonel in the Royal African Corps, and in the following year he became governor of Senegal. In Sierra Leone in 1822, M'Carthy attempted to encourage trade with goods rather than with slaves. He was fatally wounded during an 1824 campaign.

Humboldt's "Chevalier Mackintosh" was probably the Scottish political writer and politician Sir James MACKINTOSH of Kyllachy (1765–1832), who had served as a judge in India since 1804, returning to Britain in 1811. He was elected to the Royal Society in 1813. Mackintosh was granted honorary French citizenship during the French Revolution. Alongside **Huskisson**, he engaged in parliamentary debates about legislation with regard to colonial slave trade and slavery and also supported Wilberforce. While visiting Paris, Duke Arthur Wellesley of Wellington (1769–1852) consulted him on the matter of the slave trade.

Thomas CLARKSON (1760–1846) was a British abolitionist and author of *An Essay of the Slavery and Commerce of the Human Species, Particularly the African* (1786).

This publication caught the interest of Charles James Fox and led to the creation of an informal committee that succeeded in recruiting Wilberforce. Soon thereafter, the committee for effecting the abolition of slavery was formally set up. Thomas's brother John Clarkson (1764–1828) joined their crusade after his return from war in the West Indies. He acted as his brother's secretary and one of his agents. John also recruited former American slaves in Halifax for an abolitionist colony in Sierra Leone, an ill-fated project. Thomas continued to devote time to the abolitionist cause, writing persistently until, in 1833, the act abolishing slavery in the British Empire was finally passed.

David HODGSON had been a business partner of two generations of Croppers and Bensons in Liverpool. **James Cropper** was a close friend of Adam Hodgson's. Cropper's wealth allowed him to engage in antislavery activism.

II.381 table. George CHALMERS (1742–1825), a Scottish historian educated at King's College in Aberdeen and at the University of Edinburgh, moved to Maryland in 1763 to practice law in Baltimore. A devout loyalist, he returned to London in September of 1775, when revolutionary discontent grew in the American colonies. He was appointed chief clerk of the committee of the Privy Council for trade and foreign plantations, a post he held for forty years; it allowed for abundant time to devote to his studies and writings, mainly about Ireland and the affairs of America and the British monarchy. In the process, he collected a vast library of books, original manuscripts, and notes.

II.383 table. Pierre Franc MACCALLUM was a British ghost writer, scandal monger, and journalist. Humboldt spelled the name "Maculhum." After a quarrel with the governor of Halifax, Nova Scotia, in 1800, MacCallum moved to Philadelphia and New York, eventually ending up in **Saint-Domingue** during the slave insurrection. Upon the arrival of 25,000 French soldiers, he retreated to the mountains with rebel forces. Humboldt had probably read MacCallum's 1805 *Travels in Trinidad during the Months of February, March, and April, 1803*.

General Joseph François Pamphile de LACROIX (1774–1841) came to **Saint-Domingue** in 1802 as part of the expedition to reclaim Saint-Domingue from Toussaint l'Ouverture. His *Memoirs pour servir à l'histoire de la révolution de Saint-Domingue* (1819) were an indispensable source for later historians of the Saint-Domingue Revolution, notably for C. L. R. James (1901–89), whose *The Black Jacobins* appeared in 1938.

II.387 table. A friend of **Fernández de Navarrete's**, José de VARGAS Ponce (1760–1821) was a Spanish naval captain, scholar, poet, and politician. In 1782, he entered the Spanish navy, where he had a brilliant career. In that same year, Vargas won an award of the Royal Spanish Academy for his *Elogio del Rey Don Alfonso el Sabio* [In praise of King Alfonso the Wise]. A member of the Royal Academy of History in Spain since 1786, he became its dean and served twice as its director. Vargas was an avid writer, authoring scores of essays, reports, and works of fiction (many yet to be published).

Among his most famous publications are his travel narrative *Relación del último viaje al Estrecho de Magallanes* [Report of the last voyage to the Straits of Magellan] (1788), *Abdalaziz y Egilona: Tragedia* (1804), and *Importancia de la historia de la marina española* [Importance of the history of the Spanish fleet] (1807). By request of the Academy of History, Francisco de Goya (1746–1828) painted a portrait of Vargas in 1805.

II.390n. Jean Baptiste SAY (1767–1832) was a French classical economist. Under the influence of Adam Smith (1723–90), he is remembered primarily for Say's law, or the law of markets, one of the cornerstones of classical economics. In 1830, he was the first professor of political economy at the Collège de France. His main work, *Traité d'économie politique* (1803), translated as *A Treatise on Political Economy*, was suppressed by Napoleon for its views on free trade but had a tremendous influence on the study and teaching of economics in the United States throughout the ninetheenth century.

George Wilson BRIDGES (1788–1863) was a British minister who settled in Jamaica. A graduate of Trinity College, Oxford, Bridges moved to Jamaica in 1816 to become rector of the Parish of Manchester (1817–23) and then rector of the Parish of St. Ann (1823–37). A vocal and ardent supporter of slavery, Bridges wrote *A Voice from Jamaica; in Reply to William Wilberforce* (1823), a proslavery response to Wilberforce's *Appeal to the Religion, Justice, and Humanity of the Inhabitants of the British Empire, in Behalf of the Negro Slaves in the West Indies*. Serving as a proslavery propagandist tool, Bridges antiabolitionist response was reprinted four times in 1823 and 1824. Bridges expanded on his proslavery views in his *Annals of Jamaica* (1828).

II.393. Pierre François PAGE (1764–1805) wrote about **Saint-Domingue**, where he had bought planatations and various tracts of land in early 1790. In 1795, he appeared in court in Paris on behalf of other colonists, who were accusing two individuals of having devastated the French part of the island.

II.394. JAMAICA'S original MAROON POPULATION consisted of the African slaves whom the Spanish colonists had left behind when the British captured the island in 1655. These slaves escaped into the mountains rather than be reenslaved by the British, joining the Africans who had previously escaped from the Spanish to live and intermarry with the Taínos. To this population, which survived through subsistence farming and by raiding plantations, were later added slaves from different parts of Africa, mainly Ghana, who had been imported by the British. The Maroons came to control large areas of the Jamaican interior and rebelled against the colonial government. As a result, about six hundred Jamaican Maroons from Trelawney Town were shipped to Nova Scotia in 1796; others were deported to Sierra Leone. See also Cimarrón.

II.395. Jedidiah MORSE (1761–1826) was a Congregational minister and geographer from the United States, who authored the first textbook on American geography, *Geography Made Easy* (1784). He had the encouragement of, among others, Benjamin

Franklin and James Madison. Morse published his first geographical dictionary, *The American Gazetteer*, in 1797; Noah Webster (1758–1843), who would later work on his own dictionary, had initially offered his assistance but had to withdraw. Morse's sons, together with Serano Edwards Dwight (1786–1850), son of Yale president Timothy Dwight (1752–1817), assisted him in his 1810 *American Universal Geography*, whose final revision was completed in 1819 by his son, Sidney Edwards Morse (1794–1871). **Arrowsmith** and Samuel Lewis created sixty-three maps for a separate atlas. Until his death, Morse dominated the field of geography in the United States. His eldest son was Samuel Finley Breese Morse (1791–1872), the inventor of the telegraph.

II.395. Henry Charles CAREY (1793–1879) was a U.S. economist, publisher, and sociologist of Irish extraction; he is often called the founder of the American school of economics. A partner and later president of the Philadelphia publishing house Carey, Lea & Carey, he was knowledgable in a broad range of subjects. After retiring from publishing in 1835, he devoted the rest of his life to political economy. ISAAC LEA (1792–1886) was a U.S. naturalist and publisher. Lacking opportunities to study geology in the United States at the time, Lea befriended **Adam Seybert** of Philadelphia, who owned a significant collection of rock specimens. In 1821, Lea married Frances Anne Carey (1799–1874), the daughter of publisher Mathew Carey (1760–1839), and a sister of Henry Charles. In the following year, the younger Carey and Lea published a complete historical, chronological, and geographical *Atlas of the West Indies*. The elder Carey retired in 1821, and his son and Lea together proceeded to run the largest American publishing house of the nineteenth century, although H. C. Carey confided to his father that Lea was too much of a naturalist to make a good bookseller. Lea retired in 1851. In 1894, Lea's daughter gave his gem and mineral collection to the Smithsonian; it became the nucleus of their gem collection.

II.400. Johannes van den BOSCH (1780–1844) was a Dutch statesman, who expanded the poor-relief system. From 1798 to 1810, Bosch served with the army in the Dutch East Indies and wrote a book about the region, from which Humboldt quotes. As governor general of the Dutch East Indies (1828–33), he instituted the paternalistic Dutch East Indies Culture System, which forced farmers to pay revenue to the treasury of the Netherlands in the form of export crops or compulsory labor. He was minister of the colonies from 1834 to 1839.

II.402n. Charles COQUEREL (1797–1851) was the nephew of British novelist and poet Helen Maria Williams (c. 1762–1827), who translated many of Humboldt's works. She met Humboldt in Paris, where she continued to spend much time despite having been imprisoned during the **Reign of Terror**; she even became a naturalized French citizen. After her death, Charles completed her translation of Humboldt's *Relation historique*. Later that year, he translated her memoirs, *Souvenirs de la révolution française* [Remembrances of the French Revolution], into French.

ALEXANDER VON HUMBOLDT'S LIBRARY

Most of Humboldt's extensive library, which consisted of nearly seventeen thousand volumes, was destroyed in a warehouse fire at Sotheby's in London in 1865, so we have no complete record of the books he had actually owned. Although Henry Stevens's sales inventory from 1863—*The Humboldt Library: A Catalogue of the Library of Alexander von Humboldt*—predates this unfortunate event, it is also quite incomplete. Stevens, who had bought Humboldt's library in 1860, had already sold off many items prior to the scheduled auction in 1865, which the fire prevented.

The following list of the sources Humboldt directly or indirectly referenced in the *Political Essay on the Island of Cuba* is a modest attempt at a partial reconstruction of his library. We have checked all references to the extent possible and have indicated variations and other noteworthy information in the bracketed comments that follow many of the entries. Humboldt's references in the text of the *Political Essay on the Island of Cuba* have been updated to supply missing information and to help readers locate each item in this bibliography.

Abad y Queipo, Manuel de. "Representación sobre la inmunidad personal del clero, reducida por las leyes del nuevo código, en la cual se propuso al rey el asunto de diferentes leyes que establecidas harían la base principal de un gobierno liberal y benéfico para las Américas y para su metrópoli [fragmento]." In José María Luis Mora, *Obras Selectas*, 123–35. México: Porrúa, 1963.

La Abeja Argentina. Buenos Aires, 1822–23.

Acuerdos provisionales sobre arreglo de derechos y establecimiento de almacenes de depósito, sancionados por la E.ma diputación provincial, bajo la presidencia del señor Jefe superior político interino, para gobierno de esta administración general de rentas, y demás subalternas a quienes corresponda; formados con presencia del sistema general de aduanas que aprobaron las Cortes ordinarias de la monarquía en el año de 1820 y del decreto sancionado por el rey el 4 de febrero del presente comunicado con real orden de 20 del mismo a esta intendencia de ejército: de cuyo mandato se imprimen para conocimiento del público. Habana: Oficina de Arazoza y Soler, impresores del Gobierno Constitucional, 1822.

African Institution (London, England). *Foreign Slave Trade: Abstract of the Information Recently Laid on the Table of the House of Commons on the Subject of the Slave Trade: Being a Report Made by a Committee Specially Appointed for the Purpose, to the Directors of the African Institution on the 8th of May, 1821*. London: Ellerton and Henderson, 1821.

———. *A Review of the Colonial Slave Registration Acts: In a Report of a Committee of the Board of Directors of the African Institution, Made on the 22nd of February,*

1820, and Published by Order of That Board. London: Ellerton and Henderson, 1820.

———. *Sixteenth Report of the Directors of the African Institution: Read at the Annual General Meeting, Held on the 10th Day of May, 1822: With an Account of the Proceedings of the Annual Meeting, and an Appendix*. London, Ellerton, and Henderson, 1822.

Andréossy, Antoine François. *Histoire du canal du Midi, ou canal de Languedoc considéré sous les rapports d'invention, d'art, d'administration, d'irrigation, et dans ses relations avec les étangs de l'intérieur des terres qui l'avoisinent: Avec les cartes générales et particulières, ainsi que les plans, coupes et profils des principaux ouvrages*. 2 vols. New and exp. ed. Paris: Crapelet, 1804 (in year 13 of the French Revolutionary Calendar). [There is also a 1799 edition, but Humboldt tended to consult updated versions. Humboldt referenced the book variably as *Histoire du canal de Languedoc* and *Histoire du canal du Midi*.]

Anghiera, Pietro (or Petrus) Martyr. *De Orbe Novo Petri Martyris ab Angleria, mediolanensis protonarii Cæsaris senatoris Decades*. Basileae: Bebelius, 1530. [This book was originally written in Latin; work on it began in 1501, and the first installment was published ten years later. The eight decades were published together in 1530. Humboldt also references this book as "Oceanica," which refers to the title of a 1574 edition.]

———. *Opus Epistolarum Petri Martyris Anglerii Mediolanesis Protonotarii Aplici atq a cosiliis reru Indicaru*. Compluti: Eguia, 1530.

Antillón y Marzo, Isidoro de. *Elementos de la geografía astronómica, natural y política de España y Portugal*. 2nd corrected and exp. ed. by Valenci, 1815 [1808]. [The book is dedicated to Humboldt.]

Arago, François. "Notices scientifiques. Pluies des tropiques." Annuaire pour l'an 1824, presenté au Roi par le Bureau des Longitudes, 165–67. Paris: Bachelier, 1823. [Humboldt cites this yearbook as Annales du Bureau des Longitudes.]

Arango y Parreño, Francisco. "Continúan las memorias para la historia de la isla de Cuba." *El Patriota Americano* 20 (1812): 287–94.

———. *Expediente instruido por el consulado de la Habana, sobre los medios que convenga proponer para sacar la agricultura y comercio de esta isla del apuro en que se hallan*. Havana: Oficina del Gobierno y Capitanía General, 1808. [Reprinted in 1889 as "Informe del Síndico en el expediente instruido por el Consulado de la Habana sobre los medios que conviene proponer para sacar la Agricultura y Comercio de esta Isla del apuro en que se hallan," in *Obras de D. Francisco de Arango y Parreño* (Havana: Howson y Heinen), 2:17–64. This text is an updated version of the 1798 manuscript to which Humboldt refers.]

———. *Informe que se presentó en 9 de junio de 1796. A la Junta de Gobierno del Real Consulado de Agricultura y Comercio de esta ciudad e isla, por los Sres. José Manuel Torrontégui, Síndico procurador general del común y D. Francisco Arango y Parreño, Oidor Honorario de la Audiencia del Distrito y Síndico de dicho Real Consulado, cuando examinó la mencionada Real Junta el Reglamento y Arancel de capturas de esclavos cimarrones y propuso al Rey su reforma*. Havana: Imprenta de la

Capitanía General, 1796. [Reprinted in *Obras de D. Francisco de Arango y Parreño* (Havana: Dirección de Cultura, Ministerio de Educación, 1952), 1:256–74.]

———. *Reflexiones de un habanero sobre la independencia de esta isla*. 2nd Ed. Havana: Oficina de Arazoza y Soler, impresores del Gobierno Constitucional y Capitanía General por S.M., 1823. [An expanded version with notes on its publication history appeared in *Obras de D. Francisco de Arango y Parreño* (1889), 2:423–64. *See also* P. Q. S., *Contestación al indecente folleto y aviso de D. José de Arango, sobre independencia de la isla de Cuba* (Havana: Impr. Imparcial, 1821).]

———. "Representación de la ciudad de la Habana a las Cortes el 20 de julio de 1811, con motivo de las proposiciones hechas por D. José Guridi Alcocer y D. Agustín de Argüelles, sobre el tráfico y la esclavitud de los negros; extendida por el Alférez Mayor de la ciudad, D. Francisco de Arango, por encargo del ayuntamiento, consulado y Sociedad patriótica de la Habana." In *Obras* 1:145–237 (repr. 1952). Havana: Dirección de Cultura, Ministerio de Educación, 1811.

Arango y Parreño, Francisco, and José Manuel Torrontégui. *Documentos de que hasta ahora se compone el expediente que principiaron las Cortes extraordinarias sobre el tráfico y esclavitud de los negros*. Madrid: Imprenta de Repulles, 1814. [Humboldt gives the publication year as 1817. He references it variously as *Documentos sobre el tráfico de los negros* or simply as *Documentos*.]

El Argos de Buenos Aires. Buenos Aires: Imprenta de la Independencia, 1821. [This periodical began with number 1 on May 12, 1821, and folded around 1823. Weekly continued in 1824 under the name *El Argos de Buenos Aires y avisador universal*.]

Arrate y Acosta, José Martín Félix. *Llave del Nuevo Mundo Antemural de las Indias Occidentales. La Habana descripta: Noticias de su Fundación, Aumento y Estado*. 1756. [Humboldt quotes twice from "Arrate's manuscripts," which can be found at the Lilly Library at Indiana University, Bloomington.]

Atkinson, Henry. "XVI. On Astronomical and other Refractions; with a Connected Inquiry into the Law of Temperature in Different Latitudes and at Different Altitudes. Read January 14, April 8 and May 13, 1825." *Memoirs of the [Royal] Astronomical Society of London* 2:137–260. London: Baldwin, Cradock, and Joy, 1826.

Avendaño, Diego de. *Thesaurus indicus seu generalis instructor: Pro regimine conscientiæ, in iis quæ ad Indias spectant*. 2 vols. Anveres: Iacobum Meursium, 1668–78. [Reprinted: Avendaño and Angel Múñoz García, *Thesaurus indicus (1668)* (Pamplona: EUNSA, 2001).]

Bacon, Francis. *Opera omnia*. Ed. Simon Johann Arnold. Amsterdam: R. & J. Wetstenios & G. Smith, 1730. [The 1730 edition has 4 volumes; the essays are in volume 3, starting on p. 299. Humboldt quoted from chapter 25, "Of Innovations." Reprinted as *The Essays or Counsels, Civil and Moral*, ed. with an introduction and notes by Brian Vickers (Oxford and New York: Oxford University Press, 1999).]

Balanza del commercio marítimo de Veracruz correspondiente al año de 1811, formada por el consulado en cumplimiento de las órdenes del rey. [Reprinted: Miguel Lerdo de Tejada, *Comercio exterior de México desde la conquista hasta hoy* (Mexico: Banco Nacional de Comercio Exterior, [1853] 1967). *Balanzas del comercio marítimo de Vera-Cruz* for the years from 1784 to 1824 may also be found in vol.

3 of *Apuntes históricos de Vera-Cruz* by Miguel M. Lerdo de Tejada (Mexico: Imprenta de Vicente García Torres, 1858).]

Balanzas y estados de comercio. See *Estado demonstrativo del comercio*.

Balbi, Adriano. *Compendio di geografia universale conforme alle ultime politiche transazioni e più recenti scoperte*. 2 vols. Livorno: Glauco Masi, 1824–25. [An earlier edition appeared in Venice in 1819.]

———. *Essai statistique sur le royaume de Portugal et d'Algarve: Comparé aux autres états de l'Europe, et suivi d'un coup d'oeil sur l'état actuel des sciences, des lettres et des beaux-arts parmi les Portugais des deux hémisphères*. 2 vols. Paris: Rey et Gravier, 1822.

Barbé-Marbois, François. *Réflexions sur la colonie de Saint-Domingue, ou, examen approfondi des causes de sa ruine, et des mesures adoptées pour la rétablir; terminées par l'exposé rapide d'un plan d'organisation propre à lui rendre son ancienne splendeur; adressées au commerce et aux amis de la prospérité nationale*. Paris: Garney, 1796. [It is possible that Humboldt mixed up Charault's publications with others by that title, such as this one. *See* Charault, J. R.]

Barrow, John. "Art XI. *Narrative of a Voyage to Hudson's Bay, in His Majesty's Ship Rosamond, Containing Some Account of the North-Eastern Coast of America, and of the Tribes Inhabiting That Remote Region*. By Lieut. Chappell, R. N. 8 vols. London. 1817." *The Quarterly Review* 18, no. 35 (October 1818): 199–223. [Humboldt references this article as *Quarterly Review*. The running title is *On the Polar Ice and Northern Passage into the Pacific*; the publication date is 1818, but it was written in 1817.]

Bauzá, Felipe. *Derrotero de las islas Antillas, de las costas de tierra firme y de las del seno Méjicano*. 2nd corr. and exp. ed. Madrid: Imprenta Real, 1820.

———. *Memoria sobre la situación geográfica de La Habana, Veracruz, Puerto Rico, y otras islas y bajos del mar de las Antillas*. 1826. [Available at the British Library, London.]

Beaujeu, Du Temple de. *Mémoires de l'Académie* des sciences (August 1826). [On Beaujeu and his writings, see Adolphe Jules César Dureau de la Malle's *Description du bocage percheron, des moeurs et coutumes des habitans, et de l'agriculture de M. de Beaujeu* (extrait des annales de l'industrie nationale et étrangère) (Paris: de Fain, 1823).]

Beckford, William. *A Descriptive Account of the Island of Jamaica with Remarks upon the Cultivation of the Sugar-Cane, Throughout the Different Seasons of the Year, and Chiefly Considered in a Picturesque Point of View: Also Observations and Reflections upon What Would Probably Be the Consequences of an Abolition of the Slave-Trade, and of the Emancipation of the Slaves*. 2 vols. London: T. and J. Egerton, 1790. [Humboldt refers to Beckford without providing a specific source; he must have had this book in mind.]

Benzoni, Girolamo. *La Historia del Mondo Nuovo*. Venice: Appresso Francesco Rampazetto, 1565.

La Biblioteca americana, o miscelánea de literatura, artes y ciencias: Por una sociedad de americanos. Vol 1: 115–29. Published by Andrés Bello and Juan García del Río.

London: G. Marchant, 1823. [This volume contains an excerpt from Humboldt's *Vues des Cordillères*, entitled "El Chimborazo" (108–15). Along with García del Río, Bello published *Biblioteca Americana* in 1823 and *El Repertorio Americano* in 1826. They were two influential journals that featured Bello's own work, which included grammar, poetry, scientific investigations, philosophy, translations, and literary criticism.]

Bibliothèque universelle des sciences, belles-lettres, et arts, faisant suite à la Bibliothèque Britannique, rédigée à Génève par les auteurs de ce dernier recueil. Book 22. Geneva, 1823. [*Biblioteca Americana* reproduced an article from this source; see *Biblioteca Americana*.]

Biot, Jean-Baptiste. *Traité élémentaire d'astronomie physique.* Paris: Bernard, 1805.

———. *Traité élémentaire d'astronomie physique, avec des additions relatives à l'astronomie nautique, par M. de Rossel.* Paris: Klostermann, 1810.

Biot, Jean-Baptiste, Francisco Arago, and François André Méchain. *Recueil d'observations géodésiques, astronomiques et physiques, exécutées par ordre du Bureau des Longitudes de France, en Espagne, en France, en Angleterre et en Écosse, pour déterminer la variation de la pésanteur et des degrés terrestres sur le prolongement du Méridien de Paris, faisant suite au troisième volume de la base du système métrique.* Paris: Courcier, 1821.

Bolingbroke, Henry. *A Voyage to the Demerary Containing a Statistical Account of the Settlements There, and of Those on the Essequebo, the Berbice, and Other Contiguous Rivers of Guyana.* London: Richard Phillips, 1807. [There are also editions from 1809 and 1813. Humboldt did not specify which one he used.]

Bosch, Johannes van den. *Nederlandsche bezittingen in Azia, Amerika en Afrika, in derzelver toestand en aangelegenheid voor dit Rijk, wijsgeerig, staatshuishoudkundig en geographisch beschouwd, met bijvoeging der noodige tabellen, en eenen atlas nieuwe kaarten.* 2 vols. Gravenhage en Amsterdam: Gebroeders Van Cleef, 1818.

Brackenridge, Henry Mary. *Voyage to South America, Performed by Order of the American Government, in the Years 1817 and 1818, in the Frigate Congress.* 2 vols. Baltimore: published by the author, John D. Toy, printer, 1819. [There is also an 1820 British edition and an 1821 German translation.]

Brewster, David. "Meteorology." *The Edinburgh Journal of Science* (Edinburgh: John Thomson; London: T. Cadell) 4, no. 7 (Nov.–Apr. 1826): 180. [Humboldt gives the publication year as 1829.]

Bridges, George Wilson. "The Rev. G. W. Bridges on the Effects of Manumission." *Negro Slavery* (London: Ellerton and Henderson), no. 4 (1823).

Broglie, Achille Charles Léonce Victor. *Développements d'une proposition faite à la chambre et relative à exécution des lois prohibitives de la traite des noirs. Session de 1821; Séance du jeudi 28 mars 1822.* Paris: Hachette, 1822.

Brown, Charles. *Narrative of the Expedition to South America, Which Sailed from England at the Close of 1817, for the Service of the Spanish Patriots: Including the Military and Naval Transactions, and Ultimate Fate of That Expedition: Also the Arrival of Colonels Blosset and English.* London: J. Booth, 1819.

Bruncrona, Nils Abraham, and Carl Peter Hällström. "VII. Beobachtungen und Angaben über die Verminderung des Wassers an der schwedischen Küste." *Annalen der Physik und Chemie* (ed. J. C. Poggendorff), series 2, vol. 2, part 3, item 7 (1824): 308–28. [Originally in Swedish.]

Buache, Jean-Nicolas, and Ciriaco Cevallos. *Disertaciones sobre la navegación a las Indias Orientales por el norte de la Europa*. Isla de León: Y.D.C.G.M, 1798.

Buch, Leopold von. "Observations on the Climate of the Canary Islands." *Edinburgh New Philosophical Journal, Exhibiting a View of the Progressive Improvements and Discoveries in the Sciences and the Arts* (April–October 1826): 92–104.

———. *Physicalische Beschreibung der Canarischen Inseln*. Berlin: Kgl. Akad. d. Wissenschaften, 1825. [Humboldt, who highly esteemed Buch, owned this book in various editions. Also in Buch, *Gesammelte Schriften* (1967), vol. 3.]

———. *Reise durch Norwegen und Lappland*. Berlin: G.C. Nauck, 1810.

Budget et comptes de la ville de Paris pour 1825. Full title: *Présentation des comptes de finances: Réglement définitif du budget de 1824: Ouverture de crédits supplém. sur l'exercice 1825; fixation du budget des d'épenses et des recettes de 1827; Session de 1826*.

Bureau des Longitudes. *Connaissance des Tems [sic] ou des mouvements célestes a la usage des astronomes ou des navigateurs pour l'an 1810*. Paris: L'Imprimerie royale (August 1808).

Caballero y Góngora, Antonio. *Relación del estado del Nuevo Reino de Granada; que hace el Arzobispo obispo de Cordova, a su sucesor el excelentísimo señor Don Francisco Guyl [Gil] de Lemos año de 1789*. C. 1789. [Caballero y Góngora (1723–96) was viceroy of New Granada from 1782 to 1788. Manuscript, 207 leaves. Available at UCLA Library, Special Collections, Charles E. Young Research Library.]

Caldas, Francisco José de. *Semanario del nuevo reino de Granada*. Bogotá, Colombia: s.n., 1808–10. Repr. by Biblioteca de Cultura Colombiana, 3 vols., 1942.

Caldcleugh, Alexander. *Travels in South America, during the Years, 1819–20–21; Containing an Account of the Present State of Brazil, Buenos Ayres, and Chile*. 2 vols. London: J. Murray, 1825.

Carlisle, Anthony. "Letter from A. Carlisle Esq. on the Temperature of the Sea, to Mr. Nicholson, May 28, 1804." *Journal of Natural Philosophy, Chemistry, and the Arts* 8 (May 1804): 131–33.

Cartas de los Rev. Padres Observantes, no. 7 (unpublished manuscript).

Casas, Bartolomé de las. *Historia de las Indias publicada ahora por vez primera, conforme á los originales del autor, que se custodian en la biblioteca de la Academia de historia y en el Nacional de esta corte*. 5 vols. Madrid: M. Ginesta, 1875–76. [This book was still an unpublished manuscript at the time Humboldt wrote. He took his information from Herrera y Tordesillas, who quoted from Las Casas's manuscript. Critical edition: Miguel Angel Medina, Jesús Angel Barreda, and Isacio Pérez Fernández, eds., *Historia de las Indias by Bartolomé de las Casas*, 3 vols. (Madrid: Alianza, 1994). English editions, selections: Nigel Griffin, ed. and trans., *Las Casas on Columbus: Background and the Second and Fourth Voyages* (Turnhout, Belgium: Brepols, 1999); and Geoffrey Symcox and Jesús Carrillo, eds., *Las Casas on Columbus: The Third Voyage by Bartolomé de las Casas*, trans. Michael Hammer and Blair Sullivan (Turnhout: Brepols, 2001).]

Cassini, Jacques. "Observations astronomiques faites en divers lieux de l'Amérique méridionale; comparée avec celles qui ont été faites en Frances." *Histoire de l'Académie Royale des Sciences. Année MDCCXXIX. Avec les mémoires de mathématique & de physique, pour la même année. Tirés des registres de cette Académie* 31 (1729): 361–84. Paris: L'Impримérie Royale. [At the French Natonal Library. *See* http://www.bnf.fr.]

Castillo, José del, and Bergaño Simon, eds. *El Patriota Americano*. 2 vols. La Habana: D. Pedro Nolasco Palmer, 1811–12. [Contributions from Francisco Arango y Parreño and José Arango. The second volume contains the article "Noticia mineralógica del cerro de Guanabacoa comunicada al Excmo. señor marqués de Someruelos por el barón de Humboldt en el año de 1804," pp. 29–32. Index published in *Los antiguos diputados de Cuba y apuntes para la historia constitucional de esta isla* by Eusebio Domínguez Valdés (Habana: "El Telegrafo," 1879). The Biblioteca Nacional in Madrid, Spain, has 1812, nos. 4–32.]

Castillo y Rada, José María del. "Memoria que el secretario de estado y del despacho de hacienda presenta al congreso de Colombia sobre los negocios de su departamento. Año 1823.-13°." *Memorias de Hacienda. 1823–1826–1827*, 5–22. Archivo de la economía nacional. Bogotá: Banco de la República, 1952. [The report Humboldt mentions is reprinted here; it is dated "Bogotá mayo 5 de 1823."]

Caulín, Antonio. *Historia corográfica, natural y evangélica de la Nueva Andalucía, provincias de Cumaná, Guayana y Vertientes del Río Orinoco; dedicada al rey n. s. d. Carlos III*. Granada or Madrid: S.M., 1779. Repr. in Caracas: Academia Nacional de la Historia, 1966.

Cavo, Andrés. *De vita Josephi Juliani Parrenni Havanensis ab Andrea Cavo sacerdote Guadalaxarensi Mexicano*. Rome: Officina Salomoniana, 1792.

El Centinela de la Plata. No. 8 (September 15 1822): 101–3. In *Biblioteca de mayo: Colección de obras y documentos para la historia Argentina*. Vol. 9, *Periodismo*. Buenos Aires: Imprenta del Congreso de la Nación, 1960, 8019–20. [Part of the Library of May journalism collection, Buenos Aires, Argentina.]

Chabrol de Volvic, Gilbert Joseph Gaspard, Claude Philibert Barthelot Rambuteau, and Georges Eugène Haussmann. *Récherches statistiques sur la ville de Paris et le département de la Seine recueil de tableaux dressés et réunis d'après les ordres de monsieur le comte de Chabrol, conseiller d'état, préfet du département*. Paris: Impr. Royale, 1823–60.

Chalmers, George. *An Official Letter from the Commissioners of Correspondence of the Bahama Islands, to George Chalmers, Esq., Colonial Agent, Concerning the Proposed Abolition of Slavery in the West Indies*. Nassau, New-Providence: Printed at the Royal Gazette Office, 1823. [The letter is not by but *to* Chalmers. Humboldt mentioned Chalmers once in his population tables on Nevis for 1809.]

Chamisso, Adalbert von. *See* Kotzebue, Otto von.

Chappe d'Auteroche, Jean. *Voyage en Californie pour l'observation du passage de Vénus sur le disque du soleil, le 3 Juin 1769, contenant les observations de ce phénomene, & la déscription historique de la route de L'Auteur à travers le Mexique*. Ed. and published by Cassini. Paris: Chez Charles-Antoine Jombert, 1772.

Chaptal, Jean Antoine Claude. *Chimie appliquée à l'agriculture*. 2 vols. Paris: Mme

Huzard, 1823. [The first U.S. translation into English on the basis of the second French edition was Chaptal and Charles Folsom, *Chymistry Applied to Agriculture* (Boston: Hilliard, Gray, and Co., 1835).]

———. *Mémoire sur le sucre de betterave.* 3rd ed., corr. and exp. Paris: Mme Huzard, 1821.

Charault, J. R. *Coup-d'oeil sur Saint-Domingue; observations sur le caractère des nègres et sur la fièvre jaune; moyens de recouvrer cette colonie, et de se préserver des maladies qui y règnent.* Paris: Panckoucke, 1814. [Humboldt may have combined elements from several books here; see Barbé-Marbois, François. There is no 1806 edition of any of them. We traced some of Humboldt's references to "Charault, *Réflexions sur Saint-Domingue*" to Charault's *Coup-d'oeil sur Saint-Domingue.*]

Chateauneuf, Benoiston de. *Recherches sur les consommations de tout genre de la ville de Paris en 1817 comparées a ce qu'elles étaient en 1789.* 2nd corr. and exp. ed. Paris: Chez l'Auteur, [1820] 1821.

Churchill, Awnsham, and John Churchill. *A Collection of Voyages and Travels Some Now First Printed from Original Manuscripts, Others Now First Published in English: In Six Volumes with a General Preface Giving an Account of the Progress of Navigation from Its First Beginning.* London: Printed by assignment from Messrs. Churchill for John Walthoe, 1732. [Includes Fernando Colombo, "The Discovery of the West-Indies, by Christopher Columbus, together with his Life and Actions, etc.," 2:499–628. There is also a 1704 edition in 4 vols.]

Churruca, Cosme Damián. *Apéndice que contiene el de los paquebotes Santa Casilda y Santa Eulalia para completar el reconocimiento del estrecho en 1788 y 1789.* Madrid, 1793.

Clarkson, Thomas. *An Essay on the Slavery and Commerce of the Human Species, Particularly the African: Translated from a Latin Dissertation, which was Honoured with the First Prize in the University of Cambridge, for the Year 1785, with Additions.* London: T. Cadell; J. Phillips, 1786.

Colombia from Humboldt and Other Recent Authorities. See Neele, Robert Madie.

Colquhoun, Patrick. *Treatise on the Wealth, Power, and Resources, of the British Empire, in Every Quarter of the World, Including the East Indies: The Rise and Progress of the Funding System Explained; With Observations on the National Resources for the Beneficial Employment of a Redundant Population, and for Rewarding the Military and Naval Officers, Soldiers, and Seamen, for Their Services to Their Country during the Late War. Illustrated by Copious Statistical Tables, Constructed on a New Plan, and Exhibiting a Collected View of the Different Subjects Discussed in This Work.* 2nd ed. London: J. Mawman, [1814] 1815.

Columbus, Christopher. *Lettera rarissima* (Jamaica, July 7, 1503), reproduced and illustrated by Cav. Ab. Morelli. Bassano, 1810. [Humboldt owned a copy of this book. A Spanish retranslation was published by Martín Fernández de Navarrete in his *Colección de los viajes y descubrimientos, que hicieron por mar los Españoles desde fines del siglo XV, con varios documentos inéditos concernientes a la historia de la marina Castellana y de los establecimientos Españoles en [las] Indias* (Madrid: Imprenta Real, 1825), 5 vols. In 1903, John Boyd Thatcher reproduced the early Italian translation together with an English translation; see *Christopher Columbus, His Life, His Works, His Remains, Together with an Essay on Peter Martyr of Anghiera and Bartolomé de las Casas, the First Historians of America*, 2 vols. *See also* J. Franklin Jameson, ed., *Original Narratives of Early American History. Reproduced under the Auspices of the American Historical Association. The Northmen, Columbus, and Cabot 985–1503* (New York: Charles Scribner's Sons, 1906), 387–418. Richard Henry Major revised *Select Letters of Columbus* and made some changes; see *Select Letters of Christopher Columbus with Other Original Documents, Relating to His Four Voyages to the New World*, works issued by the Hakluyt Society, no. 2 (London: Printed for the Hakluyt Society). For a more recent translation, see John Michael Cohen, *The Four Voyages of Christopher Columbus; Being His Own Log-Book, Letters and Dispatches with Connecting Narrative Drawn from the Life of the Admiral by His Son Hernando Colón and other Contemporary Historians*, (Harmondsworth: Penguin, 1969).]

Comisión para el fomento de la Isla de Cuba, 1799. *Dictamen sobre las ventajas que pueden sacarse para el mejor fomento de la Isla de Cuba*, 3 folders. Report concerning the history of Cuba and the importance of developing commerce, industry, and military defense. Sewn, in 2 parts. [Available at Harvard University's Houghton Library.]

Crawfurd, John. *History of the Indian Archipelago Containing an Account of the Manners, Arts, Languages, Religions, Institutions, and Commerce of its Inhabitants*. 3 vols. Edinburgh: Archibald Constable and Co., 1820.

Cropper, James, and William Wilberforce. *A Letter Addressed to the Liverpool Society for Promoting the Abolition of Slavery on the Injurious Effects of High Prices of Produce, and the Beneficial Effects of Low Prices, on the Condition of Slaves*. London: Hatchard, 1823.

———. *Letters Addressed to William Wilberforce, M.P. Recommending the Encouragement of the Cultivation of Sugar in Our Dominions in the East Indies, as the Natural and Certain Means of Effecting the Total and General Abolition of the Slave-Trade*. London: Longman, Hurst, and Co., 1822.

———. *Relief for West-Indian Distress Shewing the Inefficiency of Protecting Duties on East-India Sugar, and Pointing out Other Modes of Certain Relief*. London: Printed by Ellerton and Henderson, Gough Square, and sold by Hatchard and Son, Piccadilly; J.& J. Arch, Cornhill, London; G. & J. Robinson, Castle Street; W. Grapel, Lord Street, Liverpool, 1823.

Cuvier, Georges. *Recherches sur les ossemens fossiles de quadrupèdes, où l'on rétablit les caractères de plusieurs espèces d'animaux que les révolutions du globe paroissent avoir détruites, &c*. 3rd ed. Paris: Dufour et d'Ocagne, [1812] 1825. [A second edition appeared between 1821 and 1824: G. Dufour et E. d'Ocagne, 5 vols.]

Cuvier, Georges, and Alexandre Brongniart. *Déscription géologique des environs de Paris*. Paris: G. Dufour et E. d'Ocagne, 1822. [Includes "Description des végétaux fossiles du terrain de sédiment supérieur, cités dans la description géologique du bassin de Paris" by Adolphe Brongniart, pp. 353–71.]

Dampier, William. *Voyages and Descriptions. Vol. II. In Three Parts, viz. 1. A Supple-*

ment of the Voyage Round the World, Describing the Countreys of Tonquin, Achin, Malacca, &c. Their Product, Inhabitants, Manners, Trade, Policy, &c. 2. Two Voyages to Campeachy; with a Description of the Coasts, Product, Inhabitants, Logwood-Cutting, Trade, &c. of Jucatan, Campeachy, New-Spain, &c. 3. A Discourse of Trade-Winds, Breezes, Storms, Seasons of the Year, Tides and Currents of the Torrid Zone throughout the World: With an Account of Natal in Africk, its Product, Negros, &c. By Captain William Dampier. Illustrated with particular Maps and Draughts. To which is Added, a General Index to both Volumes. London: James Knapton, 1699. [Humboldt apparently used the first edition of the second volume (1699). The first volume, *A New* Voyage *round the World,* had been published in 1697; all of Humboldt's references can be traced to the second edition from 1700. This volume comes in three parts, each of which starts with page 1.]

Dauxion Lavaysse, Juan-José. *Voyage aux îles de Trinidad, de Tabago, de la Marguerite, et dans diverses parties de Vénézuéla, dans l'Amérique Méridionale.* Paris: F. Schoell, 1812–13. [Some sources have "Jean François Dauxion Lavaysse," others "Juan José Dauxion Lavaysse." For an English translation, see Dauxion-Lavaysse and Edward Blaquiere, *A Statistical, Commercial, and Political Description of Venezuela, Trinidad, Margarita, and Tobago: Containing Various Anecdotes and Observations, Illustrative of the Past and Present State of These Interesting Countries; from the French of M. Lavaysse: With a Preface and Explanatory Notes,* 2nd ed. (London: Printed for G. and W. B. Whittaker, 1821).

Davy, John. *An Account of the Interior of Ceylon, and of Its Inhabitants with Travels in That Island.* London: Printed for Longman, Hurst, Rees, Orme, and Brown, 1821.

———. "Observations on the Temperature of the Ocean and Atmosphere, and on the Density of Sea-Water, Made during a Voyage to Ceylon." *Philosophical Transactions of the Royal Society of London* 107 (January 1, 1817): 275–92. [Davy included these *Observations* in a letter addressed to his brother Sir Humphry Davy, dated November 3, 1816, in Colombo.]

De la situación actual de la real factoría de tabacos de la Habana en Abril 1804. Official handwritten document.

Denham, Dixon, Hugh Clapperton, and Walter Oudney. *Narrative of Travels and Discoveries in Northern and Central Africa: In the Years 1822, 1823 and 1824: Extending across the Great Desert to the Tenth Degree of Northern Latitude, and from Kouka in Bornou, to Sackatoo, the Capital of the Felatah Empire.* 2 vols. London: J. Murray, 1826. [Reprinted 1985]

Descourtilz, Michel Étienne. *Voyages d'un naturaliste, et ses observations faites sur les trois règnes de la nature dans plusieurs ports de mer français, en Espagne, au continent de l'Amérique septentrionale, à Saint-Yago de Cuba, et à St.-Domingue, où l'auteur devenu le prisonnier de 40,000 noirs révoltés, et parsuite mis en liberté par une colonne de l'armée française, donne des détails circonstanciés sur l'expédition du général Leclerc* [1799–1803]. 21 vols. Paris: Dufart, 1809.

Dobrizhoffer, Martin. *Historia de Abiponibus equestri, bellicosaque Paraquariae natione.* Viennae: J. Nob de Kurzbek, 1784. [For an English version, see Dobrizhoffer and Sara Coleridge, *An Account of the Abipones, an Equestrian People of Paraguay. From the Latin of Martin Dobrizhofer* (London: J. Murray, 1822) .]

Dorta, Bento Sanches. "Observações Astronomicas. Feitas junto ao Castello da Cidade do Rio de Janeiro para de ser minar a Latitude e Longitude da dita Cidade." *Memorias da Academia Real das Sciencias de Lisboa* 1 (1780–88): 325–44, 345–78. Lisboa: Na Typografia da Academia, 1797.

———. "Observações Astronomicas, e Meteorologicas. Feitas na Cidade do Rio de Janeiro no anno de 1784." *Memorias da Academia Real das Sciencias de Lisboa* 2: 347–68. Lisboa: Na Typografia da Academia, 1799.

Ducros, J. A., and Gaspar Clair François Marie Riche de Prony. *Mémoires sur les quantités d'eau, qu'exigent les canaux de navigation.* Paris: Goujon, 1800. [There are 1801 and 1802 editions that give the publication year as "IX" on the French Revolutionary Calendar, which started in 1792 with the year "I." A German edition was published in 1810.]

Dupin, Pierre Charles François. *Mémoires sur la marine et les ponts et chaussées de France et d'Angleterre, contenant deux relations de voyages faits par l'auteur dans les ports d'Angleterre, d'Écosse et d'Irlande dans les années 1816, 1817 et 1818; la description de la jetée de Plymouth, du canal Calédonien, etc.* Paris: Bachelier, 1818.

Dupuget, Edme Jean Antoine d'Orval. "Extrait d'un mémoire du citoyen Dupuget, intitulé: Coup-d'œil rapide sur la physique générale et la minéralogie des Antilles." *Journal des mines* 3, no. 18 (1797): 43–60.

Dutens, Joseph. *Mémoires sur les travaux publics de l'Angleterre suivis d'un mémoire sur l'esprit d'association et sur les différens modes de concession, et de quinze planches avec une carte générale de la navigation intérieure, indiquant les deux systèmes des grands et des petits canaux de ce pays.* Paris: Imprimerie Royale, 1819. [For a description of the Caledonian Canal see pp. 330ff. Dutens acknowledges Robert Stevenson, a civil engineer, as the author, which would make him the translator of this particular passage.]

East and West India Sugar. See *Substance of a Debate in the House of Commons, on the 22nd of May 1823.*

Edwards, Bryan. *The History, Civil and Commercial, of the British Colonies in the West Indies: In Two Volumes.* Dublin: Luke White, 1793. [This book has a complicated publication history, which Humboldt illustrates by referencing different editions in successive parts of his narrative. Another edition: *The History, Civil and Commercial, of the West Indies: With a Continuation to the Present Time,* 5th ed. (London: Whittaker, etc., 1819).]

Elphinstone, Mountstuart. *An Account of the Kingdom of Caubul, and Its Dependencies in Persia, Tartary, and India. Comprising a View of the Afghaun Nation, and a History of the Dooraunee Monarchy.* 2 vols. London: Longman, Hurst, Rees, Orme, and Brown, and J. Murry, [1815] 1819.

Empréstito de la Intendencia de la Habana. November 5, 1804.

Entrecasteaux, Antoine Raymond, and Joseph de Bruni. *The Voyage of d'Entrecasteaux.* Ed. É. P. É. de Rossel, trans. John Joseph Smith. Paris: Imperial Press, 1808. [It is a rare book.]

Espinosa y Tello, José. *Memorias sobre las observaciones astronómicas, hechas por los navegantes españoles en distintos lugares del globo; las cuales han servido de fundamento para la formación de las cartas de marear publicadas por la dirección de trabajos hidrográficos de Madrid.* Vol. 2. Madrid: Imprenta Real, 1809. [Vols. 1 and 2 have two and three different parts respectively, each of which starts with page 1. Humboldt quotes from vol. 2.]

Estado demonstrativo del comercio de importación y exportación que se ha hecho por el Puerto de la Habana en todo el año 1816, con distinción de buques nacionales y extranjeros, efectos introducidos y extraídos, sus valores por aforos y derechos reales y municipales que han adeudado. Havana: Oficina del Gobierno y Capitanía General. [Trade balances were published annually. A few of them, including the one for 1816, which Humboldt mentions, may be found in the Alexander von Humboldt Papers, Biblioteka Jagiellónska, Krakow, Poland.]

Fajardo, Manuel Palacio. *Outline of the Revolution in Spanish America, or, an Account of the Origin, Progress and Actual State of the War Carried on between Spain and Spanish America Containing the Principal Facts Which Have Marked the Struggle.* New York: James Eastburn & Co., 1817. [There is also a British edition from 1817.]

Farquhar, Robert Townsend. *Suggestions, Arising from the Abolition of the African Slave Trade for Supplying the Demands of the West India Colonies with Agricultural Labourers.* London: Printed for John Stockdale, 1807.

Fernández de Oviedo y Valdés, Gonzalo. *L'Histoire naturelle et générale des Indes, isles et terre ferme de la grand mer océane.* Paris: Imp. de Michel de Vascosan, 1556.

———. *La Historia general de las Indias.* Sevilla: en la Imprenta de Cromberger, 1535.

Ferrer, José Joaquín de. "No. XXIX. Astronomical Observations Made by José Joaquín de Ferrer, Chiefly for the Purpose of Determining the Geographical Position of Various Places in the United States, and Other Parts of North America. Communicated by the Author." Translated from Spanish, and read at different times. *Transactions of the American Philosophical Society* 6 (1809): 158–61. [Discrepancy with Humboldt's page numbers.]

———. "I. Observations Made in the Island of Cuba. By the Late Don José Joaquín de Ferrer, of Cadiz, Member of the Phil. Acad. Boston. Communicated by H. T. Colebrooke, Esq. Translated from the Spanish: and Comprehending 1. Observations of the Comet of 1807, with the Determination of the Elements of Its Orbit. 2. Observations of the Lunar Eclipse, Nov. 14, 1807. 3. Observations of the Comet of 1813, with the Determination of the Elements of Its Orbit; Together with Remarks on Its Magnitude, and That of the Comet of 1811. 4. Observations, and Computations of the Elliptic Orbit, of the Comet of 1811." *Memoirs of the Astronomical Society of London* 3, part 1 (1827): 1–38. Read January 9, 1824. [This publication points out some of Ferrer's major contributions to science.]

———. "Occultations d'étoiles observées à la Havane, et qui peuvent server à déterminer l'inflexion du demi-diamètre de la lune." In *Connaissance des temps ou des mouvements célèstes, à l'usage des astronomes et des navigateurs,* 318–37. Paris: Bureau des Longitudes, 1817. [The pages leading up to p. 318 are also filled with Ferrer's observations.]

Fleming, John. *A Catalogue of Indian Medicinal Plants and Drugs.* Calcutta: Hindustani Press, 1810. [This might be the source to which Humboldt referred as "Calcutta 1810." An 1812 edition was also published in London.]

France. *Déscription de l'Égypte, ou recueil des observations et des récherches qui ont été faites en Égypte pendant l'expédition de l'armée française, publié par les ordres de sa majesté l'empereur Napoléon le grand.* 21 vols. Paris: Imprimerie Impériale, 1809–28. [Humboldt cites 1808. Full digital versions of the 21 volumes are available from the Bibliotheka Alexandrina at http://descegy.bibalex.org].

Gaimard, Joseph Paul, and Jean-René Constant Quoy. *Mémoire sur l'accroissement des polypes lithophytes considéré géologiquement: Lu à l'Académie des sciences d'institute, le 14 juillet 1823.* Paris: s.n, 1825.

Gallesio, Giorgio. *Traité du citrus.* Paris: L. Fantin, 1811.

Gardner, Edmund. *Logbook of the Ship South America* (1823–25). 1823.

Gaspari, Adam Christian, and Johann Georg Heinrich Hassel, et al. *Vollständige und neueste Erdbeschreibung des Russischen Reichs in Europa nebst Polen: Mit einer Einl. zur Statistik des ganzen Russischen Reichs.* Weimar: Geographisches Institut, 1821. [Humboldt owned this book. It was first published by Gaspari in 1797 and revised in 1802. Gaspari and Hassel are usually cited as authors; some listings include other names.]

La Gazeta de Colombia. Also: *Gazeta de Santafé, capital del N.R. de Granada.* Santafé de Bogotá: Imp. del Gobierno, 1816.

Girard, Pierre Simon. "Du Trosième mémoire sur les canaux de navigation considerés sous le rapport de la chute et de la distribution de leurs écluses." *Annales de la Chimie et de Physique* 24 (1823, par MM. Gay-Lussac et Arago): 113–39. Paris: Crochard.

Gómara, Francisco López de. *Historia general de las Indias: Y todo lo acaecido en ellas donde que se ganaron hasta agora y la conquista de México, y de la Nueva España.* Vols. 1 and 2. Madrid: CALPE, [1553] 1922. [Humboldt's references are traceable in this facsimile, but the section numbering differs.]

Guía de forasteros de la siempre fiel isla de Cuba. Habana: Imprenta del Gobierno y Capitanía General por S.M. [Edited by Antonio Robredo. Full title of this periodical is *Calendario Manual y Guía de Forasteros de la Isla de Cuba.* It ran from 1793 to 1884.]

Hall, Basil. *Extracts from a Journal Written on the Coasts of Chili, Peru and Mexico, in the Years 1820, 1821, 1822.* 2 vols. Endinburgh: Printed for A. Constable, 1824.

———. *Travels in North America in the Years 1827 and 1828.* 3 vols. Edinburgh: Printed for Cadell and Co., 1829.

Hansteen, Christopher, and Peter Treschow Hanson. *Untersuchungen über den Magnetismus der Erde. Erster Teil. Die mechanischen Erscheinungen des Magneten.* Christiania: J. Lehmann und C. Gröndahl, 1819. [Hanson is the translator.]

Harvey, George. "Art. IX.—Remarks on the Increase of the Population of the United States, and Territories of North America, with Original Tables Deduced from the

American Population Returns, to Illustrate the Various Rates of Increase in the White Population and Slaves, and also the Comparative Degrees in which Agriculture, Commerce, and Manufactures Prevail." *Edinburgh Philosophical Journal* (Edinburgh: Archibald Constable and Co.; London: Hurst, Robinson & Co.) 8 (October 1, 1822, to April 1, 1823, conducted by Dr. Brewster and Professor Jameson): 41–55.

Hassel, Georg. *Statistischer Umriss der sämtlichen Europäischen Staaten in Hinsicht ihrer Grösse, Bevölkerung, Kulturverhältnisse, Handlung, Finanz- und Militärverfassung und ihrer aussereuropäischen Besitzungen.* Braunschweig: Vieweg, 1805. [Humboldt is likely to have used the more recent edition from 1824.]

———. *Statistischer Umriss sämmtlichen Europäischen und der vornehmsten aussereuropäischen Staaten in Hinsicht ihrer Entwickelung, Größe, Volksmenge, Finanz- und Militärverfassung 3. Dritter Heft, welcher das Osmanische Reich und die außereuropäischen Staaten darstellt.* Weimar: Geographisches Institut, 1824.

Hassel, Georg, F. W. Beniken, and J. C. Schmidt, eds. *Genealogisch-historisch-statistischer Almanach.* Weimar: Landes-Industrie-Comptoir, 1824.

Hatchard, John. *Review of Registry Laws. See* African Institution, *A Review of the Colonial Slave Registration Acts.*

Herrera y Tordesillas, Antonio. *Historia general de los hechos de los Castellanos en las islas y tierra firme del Mar Océano.* Folio. Madrid: Franco, [1601–15] 1730. [This book is better known as *Historia de las Indias occidentales,* the title Humboldt uses. In 1826, the 1730 edition would have been the most recent Spanish edition; later editions were revised and expanded by various editors.]

Higgins, Bry. *Observations and Advices for the Improvement of the Manufacture of Muscovado Sugar and Rum. By Bryan Higgins, M.D.* St. Jago de la Vega: Alexander Aikman, 1797[–1803].

Hodgson, Adam, and Jean Baptiste Say. *A Letter to Jean-Baptiste Say, on the Comparative Expense of Free and Slave Labour.* Liverpool: Smith, 1823.

Hoff, Karl Ernst Adolf von. *Geschichte der durch Überlieferung nachgewiesenen natürlichen Veränderungen der Erdoberfläche. Ein Versuch. I. Theil. Eine von der Kön. Gesellschaft der Wissensch. zu Göttingen gekrönte Preisschrift. Mit einer Charte von Helgoland.* 5 vols. Gotha: Justus Perthes, 1822–41.

———. *Geschichte der durch Überlieferung nachgewiesenen natürlichen Veränderungen der Erdoberfläche: ein Versuch 2 Geschichte der Vulcane und der Erdbeben.* Gotha: Perthes, 1824.

Huber, B., and Robert Francis Jameson. *Aperçu statistique de l'île de Cuba, précédé de quelques lettres sur la Havane, et suivi de tableaux synoptiques, d'une carte de l'île, et du tracé des côtes depuis la Havane jusqu'à Matanzas.* Paris: P. Dufart, 1826. [See also the original English version by Jameson below.]

Huerne de Pommeuse, Michel Louis François. *Des Canaux navigables considérés d'une manière générale: Avec des récherches comparatives sur la navigation intérieure de la France et celle de l'Angleterre: Avec atlas.* Paris: Bachelier et al., 1822.

Humboldt, Alexander von. *De distributione geographica plantarum secundum coeli temperiem at altitudinem montium: Prolegomena.* Lutetiae Parisiorum: Libraria Graeco-Latino-Germanica, 1817.

———. "De la température des différentes parties de la zone torride au niveau des mers." *Annales de chimie et de physique* (ed. Louis Bernard Guyton de Morveau and Joseph-Louis Gay-Lussac) 33 (1826): 29–48. For the German version, see Poggendorff's *Annalen der Physik und Chemie* 8, part 2, item 3: 165–75. [See also "*On the Temperature in Different Parts of the Torrid Zone at Sea Level,*" in the *Political Essay on the Island of Cuba*, where it immediately precedes the supplement. This section is part of the Atkinson exchange regarding the equator's mean temperature.]

———. *Des lignes isothermes et la distribution de la chaleur sur le globe.* Paris: V.H. Perronneau, 1817. [English: "On Isothermal Lines, and the Distribution of Heat over the Globe," *Edinburgh Philosophical Journal* 3–5 (1820–21, nos. 5–9): 1–20, 256–74, 23–37, 262–81, and 28–39.]

———. *Essai géognostique sur le gisement des roches dans les deux hemispheres.* Paris: F.G. Levrault, 1823.

———. *Essai politique sur le royaume de la Nouvelle-Espagne avec un atlas physique et géographique, fondé sur des observations astronomiques, de mesures trigonométriques et des nivellemens barométriques.* 5 vols (octavo). Paris: F. Schoell, 1808–11.

———. "Evaluation numérique de la population du nouveau continent, considéré sous les rapports de la difference des cultes, des races et des idioms. Éxtrait d'une letter addressee à M. Charles Coquerel." *Révue Protestante* (Jan. 1825): 81–85.

———. *Vues des Cordillères et Monumens des Peuples Indigènes de l'Amérique.* Folio. Paris: F. Schoell, 1810–13.

Humboldt, Alexander von, and Aimé Bonpland. *Essai sur la géographie des plantes: Accompagné d'un tableau physique des régions équinoxiales, fondé sur des mesures exécutées, depuis le dixième degré de latitude boréale jusqu'au dixième degré de latitude australe, pendant les années 1799, 1800, 1801, 1802 et 1803.* Paris: Chez Levrault, Schoell et compagnie, libraires, 1805.

———. *Voyage aux regions équinoxiales du nouveau continent, fait en 1799, 1800, 1801, 1802, 1803, et 1804.* Quarto: 3 vols., 1814–31. Vol. 1, Paris: F. Schoell; vol. 2, Paris: N. Maze; vol. 3, Paris: Gide Fils. Octavo: 13 vols., 1816–31. Vols. 1–4, Paris: Libraria Graeco-Latino-Allemande; vols. 5–8, Paris: N. Maze; vols. 9–13, Paris: Smith/Gide.

Humboldt, Alexander von, Aimé Bonpland, and Karl Sigismund Kunth. *Nova genera et species plantarum: Quas in peregrinatione orbis novi collegerunt, descripserunt, partim adumbraverunt Amat. Bonpland et Alex. de Humboldt.* 7 vols. Paris: Librariae Graeco-Latino-Germanicae, 1815–25.

Humboldt, Alexander von, P. A. Latreille, and Georges Cuvier. *Recueil d'observations de zoologie et d'anatomie comparée: Faites dans l'océan Atlantique, dans l'intérieur du nouveau continent et dans la mer du sud, pendant les années 1799, 1800, 1801, 1802 et 1803.* 2 vols. Paris: Levrault, Schoell, 1812.

Humboldt, Alexander von, and Jabbo Oltmanns. *Recueil d'observations astronomiques, d'opérations trigonométriques et de mesures barométriques, faites pendant le cours d'un voyage aux régions équinoxiales du nouveau continent, depuis 1799 jusqu'en 1803, par Alexandre de Humboldt, redigée et calculées, d'après les tables les plus exactes, par Jabbo Oltmanns.* 2 vols. Paris: F. Schoell, 1810.

Humboldt, Alexander von, and Heinrich Karl Wilhelm Berghaus. *Briefwechsel Alexander von Humboldt's mit Heinrich Berghaus aus den Jahren 1825 bis 1858.* 3 vols. Leipzig: Hermann Costenoble, 1863.

Hüne, Albert. *Vollständige historisch-philosophische Darstellung aller Veränderungen des Negersclavenhandels, von dessen Ursprunge an bis zu seiner gänzlichen Aufhebung.* 2 vols. Göttingen: J.F. Roewer, 1820.

Huskisson, William. *The Speeches of the Right Honourable William Huskisson: With a Biographical Memoir Supplied to the Editor from Authentic Sources.* 3 vols. London: J. Murray, 1831. [These are summaries rather than the actual speeches.]

Jameson, Robert Francis. *Letters from the Havana, during the Year 1820 Containing an Account of the Present State of the Island of Cuba, and Observations on the Slave Trade.* London: Printed for John Miller, 1821. [See also the French translation by Huber above.]

Jeffrey, Francis, ed. "Art. VI. Present State of the Spanish Colonies, Including a Particular Report of Hispaniola, or the Spanish Part of St Domingo. By William Walton Junior. Secretary to the Expedition which Captured the City of Santo Domingo from the French, and Resident British Agent There. Longman & Co. London, 1810." *The Edinburgh Review or Critical Journal* 16 (Nov. 1810–Feb. 1811): 372–81.

Juan, Jorge, Antonio de Ulloa, and Garcilaso de la Vega. *Relación histórica del viaje a la América Meridional: Hecho de orden de S. Mag. para medir algunos grados de meridiano terrestre, y venir por ellos en conocimiento de la verdadera figura, y magnitud de la tierra, con otras varias observaciones astronómicas y físicas.* 2 vols. Madrid: Antonio Marín, 1748. [Humboldt uses the title in French: *Voyage historique de l'Amérique méridionale fait par ordre du roi d'Espagne par don George Juan, commandeur d'Aliaga dans l'ordre de Malthe, et commandant de la compagnie des gentils-hommes grandes de la marine, et par don Antoine de Ulloa, lieutenant de la même compagnie, tous deux capitaines de haut-bord de l'Armée navale du roi d'Espagne, membres des Sociétés royales de Londres & de Berlin, & correspondans de l'Académie des sciences de Paris. Ouvrage orné des figures, plans et cartes necessaires. Et qui contient une histoire des Yncas du Pérou, et les observations astronomiques & physiques, faites pour déterminer la figure & la grandeur de la terre* (Paris: Charles-Antoine Jombert, 1752). English: Antonio de Ulloa and John Adams, *A Voyage to South America* (New York: Knopf, 1964).]

Juarros y Montúfar, Juan Domingo. *Compendio de la historia de la ciudad de Guatemala.* Vols. 1 and 2. Guatemala: Beteta, 1810–18. [Humboldt gives the publication year as 1809. English: *A Statistical History of the Kingdom of Guatemala in Spanish America,* trans. John Baily (London: Printed for John Hearne, 1823).]

King, Rufus. *The True Advocate of Free Soil [the Substance of Two Speeches on the Missouri Bill Delivered in the Senate of the United States].* Newark: Printed by Steam Power Press, 1819. *See also* [Shaw, Samuel.]

Kirwan, Richard. *An Estimate of the Temperature of Different Latitudes.* London, 1787.

Kotzebue, Otto von, and Adalbert von Chamisso, Ivan Fedorovich Kruzenshtern, Johann Caspar Horner, and Johann Friedrich Eschscholtz. *Entdeckungsreise in die*

Südsee und nach der Berings-Strasse zur Erforschung einer nordöstlichen Durchfahrt. Unternommen in den Jahren 1815, 1816, 1817 und 1818 auf Kosten Sr. erlaucht des Herrn Reichs-Kanzlers Grafen Rumanzoff auf dem Schiffe Rurick unter dem Befehle des Lieutenants der Russisch-Kaiserlichen Marine Otto von Kotzebue. 3 vols. Wien: Kaulfuss & Krammer, [1821] 1825.

La Bastide, Martin de. *Mémoire sur un nouveau passage de la mer du Nord à la mer du Sud.* Paris: Didot fils, 1791. [Humboldt reverses South and North in the title.]

Lacépède, Comte de (Bernard Germain Étienne de La Ville sur Illon). *Histoire naturelle des poissons.* Paris: Plassau, 1797 (l'an IV–XI). [The publication years correspond to the French revolutionary calendar. *See also* Lacépède and Georges Louis Leclerc Buffon, *Histoire naturelle des poissons* (Paris: P. Didot, 1798–1803).]

Lachenaie, Thomas Luc Augustin Hapel. "A New Process for Claying Sugars." *Journal of Natural Philosophy, Chemistry, and the Arts* (by William Nicholson, London; concluded from vol. 2, 1798) 3 (1802): 41–48.

La Condamine, Charles Marie de. *Relation abrégée d'un voyage fait dans l'intérieur de l'Amérique Méridionale. Depuis la côte de la Mer du Sud, jusqu'aux côtes du Brésil & de la Guiane, en descendant la rivière des Amazones.* Paris: Veuve Pissot, 1745. [Other editions: Maestricht: J. E. Dufour & P. Roux, 1778; *Voyage sur l'Amazone* (Paris: La Découverte, 2000). This book has never been translated into English.]

Lacroix, Francois Joseph Pamphile de. *Mémoires pour servir à l'histoire de la révolution de Saint-Domingue: Avec une carte nouvelle de l'île et un plan topographique de la Crête-a-Pierrot.* Paris: Chez Pillet Aine, 1819. [A recent edition of this work is Pierre Pluchon's *La Révolution de Haïti* (Paris: Karthala, 1995).]

Laet, Joannes de. *Nieuvve wereldt, ofte, Beschrijvinghe van West-Indien wt veelderhande schriften ende aen-teeckeninghen van verscheyden natien.* Leyden: Isaack Elzeviet, 1625.

———. *Novus orbis seu descriptionis Indiae Occidentalis libri XVIII Novis tabulis geographicis et variis animantium plantarum fructuumque iconibus illustrati.* Leiden: Elzevirios, 1633. [Humboldt mentions the year 1633, which suggests that he used the first and expanded Latin translation. The book was first published in Dutch in 1625.]

Lalande, Joseph-Jérôme Lefrançais de. *Histoire céleste française, contenant les observations faites par plusieurs astronomes français.* Paris: Imprimerie de la république, 1801. [In *Cosmos,* Humboldt mentions Heinrich Christian Schumacher and Francis Baily's edition *A Catalogue of Those Stars in the Histoire Céleste Française of Jérôme Delalande, for Which Tables of Reduction to the Epoch 1800 Have Been Published by Professor Schumacher* (London: R. and J.E. Taylor, 1847).]

Lambert, Aylmer Bourke. *A Description of the Genus Cinchona, Comprehending the Various Species of Vegetables from which the Peruvian and Other Barks of a Similar Quality are Taken. Illustrated by Figures of All the Species Hitherto Discovered. To which is Prefixed Professor Vahl's Dissertation on this Genus, Read Before the Society of Natural History at Copenhagen. Also a Description, Accompanied by Figures, of a New Genus Named Hyænanche: or, Hyæna Poison.* London: B. and J. White, 1797. [Originally published in 1797, the book was reissued in 1821 and dedicated to Humboldt. Humboldt appears to have quoted from the more recent edition, to which new findings from a number of naturalists, including Humboldt himself, had been added.]

Larpent, George Gerard de Hochepied. *On Protection to* [sic] *West-India Sugar.* London: J.M. Richardson, 1823.

Lasègue, Antoine, and Benjamin Delessert. *Musée botanique de M. Benjamin Delessert. Notices sur les collections de plantes et la bibliothèque qui le composent, contenant en outre des documents sur les principaux herbiers d'Europe et l'exposé des voyages entrepris dans l'intérêt de la botanique.* Paris: Librairie de Fortin, Masson et Cie, 1845. [Republished in 1970.]

Laurent de Lionne, Pierre-Joseph. *Discours prononcé à la séance publique de l'Académie des sciences, belles-lettres et arts d'Amiens, le 25 août 1776, par M. Laurent de Lionne, directeur des canaux de Picardie et de la Somme, sur l'utilité de ces canaux.* Paris: Impr. de Cailleau, 1781.

Laurenti, Josephus Nicolaus. *Specimen medicum, exhibens synopsin reptilium emendatam cum experimentis circa venena et antidota reptilium Austriacorum.* Vienna: Typ. Joan. Thom. nob. de Trattnern, 1768.

Ledru, André P., and Charles S. Sonnini. *Voyage aux îles de Ténériffe, la Trinité, Saint-Thomas, Sainte-Croix et Porto-Ricco: Exécuté par ordre du gouvernement français: Depuis le 30 septembre 1796 jusqu'au 7 juin 1798, sous la direction du capitaine Baudin, pour faire des recherches et des collections relatives à l'histoire naturelle; contenant des observations sur le climat, le sol, la population, l'agriculture, les productions de ces îles, le caractère, les mœurs et le commerce de leurs habitants.* 2 vols. Paris: Bertrand, 1810.

León Pinelo, Antonio de. *Epitome de la Biblioteca Oriental i Occidental, nautica i geografica.* 1629.

Léon y Gama, Antonio. *Descripción octográfica del eclipse del sol, el 24 de Junio de 1778, dedicada al sr. Joaquín Velázquez de León.* Mexico, 1778.

Lindenau, Bernhard August von. *See* Zach's articles in the *Monatliche Correspondence.*

Liverpool East India Association. *Report of a Committee of the Liverpool East India Association Appointed to Take into Consideration the Restrictions on the East India Trade, Presented to the Association at a General Meeting, 9th May 1822, and Ordered to Be Printed.* Liverpool: J. Smith, 1822. [A table on p. 40 is similar to but not identical with Humboldt's table immediately preceding his East India Association reference. Humboldt's table possibly comes from the 1825 *London Price Current.*]

Livingston, Andrew. "Remarks on Cape Antonio, the S.W. of Cuba, and the Isle of Pine." In *The Columbian Navigator*, by John Purdy, 175–76. 1824.

López Gómez, Antonio. *Historia natural y política de la isla de Cuba.* Unpublished, 1794. [According to Carlos Manuel Trelles's *Ensayo de bibliografía cubana de los siglos XVII y XVIII* (1907), 129–30, the book was commissioned by the Count of Mopox. It was sent to Spain in 1800, but it had disappeared by 1812.]

Luyando, José de, Antonio Rodríguez, Fernando Selma, and Pedro Manuel Gangoiti.

Tablas lineales para resolver los problemas del pilotage astronómico con exactitud y facilidad. Madrid: [Imprenta Real,] 1803.

Lyell, Charles. "VIII.—On a Recent Formation of Freshwater Limestone on Forfarshire, and on Some Recent Deposits of Freshwater Marl; with a Comparison of Recent with Ancient Freshwater Formations; and an Appendix on the Gyrogonite of Seed-Vessel of Chara." *Transactions of the Geological Society of London* (Second Series; London: Richard Tylor) 2 (1829): 73–96. [Read on December 17, 1824, and on January 7, 1825.]

Lyon, George Francis. *A Narrative of Travels in Northern Africa in the Years 1818, 19, and 20; Accompained by Geographical Notices of Soudan, and of the Courses of the Niger. With a Chart of the Routes, and a Variety of Coloured Plates, Illustrative of the Costumes of the Several Natives of Northern Africa.* London: J. Murray, 1821.

Macaulay, Zachary. *Negro Slavery, or, A View of Some of the More Prominent Features of That State of Society as It Exists in the United States of America and in the Colonies of the West Indies, Especially in Jamaica.* London: Hatchard and Son, 1823. [Reissued in 1824. No author is mentioned in the 1823 edition. The book was published by the Society for the Mitigation and Gradual Abolition of Slavery throughout the British Dominions, or, for short, the Anti-Slavery Society. Macaulay published the *Anti-Slavery Reporter,* which was this organization's voice.]

MacCallum, Pierre Franc. *Travels in Trinidad during the Months of February, March, and April, 1803 in a Series of Letters, Addressed to a Member of the Imperial Parliament of Great Britain. Illustrated with a Map of the Island.* Liverpool: Printed for the author by W. Jones, 1805.

MacCulloch, John. "Art. I. On Limestone of Clunie, in Perthshire, with Remarks on Trap and Serpentine." *Edinburgh Journal of Science, Exhibiting a View of the Progress of Discovery in Natural Philosophy, Chemistry, Mineralogy, Geology, Botany, Zoology, Comparative Anatomy, Practical Mechanics, Geography, Navigation, Statistics, Antiquities, and the Fine and Useful Arts* (Edinburgh: William Blackwood; London: T. Cadell) 1 (April–Oct. 1824): 1–16.

Maistre, Joseph Marie. *Les Soirées de Saint-Pétersbourg: ou, entretiens sur le gouvernement temporel de la Providence: Suivis d'un traité sur les sacrifices.* Paris: Librairie Grecque, 1821.

Malte-Brun, Conrad, and J.-J. N. Huot. *Précis de la géographie universelle: Ou déscription de toutes les parties du monde sur un plan nouveau, d'après les grandes divisions naturelles du globe; précédée de l'histoire de la géographie chez les peuples anciens et modernes, et d'une théorie générale de la géographie mathématique, physique et politique.* Rev. and augmented ed. 12 vols. Paris: André Aimé, 1831–37. [Original eight-volume edition published in 1810 by Buisson in Paris. The first English edition saw print between 1822 and 1833: *Universal Geography, or, a Description of All Parts of the World, on a New Plan, According to the Great Natural Divisions of the Globe: Accompanied with Analytical, Synoptical, and Elementary Tables*, 9 vols. (London: Printed for Adam Black; and Longman, Hurst, Rees, Orme, and Brown.]

Malthus, Thomas Robert. *An Essay on the Principle of Population; or, A View of its*

Past and Present Effects on Human Happiness: With an Inquiry Into Our Prospects Respecting the Future Removal or Mitigation of the Evils which It Occasions. 6th ed. London: J. Murray, [1798] 1826.

Maraschini, Pietro. *Sulle formazioni delle rocce del Vicentino saggio geologico.* Padova: tipogr. della Minerva, [1822] 1824. [Humboldt owned Maraschini's *Observations géognostiques sur quelques localités du Vincentin* (Paris, 1822).]

Marieta, Juan. "Libro Treze de la vida del Santo padre fray Luys Bertram Valenciano, de la Orden de Santo Domingo." In *De la Historia Eclesiástica de España, que trata de la vida de Santo Domingo, fundador de la Orden de Predicadores, y de San Vicente Ferrer, y otros Santos naturales de España de las misma Orden,* by Padre Fray Juan Marieta, segunda parte, 165–200. Cuenca: Juan Masselin, 1594/96. [Humboldt refers to "Libro VII," which would be the last book of the first volume. Book 13 deals with Fray Bertram.]

Mason, Charles. *Lunar Tables in Longitude and Latitude According to the Newtonian Laws of Gravity.* London: s.n., 1787.

Masse, Étienne-Michel. *L'îsle* [sic] *de Cuba et la Havane, ou, histoire, topographie, statistique, mœurs, usages, commerce et situation politique de cette colonie: D'Après un journal écrit sur les lieux.* Paris: Lebégue, 1825.

Mathison, Gilbert Farquhar. *Narrative of a Visit to Brazil, Chile, Peru, and the Sandwich Islands, during the Years 1821 and 1822, with Miscellaneous Remarks on the Past and Present State, and Political Prospects of Those Countries.* London: Charles Knight, 1825.

———. *Short Review of the Reports of the African Institution, and of the Controversy with Dr. Thorpe, with Some Reasons against the Registry of Slaves in the British Colonies.* London: WM. Stockdale and J. Asperne, 1816.

Medina, Pedro de. *Regimento de Navegacion. En que se contienen las reglas, declaraciones y avisos del libro del arte de navegar.* Valladolid: Francisco Fernandez de Cordova, 1545. [English edition: John Frampton, *The arte of nauigation: vvherein is contained, all the rules, declarations, secrets, and aduises, which for good nauigation are necessary, and ought to be knowen and practised, and are very profitable for all kinde of mariners. Made by M. Peter de Medina, and dedicated to the right excellent and renowned lord, Don Philip, Prince of Spaine, and of both Siciles. With the declination of the sunne newly corrected* (London: Thomas Dawson, 1595). *See also* Nicolai, Nicolas de.]

Melish, John. *A Geographical Description of the United States with the Contiguous Countries, Including Mexico and the West Indies, Intended as an Accompaniment to Melish's Map of These Countries.* Philadelphia: published by the author, 1816. Rev. ed.: Philadelphia: J. & J. Harper, 1826. [Humboldt owned the map.]

———. *Travels through the United States of America, in the Years 1806 & 1807, and 1809, 1810, & 1811; Including an Account of Passages betwixt America & Britain, and Travels through Various Parts of Britain, Ireland, and Canada. With Corrections and Improvements till 1815. Illustrated by Coloured Maps and Plans. With an Appendix, Containing a Letter from Clements Burleigh, Esq. to Irish Emigrants Removing to America, and Hints by the Shamroc Society, New-York, to Emigrants from Europe.* Philadelphia: printed for the author, 1818; London: Reprinted for G. Cowie and Co.; Belfast: Reprinted by Jos. Smith.

Memminger, Johann Daniel Georg von, August Friedrich von Pauly, Rudolph Moser, Christoph Friedrich von Stälin, Paul Friedrich Stälin, Eduard von Paulus, and Julius von Hartmann. *Beschreibung des Königreichs Württemberg.* Stuttgart: J.G. Cotta, 1824–85/86.

Memorias de la Real Sociedad Económica de la Habana. See Real Sociedad Económica de Amigos del País (Cuba).

Mendinueta y Múzquiz, Pedro de. *Relaciones de los vireyes del nuevo reino de Granada: Ahora Estados Unidos de Venezuela, Estados Unidos de Colombia y Ecuador.* [A manuscript when Humboldt referenced it as "*Relacion del Gobierno,*" this report is part of José Antonio García y García's collection edited by Ignacio Gómez and published by Hallet & Breen (New York, 1869), see 409–576.]

Mendoza y Ríos, Juan José de. *Colección de tablas para varios usos de la navegación.* 1800. [A later edition is Mendoza y Ríos, José Sánchez y Cerquero, and Juan José Martínez de Espinosa y Tacón, *Colección completa de tablas para los usos de la navegación y astronomía náutica* (Madrid: s.n., 1863)]

Mercurio Americano. [This is likely the *Mercurio Peruano* that Humboldt also cites in *Vues des Cordillères.* It was published by the Sociedad Académica de Amantes de Lima. See also Jean-Pierre Clément, *El Mercurio peruano, 1790–1795,* in Textos y estudios coloniales y de la independencia, vols. 2–3 (Frankfurt: Vervuert, 1997).]

Moniteur 22 March. *Moniteur universel.* Paris, 1811.

[Monroe, James.] *Message from the President of the United States at the Commencement of the Second Session of the Fifteenth Congress: November 17, 1818. Read, and Committed to a Committee of the Whole House, on the State of the Union. Printed by Order of the Senate of the United States.* Washington: Edward De Krafft, 1818.

———. *Message from the President of the United States: Transmitting, in Pursuance of a Resolution of the House of Representatives, of the 30th Jan. Last, Communications from the Agents of the United States with the Governments South of the U. States Which Have Declared Their Independence: And the Communications from the Agents of Such Governments in the United States with the Secretary of State, as Tend to Shew the Political Condition of Their Governments, and the State of War between Them and Spain: March 8, 1822: Read, and Referred to the Committee on Foreign Relations.* Washington, 1822.

Moreau de Jonnès, Alexandre. *Le Commerce au dix-neuvième siècle: État actuel de ses transactions dans les principales contrées des deux hémisphères, causes et effets de son agrandissement et sa décadence, et moyens d'accroître et de consolider la prospérité agricole, industrielle, coloniale, et commerciale de la France.* 2 vols. Paris: L'auteur, Migneret, 1825.

———. *Histoire physique des Antilles françaises; savoir: La Martinique et les îles de la Guadeloupe.* Paris: Impr. de Migneret, 1822.

Morse, Sidney Edward, and Jedidiah Morse. *A New System of Modern Geography, or, A View of the Present State of the World: With an Appendix, Containing Statistical Tables of the Population, Commerce, Revenue, Expenditure, Debt, and Various In-*

stitutions of the United States, and General Views of Europe and the World. Boston: G. Clark; New Haven, Conn.: Howe & Spalding, 1822.

Muñoz, Juan Bautista. *Historia del Nuevo-Mundo.* 6 vols. Madrid: La viuda de Ibarra, 1793. [First edition in 1794.]

Navarrete, Manuel. *Informe de Don Manuel Navarrete, tesorero de la real hacienda en Cumaná, sobre el estanco de tabaco y los medios de su abolición total.* Manuscript.

Navarrete, Martín Fernández de, trans. "Un Discours sur les progrès, et l'état actuel de l'hydrographie en Espagne, par Don Louis Marie de Salazar, ci-devant intendant général de marine." *Correspondance astronomique, géographique, hydrographique et statistique du Baron de Zach* 13, no. 1 (1825): 50–74.

Navarro y Noriega, Fernando. *Catálogo de los curatos y misiones que tiene la Nueva España en cada una de sus diócesis, o sea la división eclesiástica de este reino, que ha sacado de las constancias mas auténticas y modernas.* Mexico: Impreso en casa de Arizpe, 1813.

———. *Memoria sobre la población del reino de Nueva España.* México: En la oficina de D. Juan Bautista de Arizpe, 1820. New ed., Llanes: J. Porrúa, 1954. [Humboldt gives the year as 1814; he also refers to this text as *Rispuesta de un Mexicano al no. 200, del Universal.*]

———. *Semanario político y literario.* Mexico: Mariano de Zúñiga y Ontiveros, 1820–22. [Some of the *Semanarios*' issues are in the Alexander von Humboldt Papers, Biblioteka Jagiellónska, Krakow, Poland.]

New Monthly Magazine. "Empire of Hayti." *New Monthly Magazine and Literary Journal* (London: Henry Colburn and Co.) (Feb. 1825): 69.

Nicolai, Nicolas de, trans. *L'art de naviguer de m. Pierre de Medine, espagnol: Contenant toutes les reigles, secrets, et enseignemens necessaires à la bonne navigation.* Lyon: Guillaume Rouille, 1569. [First edition in 1554; includes the map Nouveau Monde inserted between pages 48–49.]

Norris, Robert. *Memoirs of the Reign of Bossa Ahádee, King of Dahomy an Inland Country of Guiney: To which are Added, the Author's Journey to Abomey, the Capital; and a Short Account of the African Slave Trade.* London: Printed for W. Lowndes, 1789.

North American Review. "Art. VI.—1. *Réflexions politiques sur quelques ouvrages et journaux français concernant Hayti,* par M. le Baron de Vastey, sécretaire du roi, chevalier de l'Ordre royal et militaire de Saint Henry. Précepteur de son altesse royale Monsieur le Prince Royale d'Hayti &c. À sans-souci, de l'Imprimerie royale, 1817, 8 vo. pp. xx. 206. . . . *North American Review* 12, no. 30 (January 1821): 112–34. [The review essay is in English; no author given. Humboldt refers to p. 116, which contains a lengthy quote from De Vastey's writings].

Nuix, Juan. *Riflessioni imparziali sopra l'umanità degli Spagnuoli nell'Indie contro i pretesi filosofi e politici: Per servire di lume alle storie de' Signori Raynal e Robertson.* Venezia: F. Pezzana, 1780. [A Spanish translation by Juan Pedro y Ulloa appeared in 1782.]

Nuix, Juan, and Pedro Varela y Ulloa. *Reflexiones imparciales sobre la humanidad de los españoles en las indias, contra los pretendidos filósofos y políticos: Para ilustrar las historias de MM. Raynal y Robertson.* Madrid: J. Ibarra, 1782.

Núñez, Ignacio Benito. *An Account, Historical, Political, and Statistical, of the United Provinces of Rio de la Plata: With an Appendix, Concerning the Usurpation of Monte Video by the Portuguese and Brazilian Governments.* London: Printed for R. Ackermann, 1825. [Spanish: *Noticias históricas, políticas y estadísticas, de las provincias Unidas del Rio de la Plata.*]

Omalius d'Halloy, Jean Baptiste Julien d'. "Observations sur un essai de carte géologique de la France, des Pays-Bas, et de quelques contrées voisines." *Annales des Mines* 7 (1822): 353–76.

Page, Pierre François. *Discours historique sur la cause des désastres de la partie française de Saint-Domingue, établi sur pièces probantes, déposées au Comité colonial, & dans les archives de la commission de l'Assemblée coloniale de Saint-Domingue, auprès de la Convention nationale. Adresse à la Convention nationale, expositive des droits de la partie françoise de Saint-Domingue, & du seul moyen convenable de procéder au jugement des incendiaires de cette contrée.* Paris: L. Potier, 1793.

———. *Traité d'économie politique et de commerce des colonies.* 2 vols. Paris: Brochot, 1801. [Publication year is IX, according to the French revolutionary calendar. Vol. 6 does not exist; Humboldt's reference may be to a section or chapter. A German edition dates from 1802.]

Papel periódico de la Habana. See *El sesquicentenario del Papel periódico de la Habana.*

O Patriota, journal litterario, politico, mercantil, etc. do Rio de Janeiro. No. 1–6, janeiro-junho, 1813. Rio de Janeiro: Regia, 1813.

El Patriota Americano. See Castillo, José del, and Bergaño Simon.

Pereira Lago, Antonio Bernardino. "Observações meterologicas, feitas na cidade de São Luiz do Maranhão pelo coronel do corpo de engenheiros Antonio Bernardino Pereira Lago." *Annaes das sciencias das artes e das letras; por huma sociedade de portugueses residentes em Paris* 16 (April, fourth year): 55–80. Paris: A. Bobée, 1822.

Periódico de la Sociedad económica de Guatemala. See Valle, José Cecilio del.

Perrins, R. *See* Carlisle, Anthony.

———. "VIII. Temperaturen des Meerwassers, beobachtet auf einer Reise von England nach Bombay im Jahre 1800." *Annalen der Physik* 19 (by Ludwig Wilhelm Gilbert, 1805): 447–49.

Perronet, Jean-Rodolphe. *Description des projets et de la construction des ponts de Neuilly, de Mantes, d'Orléans & autres du projet du canal de Bourgogne, pour la communication des deux mers par Dijon, et de celui de la conduite des eaux de l'Yvette et de Bièvre à Paris: En soixante-sept planches.* Paris: Imprimerie royale, 1782.

Peuchet, Jacques. *Statistique élémentaire de la France contenant les principes de cette science et leur application à l'analyse de la richesse, des forces et de la puissance de l'empire français.* Paris: Chez Gilbert et co., 1805.

Phipps, John. *A Guide to the Commerce of Bengal for the Use of Merchants, Ship Owners, Commanders, Officers, Pursers and Others, Resorting to the East Indies, but Particularly of Those Connected with the Shipping and Commerce of Calcutta: Containing a View of the Shipping, and External Commerce of Bengal: In Three Parts: I. Port Rules and Regulations. II. Shipping. III. Commercial Statements, &c.: With *a Copious Appendix, Comprehending Various Details and Statements, Relative to the Shipping and Commerce of Countries Connected with British India and China: The Whole Compiled from Authentic Sources.* Calcutta: s.n., 1823.

Pitkin, Timothy. *A Statistical View of the Commerce of the United States of America: Its Connection with Agriculture and Manufactures: And an Account of the Public Debt, Revenues, and Expenditures of the United States. With a Brief Review of the Trade, Agriculture, and Manufactures of the Colonies, Previous to Their Independence. Accompanied with Tables, Illustrative of the Principles and Objects of the Work.* Hartford: Charles Hosmer, 1816.

Poinsett, Joel Roberts. *Notes on Mexico, Made in the Autumn of 1822.* Philadelphia: H. C. Carey and I. Lea, 1824. [In a note on the first page, Poinsett points out that he drew heavily on Humboldt.]

Pombo, José Ignacio. "Noticias varias sobre las quinas oficinales, sus especies, virtudes, usos, comercio, acopios y su descripción botánica" 1814. [This text is illustrated with watercolor drawings of the quina plant. Humboldt owned a copy of it; he nonetheless gave the year 1817 as the publication date. The *Yale University Collection of Latin American Manuscripts* (Andean Collection, 26–27, reel 8, box 16, folder 164) identifies this text as part of Pombo and Martín Melgar's *Noticias de las riquezas, minas, plantas y otros productos naturales de la América del Sur* (1783–94). But there is a significant discrepancy in publication years; Pombo might not have started working on the *Noticias varias* until after the turn of the century.]

———. *Representación que dirigió Don José Ignacio Pombo al consulado de Cartagena el 14 de Mayo 1807 sobre el reconocimiento del Atrato, Zenú y San Juan.* Fol. 38 (manuscript). 1807.

Pons, François Joseph de. *Voyage à la partie orientale de la terre-ferme, dans l'Amérique méridionale, fait pendant les années 1801, 1802, 1803 et 1804 contenant la description de la capitainerie générale de Caracas, composée des provinces de Vénézuéla, Maracaibo, Varinas, la Guiane Espagnole, Cumaná, et de l'île de la Marguerite et renfermant tout ce qui à rapport à la découverte, à la conquéte, à la topographie, à la législation, au commerce, aux finances, aux habitans et aux productions de ces provinces, avec un aperçu des mœurs et usages des Espagnols et des Indiens sauvages et civilisés.* Paris: Colnet, 1806. [Translated as *A Voyage to the Eastern Part of Terra Firma, or the Spanish Main, in South-America, during the Years 1801, 1802, 1803, and 1804 Containing a Description of the Territory under the Jurisdiction of the Captain General of Caraccas, Composed of the Provinces of Venezuela, Maracaibo, Varinas, Spanish Guiana, Cumana, and the Island of Margaretta: And Embracing Every Thing Relative to the Discovery, Conquest, Topography, Legislation, Commerce, Finance, Inhabitants and Productions of the Provinces: Together with a View of the Manners and Customs of the Spaniards, and the Savage as Well as Civilized Indians,* 3 vols. (New York: I. Riley and Co., 1806).]

Pope Alexander VI. *Inter Caetera.* Papal Bull of May 4, 1493. [*See* Frances Gardiner Davenport, *European Treaties Bearing on the History of the United States and Its Dependencies* (Washington, D.C.: Carnegie Institution, 1917).]

Powell, John. *Statistical Illustrations of the Territorial Extent and Population, Com-*

merce Taxation, Consumption, Insolvency, Pauperism, and Crime, of the British Empire. London: J. Miller, 1825.

Pownall, Thomas. *The Administration of the Colonies.* London: J. Wilkie, 1764.

Prince's London Price Current. London: Printed for the Proprietor by S. Wright, 1825. [Humboldt attributes the *London Price Current* to Nichols, a connection we have been unable to verify. *Prince's London Price Current* was available from 1796 until at least 1820. A *London Price Current* existed from 1822 until 1826 and was continued until the early 1830s as the *British and Foreign Price Current.*]

Prony, Gaspar Clair François Marie Riche de. *Description hydrographique et historique des marais pontins.* Paris: Firmin Didot, père et fils, 1822.

Purdy, John. *The Columbian Navigator: Or, Sailing Directory for the American Coasts and the West-Indies: Volume the Second.* London: R. H. Laurie, [1817] 1824. [Reprinted and revised repeatedly since at least 1817; 2 vols. in 1823 and 1824, respectively.]

Puységur, Antoine Hyacinth Anne de Chastenet, Count of, and François Cyprien Antoine Lieudé de Sepmanville. *A Treatise upon the Navigation of St. Domingo with Sailing Directions, for the Whole Extent of Its Coasts, Channels, Bays and Harbours: Undertaken by Order of the King.* Trans. Charles de Monmonier. Baltimore: Printed for the translator, by W. Pechin, 1802. [First published in French in 1787–88 under the title *Le pilote de l'Isle de Saint-Domingue et des débouquemens de cette isle, comprenant une carte de l'Isle de Saint-Domingue et une carte mens, depuis la Caye d'Argent jusqu'à la partie ouest du places des Isles Lucayes des débouque*(Paris: De l'Imprimerie Royale).]

Raynal, Abbé Guillaume Thomas François. *Histoire philosophique et politique des établissemens et du commerce des européens dans les deux Indes.* 3rd ed. 10 vols. Neuchatel: Libraires associés, 1783. [The most extensive discussion of New World slavery is in volume 6, book 11.]

Real Sociedad Económica de Amigos del País (Cuba). *Memorias de la Real Sociedad Económica de la Habana.* Habana: Oficina del Gobierno y de la Real Sociedad Patriótica por S.M., 1793–1845.

Redactor general. Guatemala: Imprenta de la Union, 1825–26.

Redhead, Joseph James Thomas. *Memoria sobre la dilatación progresiva del aire atmosférico.* Buenos Aires: Imp. de la Independencia, 1819. [Available at the Biblioteca Nacional in Madrid, Spain.]

Relación del estado del nuevo reino de Granada que hace el Arzobispo Obispo de Cordova. See Caballero y Góngora, Antonio.

Report of a Committee of the Liverpool East India Association. See Liverpool East India Association.

Representación dirigida por el Real Consulado de la Habana al ministro de hacienda en 10 de Julio de 1799. Reprinted in vol. 5 of *Historia de la esclavitud desde los tiempos más remotos hasta nuestros días,* by José Antonio Saco, 131–49. 2nd ed. Havana: Editorial ALFA, [1879] 1944.

*Representación del 16 de Agosto 1811, que por encargo del Ayuntamiento, consulado y sociedad patriótica de la Habana, hizo el Alférez mayor de aquella ciudad, y se elevó *a las Cortes por los expresados cuerpos.* [*See* Arango, *Documentos sobre el tráfico y esclavitud de negros,* 1–86.]

Restrepo, José Manuel. "Memoria que el secretario de estado y del despacho del interior presentó al congreso de Colombia sobre los negocios de su departamento: Año de 1823–13." Biblioteca Nacional de Colombia, Miscelánea no. 1,160, 1823.

Rigada, Juan Enrique de la. MSS. "Adjunta la nueva derrota de la canal vieja al norte de la isla de Cuba." In *Correspondencia y expedientes de la Administración de Correos de La Coruña* (1796–98). Seville, Spain: Archivo General de Indias, 1793.

Robertson, William. *The History of America.* London: W. Strahan; T. Cadell; Edinburgh: J. Balfour; Dublin: Whitestone et al., 1777.

Robinson, James H. and John Robertson. *Journal of an Expedition 1400 Miles up the Orinoco and 300 up the Arauca with an Account of the Country, the Manners of the People, Military Operations, &c.* London: Printed for Black, Young, and Young, 1822.

Robinson, William Davis. *Memoirs of the Mexican Revolution Including a Narrative of the Expedition of General Xavier Mina; to which Are Annexed Some Observations on the Practicability of Opening a Commerce between the Pacific and Atlantic Oceans, through the Mexican Isthmus, in the Province of Oaxaca, and at the Lake of Nicaragua; and the Vast Importance of Such Commerce to the Civilized World.* 2 vols. London: Lackington, Hughes, Harding, Mavor, & Lepard, 1821. [There is also a one-volume edition from 1820.]

Robredo, Antonio. "Eclipse del sol observado en la Havana el 21 de Febrero de 1803." Unpublished manuscript. [In the Humboldt Papers at the Staatsuniversität Berlin.]

Rodet, D.-L. *Du Commerce extérieur et de la question d'un entrepôt a Paris.* Paris: Renard, 1825.

Rodman, Samuel. "Of the Comparative Temperature of the Air and of the Water of the Ocean, as Indicated by Fahrenheit's Thermometer, at 12 O'clock at Noon, in a Voyage from Philadelphia to Batavia." *Philadelphia Medical Museum* (1800). [Dr. Thomas Horsfield published Rodman's table in his "An Account of a Voyage to Batavia, in the Year 1800," *Philadelphia Medical Museum* 1 (1805): 75–82.]

Rogers, Woodes. *A Cruising Voyage Round the World First to the South-Seas, thence to the East-Indies, and Homewards by the Cape of Good Hope. Begun in 1708, and Finish'd in 1711. Containing a Journal of all the Remarkable Transactions; Particularly, of the Taking of Puna and Guiaquil, of the Acapulco Ship, and Other Prizes; an Account of Alexander Selkirk's Living Alone Four Years and Four Months in an Island; and a Brief Description of Several Countries in our Course Noted for Trade, Especially in the South-Sea. With Maps of All the Coast, from the Best Spanish Manuscript Draughts. And an Introduction Relating to the South-Sea Trade.* London: Printed for A. Bell and B. Lintot, 1712.

Romay Chacón, Tomás, and Juan González. *Elogios fúnebres del excelentísimo señor D. Luis de las Casas y Aragorri, teniente general de los rles. Exércitos, gobernador político y militar de la ciudad de Cádiz, y capitán general honorario de provincia: Hechos y publicados por la Real Sociedad Económica de la Havana, y por el tribunal del consulado de la misma ciudad.* Sociedad Económica de Amigos del País (Cuba). Havana: Impr. de la Capitanía general, 1802.

Rossel, Élisabeth Paul Édouard de. "Practical Treatise on Finding the Latitude and Longitude." *Monthly Review or Literary Journal, Enlarged: From May to August, Inclusive* (Ralph Griffiths and George Edward Griffiths) 83 (1817): 412–17.

Roxburgh, William. "An Account of the Hindoo Method of Cultivating the Sugar Cane, and Manufacturing the Sugar and Jagary in the Rajahmundry Circar; Interspersed with such Remarks, as Tend to Point Out the Great Benefit that Might be Expected from Increasing this Branch of Agriculture, and Improving the Quality of the Sugar; also the Process Observed, by the Natives of the Ganjam District, in Making the Sugars of Barrampore." *Oriental Repertory* (London: G. Bigg) 2 (1793): 497–514, ed. Alexander Dalrymple (East India Company). [The *Oriental Repertory* was published between 1791 and 1797. *See also* Tennant, William.]

Sabine, Edward. *An Account of Experiments to Determine the Figure of the Earth by Means of the Pendulum Vibrating Seconds in Different Latitudes, as Well as on Various Other Subjects of Philosophical Inquiry.* London: John Murray, 1825. [Humboldt twice mentions 1826 as the year of publication.]

Saint-Cricq, Pierre Laurent Barthélemy. *Documens relatifs au commerce des nouveaux états de l'Amérique: Communiqués par le Bureau de commerce et des colonies, aux principales chambres de commerce de France.* Paris: Libraire de l'industrie, et chez Renard, 1825.

Saint-Hilaire, Auguste de. *Aperçu d'un voyage dans l'intérieur du Brésil, la province cisplatine et les missions dites du Paraguay.* 2 vols. Paris: A. Belin, 1823.

Saint-Hilaire, Étienne Geoffrey de. "Notice sur une nouvelle espèce de crocodile de l'Amérique." *Annales du museum national d'histoire naturelle* (Paris: Levrault) 2 (1803): 53–56.

———. "Observations anatomiques sur le crocodile du Nil." *Annales du museum national d'histoire naturelle* (Paris: Levrault) 2 (1803): 37–52.

Seba, Albertus. *Locupletissimi rerum naturalium thesauri accurata descriptio, et iconibus artificiosissimis expressio, per universam physiees historiam. Opus, cui, in hoc rerum genere, nullum par exstitit.* Amsterdam: Janssonio-Waesbergios, 1734–65.

Sedano, Diego José de. *Dos preguntas que el capitán general de la isla de Cuba, como presidente de la junta económica de aquel consulado, hizo a don Diego José de Sedano, y su respuesta sobre la decadencia del ramo de azúcar.* London: C. Wood, 1812. [Humboldt owned this book.]

Seijas y Lobera, Francisco de. *Descripción geográfica y derrotero de la región austral Magallánica.* Madrid, 1690.

———. *Teatro naval hidrográfico, de los flujos y reflujos y de las corrientes de los mares, estrechos, archipiélagos y pasajes aquales del mundo, y de las diferencias de las variaciones de la ahuja de marear y efectos de la luna con los vientos generales y particulares que reinan et las cuatro regiones marítimas del Orbe. Dirigido al Rey nuestro Señor.* Madrid: Antonio de Zafra, 1688.

El sesquicentenario del Papel periódico de la Habana, 1790—24 de octubre—1940. Habana: Municipio de la Habana. [This item is available in the Alexander von Humboldt Papers, Biblioteka Jagiellónska, Krakow, Poland.]

Seybert, Adam. *Statistical Annals: Embracing Views of the Population, Commerce, Navigation, Fisheries, Public Lands, Post-Office Establishment, Revenues, Mint,*

Military and Naval Establishments, Expenditures, Public Debt and Sinking Fund, of the United States of America: Founded on Official Documents: Commencing on the Fourth of March Seventeen Hundred and Eighty-Nine and Ending on the Twentieth of April Eighteen Hundred and Eighteen. Philadelphia: Thomas Dobson & Son, 1818. [Humboldt provides references in French. There was an 1820 French translation: *Annales statistiques des États-Unis.*]

[Shaw, Samuel.] "Art. VIII—1. *Substance of Two Speeches, Delivered in the Senate of the United States, on the Subject of the Missouri Bill. By the Hon. Rufus King of New York.* New York, 1819. pp. 32. 2. *A Charge Delivered [by Mr. Justice Story] to the Grand Juries of the Circuit Court at October Term 1819, in Boston, and at November Term 1819, at Providence, and Published at their Unanimous Request.* Boston, 1819. pp. 8." *North American Review* 10, no. 1, issue 26 (Jan. 1820): 137–68. [No author given. The facsimile has a handwritten "Samuel Shaw" at the end of the article.]

Spain. *Real cédula de Su Majestad concediendo libertad para el comercio de negros con las islas de Cuba, Santo Domingo, Puerto Rico y provincia de Caracas, a españoles y extranjeros, baxo las reglas que se expresan.* Madrid: en la imprenta de la vuida de Ibarra, 1789. [Issued on February 28, 1789. A copy is available at the Archivo Histórico Nacional (AHN) in Madrid.]

Steetz, William. *Instruction nautique sur les passages à l'île de Cuba et au golfe du Mexique, par le canal de la Providence et le grand banc de Bahama.* Paris: Béchet, 1825. [Humboldt owned Steetz's *Cartes pour l'instruction sur les passages à l'île de Cuba* (1824).]

Stewart, John. *A View of the Past and Present State of the Island of Jamaica with Remarks on the Moral and Physical Conditions of the Slaves, and on the Abolition of Slavery in the Colonies.* Edinburgh: Oliver & Boyd; London: G. & W.B. Whittaker, 1823. [Humboldt gives the year as 1825.]

Strabo. *Géographie de Strabon. Traduite du Grec en Français par MM. Coray et Letronne, avec de notes par M. Gosselin, tome V.* Paris: Imprimerie Royale, 1819. [The work consists of seventeen books; book 16 is on Assyria, Babylonia, Syria, and Arabia. Humboldt owned several editions in Latin and French and a German edition from 1831–34.]

Substance of the Debate of the House of Commons on the 15th May, 1823, on a Motion for the Mitigation and Gradual Abolition of Slavery throughout the British Dominions. With a Preface and Appendixes, Containing Facts and Reasoning Illustrative of Colonial Bondage. London: Ellerton and Henderson.

Substance of a Debate in the House of Commons, on the 22nd of May 1823 on the Motion of Mr. W. W. Whitmore, "That a Select Committee be Appointed, to Inquire into the Duties Payable on East and West India Sugar." England: s.n. [Each contribution to the debate starts with p. 1. Section H: Mr. Wilberforce, *East and West India Sugar.*]

Sucinta noticia de la situación presente de la Habana. Unpublished, 1800. *See* Valle Hernández, Antonio del.

Tennant, William. "Sect. I. V. The Hindoo Method of Cultivating the Sugar Cane."

Indian Recreations; Consisting of Thoughts on the Effects of the British Government on the State of India, Accompanied with Hints Concerning the Means of Improving the Condition of the Natives of That Country (Edinburgh: University Press; sold by Longman, Hurst, Rees & Orme) 2 ([1803] 1804): 31–44. [Tennant quotes amply from Roxburgh; Humboldt's footnote only references Roxburgh, not Beckford. "Calcutta 1810" appears to be a separate source; *see* Fleming, John.]

Thomson, Thomas. *Chimie générale. Systéme de chimie de M. Th. Thomson; traduit de l'anglais sur la derniére éd. de 1807 par M. J. Riffault; précedé d'une introduction de M. C. L. Berthollet, vol. 2.* Paris: Bernard, 1809. [English version on which the French is based: Thomas Thomson, *A System of Chemistry in Five Volumes* (Edinburgh: John Brown for Bell & Bradfute, 1807).]

Tooke, Thomas. *Thoughts and Details on the High and Low Prices of the Thirty Years, from 1793 to 1822.* 2nd ed. London: J. Murray, [1823] 1824.

Tuckey, James Hingston. *Maritime Geography and Statistics, or a Description of the Ocean and Its Coasts, Maritime Commerce, Navigation etc.: In 4 Volumes.* London: Black, 1815. [Older sources give his middle name as "Hingston" rather than "Kingston."]

Unanue y Pavón, Hipólito José. *Guía Política, Eclesiástica y Militar del Vireynato del Perú.* Lima: Sociedad Académica de Amantes del País de Lima, 1793.

Urquinaona y Pardo, Pedro de. *Relación documentada del orígen y progresos del trastorno de las provincias de Venezuela hasta la exoneración del capitán general Don Domingo Monteverde, hecha en el mes de diciembre de 1813 por la guarnición de la plaza de Puerto Cabello.* Madrid: En la imprenta nueva, 1820. [Reissued as *Memorias de Urquinaona (comisionado de la regencia española para la pacificación del Nuevo reino de Granada)* (Madrid: Editorial América, 1917).]

Valle, José Cecilio del, et al., ed. *Periódico de la Sociedad Económica de Guatemala.* Guatemala: Beteta, 1815–16.

———, ed. *El Amigo de la Patria.* Guatemala: Manuel Arévalo, 1820–21. [Twenty-four issues, no. 1 from October 16, 1820; no. 24 from April 30, 1821.]

Valle Hernández, Antonio del. "Sucinta noticia de la situación presente de esta colonia, 1800." *Boletín del Archivo Nacional* (Havana: s.n.) 17 (1918): 171–90.

Vargas Ponce, José de. *Descripciones de las islas Pithiusas y Baleares. Viajeros y filósofos.* Madrid: Vda. de Ibarra, 1787. [Repr. Palma de Mallorca: J.J. de Olañeta, 1983.]

Vargas Ponce, José, Antonio de Córdoba, Dionisio Alcalá Galiano, and Alejandro Belmonte. *Relación del último viage al Estrecho de Magallanes: de la fragata de S.M. Santa María de la Cabeza en los años de 1785 y 1786: Extracto de todos los anteriores desde su descubrimiento impresos y MSS. y noticia de los habitantes, suelo, clima y producciones del estrecho.* Madrid: Por la Viuda de Ibarra, Hijos y Compañía, 1788. [English: *A Voyage of Discovery to the Strait of Magellan: with an account of the manners and customs of the inhabitants; and of the natural productions of Patagonia* (London: Printed for R. Phillips, 1820/1973).]

Vastey, Pompée-Valentin de. *Réflexions politiques sur quelques ouvrages et journaux français concernant Hayti.* Sans-Souci, Haiti: L'Imprimerie royale, 1817.

———. *Réflexions politiques sur quelques ouvrages et journaux français concernant Haïti, essai sur les causes de la révolution et des guerres civiles d'Haïti.* 1819. [English translation from 1823: *An Essay on the Causes of the Revolution and Civil Wars of Hayti, Being a Sequel to the Political Remarks upon Certain French Publications and Journals Concerning Hayti* (reprint, New York: Negro Universities Press, 1969). See also *North American Review.*]

———. *Réflexions sur une lettre de Mazères, ex-colon français, adressée à m. J .C. L. Sismonde de Sismondi, sur les noirs et les blancs, la civilisation de l'Afrique, le royaume d'Hayti, etc.* Cap-Henry, Haiti: P. Roux, Imprimeur du Roi, 1816. [English: *Reflexions on the Blacks and Whites: Remarks upon a Letter Addressed by M. Mazères, a French Ex-Colonist, to J. C. L. Sismonde De Sismondi, Containing Observations on the Blacks and Whites, the Civilization of Africa, the Kingdom of Hayti, Etc.* (London: Sold by J. Hatchard, 1817).]

———. *Le Système colonial dévoilé [1. ptie.].* Cap-Henry, Haiti: P. Roux, Imprimeur du Roi, 1814.

Vega, Garcilaso de la, El Inca. *Historia general del Perú trata el descubrimiento de él; y como lo ganaron los Españoles. Las guerras civiles que hubo entre Piçarros, y Almagros, sobre la partija de la tierra. Castigo y levantamieto de tiranos: y otros sucesos particulares que en la historia se contienen.* Córdoba: La viuda de Andrés Barrera, 1617.

———. *Primera parte de los Comentarios reales: que tratan, del origen de los Incas, reyes, que fueron del Perú, de su idolatría, leyes y gobierno, en paz y en guerra: de sus vidas, y conquistas, y de todo lo que fue aquel imperio y su república, antes que los españoles pasaran a él.* Lisbon: Pedro Crasbeeck, 1609. [For an English version see *Royal Commentaries of the Incas and General History of Peru,* 2 vols., trans. Harvold V. Livermore (Austin: University of Texas Press, 1966).]

Veloso de Oliveira, Antonio Rodrigues. "A Igreja do Brasil, ou informação para servir de base à divisam dos Bispados projectada no anno de 1819, com a Statistica da populaçam do Brasil, considerada en tõdas as diferentes classes na conformidade dos mapas das respectivas provincias e número de seus habitantes." *Anais Fluminenses de Ciências, Artes e Literatura* 1 (1822): 57–115. [Also published as "A Igreja do Brasil," *Revista do Instituto Histórico e Geográfico Brasileiro* 29 (1866): 159–99.]

Wafer, Lionel. *A New Voyage and Description of the Isthmus of America: Illustrated with Several Copper-Plates.* 3rd ed. London: Knapton, [1699] 1729. [This edition was included in the second volume of William Dampier's 4th edition of his voyages; the two men traveled together.]

Walker, Alexander. *Colombia: Being a Geographical, Statistical, Agricultural, Commercial, and Political Account of That Country.* 2 vols. London: Baldwin, Cradock & Joy, 1822.

Walsh, Robert. *An Appeal from the Judgments of Great Britain Respecting the United States of America, First Part, Containing an Historical Outline of Their Merits and Wrongs as Colonies.* Philadelphia: Mitchell, Ames, and White, 1819.

Walton Jr., William. *Present State of the Spanish Colonies, Including a Particular Re-*

port of Hispaniola, or the Spanish Part of St Domingo. 8 vols. London: Longman & Co., 1810. [*See also* Jeffery, Francis.]

———. "The Isthmus of Panama. Considered as Affording a Passage to Unite the Pacific with the Atlantic or Western Ocean; and This Passage (if Practicable) Compared with the Land Route, over the Buenos Ayres Plaines." *Colonial Journal* 5 (March 1817): 86–101. [Written as a letter to the editor in two parts.]

———. "Route to the Pacific. Concluded from Page 101." *Colonial Journal* 6 (July 1817): 331–44. [Second part of the letter to the editor.]

Warden, David Bailie. *A Statistical, Political, and Historical Account of the United States of North America; from the Period of Their First Colonization to the Present Day.* 3 vols. Edinburgh: A. Constable and Co., 1819. [French translation: *Description statistique, historique et politique des États-Unis de l'Amérique septentrionale: Depuis l'époque des premiers établissemens jusqu'à nos jours,* 5 vols. Paris: Rey et Gravier, 1820. Humboldt referred to the French edition.]

Wells, William Charles. *An Essay on Dew, and Several Appearances Connected with It.* 2nd ed. London: Taylor and Hessey, 1815. [Humboldt owned an 1817 French version of Wells's essay in a translation by A. J. Tordeux.]

Wilberforce, William. *An Appeal to the Religion, Justice, and Humanity of the Inhabitants of the British Empire, in Behalf of the Negro Slave in the West Indies.* London: J. Hatchard, 1823.

Wilmot Horton, Robert. *Letters on the Necessity of a Prompt Extinction of British Colonial Slavery: Chiefly Addressed to the More Influential Classes. "Whatsoever Thy Hand Findeth to Do, Do It with All Thy Might." To Which Are Added, Thoughts on Compensation.* London: Sold by Hatchard and Son, 1826. [Humboldt refers to him as "Mr. Wilmot," his birth name. It is unclear what documents Humboldt received from Wilmot Horton. This text is an example of Wilmot Horton's writings.]

Wilson, Patrick. "An Account of the Most Extraordinary Degree of Cold at Glasgow in January Last, Together with Some New Experiments and Observations on the Comparative Temperature of Hoarfrost and the Air Near to It, Made at the Macfarlane Observatory, Belonging to the College. In a Letter from Patrick Wilson, M.A., to the Rev. Nevil Maskeleyne, D.D., F.R.S. and Astronomer-Royal." *Transactions of the Royal Society of London* (1780). A shorter paper on the topic followed in 1781, entitled "Further Experiments on Cold."]

Wollaston, William Hyde. "On the Water of the Mediterranean." *Philosophical Transactions of the Royal Society of London* 119 (1829): 29–31.

Zach, Franz Xaver. "XXIV. Über einige Breitenbestimmungen und über den daraus folgenden mittlern Werth eines Breiten-Grades am Aequator." *Monatliche Correspondenz zur Beförderung der Erd- und Himmels-Kunde* (Oct. 1807): 302–29. [It is possible that Humboldt mixed up the page numbers of this article with those of an article in the December issue; see below.]

———."XXX. Beyträge zur Kenntnis des Flächen-Inhaltes einiger Länder und Inseln in und ausser Europa." *Monatliche Correspondenz zur Beförderung der Erd- und Himmels-Kunde* (Nov 1807): 398–423.

———."XL. Beyträge zur Kenntniss des Flächen-Inhaltes einiger Länder und Inseln in und ausser Europa. (Fortsetz. zum Novbr. Heft, S. 425.) Angabe von sechs spanischen See-Charten, auf die sich die Berechnungen gründen. Flächen-Inhalt des Mexicanischen Meerbusens und der Antillen. Kurze Übersicht ihrer ersten Entdeckung. Geographische Ortsbestimmung." *Monatliche Correspondenz zur Beförderung der Erd- und Himmels-Kunde* (Dec. 1807): 516–38. [Part two continued from the November 1807 issue. Humboldt attributes this article to Lindenau in Zach's journal. Zach founded the journal in 1800; it existed until 1813. Starting in 1807, Lindenau edited it. No author given in the journal.]

———. "Discussion. Sur la longitude du morne (Morro) de Porto-Rico. Par Don Joseph Sánchez Cerquero, directeur provisoire de l'observatoire royal dans la ville de S. Fernando (île de Léon)." *Correspondance astronomique, géographique, hydrographique et statistique du Baron de Zach* 13, no. 1 (1825): 114–25.

———. "Du discours et des mémoires publiés par la direction hydrographique à Madrid, sur les fondemens qui l'ont guidé dans la construction des cartes marines publiées dans ce dépot depuis l'an 1797." *Correspondance astronomique, géographique, hydrographique et statistique du Baron de Zach* 13, no. 1 (1825): 49–74.

Zamora y Coronado, José María. "Reglamento y arancel de gobierno en la captura de esclavos prófugos ó cimarrones, aprobado por S. M. en real órden de 20 de diciembre de 1796: Reformado por real cédula de 7 de febrero de 1820, y real órden de 22 de abril de 1822; y que por acuerdo de la junta consular de 26 de mayo de 1824 que mandó imprimir y circular." In vol. 2 of *Biblioteca de legislación ultramarina en forma de diccionario alfabético. Contiene: El Texto de todas las leyes vigentes de Indias, y extractadas las de algún uso, aunque sólo sea para recuerdo histórico: Las dos ordenanzas de intendentes de 1786 y 1803; el código de comercio de 1829, con su ley de enjuiciamiento; las reales cédulas, ordenes, reglamentos y demás disposiciones legislativas aplicadas a cada ramo, desde 1680 hasta el día, en que se comprenden las del registro ultramariano con oportunas reformas, y agregación de acordados de audiencias, bandos y autos generales de gobierno; y cuantas noticias y datos estadísticos se han creído convenientes para marcar el progreso sucesivo de las posesiones ultramarinas, y a los fines de su más acertado régimen administrativo, mejoras que admita, y represión de abusos,* 217–20. N.p.: Alegria y Charlain, 1844.

Zayas y Chacón, José de, and José Domingo Benítez. *Reclamación hecha por los representantes de la isla de Cuba contra la ley de aranceles sobre las restricciones que ésta impone al comercio de dicha isla.* Madrid: J. del Collado, 1821.

Humboldt's Maps

Arrowsmith, Aaron. *Abrégé de la nouvelle géographie universelle, physique, politique et historique.* Ed. William Guthrie. Paris: Chez Hyacinthe Langlois, 1811. [Most of Arrowsmith's maps (41 of them) are at the David Rumsey Map Collection and are accessible online.]

———. *Chart of Rio De La Plata, from an Actual Survey.* London: A. Arrowsmith, 1798.

———. *Chart of the West Coast of South America from the Gulf of Dulce to Point Aguja.* London: A. Arrowsmith, 1809.
———. *A Map of Part of the Viceroyalty of Buenos Ayres 1806.* London: Published by A. Arrowsmith, 1806.
———. *Sketch of the Isthmus of Darien.* London: A. Arrowsmith, 1803.
———. *South America.* Philadelphia: John Conrad & Co. 1804.
———. *Survey of the Harbour of Panama by the Sloops Descubierta & Atrevida by Order of His Catholic Majecty in the Year 1791.* London: A. Arrowsmith, 1800.
Arrowsmith, Aaron, and J. Bapt. Bourguignon d'Anville. *Nouvel atlas universal-portatif de géographie ancienne & moderne, contenant 39 cartes dont 33 pour la partie moderne.* 1817.
Arrowsmith, Aaron, Jean-Nicolas Buache, and Jean-Gabriel Dentu. *Amérique Méridionale. Amérique Septentrionale. Indes Occidentales.* Paris: Dentu, 1805.
Arrowsmith, Aaron, et al. *A New and Elegant General Atlas Comprising All the New Discoveries, to the Present Time: Containing Sixty-Three Maps.* Boston: Thomas & Andrews, 1812.
———. *Géographie moderne rédigée sur un nouveau plan, ou description historique, politique, civile et naturelle des empires, royaumes, états et leurs colonies. . . .* Paris: Dentu, 1804.
Barcaíztegui, Ventura de, and José del Río. *Carta esférica que comprende la Cista Meridional, parte setentrional e islas adyacentes de la isla de Cuba, desde la Punta Maisí hasta Cabo San Antonio.* Madrid: Dirección Hidrográfica, 1821. Inserte siete vistas de costa—(Con la mitad occidental de la costa norte de la isla en blanco).—Longitud occidental de Cádiz.—Grabada por Félipe Cardano.—Caducó con la publicación de la carta de 1866. [This map became obsolete in 1866. *See* Roussin, Albin Reine, and Alexandre P. Givry.]
Bauzá, Felipe. *Carta esférica que comprende las costas del seno Mexicano: Construida de orden del Rey en el Depósitio hidrográfico de Marina.* Madrid: s.n., 1799. Longitud occidental de Cádiz.—El 26 de noviembre de 1803 se insertó una corrección con una nota sobre la situación de los Caymanes Chicos, que fue observada por el C. de N. Ciriaco de Zeballos en 1802, y las del Triángulo, el Obispo y las Arcas en 1803.—Corregida en 1805.—Dibujada por Felipe Bauzá. Grabada por Selma. [Corrected in 1803 and in 1805. The map could be based on sketches that José Antonio de Evia made during José Espinosa's directorship. *See also* Selma, *Carta particular*; Suárez, *Mapa y plano del Seno Mexicano.*]
Bauzá, Felipe, and Juan de Lángara. *Carta esférica que comprende una parte de las islas Antillas: Las de Puerto Rico, Santo Domingo, Jamaica y Cuba con los bancos y canales adyacentes.* Madrid: Depósito Hidrográfico de Marina, 1815.
Bauzá, Felipe, Juan de Lángara, and Fernando Selma Moreno. *Carta esférica que comprende una parte de las Islas Antillas, las de Puerto Rico, Santo Domingo, Jamaica y Cuba, con los bancos y canales adyacentes.* (Construida de orden del Rey en el Dep. H. de Marina por disposición del Exmo. Sr. Don Juan de Lángara, secretario de estado y del despacho universal de ella. Año de 1799.) Madrid: Depósito Hidrográfico de Marina, 1799. Longitud occidental de Cádiz.—Corregida en 1804,

con siete vistas de costa.—Corregida en 1815, sin vistas.—Sondas en brazas de a dos varas castellanas.—Delineada por Felipe Bauzá. Grabada por Fernando Selma. [Corrected in 1804 and 1815.]

Buchenröder, Friedrich von, and Jakob Turpin. *Carte générale & particulière de la Colonie d'Essequebe & Demerarie: Située dans la Guiane, en Amérique*. Amsterdam: Wouter Brave, 1798. [At Stanford University's Rare Book Collection.]

Carey, Henry Charles, Isaac Lea, Emmanuel-Auguste-Dieudonné Las Cases, and C. V. Lavoisne. *A Complete Historical, Chronological, and Geographical American Atlas, Being a Guide to the History of North and South America, and the West Indies: Exhibiting an Accurate Account of the Discovery, Settlement, and Progress of Their Various Kingdoms, States, Provinces, &c. Together with the Wars, Celebrated Battles, and Remarkable Events, to the Year 1822. According to the Plan of Le Sage's Atlas, and Intended as a Companion to Lavoisne's Improvement of that Celebrated Work*. Philadelphia, PA: H. C. Carey & I. Lea, 1822.

Carta de la Isla de Santo-Domingo y parte oriental del Canal Viejo de Bahama. See Ferrer, *Carta esférica*; Selma, *Nueva carta del Canal de Bahama.*

Carta de una parte de las Islas Antillas. 1799 (corr. 1805). *See* Bauzá and Lángara; Bauzá, Lángara, and Moreno; Churruca y Elorza.

Carte réduite des côtes de la Guyane comprise entre les bouches de la Rivière des Amazones et celles du Maroni d'après les plans levés par les ingénieurs français et portugais. Publiée par ordre du roi sous le ministère de son excellence le Vicomte Dubouchage, etc. Au dépôt de la Marine, 1817. [Humboldt refers to this map as *Carte des côtes de la Guyane*.]

Cevallos, Ciriaco, and F. Cabarcos. *Plano de la costa y sonda de Campeche*. Nuevamente levantado y enmendado pr. el capitán de navío Dn. Ciriaco Cevallos, que para el uso de esta academia se copio de su original, Ferrol 8 de sept. de 1807. F. Cabarcos, 1807.

Churruca y Elorza, Cosme Damián de, Felipe Bauzá, Fernando Selma, and Joaquín F. Fidalgo. *Carta esférica de las islas Antillas con parte de la costa del continente de América*. Madrid: Dirección Hidrográfica, 1802. Corr. 1825.

D'Anville, Jean Baptiste Bourguignon. *Nouvel atlas de la Chine de la tartarie chinoise, et du Thibet*. La Haye: Scheurleer, 1737.

———. *Orbis veteribus notus auspiciis serenissimi principis Ludovici Philippi Aurelianorum ducis publici jurs factus*. 1763.

D'Anville, Jean Baptiste Bourguignon, Charles-Marie de La Condamine, and Pedro Vicente Maldonado. *Carte de la province de Quito au Perou: dressée sur les observations astromoniques, mesures géographiques, journaux de route et mémoires de M. de la Condamine et sur ceux de Don Pedro Maldonado*. France: s.n., 1751.

Espinosa y Tello, José. *Carta esférica del mar de las Antillas y de las costas de tierra firme, desde las Bocas del Río Orinoco hasta el Golfo de Honduras*. London, 1810. *See also* Churruca y Elorza; Selma Moreno, *Carta esférica*.

Ferrer, Juan. *Carta esférica que comprende los desemboques al norte de la isla de Sto. Domingo y la parte oriental del canal viejo de Bahama*. Madrid: Dirección Hidrográfico, 1802.

Humboldt, Alexander von. *Atlas géographique et physique des régions équinoxiales du nouveau continent, fondé sur des observations astronomiques, des mesures trigonométriques et des nivellemens barométriques.* Paris: Librairie de Gide, 1814.

Hydrographic Office (Madrid). *Cuarta hoja que comprende las costas de la Provincia de Cartagena, Golfo de Darién y Provincia de Porto Velo con el Golfo de Panamá y Archipiélago de las Perlas.* Madrid, 1817. Longitud occidental de Cádiz—Sondas en brazas de seis pies de Burgos—Delineada por Clemente Noguera a partir de los borradores originales, grabada por Felipe Cardano y escrita por Juan Morata—Adicionada en 1897.

———. *Portulano de la América Septentrional: Dividido en cuatro partes.* Madrid: Dirección de Trabajos Hidrográficos, 1809. [Corr. in 1818 and 1825. At the Library of Congress].

Jefferys, Thomas, Francisco Mathias Celi, and Thomas Bully. *Plan of the Town and Harbor of San Juan de Puerto Rico.* 1768. [This map was probably published in London.]

La Cruz Cano y Olmedilla, Juan de. *Mapa Geográfico de América Meridional, Dispuesto y Grabado por D. Juan de la Cruz Cano y Olmedilla. Año de 1775.* 1776. [Humboldt owned La Cruz's *Mapa geográfico de América meridional, the original large map mounted on clith, Madrid 1775.* This valuable and very rare map has Humboldt's autograph note, stating that it cost him 15 Napoleons: "Al. Humboldt. L'original de la Cruz dont les planches ont été détruit à Madrid. Acheté à Paris 15 Napoléons."]

Lapie, Pierre, and Antoine François Tardieu. *A Map of South America—Carte de l'Amérique Méridionale.* Paris, 1814.

Lemaur, Félix, and Francisco Lemaur. *Plano de la Habana y Batabanó hasta el Cabo de S. Antonio.* 1804.

López de Vargas Machuca, Juan. *Carta marítima del Reino de Tierra Firme u Castilla del oro: comprende el istmo y provincia de Panamá; las Provincias de Veragua, Darién y Biruquete.* Madrid, 1785.

———. *Carta plana de la provincia de Caracas ó Venezuela.* Madrid, 1787.

López de Vargas Machuca, Tomás. *Mapa geográfico del reino de tierra firme y sus provincias de Veragua y Darién, nuevamente dado a luz y corregido.* 1802.

Malaspina, Alejandro. *Carta de la costa occidental de América desde el 7° N[orte] hasta las 9° S[ur].* Mardid: Depósito Hidrográfico, 1800.

———. *Carte réduite de la côte occidentale de l'Amérique: depuis 9° de latitude nord jusqu'à 7° de latitude sud dressée d'après les observations astronomiques et nautiques faites en 1791 par divers officiers de la Marine d'Espagne; gravé par Vicq; écrit par Besançon.* Paris: Département de la marine et des colonies au Dépôt-général de la marine, 1821.

Mar de las Antillas. 1809. *See* Selma Moreno, *Carta esférica.*

Mar de las Antillas. 1815. [Humboldt is likely referring to Bauzá 1815. But that map only shows a part of the Antilles rather than the Antillean Sea.] For maps of the Antillean Sea, see Espinosa; Selma, *Carta esférica.*

Mascaró, Manuel Augustin. *Mapa geográfico de una gran parte de la América septen-*

trional: comprehendido, entre los 19° y 41° de latitud norte y los 255° y 289° de longitud oriental de Tenerife, en el que se contienen las provincias de la Antigua, y Nueva California, las de Sonora, Nueva Vizcaya, Nueva Mexico, Cohahuila y Texas, erigidas en Capitanía general por S.M. en el año de 1779. 1779.

Mayne, Anthony de. *A General Chart of the West Indies and Gulf of Mexico, Describing the Gulf and Windward Passages, Coasts of Florida, Louisiana, and Mexico, Bay of Honduras, and Musqito Shore; Likewise the Coast of the Spanish Main to the Mouths of the Orinoco. Drawn chiefly from the Surveys of Anthony de Mayne, RN, the new Spanish Charts, &c. Inset: Virgin Islands. 1 inch to about 33 miles.* Published by the Admiralty, 1824. [This map is at the British National Archives.]

Misiones de Mojos de la Compania de Jesús. 1713. See *Relación de las misiones.*

Neele, Robert Madie, and Frank Howard. *Colombia tomado de Humboldt y de varias otras autoridades recientes.* London: Baldwin Cradock and Jay, 1823.

Omalius d'Halloy, Jean Baptiste Julien d'. *Essai d'une carte géologique de la France, des Pays-Bas et de quelques contrées voisines.* Paris: Gravé par Berthe, 1823.

Purdy, John, and Richard Holmes Laurie. *General Chart of the West-India Islands with the Adjacent Coasts of the Southern Continent: Including the Bay of Yucatan or Honduras.* London: Published by R.H. Laurie, chartseller to the Admiralty, &c. &c., No. 53, Fleet Street, 1823.

Relación de las misiones de los Mojos de la Compañía de Jesús en esta provincia del Perú, del P. Alonso Mejía, 1713. Fol. 175r–222r. [Humboldt references it as "*Misiones de Mojos de la Compañía de Jesús,* 1713." Located at the Archivo histórico de la Compañía de Jesús ARSI in Rome, Italy: Fondos Provincia Peruana, Peru, nos. 1–26, vol. 21, *Historia* (1689–1753).]

Requena, Francisco. *Mapa de una parte de la América meridional en que se manifiestan los paises pertenecientes al Nuevo Reino de Granada y Capitanía General de Caracas de los dominios de nuestro muy augusto soberano que confinan con los establecimientos de S.M. fidelísima para acompañar a la relación sobre las operaciones ejecutadas en la demarcación de límites por la cuarta partida de división.* 1783.

———. *Mapa de una parte del Reino Marañon o de las Amazonas: comprendida entre la boca del Caño de Avatiparana y la villa de Ega o Tefé.* 1788. [Both maps are available at the Library of Congress digital collection.]

Río, José del. *Carta de la costa de la isla de Cuba. Trazado por D.J. del Río en 1804.* Madrid: Dirección Hidrográfica, [1804]. [*See also* Barcaíztegui and del Río.]

———. *Carta esférica que comprende desde el Río Guaurabo: Hasta Boca-Grande en la parte meridional de la Isla de Cuba levantada en 1803.* Madrid: Dirección Hidrográfica, 1805.

Roussin, Albin Reine, Louis Claude Desaulses de Freycinet, and Alexandre P. Givry. *Carte de la Province de Rio de Janeiro: Rédigée d'après un manuscrit portugais inédit et les cartes nautiques de MM. Roussin et Givry.* 1824.

Roussin, Albin Reine, and Alexandre P. Givry. [*Map of Cuba's South Coast.*] French naval Depot, 1824. [The Spanish Depository did not issue a Cuba map between 1822 and 1831. In 1821, however, they issued a map of the southern coast of Cuba, which can be traced back to drawings by Ventura de Barcaíztegui in 1793 and to José del Río's drawings of 1804. *See* Barcaíztegui and del Río.]

Selma Moreno, Fernando, and Juan Morata. *Carta esférica del Mar de las Antillas y de las costas de tierra firme, desde la Isla de Trinidad, hasta el Golfo de Honduras.* Madrid: Dirección Hidrográfica, 1809.

———. *Carta particular de las costas septentrionales del seno Mexicano que comprende las de Florida Occidental las márgenes de la Luisiana y toda la rivera que sigue por la Bahía de San Bernardo y el río Bravo del Norte hasta la Laguna Madre.* Madrid: s.n., 1807. Tiene representada la derrota que realizó el capitán Ferrer en agosto de 1801.—Longitud occidental de Cádiz.—Sondas en brazas de seis pies de Burgos.—Delineada por Miguel Moreno, grabada por Fernando Selma y escrita por Juan Morata.—Caducó con la publicación de la carta de 1846. [Based on data provided by Captain Ferrer in 1801. The map became obsolete with the publication of the 1846 map. *See* Bauzá, *Carta esférica que comprende las costas del seno Mexicano.*]

———. *Nueva carta del Canal de Bahama que comprende también los de Providencia y Santaren con los bajos, islas y sondas al este y al oeste de la península de la Florida.* Madrid: Dirección Hidrográfica, 1805. Longitud occidental de Cádiz.—Sondas en brazas de a seis pies de Burgos.—Delineada por Miguel Moreno, grabada por Fernando Selma y escrita por Juan Morata.

Seno Mexicano. 1799. Corr. in 1805. *See* Bauzá, *Carta esférica que comprende las costas del seno Mexicano.*

Silva, Pedro de. *Mapa topográfico de la parte más poblada de la isla de Cuba, levantado recientemente por acuerdo y a expensas del real consulado de la misma isla establecido en la Havana y con aprobación del Superior Gobierno. Año 1805.* 1809.

Suárez, José de San Martín. *Mapa y plano del Seno Mexicano. Contodas las costas, de tierra firme e islas de barlovento con sus adyacentes, recopiladas, sus-latitudes y longitudes en el puerto de la Habana con junta de primeros y seguidos pilotos de la esquadra y según el nuevo padrón;* por Dn. Josef de Sn. Martín Suares, teniente de navio de la Real Armada, Ayudante y primer piloto mayor de Derrotas, celebrada por disposición del Exmo Sor. Dn. Josef Solano y Bote, caballero. Delineado en Cádiz por Dn. Josef Dias Portaly. Año de 1787. 1787. [The map is in the Library of Congress digital collection.]

Talledo y Rivera, Vicente. *Mapa corografica de la Provincia de Cartagena de Indias.* [This is likely the *Mapa corográfico del Nuevo Reyno de Granada, del "Golfo de Guayaquil. No. VI. Mompox, enero 13 de 1816. Copiado en 27 de octubre de 1816."* Meridiano de Cartagena. Mapa manuscrito en colores. At the Biblioteca Luis Angel Arango in Bogotá, Colombia.]

Ugarte y Liaño, Tomás. *Carta Esférica del Seno Mexicano y Canales Viejo de Bahama corregida la situación de varios puntos principales des sus navegantes. Por el Capitán de Navío y Comandante del San Lorenzo. Don Tomas de Ugarte y Liaño.* Madrid: Dirección de Hidrográfica, 1794. [The Library of Congress has one photographic copy in the Louis C. Karpinski Collection.]

Varela y Ulloa, José. *Carta Esférica de la Costa de Africa desde Cabo Espartel, a Capo Bojado, è Yslas Adjacentes.* Madrid: Dirección de Hidrografía, 1787. [This map includes the Canary Islands. He appears to have collaborated on his 1776 map of Tenerife with Luis de Arguedas.]

CHRONOLOGY

1769 Friedrich Wilhelm Heinrich Alexander von Humboldt is born in Berlin (September 14).

1779 Death of Humboldt's father, Major Alexander Georg von Humboldt (January 6).

1787–88 Studies at the University of Frankfurt/Oder; returns to Berlin for private tutoring in botany under the tutelage of Karl Ludwig Willdenow.

1790 Travels with Georg Forster, from Mayence via Cologne, Brussels, and Amsterdam to England. Returns via revolutionary Paris, where Humboldt carts sand for the "temple of freedom that is still under construction" (March to July).

1791–92 Studies at the School of Mines at Freiberg in Saxony.

1792 Returns to Berlin. Begins his career as inspector of mining operations in Prussia.

1793 *Florae Fribergensis specimen plantas cryptogamicus praesertim subterraneas exhibens* [Examples of the Flora of Freiburg, especially displaying cryptogamic underground plants].

1794 First visit with Johann Wolfgang von Goethe in Jena.

1795 Scientific travels to northern Italy, Switzerland, and the French Alps.

1796 Death of his mother Elisabeth, née Colomb (November 19). Decides to retire from his public office to focus on the preparations for his planned voyage.

1797 Humboldt's relations with Goethe and Friedrich Schiller deepen. Conducts astronomical studies based on Franz Xavier von Zach's method of geographically determining locations. Travels to Vienna at the end of May, where he prepares intensively for traveling to the tropics (August to October). Excursion to Ödenburg (Sopron) in Hungary.

1797–98 Humboldt in Salzburg: research with Leopold von Buch; numerous excursions.

1798 Leaves Salzburg for Paris. Meets the young physician and botanist Aimé Bonpland. They become friends and travel together to Marseille, where they wait in vain for transport to North Africa. Finally depart Marseille for Spain in December.

1799 Travels through Spain, from Barcelona via Valencia to Madrid. Audience with the Spanish King in Aranjuez and approval of the expedition to the Spanish colonies. Departure for the New World from La Coruña (June 5). Stopover in the Canary Islands (June 25), where Humboldt climbs the volcano Teide. Humboldt and Bonpland arrive in Cumaná on July 16 after a forty-one-day sea voyage.

1800 Travels on the Orinoco (February to July). Travels across the cataracts of Atures and Maipures up to the border of the Portuguese colonies. Return voyage up the Orinoco to Angostura (Ciudad Bolívar); through the Llanos to Nueva Barcelona and Cumaná. Crossing from Nueva Barcelona to Cuba (November 24 to December 19).

1801 First stay in Cuba (until early March), then departure from the port of Trinidad for Cartagena de Indias. Travels to Santa Fé de Bogotá, where the botanist José Celestino Mutis welcomes Humboldt and Bonpland.

1802 In Quito and the Ecuadorian Andes, where Humboldt climbs several volcanoes, among them Pichincha and Chimborazo (January 6 to October 21). Proceeds to Lima and environs; then sea voyage from Callao via Guayaquil to Acapulco.

1803 Arrives in New Spain (Mexico) on March 23. Travels from Acapulco to Mexico City.

1804 Travels from Mexico City to Veracruz, then again to Havana, where Humboldt and Bonpland arrive in April. Second stay in Cuba (April 29 to May 19). Sails from Havana to Philadelphia; meets with President Thomas Jefferson in Monticello. Leaves Philadelphia on July 9 to return to France; arrives in Bordeaux on August 3. Humboldt returns to Paris on August 27 and stays there until March of the following year.

1805 Appointed a member of the Berlin Academy of Sciences. Travels to Italy (March to October) to visit his brother Wilhelm in Rome. Travels to Naples with Joseph-Louis Gay-Lussac and Leopold von Buch; climbs to Vesuvius and conducts comparative studies in volcanology. Returns to Berlin, where he stays until November of 1807. Commences work on the *Voyage aux régions équinoxiales du nouveau continent*.

1807 *Ideen zu einer Geographie der Pflanzen, nebst einem Naturgemälde der Tropenländer*. English: *Essay on Plant Geography* (2009).

1807–27 Humboldt returns to Paris, where he concentrates on his *opus americanum* and on other publications. He also engages in collaborations with French scientists and extends his international network further.

1808 *Ansichten der Natur*. English: *Aspects of Nature* (1849) and *Views of Nature* (1850).

1808–11 *Essai politique sur le royaume de la Nouvelle-Espagne*. English: *Political Essay on the Kingdom of New Spain* (1811).

1810–13 *Vues des Cordillères et Monumens des Peuples Indigènes de l'Amérique*. English: *Researches, Concerning the Institutions & Monuments of the Ancient Inhabitants of America, with Descriptions and Views of Some of the Most Striking Scenes in the Cordilleras* (1813).

1814–31 *Relation historique*, the actual travelogue of the voyage to the Americas. English: *Personal Narrative of Travels to the Equinoctial Regions of the New Continent* (1814–31).

1814–38 *Atlas géographique et physique des régions équinoxiales du nouveau conti-*

nent [Geographical and physical atlas of the equinoctial regions of the new continent].

1817 *Des lignes isothermes et la distribution de la chaleur sur le globe.* English: "On Isothermal Lines, and the Distribution of Heat over the Globe" (1820–21).

1823 *Essai géognostique sur le gisement des roches dans les deux hémisphères.* English: *Geognostical Essay on the Superposition of Rocks in Both Hemispheres* (1823).

1827 Humboldt returns to Berlin, which will remain his permanent residence until the end of his life.

1829 Travels in Russia, to Siberia, and to the largely unknown Central Asia up to the Chinese frontier.

1834–38 *Examen critique de l'histoire de la géographie du nouveau continent* [Critical examination of the historical development of the geographical knowledge about the New World].

1835 Death of his brother Wilhelm von Humboldt (April 8).

1843 *Asie Centrale* [Central Asia].

1845 First volume of *Kosmos* (vol. 2, 1847; vol. 3, 1850; vol. 4, 1858; vol. 5, 1862 [posthumously]). English: *Kosmos: A General Survey of the Physical Phenomena of the Universe* (1845–58).

1848 Humboldt joins the funeral procession for the dead revolutionaries in Berlin.

1859 Alexander von Humboldt dies at his residence in Berlin (May 6).

EDITORIAL NOTE

From today's perspective, it is astonishing that an early nineteenth-century text, to which scholars in the Spanish-, French-, and German-speaking parts of the world have long attributed great significance and influence, has not been available in a complete and reliable English translation until now. Humboldt's text was often radically abridged in their English translations, and his language was homogenized. Translators tacitly converted currencies and other units of measure; they removed italics and foreign-language words and excised sentences, footnotes, and even entire passages; and they also added formal features, such as chapter titles and subsections. All prior English versions of the *Political Essay on the Island of Cuba* date back to the early to mid-nineteenth century; all are incomplete to varying degrees, and all have numerous transcriptional and translational errors.

About the French Editions

Alexander von Humboldt's *Essai politique sur l'île de Cuba*—the *Political Essay on the Island of Cuba*—was originally written in French and appeared in three different versions in 1825 and 1826. The first two versions were part of Humboldt's travelogue or personal travel narrative, known as the *Relation historique*. The third version was a two-volume freestanding edition. The *Relation historique* was published by at least three different publishers in Paris between 1814 and 1831 as part of the *Voyage aux régions équinoxiales du nouveau continent, fait en 1799, 1800, 1801, 1802, 1803, et 1804 par Al. de Humboldt et A. Bonpland.* The actual travelogue, then, is a fairly small part of Humboldt's monumental *Voyage*, a scientific compendium that came to encompass at least thirty volumes.

Different parts of the *Voyage* were also published in additional freestanding editions. Among these was the third version of Humboldt's *Political Essay on the Island of Cuba*, a two-volume edition that appeared in Paris on October 28, 1826, under the title *Essai politique sur l'île de Cuba par Alexandre de Humboldt. Avec une carte et un supplément que renferme des considérations sur la population, la richesse territoriale et le commerce de l'archipel des Antilles et de Colombia.* The publisher was Gide fils, the printer J. Smith. In an "Advertisement" included with this edition, the publisher explained the purpose of this separate edition: "the lively interest that M. de Humboldt's research has inspired about the territorial riches of the island of Cuba and about the population of the Antillean archipelago, compared to those of other regions in America, have led us to make available this knowledge to readers, who do not own the *Relation historique*." Price was a concern as well: depending on the quality of the paper used, the three volumes of the *Relation historique* in quarto format (25 x 34 cm) would have cost either 158 or 234 francs, whereas the two-volume *Essai politique sur l'île de Cuba* sold for 17 francs. (At today's currency values, these prices would roughly correspond to $720, $1,065, and $77.)

Volumes 11 and 12 of the thirteen-volume octavo edition (12.5 x 20 cm) of the *Relation historique* were published on the same date in 1826. These two volumes, along with volume 9, which had already appeared on October 22, 1825, contain the main portions of Alexander von Humboldt's *Political Essay on the Island of Cuba*. The less expensive octavo *Relation historique* was intended for a broader audience; its function was that of a paperback edition in today's market: each volume sold for 7 francs, the set for 90 francs. The two octavo volumes correspond roughly to the final volume 3 of the more expensive quarto edition, which had been published in installments (Livres) between June 27, 1825, and October 4, 1826. In the quarto edition, the *Political Essay on the Island of Cuba* appears in chapter 26 (Livre 9) and chapters 27–28 (Livre 10).

In addition to Humboldt's "Analyse raisonnée de la carte de l'île de Cuba," which was also printed separately in 1826 (besides being included in the quarto and octavo edition and at the beginning of the freestanding edition), the two-volume version of *The Political Essay on the Island of Cuba* includes sections on sugar consumption ("On the Consumption of Sugar in Europe") and temperature ("On the Temperature in Different Parts of the Torrid Zone at Sea Level"), as well as a two-part supplement, in which Humboldt compares Cuba to the rest of what he likes to call insular and continental America ("Venezuela") and contrasts different populations in the Americas ("Population Overview: Considered under the Rubrics of Race, Language, and Religion"). These additions, which comprise most of the second volume, are drawn from other portions of the *Relation historique*. The table below shows the correspondences among the three versions of *Essai politique sur l'île de Cuba*; it is based on the work in Horst Fiedler and Ulrike Leitner's comprehensive bibliography of Alexander von Humboldt's separately published writings (*Bibliographie der selbständig erschienenen Werke*, Berlin, 2000).

Essai politique	2-vol. Ed.	RH	Quarto	Octavo
Analyse raisonnée	1.7–36	Additions		Vol.12: 1^{2}–38^{2}*
			Vol.3:580–88	Vol.13:2–45
Essay	1.1–364	Chap. 28	Vol.3:345–469	Vol.11:176–
	2.5–40			Vol.12:125
Cons. of Sugar	2.40–79	Chap. 28	Vol.3:484–97	Vol.12:160–96
On Temperature	2.79–92	Chap. 28	Vol.3:498–501	Vol.12:199–212
Suppl.: Venezuela	2.93–376	Chap. 26	Vol.3:56–154	Vol.9:136–419
Suppl.: Population	2.377–408	Chap. 27	Vol.3:322–44	Vol.11:122–75

*This superscript refers to a second set of pagination at the end of vol. 12, which begins again with p.1.

Through line-by-line comparisons of three editions of the *Essai politique sur l'île de Cuba* we have found they are not identical, even though Fiedler and Leitner claim that the texts of the quarto and octavo editions are the same and that Humboldt made no revisions even in the final freestanding version. In this, they follow the lead of Hanno Beck, who had published a German translation of the quarto edition in 1992 (*Cuba-Werk* [Darmstadt: Wissenschaftliche Buchgesellschaft]). Irene Prüfer-Leske, the author of a 2001 German translation of Humboldt's text (*Politischer Essay über die Insel Cuba* [Alicante, Spain: ECU]), also asserts that Humboldt revised neither the octavo edition nor the two-volume edition. Taking the freestanding edition as a basis, we compared the texts of all three versions (excepting the additions in vol. 2). To make the textual variations easily visible, we have created a digital facsimile of the basis text. In this multilayered file, which shows all textual variations without interfering with Humboldt's text itself, readers can follow the textual genesis of Humboldt's two-volume edition. The textual comparison file is available at www.press.uchicago.edu/books/humboldt/, along with several tutorial videos explaining its use.

While Humboldt did revise and correct the text of the two-volume edition, he did not, however, adapt his frequent internal references to the *Relation historique*. We have adjusted these references whenever necessary. Humboldt also refers to his other writings with some frequency, notably to the *Essai politique sur le royaume de la Nouvelle-Espagne*, which John Black translated as *Political Essay on the Kingdom of New Spain* in 1811. We have checked all of Humboldt's references to this and other works for accuracy and made corrections in [square brackets] as needed.

In addition to the facsimile of the two-volume edition on the HiE website, to which page numbers in the margins of our edition refer, all thirteen volumes of the first octavo edition of the *Relation historique* are available on Google Books. The website of the French National Library (www.bnf.fr) has facsimiles of all three volumes of the quarto edition of the *Relation historique*, of which Brockhaus issued a historical reprint (edited by Hanno Beck) in 1970.

Translations

Alexander von Humboldt's travelogue has been translated into a variety of languages—German, English, Spanish, Italian, Dutch, Polish, Russian, Hungarian, and Czech—but rarely in its entirety. The figure below shows the English translations of the *Essai politique sur l'île de Cuba*.

Most nineteenth-century readers of this text in languages other than French would have encountered the travelogue in an abridged version. The only complete English version of the *Relation historique* is Helen Maria Williams's *Personal Narrative of Travels to the Equinoctial Regions of the New Continent during the Years 1799–1804 by Alexander de Humboldt and Aimé Bonpland*, published in seven volumes between 1814 and 1829 by the London house of Longman, Hurst, Rees, Orme, and Brown. Humboldt apparently kept a close eye on everything Williams sent to the London publishers. After Williams's death in 1827, her nephew, Charles Coquerel, completed the translation. Williams's very literal translation, which is fraught with problematic Latinate cognates, was certainly not unsuccessful in its time, and it is the only complete English translation of the text of the *Political Essay on the Island of Cuba* in its incarnation as part of the *Relation historique*. The English-language version of the

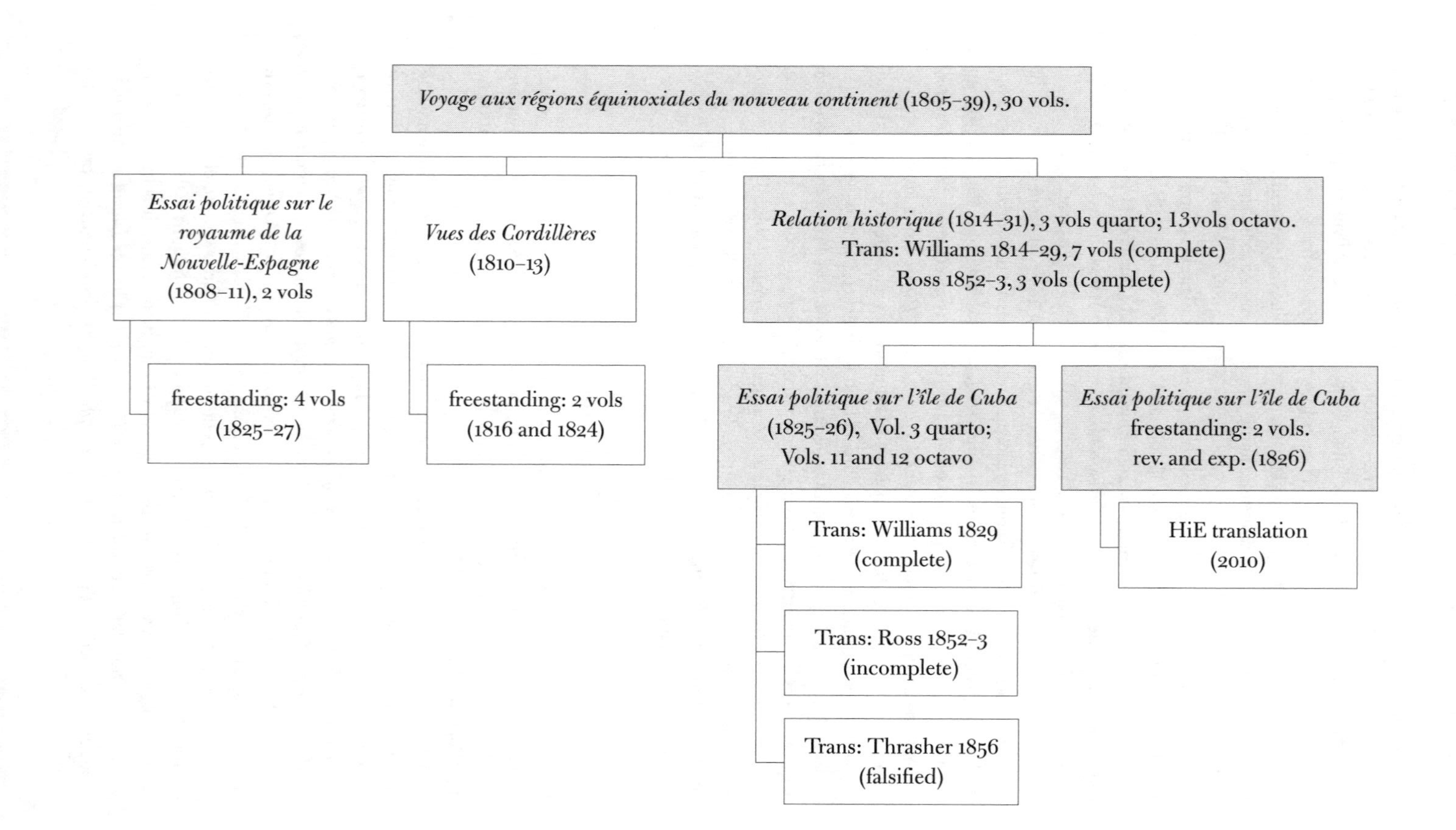
Voyage aux régions équinoxiales du nouveau continent (1805–39), 30 vols.
Essai politique sur le royaume de la Nouvelle-Espagne (1808–11), 2 vols
freestanding: 4 vols (1825–27)
Vues des Cordillères (1810–13)
freestanding: 2 vols (1816 and 1824)
Relation historique (1814–31), 3 vols quarto; 13vols octavo.
Trans: Williams 1814–29, 7 vols (complete)
Ross 1852–3, 3 vols (complete)
Essai politique sur l'île de Cuba (1825–26), Vol. 3 quarto; Vols. 11 and 12 octavo
Trans: Williams 1829 (complete)
Trans: Ross 1852–3 (incomplete)
Trans: Thrasher 1856 (falsified)
Essai politique sur l'île de Cuba freestanding: 2 vols. rev. and exp. (1826)
HiE translation (2010)

Relation historique that was more widely circulated was the abridged, three-volume translation by Thomasina Ross, which has a similar title, *Personal Narrative of Travels to the Equinoctial Regions of America, during the Year 1799–1804* (1852–53). Ross's translation had been commissioned by a competing London publishing house, Henry G. Bohn, because Williams's version was deemed too peculiar and too difficult to be popular with a broader audience. Later editions have typically either omitted the text of the *Political Essay on the Island of Cuba* included in the *Relation historique* or else reduced it drastically. Jason Wilson's translation of the *Relation historique* as *Personal Narrative of a Journey to the Equinoctial Regions of the New Continent*, published by Penguin in 1995, includes a mere fourteen pages on Cuba.

The first Spanish translation of the *Essai politique sur l'île de Cuba* from 1826 was anonymous and apparently quite inadequate. José de Bustamante (who provided only the initials D. J. B. de V. Y. M) produced a second Spanish translation in 1827, which, like the original text, was published in Paris. (Bustamante's is usually referred to as the first Spanish translation.) Humboldt approved of this Spanish translation. Bustamante's *Ensayo político sobre la isla de Cuba* has been the basis for the majority of the Spanish editions that have followed since. In 1856, John S. Thrasher prepared a third English translation based on Bustamante's Spanish text, rather than on one of the original French editions. Thrasher's *The Island of Cuba* appeared under the imprint of Derby & Jackson in New York. Though he claimed to have kept Humboldt's text intact, Thrasher actually made significant changes, omitting Humboldt's section on slavery and making countless emendations that Humboldt himself saw as "politically motivated"—unlike Humboldt, Thrasher was a supporter of slavery. (Humboldt's remark appears as a handwritten annotation in his copy of Thrasher's book.) Thrasher also added a lengthy introductory essay to update the volume, placing it squarely in the context of the United States' territorial ambitions in Cuba. Thrasher's text was reprinted twice during the twentieth century, first (and ironically) by the Negro University Press in 1969 and second by Markus Wiener and Ian Randle Publishers in 2001. The latter reprint, in an edition by Luis Martínez-Fernández, is entitled *The Island of Cuba: A Political Essay*. This only slightly corrected text restores the section on slavery Thrasher had excised, in a translation from the German (rather than from the original French); it does not, however, restore any of Thrasher's other numerous, often lengthy omissions. An annotated facsimile of Thrasher's 1856 text, which identifies all of his changes, can be found at www.press.uchicago.edu/books/humboldt/thrasher.

The first Cuban edition of the *Ensayo político sobre la isla de Cuba* was prepared by Fernando Ortiz for his prestigious Libros Cubanos series and appeared in Havana in 1930. This edition includes a substantial introductory essay and an afterword by Ortiz, entitled "El traductor de Humboldt en la historia cubana" ("Humboldt's translator in the context of Cuban history"). This afterword prefaces the only Spanish translation of Thrasher's introduction, which is included in the edition's appendix. (An English translation of Ortiz's essay can be found in *Atlantic Studies* 6.3 [Dec. 2009]: 327–44.) In Ortiz's edition, which was reprinted in Havana in 1959, 1960, 1969, and 1998, the actual text of the translation also includes Thrasher's additions, along with the annotations of the two-volume 1826 text, which Humboldt received from Francisco Arango y Parreño, both in the form of footnotes. Arango's comments, which he included in his letter to Humboldt from July 3, 1827, were published in 1839 as "Observaciones al 'Ensayo político sobre la isla de Cuba,' escritas en 1827" and reprinted in *Examen político sobre la isla de Cuba, por el B. on A. de Humboldt*, translated by Vicente González Arnao (Gerona: Imprenta de A. Oliva, 1839). Later Spanish editions of the *Ensayo político sobre la isla de Cuba* have tended to retain Arango's comments as if they were part of Humboldt's actual text. A notable example is the 1998 Doce Calles (Spain) edition by Miguel Angel Puig-Samper, Consuelo Naranjo Orovio, and Armando García González. None of these translations are based on the two-volume edition of *Essai politique sur l'île de Cuba* but either on the quarto or the octavo editions of the *Relation historique*. An English translation of Arango's comments and his letter can be found at www.press.uchicago.edu/books/humboldt/arango.

About This Edition

COMPLETENESS: This edition provides the first unabridged English translation of the two-volume *Essai politique sur l'île de Cuba* from 1826, in the exact order of the French original. We have restored Humboldt's text, adding neither chapters nor chapter titles nor subheadings. Nothing was subtracted. Any editorial changes have been marked in [square brackets]. The *Tableau statistique de l'île de Cuba* [Statistical overview of the island of Cuba], which Humboldt published separately in 1831, was not part of this edition and has not been included for this reason. Interested readers can find it at www.press.uchicago.edu/books/humboldt/tableau.

ACCURACY AND RELIABILITY: It is well worth asking what *accuracy* and *reliability* might mean, given that Humboldt revised and amended his writings almost obsessively and that his texts were issued by different publishers in different editions, often simultaneously, and over a period of almost thirty years. We decided to translate the two-volume edition of the *Essai politique sur l'île de Cuba* because Humboldt stopped revising the essay with this edition, which has never before appeared in English in its entirety (although Helen Williams had already translated earlier versions of the portions that Humboldt later added as supplements to the freestanding edition). Our goal was not to make this chronologically last edition *the* authoritative text—the one original—to which all other versions must defer. Rather than trying to immobilize Humboldt's unstable textual creations to satisfy some readers' desire for an original, we have tried to account for the historical existence of multiple originals by rendering transparent the exact nature and extent of his revisions. Technology has greatly facilitated this detailed textual comparison (see above).

Being sensitive to the historical dimensions that retrospectively render a text multiple and unstable has affected our translation in various ways. One instance in which we have not simply modernized Humboldt's language, however cautiously, is nineteenth-century racial terminology. While we have translated "nègre" as "black" throughout, we have rendered Humboldt's "gens de couleur" as "people of color" and his "mulâtres" and "mulâtresses" as men and women of "mixed-race." For any other racial or ethnic categories, we have preserved the specific French terms Humboldt employs in [square brackets].

READABILITY: In addition to completeness and accuracy, readability has been our third major concern in this edition, especially given the persistent claims that John Thrasher's mutilated translation "reads well." Our goal was to create a readable scholarly edition unencumbered by an extensive critical apparatus. In keeping with this objective, we have limited our apparatus to the annotations that follow the translation, an annotated bibliography of Humboldt's directly and indirectly referenced sources, and two indexes. Annotations are marked in the text as ▼ so as not to interfere with Humboldt's own footnote numbers. Although intended primarily for nonspecialist readers, the annotations will (we hope) also be informative for readers already familiar with Humboldt's cast of characters and the historical contexts of his writings. While they include historical and select scientific information, our annotations focus particularly on giving readers a sense of the extensive global network that Humboldt created and carefully nurtured during his lifetime. We have done so by emphasizing relevant connections between the historical personalities that populate his essay's pages.

Above all, we wanted to create a readable English version that also remains very close to Humboldt's French text. Since French has changed far less, and far less quickly, during the nineteenth and twentieth century than has English, careful modernization seemed to us the best approach to rendering Humboldt's voice. We have cautiously modernized Humboldt's language and updated his spellings, avoiding archaisms that might sound precious, without, however, collapsing entirely the inevitable distance between the twenty-first century and the time of Humboldt's writing. We have marked this distance by retaining in English Humboldt's at times elaborate linguistic formalities, especially in accounts of his collaborations, which have an endearing ring in modern English. We have also tried to retain other linguistic characteristics at the level of adjectives and adverbs. In this way, we hope to have kept Humboldt's French, which carries the unmistakable cadences of his native German, from sounding too much like colloquial American English.

ORTHOGRAPHY: In the *Essai politique sur l'île de Cuba*, Alexander von Humboldt used a plethora of terms and phrases in Spanish, Italian, and Latin either in their contemporary or historical spellings. As was typical for his time, Humboldt often eschewed accents in the Spanish names of places, persons, and idioms. In places where we did not have to interfere with Humboldt's original quotations in other languages, we have tacitly modernized his orthography. At times, Humboldt went back and forth between different spellings of the same term: for instance, *tasajo* and *tassajo* (cured meat). To increase the text's readability, we have, in such cases, created consistencies, but never at the expense of possibly important nuances or different levels of signification. Place names have been tacitly updated to current use, except in cases where a new name is radically different. In those cases, the old name was kept, followed by the new name in [square brackets]. To give the reader a strong sense of the texture of Humboldt's writing, we have kept intact Humboldt's seemingly capricious uses of capitalization, lower case, and italics.

CORRECTIONS: We have silently corrected errors that we could determine to be merely typographical. Most of those are numbers in the tables or in the pagination of references. Questionable instances, which went beyond purely formal aspects into the realm of content, have only been corrected when it was clear that they were the result of an error. All other corrections, including in Humboldt's references, appear in [square brackets].

USE OF FOREIGN LANGUAGES: In the *Essai politique sur l'île de Cuba*, as in many of his writings, Humboldt used many foreign-language terms and even quotations without offering his readers translations. To enhance our translation's readability, we have provided English translations in [square brackets] immediately after each word or citation in all cases where context does not render their meanings self-evident. In each case, we were concerned with creating linguistic and conceptual transparency, while retaining the characteristic polyvocality of Humboldt's writing.

REFERENCES: Humboldt's bibliographical references in both the footnotes and the textual parentheses (at times even directly in the text) follow his two-volume French edition, except that we have formally unified his references to volumes and other sources and have replaced his sometimes odd abbreviations of titles with fuller information. We have provided additional information in square brackets to enable readers to locate specific items in our bibliography of Alexander von Humboldt's sources, which, in addition to identifying the editions Humboldt actually used (insofar as that was possible), also contains additional information, for instance, about other editions and translations.

TABLES: This edition includes all of Humboldt tables exactly where he placed them in the two-volume 1826 edition. We have also tried to approximate the visual appearance of these tables as much as was sensible. In only one instance have we represented as a table what Humboldt wrote as continuous text to improve its readability (see 62–63; in the quarto edition of the *Relation historique*, this section is also printed in smaller font).

INDEXES: We have included two indexes to help the reader better navigate the text of the translation and the annotations: (1) a subject index that includes proper names of persons, scientific instruments and concepts, historical events, politics, and cultural topics; (2) a toponym index devoted to the many geographical names Humboldt uses. Since Humboldt himself did not provide an index specifically for the *Political Essay on the Island of Cuba*, we have followed the lead of his indexes to other works, notably *Vues des Cordillères et Monumens des Peuples Indigènes de l'Amérique—Views of the Cordilleras and Monuments of the Indigenous Peoples of the Americas*, in a forthcoming new translation from HiE—and the *Political Essay on the Kingdom of New Spain*. We have also consulted the indexes in Hanno Beck's and Ottmar Ette's respective editions of the *Relation historique*.

MAPS: Although Humboldt included only his corrected map from 1826 with the freestanding version of the *Political Essay on the Island of Cuba*, this edition has both his original 1820 map and the revised 1826 map as foldouts. We decided to include both because the differences between them are notable. The first map, "Carte de l'île de Cuba," includes the following text (1820): "based on the astronomical observations by Spanish sailors and by Mr. von Humboldt himself. By P. Lapie, Squadron Leader of the Royal Corps of the Geographer-Engineers of France, 1820. Writing by Lallemand, rue de Noyers N° 49, engraved by Falhaut." In the top-left corner, there are comments on the survey measurements: "In composing this map, Mr. von Humboldt used his own observations made east of the meridian of the Port of Trinidad and published by Mr. Oltmanns (*Recueil d'observations astronomiques*. Vol. II, pp. 13–147). He also used data from Mr. Josef Joaquín de Ferrer, Don Antonio Robredo, Don Ciriaco Cevallos, and Don Dionisio Alcala Galiano, as well as maps from the Depósito Hidrográfico in Madrid which were made under the direction of Mr. Espinoza and Mr. Bauzá from two unpublished maps put together in Havana in 1803 and 1805." Inset: Plan of the harbor of the city of Havana. The second map, "Carte de l'île de Cuba," includes this text (1826): "based on the astronomical observations by Spanish sailors and by Mr. von Humboldt himself. By P. Lapie, Squadron Leader of the Royal Corps of the Geographer-Engineers of France, 1826." In the top-left corner, there are comments on the survey measurements: "In composing this map, Mr. von Humboldt used his own observations made east of the meridian of the Port of Trinidad and published by Mr. Oltmanns (*Recueil d'observations astronomiques*. Vol. II, pp. 13–147). He also used data from Mr. Josef Joaquín de Ferrer, Don Antonio Robredo, Don Ciriaco Cevallos, Don Francisco Lemaur, and Don Dionisio Alcala Galiano, as well as maps from the Depósito Hidrográfico in Madrid which were made under the direction of Mr. Espinosa and Mr. Bauzá from two unpublished maps put together in Havana in 1803 and 1805. The southern portion of this map was corrected in 1826 in accordance with observations by Don Ventura de Barcaíztegui and Don José del Río and with a sketch that the famous geographer Don Felipe Bauzá was kind enough to sent [sic] to the author." Inset: Plan of the harbor of the city of Havana.

WEBSITE This edition, like future volumes in this series, has an accompanying website, which makes additional materials available to interested readers and researchers. Among the items included are a number of maps and first English translations of writings on Humboldt by Francisco Arango y Parreño, Fernando Ortiz, and others. The site at www.press.uchicago.edu/books/humboldt is free and open to the public.

ACKNOWLEDGMENTS: The members of the Alexander von Humboldt in English team wish to thank the staff of the Center for the Americas at Vanderbilt, particularly Vashti Wells, Janell Lees, and Lawrence Staten, and Gabriele Penquitt at the

University of Potsdam, for photocopying, scanning, and all other forms of logistical support. Many people at Vanderbilt's Heard Library have assisted us in our research in so many invaluable ways. For patiently and efficiently managing the flood of questions and requests for books and articles, for assisting in background research, and for digitalizing images, we heartily thank Peter Brush, Paula Covington, Henry Shipman, and Jim Toplon. The staff of various libraries in Berlin and Havana, Cuba, is no less deserving of our gratitude, as are friends and family members who have encouraged and helped each of us in a variety of ways: F. David T. Arens, Nara Araujo (†), María del Carmen Barcía, Adriana Bernal Calderón, Victor Casaus, Irma Castro, Detlev Eggers, Santiago Khalil, Esther Lobaina, Rosa María López González, José Matos, Daysi Rayo, Tomás Robaina, Ewa Slojka, Eduardo Torres-Cuevas, Amir Valle, Dania Vásquez, and Olga Vega. We are particularly grateful to Miguel Guzmán Stein for his support of Tobias Kraft's research in Havana; he was undeterred in his efforts even by a tropical hurricane.

For reading and commenting on different aspects of the translation, the annotations, and the afterword, we thank Pablo Gómez, Arik Ohnstadt, Daniel Spoth, and Nafissa Thompson-Spires (from the 2008–9 Center for the Americas Graduate Workshop at Vanderbilt), Chantal Bizzini, John Ochoa, John Ryan Poynter, Justin Ruaysamran, Neil Safier, and Laura Dassow Walls. Our editor, Christie Henry, has been a fount of steady encouragement and good will. We cannot thank her enough.

The work of the HiE team began in the fall of 2007. Various aspects of the work on this volume have been supported by grants from the former Center for the Americas at Vanderbilt, the Martha Rivers Ingram Chair of English at Vanderbilt, the Alexander von Humboldt Foundation, and the German Academic Exchange Service (DAAD). We are deeply grateful to them for making this volume possible.

SUBJECT INDEX

This index encompasses the major subjects that Alexander von Humboldt takes up in his text, along with the historical personalities, journals, and institutions he references. It covers both Humboldt's text and the annotations (all page numbers higher than 323 refer to the annotations). Geographical names have been reserved for the Toponym Index. While some of these names appear in both indexes, there is typically little overlap between the pages to which they point in each index.

TOPONYM INDEX

This index is devoted entirely to the names of places, including important landmarks, that Alexander von Humboldt mentions in his text. Unlike the *Subject Index*, it does not extend to the annotations. Our purpose in compiling this separate index is to show better the global range of Humboldt's geographical references. To make it easier to recognize this range and locate specifics, we have used cross-references to create some continental clusters, mainly for places outside the Americas (Africa, Asia, and Russia, for instance). Europe is represented under individual country names, as are the states of the United States of (North) America—both with some degree of caution. Confronted with historical settings in which imperial and national borders were constantly shifting as new political entities came into being during the process of decolonization, we did not attempt to assign country locations to many of the places in the continental Americas. Doing so would have added potential confusion to an already rather complicated picture. We wanted to convey a sense that these place names were, and in some cases still are, parts not of a static map but of a political geography in motion. When a particular location is not already self-explanatory, we have specified whether it refers to a district, island, province, region, river, mountain, swamp, village, etc. In not a few cases, several of these categories apply to a single name. We have also indicated some alternative spellings.